Edited by
Xiaoniu Yang

Semiconducting Polymer Composites

Related Titles

Klauk, H. (ed.)
Organic Electronics II
More Materials and Applications
2012
ISBN: 978-3-527-32647-1

Mittal, V. (ed.)
In-situ Synthesis of Polymer Nanocomposites
2012
ISBN: 978-3-527-32879-6

Knoll, W., Advincula, R. C. (eds.)
Functional Polymer Films
2 Volume Set
2011
ISBN: 978-3-527-32190-2

Mittal, V. (ed.)
Surface Modification of Nanotube Fillers
2011
ISBN: 978-3-527-32878-9

Chujo, Y. (ed.)
Conjugated Polymer Synthesis
Methods and Reactions
2011
ISBN: 978-3-527-32267-1

Kumar, C. S. S. R. (ed.)
Nanocomposites
2010
ISBN: 978-3-527-32168-1

Cosnier, S., Karyakin, A. (eds.)
Electropolymerization
Concepts, Materials and Applications
2010
ISBN: 978-3-527-32414-9

Leclerc, M., Morin, J.-F. (eds.)
Design and Synthesis of Conjugated Polymers
2010
ISBN: 978-3-527-32474-3

Guldi, D. M., Martín, N. (eds.)
Carbon Nanotubes and Related Structures
Synthesis, Characterization, Functionalization, and Applications
2010
ISBN: 978-3-527-32406-4

Mittal, V. (ed.)
Optimization of Polymer Nanocomposite Properties
2010
ISBN: 978-3-527-32521-4

Edited by Xiaoniu Yang

Semiconducting Polymer Composites

Principles, Morphologies, Properties and Applications

WILEY-VCH Verlag GmbH & Co. KGaA

The Editor

Prof. Xiaoniu Yang
Chinese Academy of Sciences
Polymer Composite Engineering
Changchun 130022
China

All books published by **Wiley-VCH** are carefully produced. Nevertheless, authors, editors, and publisher do not warrant the information contained in these books, including this book, to be free of errors. Readers are advised to keep in mind that statements, data, illustrations, procedural details or other items may inadvertently be inaccurate.

Library of Congress Card No.: applied for

British Library Cataloguing-in-Publication Data
A catalogue record for this book is available from the British Library.

Bibliographic information published by the Deutsche Nationalbibliothek
The Deutsche Nationalbibliothek lists this publication in the Deutsche Nationalbibliografie; detailed bibliographic data are available on the Internet at http://dnb.d-nb.de.

© 2012 Wiley-VCH Verlag & Co. KGaA, Boschstr. 12, 69469 Weinheim, Germany

All rights reserved (including those of translation into other languages). No part of this book may be reproduced in any form – by photoprinting, microfilm, or any other means – nor transmitted or translated into a machine language without written permission from the publishers. Registered names, trademarks, etc. used in this book, even when not specifically marked as such, are not to be considered unprotected by law.

Print ISBN: 978-3-527-33030-0
ePDF ISBN: 978-3-527-64871-9
ePub ISBN: 978-3-527-64870-2
mobi ISBN: 978-3-527-64869-6
oBook ISBN: 978-3-527-64868-9

Cover Design   Adam-Design, Weinheim; Germany

Typesetting   Thomson Digital, Noida, India

Printing and Binding   Markono Print Media Pte Ltd, Singapore

Printed in Singapore
Printed on acid-free paper

Contents

List of Contributors *XV*
Preface *XXI*

1 Solubility, Miscibility, and the Impact on Solid-State Morphology *1*
Florian Machui and Christoph J. Brabec
1.1 Introduction *1*
1.2 General Aspects *2*
1.2.1 Solubility *3*
1.2.2 Miscibility–Thermodynamic Relationships *5*
1.3 Solubility, Solvents, and Solution Formulations *6*
1.3.1 Solubility *6*
1.3.2 Solvents *9*
1.3.2.1 Impact of Different Solvents on the Solid-State Morphology *10*
1.3.2.2 Non-Halogenic Solvents *14*
1.3.2.3 Solvent Blends *15*
1.3.2.4 Addition of Poor Solvents *16*
1.3.2.5 Processing Additives *18*
1.3.2.6 Solution Concentration *21*
1.3.3 Conclusive Outlook *21*
1.4 Miscibility *22*
1.4.1 Methods *22*
1.4.1.1 Glass Transition *23*
1.4.1.2 Surface Energy *23*
1.4.1.3 Photoluminescence Quenching *24*
1.4.2 Polymer–Polymer Miscibility *26*
1.4.3 Polymer–Fullerene Miscibility *28*
1.4.4 Phase Diagrams *30*
1.5 Conclusions *32*
References *34*

2	**Nanoscale Morphological Characterization for Semiconductive Polymer Blends** 39
	Joachim Loos
2.1	Introduction 39
2.2	The Importance of Morphology Control 40
2.3	The Classic Blend: MDMO-PPV/PCBM as a Model for an Amorphous Donor System 42
2.4	Intermezzo: Morphology Imaging with Scanning Transmission Electron Microscopy 48
2.5	Volume Characterization of the Photoactive Layer: Electron Tomography 50
2.6	Measuring Nanoscale Electrical Properties: Conductive AFM 56
2.7	Current Progress and Outlook 60
	References 62
3	**Energy Level Alignment at Semiconductive Polymer Interfaces: Correlating Electronic Energy Levels and Electrical Conductivity** 65
	Nobuo Ueno
3.1	Introduction 65
3.2	General View of Electronic Structure of Organic Solids 65
3.2.1	Introduction to Correlating Electronic Structure and Electrical Conductivity 65
3.2.2	Evolution of Electronic Structure from Single Molecule to Molecular Solid 67
3.2.3	Evolution of Electronic Structure from Single Atom to Polymer Chain 70
3.2.4	Polaron 72
3.2.5	Energy Level Alignment at the Interface 73
3.3	Experimental Methods 75
3.3.1	Ultraviolet Photoelectron Spectroscopy 75
3.3.2	Penning Ionization Electron Spectroscopy 78
3.4	Valence Electronic Structure of Organic Semiconductors: Small Molecules 79
3.4.1	Energy Band Dispersion and Band Transport Mobility 79
3.4.2	Electron–Phonon Coupling and Hopping Mobility 84
3.4.2.1	Fundamental Aspects on Charge Hopping 84
3.4.2.2	Reorganization Energy and Small Polaron Binding Energy 86
3.5	Valence Electronic Structure of Polymers 90
3.5.1	Quasi-One-Dimensional Band Dispersion Along Polymer Chains 90
3.5.1.1	σ-Bond Polymer 90
3.5.1.2	π-Conjugated Polymer Chain 91
3.5.2	Pendant Group Polymers: Is the Surface of Solution-Cast Film Clean on Molecular/Atomic Scale? 92

3.5.3	P3HT: Electronic Structure and Control of π-Electron Density Distribution at the Surface for Realizing a Functional Interface	*93*
3.6	Role of the Interface Dipole Layer: Its Impact on the Energy Level Alignment *101*	
3.7	Future Prospects *103*	
	References *103*	

4 Energy and Charge Transfer   *107*
Ralf Mauer, Ian A. Howard, and Frédéric Laquai

4.1	Introduction *107*	
4.2	Energy Transfer *108*	
4.2.1	Electronic Structure and Excited States of Conjugated Polymers *108*	
4.2.1.1	Excitons: The Nature of Excited States in Conjugated Polymers	*108*
4.2.2	Excited State Dynamics in Conjugated Polymers *113*	
4.2.2.1	Role of Disorder in Energy Transfer *113*	
4.2.2.2	Singlet Exciton Energy Transfer *114*	
4.2.2.3	Triplet Exciton Dynamics *115*	
4.2.3	Energy Transfer: Relevance to Device Performance *123*	
4.3	Charge Transfer in Polymer/Fullerene Composites *125*	
4.3.1	Theoretical Background *125*	
4.3.1.1	Theory of Charge Transfer *125*	
4.3.1.2	Theory of Field and Temperature Dependence of Charge Separation *129*	
4.3.2	The Role of Charge Transfer States for Charge Separation *131*	
4.3.2.1	Parameters Influencing the Separation of Charge Transfer States *133*	
4.3.3	Charge Transfer: Relevance to Device Performance *139*	
	References *140*	

5 Percolation Theory and Its Application in Electrically Conducting Materials   *145*
Isaac Balberg

5.1	Introduction *145*
5.2	Lattice Percolation *146*
5.3	Continuum Percolation *152*
5.4	Percolation Behavior When the Interparticle Conduction Is by Tunneling *154*
5.5	The Structure of Composite Materials *156*
5.6	The Observations and Interpretations of the σ(x) Dependence in Composite Materials *159*
5.6.1	The Percolation Threshold *159*
5.6.2	The Critical Behavior of σ(x) *161*
5.7	Summary and Conclusions *165*
	References *168*

6		**Processing Technologies of Semiconducting Polymer Composite Thin Films for Photovoltaic Cell Applications** *171*
		Hui Joon Park and L. Jay Guo
6.1		Introduction *171*
6.2		Optimization of Bulk Heterojunction Composite Nanostructures *173*
6.3		Fabrication of Sub-20 nm Scale Semiconducting Polymer Nanostructure *182*
6.3.1		Nanoimprint Mold Fabrication *183*
6.4		Conclusions *186*
		References *187*
7		**Thin-Film Transistors Based on Polythiophene/Insulating Polymer Composites with Enhanced Charge Transport** *191*
		Longzhen Qiu, Xiaohong Wang, and Kilwon Cho
7.1		Introduction *191*
7.2		Fundamental Principle and Operating Mode of OTFTs *193*
7.3		Strategies for Preparing High-Performance OTFTs Based on Semiconducting/Insulating Blends *194*
7.4		Blend Films with Vertical Stratified Structure *194*
7.4.1		Phase Behavior of Polymer Blends *194*
7.4.2		One-Step Formation of Semiconducting and Insulating Layers in OTFTs *198*
7.4.3		Improved Environmental Stability *201*
7.4.4		Patterned Domains of Polymer Blends *201*
7.4.5		Improved Charge Carrier Mobility *204*
7.4.6		Crystallization-Induced Vertical Phase Segregation *206*
7.5		Blend Films with Embedded P3HT Nanowires *207*
7.5.1		P3AT Nanowires *208*
7.5.2		Polymer Blends with Embedded P3HT Nanowires *209*
7.5.3		Nanowires from Conjugated Block Copolymers *212*
7.5.4		Electrospun Nanowires from Conjugated Polymer Blends *212*
7.6		Conclusions and Outlook *214*
		References *214*
8		**Semiconducting Organic Molecule/Polymer Composites for Thin-Film Transistors** *219*
		Jeremy N. Smith, John G. Labram, and Thomas D. Anthopoulos
8.1		Introduction *219*
8.1.1		OFET Device Operation *220*
8.1.2		Small-Molecule/Polymer Film Morphology *222*
8.2		Unipolar Films for OFETs *224*
8.2.1		Oligothiophene/Polymer Blends *224*
8.2.2		Acene/Polymer Blends *227*

8.3	Polymer/Fullerene Ambipolar OFETs *232*
8.3.1	Polymer:Fullerene Blend Morphology *233*
8.3.1.1	Solvent and Polymer Molecular Weight *235*
8.3.1.2	Blend Composition *236*
8.3.1.3	Temperature- and Time-Dependent Annealing *238*
8.3.1.4	Effect of Fullerene Molecular Weight *240*
8.3.2	Polymer:Fullerene Bilayer Diffusion *241*
8.3.2.1	Modeling Fullerene Diffusion *243*
8.4	Conclusions *245*
	References *246*

9 Enhanced Electrical Conductivity of Polythiophene/Insulating Polymer Composite and Its Morphological Requirement *251*
Guanghao Lu and Xiaoniu Yang

9.1	Introduction *251*
9.2	Phase Evolution and Morphology *253*
9.3	Enhanced Conductivity of Conjugated Polymer/Insulating Polymer Composites at Low Doping Level: Interpenetrated Three-Dimensional Interfaces *258*
9.4	Conductivity of Semiconducting Polymer/Insulating Polymer Composites Doped by Molecular Dopant *260*
9.5	Mechanisms for the Enhanced Conductivity/Mobility *261*
9.5.1	Improved Crystallinity and Molecular Ordering *261*
9.5.2	"Self-Encapsulation" Effect *262*
9.5.3	Bulk 3-D Interface and Reduced Polarization of Matrix at Interface *263*
9.5.4	"Zone Refinement" Effect *264*
9.5.5	Reduced Polaron–Dopant Interaction *265*
9.6	Perspective *267*
	References *268*

10 Intrinsically Conducting Polymers and Their Composites for Anticorrosion and Antistatic Applications *269*
Yingping Li and Xianhong Wang

10.1	ICPs and Their Composites for Anticorrosion Application *269*
10.1.1	Introduction *269*
10.1.2	Protection Mechanism *270*
10.1.2.1	Anodic Protection Mechanism *270*
10.1.2.2	Inhibitory Protection Mechanism *274*
10.1.2.3	Cathodic Protection Mechanism *274*
10.1.2.4	Comprehensive Understanding on Protection Mechanism of ICPs *277*
10.1.3	Matrix Resin of Conducting Composite Coating *278*
10.1.4	Processing Methods *279*
10.1.5	Conclusions and Perspectives *280*

10.2	Antistatic Coating 282
10.2.1	Introduction 282
10.2.2	Synthesis of Processable ICPs 283
10.2.3	Processing of ICPs for Antistatic Application 284
10.2.4	Water-Based Polyaniline and Its Complex 286
10.3	Summary 288
	References 289
11	**Conjugated–Insulating Block Copolymers: Synthesis, Morphology, and Electronic Properties** 299
	Dahlia Haynes, Mihaela C. Stefan, and Richard D. McCullough
11.1	Introduction 299
11.2	Oligo- and Polythiophene Rod–Coil Block Copolymers 300
11.3	Poly(p-phenylene vinylene) Block Copolymers 308
11.4	Polyfluorenes 313
11.5	Other Semiconducting Rod–Coil Systems 319
11.6	Conjugated–Insulating Rod–Rod Block Copolymers 320
11.7	Conclusions and Outlook 322
	References 322
12	**Fullerene/Conjugated Polymer Composite for the State-of-the-Art Polymer Solar Cells** 331
	Wanli Ma
12.1	Introduction 331
12.2	Working Mechanism 332
12.2.1	Unique Properties of Organic Solar Cells 332
12.2.2	Understanding the Bulk Heterojunction Structures 332
12.2.3	Device Parameters and Theoretical Efficiency 334
12.2.3.1	Short-Circuit Current Density 335
12.2.3.2	Open-Circuit Voltage 336
12.2.3.3	Fill Factor 337
12.2.3.4	Theoretical Efficiency 338
12.3	Optimization of Fullerene/Polymer Solar Cells 338
12.3.1	Design of New Materials 339
12.3.1.1	Absorption Enhancement 340
12.3.1.2	Fine-Tuning of HOMO and LUMO Energy Levels 344
12.3.1.3	Mobility and Solubility Improvement 345
12.3.1.4	New Fullerene Derivative 345
12.3.2	Optimization of Polymer Solar Cell Devices 346
12.3.2.1	Morphology Control 346
12.3.2.2	Device Architectures 351
12.4	Outlook 354
	References 355

13	**Semiconducting Nanocrystal/Conjugated Polymer Composites for Applications in Hybrid Polymer Solar Cells** *361*	
	Michael Krueger, Michael Eck, Yunfei Zhou, and Frank-Stefan Riehle	
13.1	Introduction *361*	
13.2	Composite Materials *361*	
13.2.1	Colloidal Semiconductor Nanocrystals *361*	
13.2.2	Conjugated Polymers *366*	
13.3	Device Structure *367*	
13.3.1	Photoactive Layer *367*	
13.3.2	Device Principle *371*	
13.3.3	Band Alignment and Choice of Donor/Acceptor Pairs *373*	
13.4	State of the Art of Hybrid Solar Cells *374*	
13.5	Novel Approaches in Hybrid Solar Cell Development *381*	
13.5.1	Utilization of Less Toxic Semiconductor NCs *381*	
13.5.2	*In Situ* Synthesis of Ligand-Free Semiconductors in Conjugated Polymers *381*	
13.5.3	Utilization of One-Dimensional Structured Donor–Acceptor Nanostructures for Hybrid Film Formation *382*	
13.5.4	Toward Nanostructured Donor–Acceptor Phases *384*	
13.6	Outlook and Perspectives *390*	
13.6.1	Hybrid Solar Cells Versus Pure OPVs *390*	
13.6.2	Hybrid PV and OPV Versus Other PV Technologies *391*	
	References *393*	
14	**Conjugated Polymer Blends: Toward All-Polymer Solar Cells** *399*	
	Christopher R. McNeill	
14.1	Introduction *399*	
14.2	Review of Polymer Photophysics and Device Operation *400*	
14.3	Material Considerations *401*	
14.4	Device Achievements to Date *403*	
14.5	Key Issues Affecting All-Polymer Solar Cells *407*	
14.5.1	Interfacial Charge Separation *407*	
14.5.2	Morphology *411*	
14.5.3	Charge Transport *418*	
14.6	Summary and Outlook *421*	
	References *421*	
15	**Conjugated Polymer Composites and Copolymers for Light-Emitting Diodes and Laser** *427*	
	Thien Phap Nguyen and Pascale Jolinat	
15.1	Introduction *427*	
15.2	Properties of Organic Semiconductors *428*	
15.3	Polymer-Based Composites *429*	
15.4	Use of Polymer Composites in Photonic Applications *430*	
15.4.1	Organic Light-Emitting Diodes *430*	

15.4.1.1	Efficiency of OLEDs	432
15.4.1.2	Color Emission	433
15.4.1.3	Stability	440
15.4.2	Organic Semiconductor Lasers	441
15.4.2.1	Working Principle	441
15.4.2.2	Materials	443
15.4.2.3	Types of Resonators	446
15.4.2.4	Applications and Developments	448
15.5	Conclusions	451
	References	452

16   Semiconducting Polymer Composite Based Bipolar Transistors   457
Claudia Piliego, Krisztina Szendrei, and Maria Antonietta Loi

16.1	Introduction	457
16.2	Basics of Organic Field-Effect Transistors	458
16.2.1	Operation Principles of FETs	458
16.2.1.1	Unipolar FETs	459
16.2.1.2	Bipolar FETs	460
16.2.2	Current–Voltage Characteristics	461
16.2.2.1	Unipolar FET	461
16.2.2.2	Bipolar FETs	462
16.2.3	Device Configurations	462
16.2.4	Role of the Injecting Electrodes	463
16.2.5	Applications: Inverters and Light-Emitting Transistors	464
16.3	Bipolar Field-Effect Transistors	465
16.3.1	Single-Component Bipolar FETs	465
16.3.2	Bilayer Bipolar FETs	469
16.3.3	Bulk Heterojunction Bipolar FETs	474
16.3.3.1	Coevaporated Blends	475
16.3.3.2	Polymer–Small Molecule Blends	476
16.3.3.3	Hybrid Blends	479
16.3.3.4	Polymer–Polymer Blends	480
16.4	Perspectives	485
	References	486

17   Nanostructured Conducting Polymers for Sensor Development   489
Yen Wei, Meixiang Wan, Ten-Chin Wen, Tang-Kuei Chang, Gaoquan Shi, Hongxu Qi, Lei Tao, Ester Segal, and Moshe Narkis

17.1	Introduction	489
17.2	Conducting Polymers and Their Nanostructures	490
17.3	Synthetical Methods for Conducting Polymer Nanostructures	493
17.3.1	Hard-Template Method	494
17.3.2	Soft-Template Method	496
17.3.3	Electrospinning Technology	496
17.4	Typical Conducting Polymer Nanostructures	497

17.4.1	Polyaniline (PANI)	497
17.4.2	Poly(3,4-ethylenedioxythiophene) (PEDOT)	501
17.4.3	Polypyrrole (PPy)	502
17.5	Multifunctionality of Conducting Polymer Nanostructures	503
17.6	Conducting Polymer-Based Sensors	505
17.6.1	Gas Sensors	506
17.6.2	pH Sensors	509
17.6.3	Biosensors	510
17.6.4	Artificial Sensors	511
17.7	Summary and Outlook	512
	References	513

Index 523

List of Contributors

Thomas D. Anthopoulos
Imperial College London
Department of Physics
South Kensington Campus
London SW7 2AZ
UK

Isaac Balberg
The Hebrew University
The Racah Institute of Physics
Jerusalem 91904
Israel

Christoph J. Brabec
Friedrich-Alexander University
Department of Material Science
and Engineering
Institute of Materials for Electronics
and Energy Technology
Martensstrasse 7
91058 Erlangen
Germany

and

Bavarian Center for Applied Energy
Research (ZAE Bayern)
Am Weichselgarten 7
91058 Erlangen
Germany

Tang-Kuei Chang
National Cheng Kung University
Department of Chemical Engineering
Taiwan 70101
Taiwan

Kilwon Cho
Pohang University of Science
and Technology
Department of Chemical Engineering
San 31 Hyojia-dong, Namgu
Pohang 790-784
Korea

Michael Eck
FMF - Freiburger
Materialforschungszentrum
Institute for Microsystems Technology
Stefan-Meier-Str. 21
79104 Freiburg
Germany

L. Jay Guo
The University of Michigan
Macromolecular Science
and Engineering

and

Department of Electrical
Engineering
and Computer Science
Ann Arbor, MI 48109
USA

Dahlia Haynes
Carnegie Mellon University
The McCullough Group
5000 Forbes Avenue
Warner Hall 608
Pittsburgh, PA 15213
USA

Ian A. Howard
Max Planck Institute for Polymer Research
Ackermannweg 10
55128 Mainz
Germany

Pascale Jolinat
Université de Toulouse
Laboratoire Plasma
et Conversion d'Energie
118, Route de Narbonne
31062 Toulouse cedex 9
France

Michael Krueger
FMF - Freiburger Materialforschungszentrum
Institute for Microsystems Technology
Stefan-Meier-Str. 21
79104 Freiburg
Germany

John G. Labram
Imperial College London
Department of Physics
London SW7 2AZ
UK

Frédéric Laquai
Max Planck Institute
for Polymer Research
Ackermannweg 10
55128 Mainz
Germany

Yingping Li
Graduate School of Chinese
Academy of Sciences
Beijing 100039
China

Maria Antonietta Loi
University of Groningen
Zernike Institute
for Advanced Materials
Nijenborgh 4
Groningen 9747 AG
The Netherlands

Joachim Loos
University of Glasgow
Kelvin Nanocharacterisation Centre (KNC)
Scottish University Physics Alliance (SUPA)

and

School of Physics and Astronomy
Glasgow G12 8QQ
Scotland
UK

Guanghao Lu
Chinese Academy of Sciences
Changchun Institute
of Applied Chemistry
Changchun 130022
China

Wanli Ma
Soochow University
Institute of Functional Nano and Soft Materials
No. 1, Shizi Street
Suzhou, 215123
China

Florian Machui
Friedrich-Alexander University
Department of Material Science
and Engineering
Institute of Materials for Electronics
and Energy Technology
Martensstrasse 7
91058 Erlangen
Germany

Ralf Mauer
Max Planck Institute for Polymer
Research
Ackermannweg 10
55128 Mainz
Germany

Richard D. McCullough
Carnegie Mellon University
The McCullough Group
5000 Forbes Avenue, Warner Hall 608
Pittsburgh, PA 15213
USA

Christopher R. McNeill
Monash University
Department of Materials
Engineering
Clayton, Victoria 3800
Australia

Moshe Narkis
Technion-Israel Institute of
Technology
Department of Chemical
Engineering
Haifa 32000
Israel

Thien Phap Nguyen
Université de Nantes
Institut des Matériaux Jean Rouxel
CNRS
2, rue de la Houssinière
44322 Nantes Cedex 3
France

Hui Joon Park
The University of Michigan
Macromolecular Science and
Engineering
Ann Arbor, MI 48109
USA

Claudia Piliego
University of Groningen
Zernike Institute for
Advanced Materials
Nijenborgh 4
Groningen 9747 AG
The Netherlands

Hongxu Qi
Tsinghua University
Department of Chemistry
Beijing 100084
China

Longzhen Qiu
Hefei University of Technology
Academy of Opto-Electronic Technology
Key Lab of Special Display Technology
Ministry of Education
National Engineering Lab of Special
Display Technology
National Key Lab of Advanced
Display Technology
193 Tunxi Road
Hefei 230009
China

Frank-Stefan Riehle
FMF - Freiburger
Materialforschungszentrum
Institute for Microsystems Technology
Stefan-Meier-Str. 21
79104 Freiburg
Germany

Ester Segal
Technion-Israel Institute
of Technology
Department of Chemical Engineering
Haifa 32000
Israel

Gaoquan Shi
Tsinghua University
Department of Chemistry
Beijing 100084
China

Jeremy N. Smith
Imperial College London
Department of Physics
London SW7 2AZ
UK

Mihaela C. Stefan
Carnegie Mellon University
The McCullough Group
5000 Forbes Avenue
Warner Hall 608
Pittsburgh, PA 15213
USA

Krisztina Szendrei
University of Groningen
Zernike Institute for Advanced
Materials
Nijenborgh 4
Groningen 9747 AG
The Netherlands

Lei Tao
Tsinghua University
Department of Chemistry
Beijing 100084
China

Nobuo Ueno
Chiba University
Graduate School of Advanced
Integration Science
Inage-ku
Chiba 263-8522
Japan

Meixiang Wan
Chinese Academy of Sciences
Institute of Chemistry
Center for Molecular Science
Organic Solid Laboratory
Beijing 100080
China

Xianhong Wang
Chinese Academy of Sciences
Changchun Institute of Applied
Chemistry
Key Laboratory of Polymer
Ecomaterials
Renmin Street 5625
Changchun 130022
China

Xiaohong Wang
Hefei University of Technology
Academy of Opto-Electronic Technology
Key Lab of Special Display Technology
Ministry of Education
National Engineering Lab of Special
Display Technology
National Key Lab of Advanced
Display Technology
193 Tunxi Road
Hefei 230009
China

Yen Wei
Tsinghua University
Department of Chemistry
Beijing 100084
China

and

National Cheng Kung University
Department of Chemical
Engineering
Tainan 70101
Taiwan

Ten-Chin Wen
National Cheng Kung University
Department of Chemical
Engineering
Taiwan 70101
Taiwan

Xiaoniu Yang
Chinese Academy of Sciences
Changchun Institute of
Applied Chemistry
Renmin Street 5625
Changchun 130022
China

Yunfei Zhou
FMF - Freiburger
Materialforschungszentrum
Institute for Microsystems Technology
Stefan-Meier-Str. 21
79104 Freiburg
Germany

Preface

The research on (semi-)conducting polymers has attracted dramatically increased attention from both academic and industrial communities. The commercial products based on these new materials, for example, polymer thin-film displays and polymer solar cells, are already available on the market. Solution-based thin-film deposition technology makes it possible to carry out large-scale device fabrication with very low cost, which has been regarded as the most attractive advantage of semiconducting polymers for applications in next-generation optoelectronic devices. In most cases, a composite instead of only one polymer species is employed to realize the specific functionality of the device, which results in more scientific questions that need to be answered, for example, with respect to morphological, interfacial, and mechanical properties as well as to charge transfer mechanisms within the composite film. A book collecting the already existing knowledge on the respective topics is necessary for new researchers to become acquainted with the field as well as for giving an overview and addressing the key questions within a short time. In addition, this book aims at giving a systematic and in-depth coverage of semiconducting polymer composites from their fundamental concepts to morphology control and their applications in real devices for researchers already working in the field. Consequently, particular attention is given to the unique advantages of semiconducting polymer composites where polymers with specific functionalities are employed to form a multicomponent material with a desired morphology in order to obtain required materials properties and high-performance devices.

This book contains three parts, where the first part describes the principles and concepts of semiconducting polymer composites, including the mechanism of morphology formation, morphology characterization, energy level alignment at interfaces, energy transfer between the components, percolation theory, and processing techniques. These composites can be classified into two categories in terms of functionality of the components, mainly the matrix polymer involved, which is detailed in Parts II and III, respectively. Part II discusses the semiconducting/insulating polymer composites where a conjugated polymer or an organic semiconductor is dispersed in an insulating polymer matrix, forming a composite with exceptional properties. Part III is concerned with semiconducting/semiconducting polymer composites where conjugated polymers are used as the matrix. The applications of these composites in, for example, polymer solar cells, light-emitting diodes, transistors, and biosensors are presented.

I am greatly indebted to my colleagues who have been working in the respective fields for years and have agreed to contribute their expertise to this book. Their support made it possible to present the current state-of-the-art overview of semiconducting polymer composites in terms of both its academic value and potential applications.

I would also like to thank the people at Wiley-VCH who offered me this opportunity initially, helped me to overcome numerous difficulties, and made it become reality eventually.

Changchun, China   Xiaoniu Yang
April 2012

1
Solubility, Miscibility, and the Impact on Solid-State Morphology
Florian Machui and Christoph J. Brabec

1.1
Introduction

In recent years, organic semiconductors have been of increasing interest in academic and industrial fields. Compared to their inorganic counterparts, they offer various advantages such as ease of processing, mechanical flexibility, and potential in low-cost fabrication of large areas [1]. Furthermore, modifications of the chemical structure allow tailoring material properties and thus enhancing the applicability [2]. After the discovery of metallic conduction in polyacetylene in 1977 by Heeger, MacDiarmid, and Shirakawa, the path was paved for new material classes of electrical conductive polymers, possible due to chemical doping of conjugated polymers. This resulted in an increase of electrical conductivity by several orders of magnitude [3]. The main advantage of organic semiconductors is their processability from solution, which opens different applications such as flat panel displays and illumination, integrated circuits, and energy conversion [4–7]. Before widespread commercial application, further scientific investigations are necessary to achieve improved device performance and environmental stability.

The first organic solar cells were based on an active composite consisting of one single material between two electrodes with different work functions. Light absorption forms Coulomb-bound electron–hole pairs, so-called excitons, which have to be separated for charge generation [8]. In single-material active layers, this is possible by overcoming the exciton binding energy, either thermally or at the contacts [9]. Since both processes have rather low (<1%) efficiencies for pristine organic semiconductors, only few excitons are dissociated and recombination is very dominant. Therefore, single-layer organic solar cells exhibit device efficiencies far below 1% [10]. The first organic bilayer solar cell was presented by Tang, where copper phthalocyanine in combination with a perylene derivative is used as light absorption composite. In bilayer devices, excitons could diffuse within the donor phase toward an interface with a strongly electronegative acceptor material, which provides enough energy for exciton separation [11]. The electron gets transferred to the acceptor (i.e., lower in energy) and the hole remains on the donor. Currently, the most commonly used concept for the active layer in organic photovoltaic devices is the bulk heterojunction (BHJ), which consists of an interpenetrating network of a hole conductor and an electron acceptor, taking care of the low exciton diffusion length [12]. The main advantage of the BHJ concept is the increased interfacial surface leading to very efficient exciton dissociation within the whole active layer of the solar cell. The most commonly employed materials are conjugated polymers as donors and fullerene derivatives as acceptors [13–16]. By spontaneous phase separation, a specific nanostructure is formed that is decisive for the charge transport, since charge separation takes place at the interface. In the field of organic photovoltaics, several groups have realized devices with efficiencies over 6% [17–20]. Significant improvements have raised certified efficiencies up to 8.3% and novel concepts are under investigation to reach efficiencies beyond 10% (Konarka, http://www.konarka.com; Heliatek, http://www.heliatek.com.) [21].

For an efficient bulk heterojunction solar cell, good control of morphology is a key aspect, which is mainly influenced by the components' solubility during processing, the components' miscibility, and the formation of the resulting film. Solubility describes to what extent a substance dissolves in a particular solvent. This is the key phenomenon with regard to the design of inks and solvent systems with mutual multicomponent solubility regimes. The miscibility of several components in the film is mainly influenced by thermodynamic parameters. Film formation is additionally influenced by the surface energy differences of the substrate to the printing medium as well as by kinetic aspects.

Upscaling from lab to mass production facilities is one of the major necessities for cost optimization. In the case of organic solar cells, this is possible by large-area roll-to-roll processing, allowing throughputs of $10\ 000\ \text{m}^2\ \text{h}^{-1}$. This is orders of magnitude higher compared to silicon processing capabilities [22]. Currently, the most employed deposition method for organic solar cells is spin coating, since inherent advantages such as high film uniformity and ease of production are suitable for research activities. However, spin coating is very unfavorable for production due to its limitation in size. Doctor blading as an alternative coating technique is more suitable for larger area substrates and is easily transferred to roll-to-roll processing. For all of these techniques, it is necessary to know of the ink's solubility to adjust the formulation. Accordingly, the material parameters for ink definition are viscosity, evaporation rate of the solvent systems, and the spreading behavior. These phenomena together define the quality and the functionality of an organic semiconductor layer. Due to the high technical relevance for organic photovoltaics and, more generally, for organic electronics, the impacts of these phenomena on the performance and functionality of bulk heterojunction composite formation are the major topics in this chapter.

1.2
General Aspects

In general, chlorinated solvents are commonly used for processing in laboratories, which have restricted application in industrial operation due to safety risks and processing costs. Environment-friendly inks are therefore one decisive criterion for mass production that should provide full functionality. Since solubility is one of the determining factors for processing of the active layer in organic solar cells, several approaches are under investigation to predict solubility of the materials in question.

1.2.1
Solubility

Different approaches can be utilized to determine the solubility of a material. While simulation of solubility is a helpful tool to predict material behavior, experimental verification is of utmost importance. In order to reduce the expensive and time-consuming experimental efforts as well as frequent toxicity issues, simulations are a welcome tool to accompany experiments. One possibility to predict the material solubility is the use of solubility parameters, which was first proposed by Hildebrand and Scott and diversified by Hansen [23, 24]. In this approach, the energy of mixing is related to the vaporization energies of pure components. For liquids as well as for polymers, the solubility parameter δ was defined as the square root of the cohesive energy density (CED) with ΔE_v as energy of vaporization and V_m as average molar volume. Here the energy of mixing is related to the energies of vaporization of the pure components according to Eqs. (1.1)–(1.3). The contributions to ΔE_v in Eq. (1.2) are the difference in enthalpy of evaporation ΔH, the absolute temperature T, and the global ideal gas constant R.

$$\delta = \text{CED}^{1/2} = (\Delta E_v / V_m)^{1/2}, \tag{1.1}$$

$$\Delta E_v = \Delta H - RT. \tag{1.2}$$

Blanks, Prausnitz, and Weimer assigned the separation of vaporization energy into a nonpolar, dispersive part and a polar part [25, 26]. The polar part was further divided into dipole–dipole contribution and hydrogen bonding contribution by Hansen with δ_D as solubility parameter due to dispersion forces, δ_P as solubility parameter due to polar dipole forces, and δ_H as solubility parameter due to hydrogen bonding interactions according to Eq. (1.3) [27–29].

$$\delta^2 = \delta_D^2 + \delta_P^2 + \delta_H^2. \tag{1.3}$$

Hansen solubility properties are usually plotted in a three-dimensional coordinate system with the Hansen parameters as x, y, z axes. The coordinates of a solute can be determined by analyzing the solubility of a solute in a series of solvents with known Hansen parameters. By fitting a spheroid into the solubility space, the solubility volume of this solute can be identified. The solubility space of a solute is defined by the origin of a spheroid, resulting from the three coordinates, and the three radii in each dimension, with solvents inside the spheroid and nonsolvents outside. The radius of the sphere, R_0, indicates the maximum difference for solubility. Generally, good solvents are within the sphere, and bad ones are outside of it. Furthermore, the solubility "distance" parameter, R_a, between one solvent and one solute reflecting their respective partial solubility parameters can be defined with Eq. (1.4), with δ_{D2} as dispersive component for the solvent, δ_{D1} as dispersive component of the solute, and a, b, and c as weighting factors. Setting of $a=4$ and $b=c=1$ was suggested by Hansen based on empirical testing. To convert the Hansen spheroid into an ellipsoid, different ratios of weighting factors are used. When the scale for the dispersion parameter is doubled, the spheroidal shaped volume is converted into a spherical body [24].

$$R_a^2 = a(\delta_{D2} - \delta_{D1})^2 + b(\delta_{P2} - \delta_{P1})^2 + c(\delta_{H2} - \delta_{H1})^2. \quad (1.4)$$

Further studies by Small revealed that solubility parameters of polymers could also be calculated by using group contribution methods, which was intensified by Hoy, van Krevelen, and Coleman *et al.* [30–33]. The properties of molecules are investigated by separating them into smaller subgroups. The basic assumption is that the free energy of a molecule transfer between two phases is the sum of its individual contributions of groups, and that these group contributions are independent of the rest of the molecule. There is an obvious trade-off in group contributions. It is possible to define several groups in different ways. The more the subgroups used, the more accurate the group contributions become, but the less likely that there is sufficient statistical data to make predictions. More examples have been employed elsewhere [34–36]. For predicting solubility parameters using the group contribution method, frequently the following approach is used with F_i as molar attraction constant of a specific group i and V_m as molar volume:

$$\delta = \left| \frac{\sum F_i}{V_m} \right|. \quad (1.5)$$

Another method to predict solubility of solutes in different solvents is based on the prediction of the activity coefficient using density functional theory [37]. Molecules exhibit a rigid structure, but can possess different conformations, whose physical and chemical properties depend on their ultimate three-dimensional confirmation. Jork *et al.* showed that different conformations have different influence on the predictions of the activity coefficient [38]. Klamt *et al.* introduced a conductor-like screening model for real solvents (COSMO-RS), which allows a priori calculation of chemical potentials of one component within an arbitrary environment [39–42]. Here, modeling is realized by statistical thermodynamics where interacting molecules are substituted by corresponding pairwise interacting surface segments with densely packed contact areas. Since every segment has a constant charge density σ, the characterization of a molecule is possible by knowing the distribution function of the charge density $P(\sigma)$, the σ-profile. With that the properties of the molecules are solely dependent on the number of segments. The σ-profile of a pure component results directly from the density functional theory calculation. Further methods are based on molecular dynamics or Monte Carlo simulations but discussions are beyond the topic of this chapter.

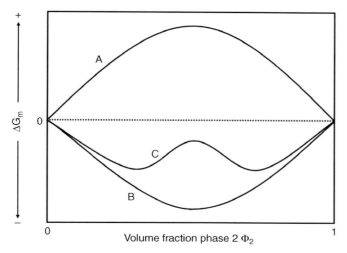

Figure 1.1 Gibbs free energy of mixing as a function of composition. (According to Refs. [43, 44].)

1.2.2
Miscibility–Thermodynamic Relationships

A decisive criterion of organic semiconductor applications is their ability of mixing. In general, blends of two or more components can be categorized according to the miscibility of their phases in one-phase or multiphase systems. Miscibility is usually defined by thermodynamic parameters. Here the Gibbs free mixing enthalpy ΔG_m is decisive for compatibility of two phases. Figure 1.1 shows the Gibbs free energy as a function of compositions. If ΔG_m is positive, the components are not miscible (A). If ΔG_m is negative and the second derivative is positive, both components are totally miscible (B). Independent of composition, a homogeneous blend is formed. If ΔG_m is negative and the second derivative is negative as well, the components are partially miscible (C). Phases with different composition are formed, which consist of both components [43, 44].

$$\Delta G_m = \Delta H_m - T\Delta S_m. \tag{1.6}$$

ΔG_m can be determined according by changes in enthalpy (ΔH_m) and entropy of mixing the components (ΔS_m). Compared to low molecular mass components, the entropy increase is low for mixing polymers. Mixing of two polymers results in a smaller increase of ΔS_m as compared to a binary blend of two low molecular weight components. Therefore, according to Eq. (1.6), the enthalpy change is the decisive parameter for thermodynamic miscibility [43]. The relatively smaller increase of entropy for polymers versus small molecules can be explained with Figure 1.2. The two-dimensional grids in Figure 1.2 represent places for molecules or for polymer segments. The number of possible configurations W is significantly higher for the

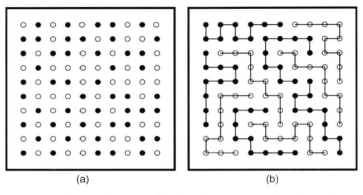

Figure 1.2 Schematic depiction of blends with components of smaller molar mass (a) and higher molar mass (a). (According to Refs [43, 44].)

arrangement with the small molecules. With $S \sim kT\ln(W)$, the lower entropy increase for polymer blends becomes obvious.

Since Gibbs free energy ΔG_m cannot be determined directly, thermodynamic models are used for the estimation. An often used model for polymer–polymer systems is the Flory–Huggins theory [45]. The Flory–Huggins definition of the Gibbs free energy and its implication on polymer blends are discussed in Eq. (1.7). It describes the free energy of binary systems, with the first two parts of the equation representing the entropic part and the third part describing the enthalpic phenomena. Here, φ_i is the volume fraction of component i, V_i is the molar volume of component i, B_{12} is the interaction parameter, and R is the ideal gas constant. In the case of polymer blends, the free energy is dominated by the enthalpic changes, which need to be negative for miscible systems. ΔH_m is directly proportional to the number of interactions between the two components, and becomes negative for strong interactions such as ion, acid–base, hydrogen bonds, or dipole–dipole interactions.

$$\Delta G_m = \left(\frac{\varphi_1}{V_1} \ln \varphi_1 + \frac{\varphi_2}{V_2} \ln \varphi_2 + \varphi_1 \varphi_2 B_{12} \right) RTV. \tag{1.7}$$

1.3
Solubility, Solvents, and Solution Formulations

1.3.1
Solubility

In this chapter, the solubility of organic semiconductors, their influence on OPV devices, and their correlation with Hansen solubility parameters (HSPs) are discussed. Before that, experimental methods to determine the absolute solubility of organic semiconductors are reviewed. High-performance liquid chromatography

(HPLC) is a separation method to identify and quantify exact concentrations of nonvolatile components. Saturated filtered solutions are analyzed and compared with standard solutions with known concentrations [48]. Spectrophotometrical measurements are also a commonly used method to determine the absolute solubility. First, saturated solutions are filtered or centrifuged to obtain true solutions. Next, these solutions are further diluted and characterized by optical absorption measurements. By comparing the optical density (OD) of the investigated solutions with the OD of calibrated master solutions, the solubility of the component in the investigated media can be determined. Examples for determination of organic semiconductor solubility measurements with this method have been reported by Walker et al. and Machui et al. [46, 47].

Ruoff et al. analyzed the solubility of pure C_{60} in different solvents [48]. HPLC was used to measure the solubility at room temperature in 47 solvents. Categorizing the solvents according to their chemical structure helped to identify good solvents such as naphthalenes and halogenated aromatics. In the first study on conjugated polymer:fullerene bulk heterojunction solar cells, the limited solubility of pure C_{60} in organic solvents and their tendency to crystallize during film formation was recognized by members of the Heeger group [12, 49]. Homogeneous stable blends with more than 80 wt% fullerene content became processable by the use of soluble C_{60} derivatives such as [6,6]-phenyl-C_{61}-butyric acid methyl ester ($PC_{61}BM$). A rough estimation of the solubility of $PC_{61}BM$ in toluene and chlorobenzene (CB) was achieved via saturated solutions by Hoppe and Sariciftci and reported with 1 wt% in toluene and 4.2 wt% in CB [64]. Kronholm and Hummelen later on published solubility values for $PC_{61}BM$ and [6,6]-phenyl-C_{71}-butyric acid methyl ester ($PC_{71}BM$) in different aromatic solvents, that is, toluene, p-xylene, o-xylene, CB, chloroform, and 1,2-dichlorobenzene (o-DCB) [50]. Solubility was determined by HPLC analysis of the liquid phase at room temperature. For both $PC_{61}BM$ and $PC_{71}BM$, highest solubility was found in o-DCB (30 mg ml^{-1} for $PC_{61}BM$), followed by CB and chloroform (each 25 mg ml^{-1}) and o-xylene, toluene, and p-xylene (<20 mg ml^{-1}). $PC_{71}BM$ was in all cases better soluble than $PC_{61}BM$. Troshin et al. analyzed the solubility of different fullerene derivatives and compared them to the resulting device performance, which is shown in Figure 1.3 [51]. Especially remarkable is a steep increase of short-circuit current density (J_{SC}), fill factor (FF), and power conversion efficiency (PCE, η) for increasing fullerene solubility in CB from 0 to about 40 mg ml^{-1}. Higher solubility values of about 60 mg ml^{-1} again resulted in a decrease of device performance. For open-circuit voltage (V_{OC}), an increase until a solubility of 30 mg ml^{-1} was recognizable. Higher solubility values did not change V_{OC}.

Hansen and Smith introduced Hansen solubility parameters for organic semiconductors and analyzed pristine C_{60} in organic solvents [52]. It was concluded that C_{60} would be soluble in polymers with aromatic rings or atoms that are significantly larger than carbon, such as sulfur or chlorine. The temperature-dependent solubility and the mutual solubility regimes for poly(3-hexylthiophene-2,5-diyl) (P3HT), $PC_{61}BM$, and small bandgap polymer-bridged bithiophene poly[2,6-(4,4-bis-(2-ethylhexyl)-4H-cyclopenta[2,1-b;3,4-b']-dithiophene)-alt-4,7-(2,1,3-

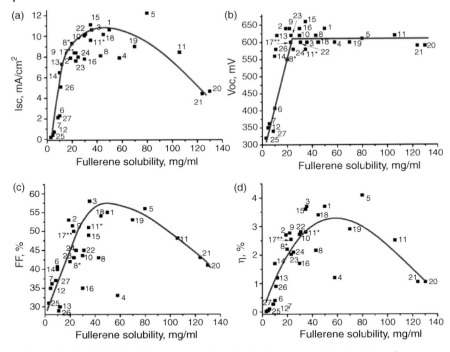

Figure 1.3 (a–d) Relationship between solar cell output parameters (I_{SC}, V_{OC}, FF, and η, respectively) and solubility of the fullerene derivative used as electron-acceptor material in the active layer. The lines are included as a guide for the eye. (From Ref. [51].)

benzothiadiazole)] (PCPDTBT) have been analyzed [47]. For the dominantly amorphous polymer PCPDTBT and the fullerene, results showed a good consistency over a broad temperature regime. Due to the semicrystalline character of P3HT, an exact determination of the solubility parameters was found difficult in a temperature regime of 25–140 °C. With increasing temperature, the solubility radius of P3HT increases significantly as well, which was explained by breaking of aggregates at elevated temperatures. Mutual solubility regimes for all three components have been identified as shown in Figure 1.4. For P3HT, the HSP parameters at 60 °C were reported as $\delta_D = 18.7 \, \text{MPa}^{1/2}$, $\delta_P = 1.4 \, \text{MPa}^{1/2}$, $\delta_H = 4.5 \, \text{MPa}^{1/2}$, and solubility radius $R_0 = 4.3 \, \text{MPa}^{1/2}$. The δ_D, δ_P, δ_H, and R_0 values were determined to be 17.3, 3.6, 8.7, and 8.2 $\text{MPa}^{1/2}$ for PCPDTBT and 18.7, 4.0, 6.1, and 7.0 $\text{MPa}^{1/2}$ for $PC_{61}BM$, respectively, at 60 °C.

Park et al. used Hansen solubility parameters and showed that non-halogenated solvent blends with the same Hansen parameters as o-DCB can be used to reach comparable device performance [53]. They mixed mesitylene (MS) with acetophenone (AP) in different ratios to match o-DCB Hansen parameters. Different mixtures of AP and MS were used with different ratios resulting in PCEs ranging from 1.5% (pure MS) to 3.38% (20 vol.% acetophenone) for P3HT:$PC_{61}BM$ cells with best external quantum efficiency (EQE) match with o-DCB. This has so far been

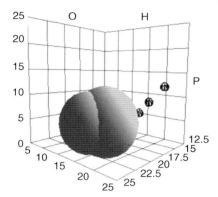
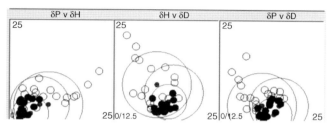

Figure 1.4 HSP diagram for solutes at 60 °C with 34 solvents, 2.5 g l^{-1} for P3HT, PCPDTBT, and PC$_{61}$BM. (From Ref. [47].)

the first combination for solvent blends and Hansen solubility parameters for organic semiconductors. Walker *et al.* analyzed a conjugated polymer 3,6-bis(5-(benzofuran-2-yl)thiophen-2-yl)-2,5-bis(2-ethylhexyl)pyrrolo[3,4-c]pyrrole-1,4-dione (DPP(TBFu)$_2$) and PC$_{71}$BM [46]. The solvents were classified into good, intermediate, and poor solvents. For PC$_{71}$BM, mostly higher solubility values were found in comparison to PC$_{61}$BM. The average δ_D, δ_P, and δ_H parameters were 19.33 ± 0.05, 4.78 ± 0.50, and 6.26 ± 0.48 MPa$^{1/2}$, respectively, for DPP(TBFu)$_2$ and 20.16 ± 0.28, 5.37 ± 0.80, and 4.49 ± 0.57 MPa$^{1/2}$, respectively, for PC$_{71}$BM. Atomic force microscopy (AFM) images of films prepared with chloroform, thiophene, trichloroethylene, and carbon disulfide were compared before and after annealing at 110 °C for 10 min. As-cast devices with the different solvents showed poor efficiencies. This is in agreement with the AFM images showing little phase separation. Annealing improves the PCE with efficiencies of up to 4.2% for carbon disulfide and 4.3% for chloroform. It was concluded that good solvents for both components result in optimal phase separation after annealing and HSPs could be used as a general tool for designing and understanding of solution-processed devices.

1.3.2
Solvents

As the active layer of organic solar cells is typically processed from solution, morphology is mainly determined by interactions between the used semiconductor solutes and the solvent during film formation. In this chapter, the influence of the used solvent on the resulting morphology and thus device performance is discussed. Different approaches to manipulate the morphology by solution processing methods are introduced.

1.3.2.1 Impact of Different Solvents on the Solid-State Morphology

Generally, good device efficiencies require the use of solvents that contain halogens (e.g., chloroform (CF), CB, o-DCB, and 1,2,4-trichlorobenzene (TCB)), whose toxicity poses potential problems for manufacturing [54–60]. Dang et al. compared different publications of P3HT:$PC_{61}BM$ analyzing material parameters and resulting device efficiencies including a comparison of different solvents and device performance [61]. For the most popular solvents such as CB and o-DCB, most PCEs were in the range of 2.5–4%. However, reports for other solvents for device processing such as CF, toluene, xylene, and tetrahydronaphthalene with also high efficiencies were found.

The choice of solvent has a great influence on the resulting morphology and thus on the device performance. This phenomenon was observed in case of poly [2-methoxy-5-(3,7-dimethyloctyloxy)]-1,4-phenylenevinylene (MDMO-PPV) blended with $PC_{61}BM$ by Shaheen et al., who compared toluene and CB as processing solvents [62]. They found a threefold better device performance for CB-processed cells, mainly attributed to higher short-circuit current density and better fill factor due to the better solubility of both components in CB (Figure 1.5). AFM images showed a smaller scale of phase separation, that is, smaller $PC_{61}BM$-rich domains in MDMO-PPV-rich matrix suppressing phase segregation of $PC_{61}BM$ molecules into clusters, and smoother surface roughness that improved the interface contacts to the cathode.

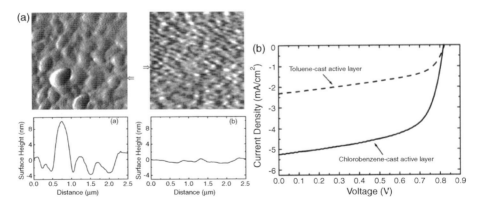

Figure 1.5 (a) AFM images showing the surface morphology of MDMO-PPV:$PC_{61}BM$ blend films when spin coated from a toluene solution (left) and from a CB solution (right). The images show the first derivative of the actual surface heights. The cross sections of the true surface heights for the films were taken horizontally from the points indicated by the arrow. (b) Characteristics for devices with an active layer that is spin coated from a toluene solution (dashed line) and from a CB solution (full line). (From Ref. [62].)

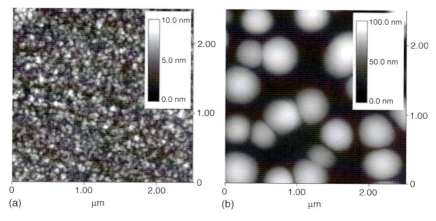

Figure 1.6 Tapping mode AFM topography scans of MDMO-PPV:PC$_{61}$BM 1:4 (by weight) blended films, spin cast from CB (a) and toluene (b) solution. The toluene-cast film exhibits height variations that are one order of magnitude larger than those on CB-cast films. Features of a few hundred nanometers in width are visible in (a), while features in (b) are around 50 nm. (Reproduced from Ref. [63].)

Hoppe *et al.* further investigated the influence of various solvents on the morphology [63, 64]. For toluene as processing solvent, photoluminescence (PL) measurements indicated pure PC$_{61}$BM clusters with larger extent than exciton diffusion range. PL measurements also showed increased material phase separation after annealing and a photocurrent loss due to PC$_{61}$BM clusters. Phase separation for toluene was dependent on blend ratio and solute concentration. The comparison of PC$_{61}$BM in toluene and CB showed larger PC$_{61}$BM clusters in case of using the poorer solvent toluene as shown in Figure 1.6 [63].

For MDMO-PPV:PC$_{61}$BM, Rispens *et al.* analyzed the influence of solvents on crystal structure of PC$_{61}$BM as shown in Figure 1.7 [65]. A comparison of *o*-DCB, CB, and xylene as spin casting solvent showed that CB was the best choice as processing solvent. Single PC$_{61}$BM crystals were obtained from CB resulting in

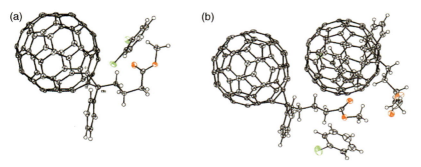

Figure 1.7 Molecular structure of PC$_{61}$BM, crystallized from (a) *o*-DCB and (b) CB (red = oxygen; green = chlorine). (From Ref. [65].)

significantly higher charge mobility than from other solvents resulting in amorphous confirmations.

The influence of different solvents on morphology was also investigated by Ruderer et al. for P3HT and $PC_{61}BM$ [66]. Spin-coated films with processing solvents such as CF, toluene, CB, and xylene were investigated by optical microscopy, grazing incidence wide-angle X-ray scattering (GIWAXS), AFM, X-ray reflectivity (XRR), and grazing incidence small-angle X-ray scattering (GISAXS) investigations. Using this wide range of investigation tools led to good understanding of how processing solvents can manipulate the lateral and vertical phase separation. Major influence on device performance resulted from vertical phase separation. $PC_{61}BM$ clusters were formed for low-solubility solvents. P3HT crystallinity was mainly influenced by annealing, and increased with higher boiling point of the solvent attributed to longer drying time during spin coating. The lattice constants were independent for the used solvents. Figure 1.8 shows the schematic vertical morphology resulting from different solvents for P3HT (white areas) and $PC_{61}BM$ (black areas), neglecting phases containing both components. These structures were reconstructed representing the findings with aforementioned methods, suggesting vertical nanostructures for CF-, toluene-, CB-, and xylene-processed films. For toluene-, CB-, and xylene-processed films, lateral nanostructures were found. P3HT accumulation at the bottom was found for toluene- and CB-processed films, while $PC_{61}BM$ accumulation at the bottom was found for chloroform and xylene. P3HT enrichment at the bottom and $PC_{61}BM$ accumulation at the top are considered as advantageous for the "normal" device architecture. Nevertheless, there was no great difference in device performance for all four solvents. It was concluded that lateral and vertical structures are not the only determining factors as long as the phase separation and the material distribution are in the range of the exciton diffusion length (here from 35 to 65 nm) and percolation paths are recognizable.

Yu compared the influence of different solvents on device performance [67]. As processing solvents, CF, CB, o-DCB, and TCB were used. According to absorption and PL measurements, charge transport dark current density–voltage (j–V) curve, XRD pattern, and AFM images, a higher P3HT crystallinity for higher boiling point solvents was concluded, since polymer chains have longer time for self-organization. This resulted in increased absorption and charge carrier mobility leading to

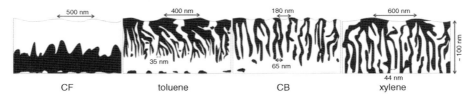

Figure 1.8 Black and white schematic morphology of annealed P3HT:$PC_{61}BM$ films made using CF, toluene, CB, and xylene solutions, as reconstructed from the results of AFM, XRR, and GISAXS investigations. Black areas correspond to pure $PC_{61}BM$ phases and white to pure P3HT phases. Characteristic lengths are indicated. (Reprinted from Ref. [66].)

higher device performance for high boiling point solvent-processed devices. Kwong *et al.* processed P3HT:TiO$_2$ nanocomposite solar cells using different solvents for spin coating the active layer [68]. A comparison of tetrahydrofuran (THF), CB, CF, and xylene showed that device performance can be strongly influenced by the used solvent. Best cells in this case were achieved with xylene. It was concluded that a good solvent for P3HT with a low evaporation rate may improve the mixing of the components resulting in better exciton dissociation and short-circuit current density. AFM studies showed that the roughest surface was obtained for films spin coated from xylene. Park *et al.* compared the influence of different solvents on solar cells made of the copolymer poly[*N*-9″-heptadecanyl-2,7-carbazole-*alt*-5,5-(4′,7′-di-2-thienyl-2′,1′,3′-benzothiadiazole)] (PCDTBT) in bulk heterojunction composites with the fullerene derivative PC$_{71}$BM [69]. Transmission electron microscopy (TEM) and AFM comparison for CF-, CB-, and *o*-DCB-processed films showed a decreased phase separation with decreasing volatility of solvents and a higher incident photon to electron conversion efficiency (IPCE).

Jaczewska *et al.* presented a polymer–solvent diagram including film structures for polystyrene (PS):polythiophene blends (1 : 1, w/w) for different processing solvents [70]. Structures were observed with different microscopic techniques, and solubility parameters were used for establishing a polymer–solubility versus solvent–solubility relation. A relation between film morphology and stability of the layers showed a dependence on the surface energy. Dewetting effects could be inhibited by decreasing polythiophene content. Furthermore, a ternary phase diagram was developed for the system polymer, fullerene, and solvent [64]. At constant temperature and pressure, a schematic diagram is shown in Figure 1.9. A decreasing amount of solvent leads to higher repulsive interactions between polymer and

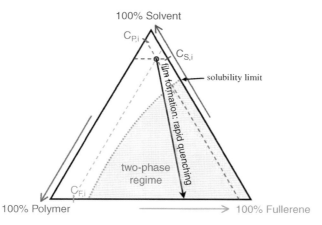

Figure 1.9 Schematic ternary phase diagram of a polymer–fullerene–solvent system at constant temperature *T* and constant pressure *p*. The arrows indicate the direction of increasing concentration; $C_{S,i}$, $C_{P,i}$, and $C_{F,i}$ are the initial concentrations of solvent, polymer, and fullerene in the solution, respectively. During film formation, a more or less rapid quenching of the solution toward a solid-state blend takes place upon extraction of the solvent. (Adapted from Ref. [64].)

fullerene molecules. Removing the solvent quickly enough can freeze the blend morphology of the polymer and the fullerene, because phase separation is a temperature-dependent process in time and the system is quenched in a metastable state. Thermal annealing does reactivate the molecule mobility and allows a reorientation and eventual recrystallization of the polymer chains within the composite. In case of very slow drying (i.e., for high boiling point solvents), molecules have more time to orient resulting in higher phase separation and larger domains.

1.3.2.2 Non-Halogenic Solvents

Xylene is an often used non-halogenic solvent that frequently offers comparable device performance as with halogenated solvents [61]. *p*-Xylene was used by Berson *et al.* to form P3HT nanofibers [71]. P3HT was previously dissolved in *p*-xylene at elevated temperatures. Nanofibers are formed after cooling to room temperature without precipitations for concentrations in a range of 0.5–3 wt% as shown in Figure 1.10. The time-dependent formation of nanofibers is monitored for a solution of P3HT in *p*-xylene. With more concentrated *p*-xylene solutions, a homogeneous thick film is obtained, which is crucial for the fabrication of photovoltaic active layers. The dimensions of the nanostructures have been determined from the AFM images; the nanofibers had lengths ranging from 0.5 to 5 μm, thicknesses ranging from 5 to 15 nm, and widths ranging from 30 to 50 nm. Moreover, cyclohexanone was also used for fiber formation. A network of fibers was obtained by using a dilute solution in cyclohexanone. Overall, this method resulted in device efficiencies of 3.4% for P3HT:PC$_{61}$BM blends with no further need of annealing.

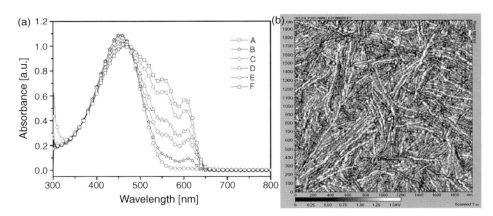

Figure 1.10 (a) Absorption spectra of a 1 wt% solution of P3HT in *p*-xylene heated at 80 °C to ensure complete dissolution of the polymeric material and then allowed to evolve after cooling to room temperature for (A) 2, (B) 4, (C) 6, (D) 21, (E) 28, and (F) 48 h. The solutions are cooled at a rate of 20 °C h^{-1}. (b) AFM phase images of a pristine nanostructured P3HT:PC$_{61}$BM film deposited on glass from a *p*-xylene solution containing 1 wt% P3HT and 1 wt% PC$_{61}$BM. (From Ref. [71].)

Due to the toxicity and processing problems of halogenic solvents, replacements with non-halogenic solvents gained interest as a concept to lower safety risks and processing costs while keeping the device performance high. Tetralene (1,2,3,4-tetrahydronaphthalene) was first suggested by Hoth *et al.* [72]. Tetralene is a high-boiling solvent showing a lower surface tension compared to *o*-DCB. The tetralene formulation provided reliable inkjet printing, but suffered from poor morphology and significantly rougher surfaces demonstrated in AFM images. This is specific to the inkjet-printed trials since doctor-bladed cells fabricated using tetralene produced cells with PCE of 3.3% for P3HT:PC$_{61}$BM [73]. Furthermore, toluene was used as a processing solvent. As mentioned previously, device performance is limited due to the lower solubility compared to halogenated solvents resulting in the formation of PC$_{61}$BM clusters, which restrain the charge separation [62–64].

1.3.2.3 Solvent Blends

Solvent blends can be used for device fabrication since they offer the possibility to adjust the morphology via the different solubility of the solutes in the various systems. For devices containing P3HT and PC$_{61}$BM, different groups investigated the influence of solvent mixtures. Kawano *et al.* reported that cells processed with a solvent mixture of *o*-DCB/CF in 60/40 (v/v) ratio had a better performance than the cells prepared from CB [74]. After annealing at 150 °C for 5 min, the cosolvent system achieved an efficiency of 3.73% compared to 3.34% for CB cells. Short-circuit current density and fill factor increased for the cosolvent system due to larger interfacial area between P3HT and PC$_{61}$BM. It was found that the cell efficiency improved by adding moderate amount of CF. The highest cell efficiency was obtained, when 40 vol% CF was added into *o*-DCB. Higher amounts of CF led to a drop in PCE. Furthermore, surface morphology was investigated, which showed that surface roughness was higher for the cosolvent system indicating a higher P3HT chain ordering. Lange *et al.* investigated the influence of adding TCB to CB as processing solvent [75]. Changes in the absorption spectra compared to the pure solvents where the P3HT absorption maximum occurred between the maxima of the two pure solvents. Therefore, it was concluded that adding TCB with a higher boiling point provides P3HT chains more time to form higher crystalline parts. Chen *et al.* mixed *o*-DCB with 1-chloronaphthalene, also providing a higher boiling point compared to *o*-DCB [76]. Absorption spectra showed again a redshift upon addition of the high boiling point solvent 1-chloronaphthalene indicating higher order due to longer time for self-organization. The cell efficiency peaked for 5 vol% 1-chloronaphthalene at 4.3%. *o*-DCB and mesitylene formulations (ratio 68 : 32) were used by Hoth *et al.* for inkjet-printed solar cells achieving PCE of 3% [72]. The solvent blend ratio was chosen to optimize droplet formation properties according to drop volume, velocity, and angularity of the inkjet print head. The combination of *o*-DCB and mesitylene served two purposes: *o*-DCB with the higher boiling point of 180 °C was used to prevent nozzle clogging and provided a reliable jetting of the print head, and mesitylene had a lower surface tension and was used to achieve optimum wetting and spreading of the solution on the substrate. Furthermore, it offered a higher vapor pressure and a lower boiling point compared to *o*-DCB and increased the drying rate of the solvent mixture, which is a critical parameter for phase separation. High-boiling solvents such as o-DCB or tetralene are an essential concept to develop inks for inkjet printing.

Influence of solvent blends was also investigated for blends of a polyfluorene copolymer poly(2,7-(9,9-dioctylfluorene)-*alt*-5,5-(4′,7′-di-2-thienyl-2′,1′,3′-benzothiadiazole)) (APFO-3) with $PC_{61}BM$ for CF, as well as solvent mixtures containing 1.2% CB, xylene, and toluene offering lower vapor pressure as compared to CF [77]. An increase in photocurrent for CF/CB blends was correlated with a finer phase separation, and a decrease in photocurrent for CF/toluene and CF/xylene was attributed to rougher surface morphologies. Furthermore, time-resolved spectroscopy supported morphological results. Wang *et al.* used blends of o-DCB and toluene for poly[2,3-bis-(3-octyloxyphenyl)quinoxaline-5,8-diyl-*alt*-thiophene-2,5-diyl] (TQ1) mixed with $PC_{71}BM$ resulting in PCEs of 4.5% [78]. Results suggest that only 5–20 vol.% o-DCB in the solvent blend system significantly increased J_{SC}, FF, and PCE. The differences to pure o-DCB were quite small. AFM topography images of the spin-coated films showed big grains in the range of 1 μm for films from pure toluene. Grain size decreased to 100 nm by adding 5 vol.% o-DCB, which corresponded well to the improved device performance. Smallest grain size resulted from pure o-DCB with the best efficiencies. The effect of mixed solvent was also studied for PCDTBT and $PC_{71}BM$ by Alem *et al.* [79]. CF and o-DCB were used as good solvents for these two materials. CF-processed films exhibited larger domains, showing increasing size with higher $PC_{71}BM$ content. The 1 : 1 mixing of CF and o-DCB was used for realizing optimum domain size resulting in power conversion efficiencies of up to 6.1%. Solvent blends containing CB and o-DCB were used for devices made from PCDTBT:$PC_{71}BM$ [69]. Increasing the amount of o-DCB in the CB/o-DCB mixture increased the contribution from $PC_{71}BM$ to the IPCE, showing pronounced peaks around 400 and 450 nm. o-DCB films showed significantly smaller phase separation. Overall, the increased IPCE could be correlated with the nanoscale phase separation.

1.3.2.4 Addition of Poor Solvents

Addition of nonsolvents to solvents can result in aggregate formation, which enhances the field-effect mobility of conjugated semiconductors. Park *et al.* added acetonitrile to chloroform and changed the P3HT organization from random coil conformation to an ordered aggregate structure [80]. Besides acetonitrile, different solvents such as hexane, acetone, ethanol, and dimethylformamide were added to chloroform-based P3HT inks as conformation modifiers. P3HT aggregation occurred at a certain solvent ratio and an additional redshifted absorption band appeared. Pristine P3HT–chloroform solution contained one peak at 455 nm, which was associated with intrachain π–π^* transition. For good solvents such as chloroform, P3HT chains were well dissolved, so no sign of molecular ordering occurred. Redshift of the absorption maximum and additional absorption bands was usually associated with ordered aggregates and interchain π–π stacking of P3HT. Both were related to an increased effective conjugation length of the chain segments in the P3HT solution, thereby decreasing energies. Moulé and Meerholz used nitrobenzene (NtB) as nonsolvent for P3HT:$PC_{61}BM$ in CB-based inks [81].

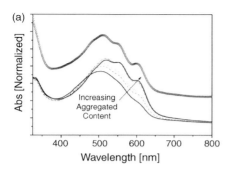
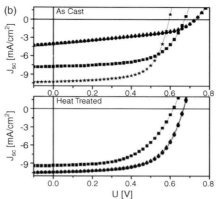

Figure 1.11 (a) UV–Vis spectra of 3:2 P3HT:$PC_{61}BM$ as-cast PV devices with 0% (solid line), 0.33% (dashed line), 0.67% (dotted line), 1.6% (dashed–dotted line), 3.2% (short dashed line), and 6.3% (solid line) nitrobenzene added into the CB solvent. Offset from the other spectra is the as-cast PV device from the o-xylene dispersion (triangles). (b) j–V curves of as-cast (upper) and heat-treated (lower) 3:2 P3HT:$PC_{61}BM$ devices. The devices were cast from CB-amorph (triangles), o-xylene-amorph (circles), o-xylene-np (squares), and CB/NtB (stars). (From Ref. [81].)

The volume fraction of P3HT aggregates in a P3HT:$PC_{61}BM$ solution could be increased from 60 up to 100% with increasing NtB content (Figure 1.11). Photovoltaic devices from P3HT:$PC_{61}BM$ mixtures with NtB addition resulted in device efficiencies of 4% without further thermal annealing. These experiments proved that a good part of the thin-film morphology can already be introduced on the solution level.

A further example was presented by Park *et al.* using blends of acetophenone and mesitylene [53]. The boiling point difference of MS (165 °C) and AP (202 °C) resulted in an increase of concentration of AP during solvent evaporation. The external quantum efficiency nearly doubled from a maximum of 35% at 500 nm for pure MS to 69% for the solvent blend as can be seen in Figure 1.12. One of the difficulties was obtaining the same drying conditions as o-DCB, which limited the ability to fully match the film thickness for different solvent blends. The better device performance of the solvent blended systems was assumed to result from lower series resistance and a superior morphology improving phase separation of P3HT and $PC_{61}BM$ that was analyzed by AFM measurements. The higher boiling point and lower evaporation rate of AP could facilitate reorganization or increase crystallinity.

Oleic acid (OA) was also reported to improve microstructure and device performance of P3HT:$PC_{61}BM$ devices [82]. After thermal annealing, the P3HT:$PC_{61}BM$ blend film with OA showed bigger domain sizes and roughness compared to films without OA. This is a result of enlarged P3HT domains with higher crystallinity analyzed by AFM and XRD measurements. The addition of OA improves the heteromolecular mixture in the solution and induces molecular local ordering in the resulting film. This allowed the formation of well-organized films with high mobility, resulting in high device performance up to 4.3%.

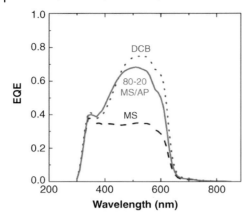

Figure 1.12 EQE measurement data for devices fabricated from o-DCB, MS, and 80 vol.% MS–20 vol.% AP mixture. (From Ref. [53].)

1.3.2.5 Processing Additives

One approach to control the morphology is the addition of small amounts of a high boiling point solvent with selected solubility into a host solvent. The advantages for this type of processing additives are the easy application to polymers with high and low solubility and the fact that no additional processing step is necessary [83]. For example, additives such as alkylthiols or diiodoalkanes are known to selectively help fullerene aggregation due to a better fullerene solubility compared to polymers [17, 84].

Peet et al. analyzed the influence of chain length of different alkane dithiols on the efficiency of P3HT:$PC_{61}BM$ and PCPDTBT:$PC_{71}BM$ solar cells [17]. Small concentrations of alkanethiols formed P3HT aggregations and modified the P3HT:$PC_{61}BM$ phase separation [85]. Since for PCPDTBT thermal or solvent annealing was not successful, additives were used. Addition of 1,8-octanedithiol into CB led to a redshift film absorption peak around 800 nm. This shift to lower energies was associated with enhanced $\pi-\pi^*$ stacking and indicated a PCPDTBT phase with more strongly and improved local structural order as compared to films processed from pure CB. Different chain lengths of alkane dithiols were analyzed. Best cell performance was achievable for the longest alkyl chain, 1,8-octanedithiol, resulting in cell efficiencies of up to 5.5%. AFM pictures showed that a specific chain length is necessary for morphological differences. While for butanedithiol no changes were recognizable compared to no additive processing, hexanedithiol addition showed larger domains.

Lee et al. investigated the use of processing additives on PCPDTBT:$PC_{71}BM$ organic solar cells [84]. Morphological control could be achieved with the criteria of a selective, differential solubility of the fullerene component and a higher boiling point compared to the host solvent. For the additive, different functional end groups of a 1,8-di(R)octane were used, achieving best results of 5.12 and 4.66% for R = I or Br, respectively. Figures 1.13 and 1.14 show the j–V curves of the devices with different additives and the schematic depiction of the role of additives, respectively.

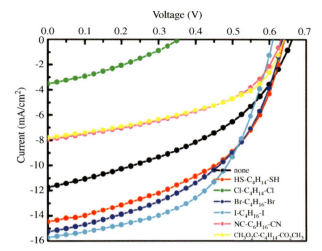

Figure 1.13 j–V characteristics of PCPDTBT/PC$_{71}$BM composite films with various additives: none (black), 1,8-octanedithiol (red), 1,8-dicholorooctane (green), 1,8-dibromooctane (blue), 1,8-diiodooctane (cyan), 1,8-dicyanooctane (magenta), and 1,8-octanediacetate (yellow). (Adapted from Ref. [84].)

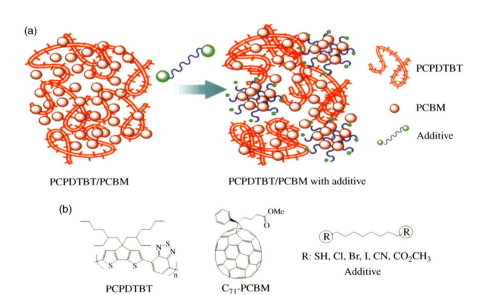

Figure 1.14 Schematic depiction of the role of the processing additive in the self-assembly of bulk heterojunction blend materials (a) and structures of PCPDTBT, PC$_{71}$BM, and additives (b). (Adapted from Ref. [84].)

Moet et al. found by modeling the photocurrent that the use of 1,8-octanedithiol can prevent recombination-limited photocurrent in PCPDTBT:$PC_{61}BM$ solar cells [86]. Modeling showed that the decay rate of bound electron–hole pairs is reduced by additive addition resulting in dissociation probability of 70% at short-circuit current.

The use of processing additives was further investigated by Su et al. for the polymer poly-{bi(dodecyl)thiophene-thieno[3,4-c]pyrrole-4,6-dione} (PBTTPD) in the system PBTTPD:$PC_{71}BM$ [87]. Diiodoalkanes with different chain lengths were added to chloroform solutions and analyzed with GISAXS and GIWAXS measurements. It was concluded that addition of the diiodoalkanes led to an improved dispersion of the $PC_{71}BM$ domains and, therefore, a better network morphology by reducing the grain boundaries of the $PC_{71}BM$-rich phases. Diiodohexane (DIH) provided the finest dispersion of $PC_{71}BM$, due to a balance of solubility for $PC_{71}BM$ and the interactions between additive and the polymer molecules. By using DIH, the polymer crystallinity could be increased and the device performance was improved from 5 to 7.3%.

Further GIWAXS measurements including the use of additives were investigated by Rogers et al. who used PCPDTBT in combination with $PC_{71}BM$ and diiodooctane or octanedithiol [83]. By using additives, the device performance could be increased from 3.2 to 5.5%. Both additives have a higher boiling point compared to the host solvent CB and the ability to solvate $PC_{71}BM$. Absorption measurements suggested increased chain aggregation and improved electrical properties were suggested from mobility and photoresponsivity measurements.

The role of additives in polymer crystallinity was further investigated by Agostinelli et al. using octanedithiol (ODT) for PCPDTBT:$PC_{71}BM$ films [88]. By using GIXRD, absorption spectroscopy, variable angle spectroscopic ellipsometry (VASE), and time-of-flight (TOF) hole mobility measurements, the degree of order was analyzed and accompanied by transient photovoltage (TPV) measurements changes in device performance were monitored. Upon addition of ODT, the polymer crystallinity was increased, resulting in higher charge pair generation efficiency. A series of polymers with alternating thieno[3,4-b]thiophene and benzodithiophene units was investigated by Liang et al. [89]. By using o-DCB/1,8-diiodooctane (97/3, v/v) as solvent, a more finely distributed polymer/fullerene interpenetrating network was obtained and a significantly enhanced solar cell conversion efficiency of up over 6% was achieved.

Chu et al. used a low-bandgap alternating copolymer of 4,4-bis(2-ethylhexyl)dithieno[3,2-b:2′,3′-d]silole and N-octylthieno[3,4-c]pyrrole-4,6-dione (PDTSTPD) and $PC_{71}BM$ as active layer with and without addition of 3% 1,8-diiodooctane (DIO) [90]. Without additive, the device performance dropped significantly below 1.0%. AFM studies of the film morphology showed that $PC_{71}BM$ formed too large isolated domains in the blend film prepared without using DIO. As a result, the J_{SC} dropped from 12.2 to 2.6 mA cm^{-2} and the V_{OC} and FF also decreased significantly. Addition of DIO to the solution resulted in much more uniform and finer domain structure, ideal for an effective polymer:$PC_{71}BM$ interpenetrating network. As a result, the device performance was greatly improved up to 6.7%. This finding highlights the importance of morphology control for high-performance solar cells.

Morana *et al.* investigated the effect of ODT on the formation of the charge transfer complex (CTC) for C-PCPDTBT and Si-PCPDTBT [91]. Despite the pristine C-PCPDTBT, no changes were observed in the absorption spectrum of the Si-PCPDTBT films prepared with ODT. Enhanced phase segregation in the C-PCPDTBT films upon addition of ODT caused increase in the molecular luminescence to CT luminescence ratio. This is due to the reduced concentration of CT complexes by a decrease in the contact area between the polymer and the fullerene because of phase separation.

1.3.2.6 Solution Concentration

The influence of solution concentration was investigated by Hoppe and Sariciftci with constant mixing ratio of MDMO-PPV and $PC_{61}BM$ [64]. Besides an increase in layer thickness with increasing concentration, also the fullerene cluster size detected by AFM analysis was increased. Further investigations have been performed by Baek *et al.* varying the solution concentration from 1 to 3 wt%. All solid film properties such as the crystalline structure formation, the interchain interaction, and the morphology were influenced [92]. $P3HT:PC_{61}BM$ absorption spectra for as-cast and annealed (150 °C for 10 min) films showed decreasing absorption with increasing concentration. Slower evaporation of the solvent at lower concentration of $P3HT:PC_{61}BM$ leads to better crystallization, stronger interchain interaction, and more ordered phase separation of P3HT. This holds for as-cast as well as for thermally annealed films.

1.3.3
Conclusive Outlook

Several approaches have been discussed how the solid-state microstructure of bulk heterojunction composites can be controlled by the design of intelligent solvent systems. Besides the choice of the right solvent, (i) addition of additional good solvents with differing drying properties has been demonstrated to control the domain size of either component, (ii) addition of nonsolvents was shown to trigger the nucleation and subsequent aggregation of individual components, and (iii) addition of processing additives was used to cause a coarsening of the microstructure.

The general ink design for organic semiconductor multicomponent composites is based on a few rules. Generally, the processing solvent has to supply a sufficient solubility, which is typically guaranteed by using halogenated aromatic solvent systems. The processing solvent mainly influences the active layer microstructure. Different $PC_{61}BM$ crystal structures were obtained by using CB, *o*-DCB, or xylene. Low-solubility solvents, in combination with a gradual variation of the surface energy, allow to control a gradient in the vertical phase separation of the two components. The kinetics of drying does impact the size of the aggregates. Slow drying (i.e., high boiling point solvents such as *o*-DCB) creates microstructures with an increased crystallinity as compared to lower boiling point solvents due to enhanced reorganization. Multicomponent solvent systems offer significantly more freedom:

Table 1.1 Solvent parameters of different key solvents for OPV.

Solvent	Hansen solubility parameters, $\delta_D + \delta_P + \delta_H$ (MPa$^{1/2}$)[a]	Molar volume (m^3 mol^{-1})[a]	Boiling point (°C)[b]	Density (g cm^{-3})[b]	Vapor pressure at 25 °C (kPa)[b]
Chlorobenzene	19.0 + 4.3 + 2.0	102.1	131.72	1.1058	1.6
o-Dichlorobenzene	19.2 + 6.3 + 3.3	112.8	180	1.3059	0.18
Chloroform	17.8 + 3.1 + 5.7	80.7	61.17	1.4788	26.2
o-Xylene	17.8 + 1.0 + 3.1	121.2	144.5	0.8802	0.88
Toluene	18.0 + 1.4 + 2.0	106.8	110.63	0.8668	3.79
1,2,4-Trichlorobenzene	20.2 + 6.0 + 3.2	125.5	213.5	1.459	0.057
Cyclohexanone	17.8 + 6.3 + 5.1	104	155.43	0.9478	0.53
Nitrobenzene	20.0 + 8.6 + 4.1	102.7	210.8	1.2037	0.03
1,8-Octanedithiol	17.2 + 6.8 + 6.4[c]	185.6[c]	269[d]	0.97[d]	0.012[d]
1,8-Dibromooctane	17.6 + 4.3 + 2.7[c]	188.6[c]	270[d]	1.477[d]	—

a) Ref. [24].
b) Laboratory solvents and other liquid reagents, in Ref. [93].
c) Ref. [94].
d) Material Safety Data Sheet Sigma–Aldrich.

solvent blends can be used to mimic solubility parameters of a good solvent by using nonhazardous solvents. Furthermore, solvent blends containing high and low vapor pressure solvents allow additional control over the degree of phase separation and interfacial area. Finally, high boiling point additives with selected solubility for one component over the other can trigger more finely distributed microstructures, preventing the aggregation of fullerene clusters. Table 1.1 summarizes the essential parameters for the most frequently used single solvents that are used for processing of organic electronic systems.

1.4
Miscibility

1.4.1
Methods

Miscibility is one essential concept in polymer science, since blended systems are commonly used to address multiple property optimizations as typical for many applications. Several methods can be used to determine the miscibility of two- and more component systems. Morphological investigations of blend systems can easily been done with microscopic methods. Several electron microscopic techniques are used, that is, scanning electron microscopy (SEM), TEM, AFM, and scanning tunneling microscopy (STM). Inhomogeneities are typically identified by scattering methods. Depending on the required resolution, visual light, X-ray, or neutron scattering methods are used. Neutron scattering is used for investigations in the nanometer domain, X-ray for structures below 1 nm and up to 40 nm, and visual light is used for structures between 100 nm and several microns. The advantage of neutron scattering is the possibility of analyzing also light elements. In the following, we will describe various miscibility aspects for polymer–polymer or polymer–fullerene-based bulk heterojunction composites.

1.4.1.1 Glass Transition

One criterion to distinguish the miscibility of blends is the glass transition temperature (T_g) that can be measured with different calorimetric methods [95]. T_g is the characteristic transition of the amorphous phase in polymers. Below T_g, polymer chains are fixed by intermolecular interactions, no diffusion is possible, and the polymer is rigid. At temperatures higher than T_g, kinetic forces are stronger than molecular interactions and polymer chain diffusion is likely. In binary or multicomponent miscible one-phase systems, macromolecules are statistically distributed on a molecular level. Therefore, only one glass transition occurs, which normally lies between the glass transition temperatures of the pure components. In partly miscible systems, interactions cause a glass transition shift of the pure components toward each other. For immiscible blends, the components are completely separated in different phases and the glass transitions of the pure components remain at their original temperature. Here it is important to emphasize that the appearance of one glass transition is not a measure of complete miscibility rather than a correlation with domain sizes of less than 15 nm. Various examples were discussed elsewhere [95].

1.4.1.2 Surface Energy

The difference in surface energy between two components can be used to define miscibility, as this was identified as one of the driving forces for vertical phase separation. Honda *et al.* analyzed the surface energy to understand the miscibility of P3HT and silicon phthalocyanine derivative (SiPc) – a light-harvesting dye [96]. The surface energy was assessed by contact angle measurements with ultrapure water on spin-coated films. Other studies have suggested as well that the surface energy between P3HT and $PC_{61}BM$ is the driving force for vertical phase separation in the binary blend [97–99]. The surface energies γ were estimated to be 20 and 29 mJ m^{-2} for P3HT and $PC_{61}BM$, respectively, also reported by Jaczewska *et al.* for P3HT [100]. Contrary results were reported by Oh *et al.* and Björström *et al.* with surface energies of 25.79 mJ m^{-2} for P3HT and 39.86 and 38.2 mJ m^{-2} for $PC_{61}BM$ [101, 102]. Nevertheless, SiPc with $\gamma_{SiPc} = 23$ mJ m^{-2} is close to P3HT but has an intermediate value between P3HT and $PC_{61}BM$. Binary blends of P3HT:SiPc and $PC_{61}BM$:SiPc showed that the SiPc molecules do not phase segregate from P3HT, as the surface energy did not change up to 40 wt% SiPc addition (Figure 1.15). On the other hand, SiPc does segregate at the air/film interface from $PC_{61}BM$ as suggested by a

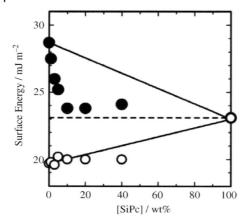

Figure 1.15 Surface energy of P3HT:SiPc (open circles) and PC$_{61}$BM:SiPc (closed circles) blend films plotted against a SiPc content. The broken line represents the surface energy of SiPc (23 mJ m^{-2}). The solid lines represent the surface energy of P3HT:SiPc or PC$_{61}$BM:SiPc predicted on the assumption of homogeneous dispersion of SiPc molecules in blend films. (Adapted from Ref. [96].)

steep decrease of the surface energy at only 10 wt% SiPc addition. Since the component with the lowest surface energy is segregated to the air/film interface, the total energy of the system becomes dominated by a surface layer of the low-energy component. For ternary blends containing P3HT:PC$_{51}$BM:SiPc, the wetting coefficient was used to predict the location of dyes in blend films. This concept was already utilized for the investigation of conductive carbon black particles, carbon nanotubes, CaCO$_3$ nanoparticles, and polymers [103–108]. In case of the ternary organic semiconductor composites, it was found that the SiPc molecules are most likely located at the P3HT:PC$_{61}$BM interface. SiPc molecules were found to be present in the disordered P3HT phases at the interface between P3HT:PC$_{61}$BM rather than in the PC$_{61}$BM and crystal P3HT domains. Thus, the addition of SiPc molecules did not impact the formation of the pristine P3HT and PC$_{61}$BM phases in the ternary blend films. This is in good agreement with the prediction based on the wetting coefficient, suggesting that the surface energy has a critical impact on such interfacial segregation.

1.4.1.3 Photoluminescence Quenching

For the determination of miscibility, the quenching effect of photoluminescence depending on the amount of quencher can be used. Quenching or intermolecular deactivation is the acceleration of the decay rate of an excited state of a material by the presence of another chemical species. The quenching effect is a reduction of PL intensity due to charge transfer to a quencher. The quantum efficiency as a function of quencher concentration can be plotted. A linear quenching represents a statistical distributed content of quencher. If the quenching centers interfere with each other, a saturation regime can be seen.

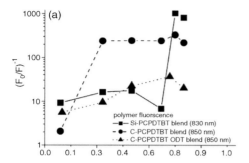 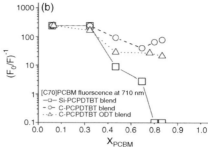

Figure 1.16 Semilogarithmic Stern–Volmer plot ($F_0/F - 1$) of (a) polymer and (b) fullerene fluorescence quenching versus the fullerene content $x_{PC_{71}BM}$ (defined as the weight ratio between the fullerene and the total solid content) for the three polymer:fullerene blends considered: Si-bridged (squares), C-bridged (circles), C-bridged/ODT (triangles). The intensity peaks we considered are positioned at 830 nm for the Si-PCPDTBT and at 850 nm for the C-PCPDTBT polymer, while the $PC_{71}BM$ emission occurs at 710 nm. Since the fullerene fluorescence at 830–850 nm is pronounced, the characteristic fullerene emission spectrum was subtracted from the total measured emission in order to isolate the polymer contribution. The measured fluorescence intensity was then normalized by the corresponding absorbance at the excitation wavelength. The maximum quenching ratio was arbitrarily set as the maximum measured signal (pristine polymer) to noise ratio. (Adapted from Ref. [91].)

According to the Stern–Volmer equation (Eq. (1.8)), the quantum efficiency can be plotted as a function of the quencher concentration, with F_0 as fluorescence intensity without quencher, F as fluorescence intensity with quencher and with the concentration [Q], and K_{SV} as Stern–Volmer constant.

$$\frac{F_0}{F} - 1 = K_{SV}[Q]. \tag{1.8}$$

The mixing behavior of the fullerene in the polymer was studied by measuring the PL of C-bridged PCPDTBT:$PC_{71}BM$ and Si-bridged PCPDTBT:$PC_{71}BM$ blends by Morana et al. as shown in Figure 1.16 [91, 109]. The polymer emission is strongly quenched upon addition of $PC_{71}BM$. Using 33 wt% $PC_{71}BM$ in the blend, the spectrally resolved PL signal intensity is decreased by a factor of 200–400 with respect to the one measured for the pure polymer. The quenching yield of $PC_{71}BM$ in PCPDTBT seems to be low compared to other conjugated polymer:fullerene systems [110, 111]. For small $PC_{71}BM$ mass fractions below 0.3 wt%, the total fluorescence intensity of all blends with Si-PCPDTBT and C-PCPDTBT with and without ODT was increasingly quenched with growing fullerene content. Since no emission from the fullerene was observed, it was concluded that the fullerene domain size is smaller than the exciton diffusion length, whereas the polymer domain size is in the range of the exciton diffusion length. The PL quenching studies suggest an increasing trend of phase segregation with growing domain sizes for both the fullerene and the polymers in the order C-PCPDTBT $<$ C-PCPDTBT with ODT $<$ Si-PCPDTBT. The larger extent of phase separation in Si-PCPDTBT was correlated to a lower solubility of the silicon-bridged polymer in the processing solvent and a stronger tendency to crystallize or aggregate compared to C-PCPDTBT.

Nismy et al. analyzed the photoluminescence quenching of carbon nanotubes in P3HT:$PC_{61}BM$ blends and speculated that acid-substituted MWNTs may react as exciton dissociation centers [112]. Excitons created within the exciton diffusion length of a donor:acceptor interface are considered to contribute to the free charge population. Therefore, the BHJ structure consisting of phase-separated donor and acceptor interfaces in the nanoscale range is important for the effective dissociation of excitons into free electron–hole pairs. Carbon nanotubes were added and acted as electron diffusion centers to the pre-existing P3HT:$PC_{61}BM$ BHJ system, for further improving the exciton dissociation by providing triple heterojunction interfaces. Therefore, more dissociated excitons and higher photocurrents were expected.

1.4.2
Polymer–Polymer Miscibility

Polymer–polymer composites are an actively researched section in the field of organic semiconductor composites. Many attempts for blending conjugated polymers for OPV applications and creating a composite microstructure with small (i.e., few nm to few tens of nm) domains were run by the Cambridge group. Granström et al. used a cyano derivative of poly(p-phenylene vinylene) (MEH-CN-PPV) as electron acceptor and a derivative of polythiophene as hole acceptor [113]. By AFM imaging, the formation of islands of the minority phase was detected, which were larger in size for the thermally treated polythiophene-rich film. In the cross section of a laminated structure, interpenetration between the two layers following the lamination and annealing procedure was recognized, with a length scale of 20–30 nm. Finer scale interpenetration was not revealed. Polyfluorene-based polymer blends consisting of poly(2,7-(9,9-di-n-octylfluorene-alt-benzothiadiazole)) (F8BT) and poly(2,7-(9,9-di-n-octylfluorene)-alt-(1,4-phenylene-((4-sec-butylphenyl)imino)-1,4-phenylene)) (TFB) were investigated by Kim et al. [114]. AFM images are shown in Figure 1.17. A microscale lateral phase separation was recognizable, but the phase-separated domains were not pure at the submicron length scale. Furthermore, a nanoscale vertical phase separation was found. McNeill et al. investigated blends of the polymers P3HT and poly((9,9-dioctylfluorene)-2,7-diyl-alt-[4,7-bis(3-hexylthiophen-5-yl)-2,1,3-benzothiadiazole]-20,200-diyl) (F8TBT) [115]. Annealing was found to be responsible for coarsening of the phase separation and increase in hole mobility of the P3HT phase, both contributing to an improved charge separation.

Different approaches to mix insulating thermoplasts with organic semiconductors have been made with the aim to combine semiconducting properties of conjugated polymers with excellent mechanical properties of commodity polymers. Goffri et al. blended P3HT with polystyrene showing a crystallinity-induced favorable phase separation. Blending P3HT and PS results in vertically stratified structures with dominantly semiconductor at the surface. Application as active layer in

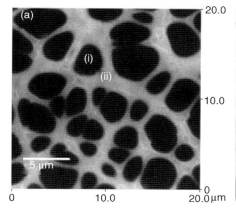 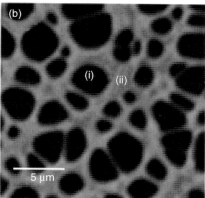

Figure 1.17 (a) AFM image with a 100 nm height scale (white: high region; black: low region) and (b) PL image of 100 nm thick F8BT:TFB (50 : 50) film (domain (i), TFB-rich region; domain (ii), F8BT-rich region). On the basis of the PL of the blend film, it is concluded that the high ridges are F8BT matrix phase (bright regions, domain (ii)) and the thin regions are TFB-rich enclosed phase (dark regions with very weak F8BT emission, domain (i)). (Reprinted with permission from Ref. [114]. Copyright 2004, American Chemical Society.)

organic field-effect transistors showed no degradation of device performance, which is a decisive advantage compared to the use of blends of P3HT with amorphous insulating polymers. Crystalline–crystalline/semiconducting–insulating multicomponent systems offer the possibility to realize high-performance semiconducting systems with reduced material cost, better mechanical properties, and improved environmental stability [116, 117].

Brabec et al. [118] blended MDMO-PPV with various nonconjugated binders such as PS, PMMA, and PC and found that addition of up to 10% of an inert polymer does not negatively influence the device performance. Ternary blends containing P3HT:PC$_{61}$BM blended with insulating polymers such as high-density polyethylene (HDPE) were also investigated by Ferenczi et al. [119]. By blending of the donor–acceptor components into the conventional polymer matrix, the percolation threshold for photovoltaic response of the three-component systems is found to be determined by percolation of the fullerene in the polymer matrix [120–122]. Up to 50 wt% of insulating semicrystalline polymers were added to a P3HT:PC$_{61}$BM blend without decreasing the device performance. The advantages of such ternary systems over the binaries without inert polymeric additives are facilitated processing, enhanced mechanical properties, and increased thickness of the active layer, which reduce defects in the films and improve large-area processing.

Spinodal decomposition has been found for spin-coated films of the ferroelectric random copolymer poly(vinylidene fluoride–trifluoroethylene) (P(VDF-TrFE)) and regioirregular (rir) P3HT. The blend separates into amorphous rir-P3HT domains embedded in a crystalline P(VDF-TrFE) matrix [123]. The number of domains decreases with increasing rir-P3HT content, indicating coarsening of morphology.

1.4.3
Polymer–Fullerene Miscibility

Intercalation of fullerene molecules in a conjugated polymer matrix was reported by Koppe et al. and was later on investigated in great detail by the Stanford group [124]. Intercalation of fullerenes in a polymer matrix results in a microstructure where individual fullerene molecules get dissolved in or close to the polymer backbone. Intercalation is dominantly facilitated by voids between the side chains along the polymer backbone and depends on the polymer structure instead of the polymer configuration. The principle of intercalation is shown in Figures 1.18 and 1.19. Intercalation has significant impact on the performance of bulk heterojunction devices, since a significant fraction of fullerenes being dissolved in the polymer is lost for electron transport. The general findings are that polymer with a tendency toward intercalation requires a significantly higher fullerene concentration to guarantee well-balanced transport. Mayer et al. have observed fullerene

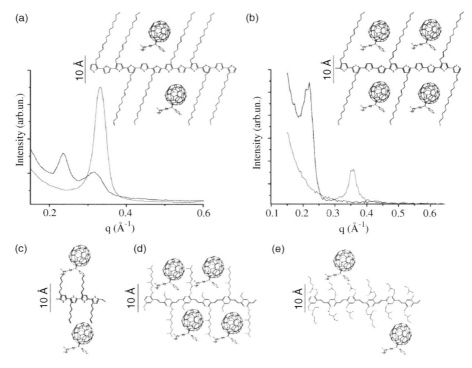

Figure 1.18 Fullerene intercalation in other polymer:fullerene systems. (a) The X-ray diffraction pattern demonstrates an expansion of the d-spacing of the pTT (gray line) upon the addition of $PC_{61}BM$ (black line) and the inset shows how the $PC_{61}BM$ fits between the side chains. (b) The same situation exists for PQT as demonstrated by the X-ray pattern. (c) There is insufficient room between the side chains of P3HT to allow for intercalation. (d) There is sufficient room for $PC_{61}BM$ intercalation between the side chains in amorphous MDMO-PPV. (e) $BisOC_{10}$-PPV, however, does not have sufficient room. (From Ref. [125].)

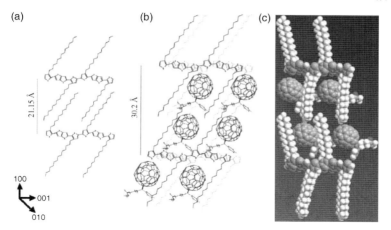

Figure 1.19 Schematic of possible structures showing the effect of $PC_{71}BM$ intercalation on the crystal lattice of pBTTT. (a) The tilt angle for the pristine pBTTT crystal and the amount of interdigitation of the side chains are set to make the d-spacing agree with X-ray diffraction. (b) The $PC_{71}BM$ is placed within the intercalated pBTTT:$PC_{71}BM$ in order to agree with the d-spacings found in X-ray scattering. (c) The total volume taken up by the electron orbitals using a space-filling routine from ChemBio3D Ultra shows that there is still sufficient room for the intercalation demonstrated in (b). The tilt of the side chains in (c) is only approximate because the simulations do not account for intermolecular interactions. The lattice axes are shown in the lower left corner for reference. (From Ref. [125].)

intercalation in blends with various amorphous and semicrystalline polymers when there is enough free volume between the side chains to accommodate the fullerene molecule [125].

Intercalation of fullerenes between side chains mostly determines the optimum polymer:fullerene blending ratios. These findings offer explanations why large-scale phase separation occurs in some polymer:fullerene blend ratios while thermodynamically stable mixing on the molecular scale occurs for others. High fullerene content is necessary to create the phase separation needed for efficient BHJ solar cells, which leads to optimum blend ratios near 1 : 3 to 1 : 4 polymer:fullerene if intercalation occurs. If no intercalation occurs, an optimum near 1 : 1 is usually found.

While no intercalation occurs in the crystalline phase of P3HT, amorphous portions of P3HT and MDMO-PPV contain significant concentrations of $PC_{61}BM$ [126]. Furthermore, depth profiles of P3HT:$PC_{61}BM$ bilayers showed interdiffusion of both materials already after short annealing times. Therefore, pure amorphous phases do not exist in BHJ or annealed bilayer devices. Energy-filtered transmission electron microscopy (EFTEM) and GISAXS measurements were used for morphological investigations, showing local P3HT concentrations in $PC_{61}BM$-rich domains [127]. This was interpreted as partial miscibility. P3HT:$PC_{61}BM$ χ parameter and Flory–Huggins phase diagram, which predicts miscibility for P3HT volume fractions above 0.42, were determined. Flory–Huggins interaction parameter enables quantifying the chemical interactions between P3HT and $PC_{61}BM$. Miscibility estimates were obtained from measurements of the melting point depression, which were analyzed with differential scanning calorimetric (DSC) experiments. Quantifying the chemical interactions between P3HT and $PC_{61}BM$ through the Flory–Huggins interaction parameter enables the determination of miscibility range for these two components as long as they are amorphous. Miscibility between P3HT and $PC_{61}BM$ suppresses fullerene crystallization. The crystallization of the polymer leads to the characteristic length scales of the mesostructure, whereby crystallization of the polymer can also lead to macroscopic phase separation by enriching the amorphous polymer phase with fullerene beyond the miscibility limit.

1.4.4
Phase Diagrams

The device performance of organic semiconducting composite devices strongly depends on the blend composition. Different approaches to analyze the phase behavior were used to correlate and improve the electric properties. Phase diagrams mainly consist of liquidus and solidus lines separating different phases. For polymers usually liquidus and solidus lines are determined using end melting temperature and peak melting temperature since all crystallites are molten and the crystalline order is broken [128, 129]. The intersection of both lines represents the eutectic point of the phase diagram, with a phase equilibrium where the degree of freedom is only selectable in a small range.

Binary organic photovoltaic blends containing poly-(3-alkylthiophene)s (P3ATs) with different side chain lengths and different fullerene derivatives were investigated by Müller et al., Zhao et al., and Kim and Frisbie [130–132]. Binary phase diagrams were reconstructed from DSC measurements, and the device performance of corresponding organic solar cells was analyzed in this phase diagram (Figure 1.20). It was suggested that all systems contain a simple eutecticum. Increasing side chain length of P3AT with poly(3-butylthiophene) (P3BT), P3HT, and poly(3-dodecylthiophene) (P3DDT) leads to a shift of the eutectic temperature T_e to higher P3AT content, accompanied with a decrease of T_e from 220 °C for P3BT to 150 °C for P3DDT. For P3HT, a T_e of 205 °C and a eutectic composition c_e of 65 wt% P3HT were found. The maximum J_{SC} of the corresponding devices was found around the eutectic composition, being slightly shifted to higher $PC_{61}BM$ ratios. The higher melting temperature of $PC_{71}BM$ compared to $PC_{61}BM$ results in a higher eutectic temperature of the binary blend with P3HT, which is also slightly shifted to higher P3HT contents compared to $PC_{61}BM$. This is also reflected in the J_{SC} maximum.

Ballantyne and coworkers investigated the blend system poly(3-hexylselenothiophene) (P3HS) with $PC_{61}BM$ by DSC and found a simple eutectic behavior with a eutectic composition c_e of 66 wt% P3HS and a T_e of 230 °C. A high crystallinity of P3HS was found by DSC, but P3HS:$PC_{61}BM$ blend films showed a lower degree of crystallinity than P3HT:$PC_{61}BM$ according to XRD measurements. Therefore, larger fractions of $PC_{61}BM$ can be dissolved in the polymer. The lower degree of

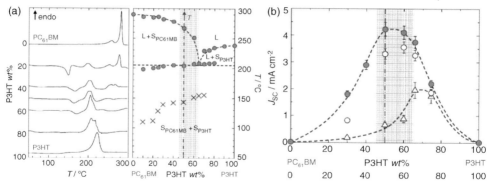

Figure 1.20 (a) Phase behavior of P3HT:PC$_{61}$BM. (a) DSC heating thermograms (left) and corresponding (nonequilibrium) temperature/composition diagram of the P3HT:PC$_{61}$BM system (right) featuring simple eutectic phase behavior (peak eutectic temperature, $T_e \sim 205\,°C$; eutectic composition, $c_e \sim 65$ wt% P3HT). Liquidus lines were constructed with end melting and end dissolution temperatures of neat components and excess component, respectively. Crosses represent the onset of recrystallization; highlighted areas in this and following part indicate the range of composition of optimum device performance. (b) Dependence of short-circuit current density J_{SC} (top panel) and power conversion efficiency (bottom panel) (under simulated solar illumination AM 1.5, 71 mW cm^{-2}) on P3HT/PC$_{61}$BM blend composition for devices thermally treated at 140 °C after spin casting (filled circles), subsequently melt quenched from 290 °C (open triangles), and then after further annealing at 140 °C (open circles). In accordance with previous reports, J_{SC} is optimized after annealing at blend compositions comprising 50–60 wt% of the polymer. Error bars represent estimated percentage error based on comparison of similar devices. (From Ref. [130].)

phase segregation is likely to contribute to the faster recombination kinetics in the P3HS:PC$_{61}$BM compared to P3HT:PC$_{61}$BM blend devices.

Müller *et al.* investigated the phase behavior of liquid-crystalline polymer:fullerene blends [133]. For poly[2,7-(9,9-dioctylfluorene)-*alt*-5,5-(40,70-di-2-thienyl-20,10,30-benzothiadiazole)] (F8TBT, also abbreviated as APFO-3 or PFDTBT) blended with PC$_{61}$BM, a eutectic phase behavior with a eutectic composition c_e of 75 wt% F8TBT at a eutectic temperature T_e of 138 °C was found. The glass transition temperature T_g was found to be independent of composition, showing good match with T_g of the pure components. Above the glass transition temperature, PC$_{61}$BM crystals tend to nucleate. Miscibility of fullerene in the polymer strongly depends on the molecular weight of the macromolecule. Molecular weight is found to have a significant influence on morphology of P3HT. For P3HT:PC$_{61}$BM blends, high molecular weight enhances intermolecular ordering (π-stacking) of P3HT. Increased molecular weight leads to extended crystallites and therefore better charge carrier mobility. A threshold for the molecular weight of P3HT was found to be necessary to guarantee sufficient device performance. This threshold was found by investigating various molecular weight fractions of a P3HT master batch [134]. On the other hand, with increasing weight average molecular weight (M_w), the crystallinity of P3HT and the crystalline orientation decreased [135]. This phenomenon was attributed to transition from a fully extended all-*trans* conformation to a semicrystalline system with crystalline lamellae and amorphous extended interlamellar zones. Huang *et al.* investigated the influence of molecular weight of poly[(4,4′-bis(2-ethylhexyl)dithieno[3,2-*b*:2′,3′-*d*]silole)-2,6-diyl-*alt*-(5,5′-thienyl-4,4′-dihexyl-2,2′-bithiazole)-2,6-diyl] (Si-PCPDTTBT) and $PC_{61}BM$ on morphology [136]. AFM and TEM images showed the increasing phase separation with increasing molecular weight leading to an interpenetrating network for carrier transport and device improvement.

Ternary systems were investigated to further enhance the spectral response of organic solar cells [137, 138]. Small amounts of PCPDTBT were added to the P3HT:$PC_{61}BM$ blend to expand the absorption spectra toward the near-infrared region. Different requirements for such ternary systems have to be fulfilled. The absorption in the near-infrared region should be complementary to the absorption spectra of P3HT. Further, the electronic levels of the sensitizer need to be aligned with respect to those of P3HT and $PC_{61}BM$ to facilitate an efficient photoinduced charge transfer between all components. Finally, the miscibility of such systems is of interest for morphological studies. Ternary phase diagrams for P3HT, PCPTDTBT, and $PC_{61}BM$ were investigated by Li *et al.* as shown in Figure 1.21 using DSC [138]. The phase diagram of the binary system P3HT:$PC_{61}BM$ showed a simple eutectic point, as already reported by Müller *et al.* and Zhao *et al.* [130, 131]. The phase diagram of the ternary system revealed that already small amounts of the dominantly amorphous polymer PCPDTBT can decrease the overall crystallinity but do not affect the position of the eutectic point in ternary blends. A comparison with cell performance showed a correlation of the phase behavior of the binary as well as the ternary blends with its electrical properties in the cells.

1.5
Conclusions

Organic semiconductor composites are a smart concept to design and customize the optoelectronic functionality of a semiconductor by simply blending multiple components with the desired individual properties. This concept is more elegant and technically easily accessible than the design of a single semiconductor compound comprehending all properties. The challenge for organic semiconductor composites is the formulation of suitable inks, the miscibility, and compatibility of the individual inks as well as the control of the solid-state microstructure.

All of these challenges need to be addressed by the formulation of semiconductor composite inks, which is also one of the key parameters for the processing of the active layer of organic photovoltaic devices. Precise information on an ink's rheological properties, the solubility of the individual components, and their influence on solid-state morphology is of highest interest for the development of coating and printing processes. The right choice of the processing solvent, offering good solubility, mainly contributes to the solid-state microstructure of the active layer. Long

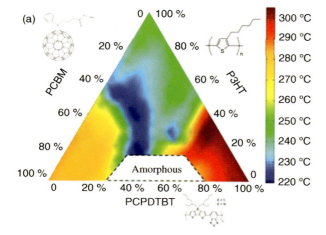

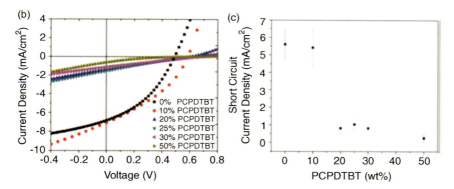

Figure 1.21 (a) Ternary phase diagram with the end melting temperature of the DSC first heating curves as liquidus lines. (b) J_{SC}–V curves of P3HT/PCPDTBT/PC$_{61}$BM ternary solar cells with a polymer:fullerene ratio of 50 : 50 (w/w). (c) J_{SC} of photovoltaic devices depends on PCPDTBT ratio for ternary blends with a constant fullerene ratio of 50 wt% (total polymer:fullerene 1 : 1). The solar cells were tested under AM 1.5, 100 mW cm^{-2} illumination. (Adapted from Ref. [138].)

drying times typically enhance the polymer crystallinity offering better possibilities for self-assembling of the polymer chains. Different approaches to control the composite's microstructure were investigated by using solvent blends. Among these concepts, the use of additional solvents with high boiling points has led to finer phase separations and therefore increasing device efficiencies. Further research activities have focused on the formation of aggregates by the addition of nonsolvents, favoring crystallinity increase in the solid state. Another task is the use of halogen-free, nonhazardous solvents for industrial processing. Here the formulation of solvent blends using Hansen solubility parameters is becoming established as a reference method.

Miscibility of the components is required to design functional semiconductor composites. Specifically, bulk heterojunction solar cells require precise control of the blend microstructure to guarantee good device efficiency. Polymer–polymer miscibility is ways more complex and difficult than polymer–fullerene miscibility. In the case of polymer–fullerene blends, the intercalation of fullerene within the side chains of conjugated polymers is found as one of the parameters defining the optimum polymer–fullerene ratio. Blending semiconducting and insulating polymers has also been proven to ease industrial production. Establishing phase diagrams and correlation with material and device properties has helped to improve the miscibility of multicomponent composites.

References

1 Brabec, C.J. (2004) *Sol. Energy Mater. Sol. Cells*, **83**, 273–292.
2 Thompson, B.C. and Fréchet, J.-M.J. (2008) *Angew. Chem.*, **120**, 62–82.
3 Chiang, C.K., Fincher, C.R., Park, Y.W., Heeger, A.J., Shirakawa, H., Louis, E.J., Gau, S.C., and MacDiarmid, A.B. (1977) *Phys. Rev. Lett.*, **39**, 1098–1101.
4 Kalinowski, J. (2004) *Organic Light-Emitting Diodes: Principles, Characteristics, and Processes*, Optical Engineering, vol. **91**, CRC Press, New York.
5 Dimitrakopoulos, C.D. and Malenfant, P.R.L. (2002) *Adv. Mater.*, **14** (2), 99–117.
6 Kroon, R., Lenes, M., Hummelen, J.C., Blom, P.W.M., and de Boer, B. (2008) *Polym. Rev.*, **48**, 531–582.
7 Dennler, G., Scharber, M.C., and Brabec, C.J. (2009) *Adv. Mater.*, **21**, 1323–1338.
8 Deibel, C. and Dyakonov, V. (2010) *Rep. Prog. Phys.*, **73**, 096401.
9 Hertel, D. and Baessler, H. (2008) *Chem. Phys. Chem.*, **9**, 666–688.
10 Chamberlain, G.A. (1983) *Sol. Cells*, **8**, 47–83.
11 Tang, C.W. (1986) *Appl. Phys. Lett.*, **48**, 183–185.
12 Yu, G., Gao, J., Hummelen, J.C., Wudl, F., and Heeger, A.J. (1995) *Science*, **270**, 1789–1791.
13 Frohne, H., Shaheen, S.E., Brabec, C.J., Müller, D.C., Sariciftci, N.S., and Meerholz, K. (2002) *Chem. Phys. Chem.*, **9**, 795–799.
14 Hoppe, H. and Sariciftci, N.S. (2004) *J. Mater. Res.*, **19**, 1924–1945.
15 McGehee, M.D. and Topinka, M.A. (2006) *Nat. Mater.*, **5**, 675–676.
16 Seong, J.Y., Chung, K.S., Kwak, S.K., Kim, Y.H., Moon, D.G., Han, J.I., and Kim, W.K. (2004) *J. Korean Phys. Soc.*, **45**, 5914.
17 Peet, J., Kim, J.Y., Coates, N.E., Ma, W.L., Moses, D., Heeger, A.J., and Bazan, G.C. (2007) *Nat. Mater.*, **6**, 497–500.
18 Green, M.A., Emery, K., Hisikawa, Y., and Warta, W. (2009) *Prog. Photovolt. Res. Appl.*, **17**, 85–94.
19 Gaudiana, R. and Brabec, C.J. (2008) *Nat. Photon.*, **2**, 287–289.
20 Park, S.H., Roy, A., Beaupré, S., Cho, S., Coates, N., Moon, J.S., Moses, D., Leclerc, M., Lee, K., and Heeger, A.J. (2009) *Nat. Photon.*, **3**, 297–303.
21 Scharber, M.C., Mühlbacher, D., Koppe, M., Denk, P., Waldauf, C., Heeger, A.J., and Brabec, C.J. (2006) *Adv. Mater.*, **18**, 789–794.
22 Brabec, C.J. and Durrant, J. (2008) *MRS Bull.*, **33**, 670.
23 Hildebrand, J.H. and Scott, R.L. (1952) *J. Chem. Phys.*, **20** (10), 1520–1521.
24 Hansen, C.M. (2007) Chapter 1, in *Hansen Solubility Parameters – A User's Handbook*, 2nd edn, CRC Press, Boca Raton, FL.
25 Blanks, R.F. and Prausnitz, J.M. (1964) *Ind. Eng. Chem. Fundam.*, **3**, 1–8.
26 Weimer, R.F. and Prausnitz, J.M. (1965) *Petrol. Refiner*, **44**, 237–242.
27 Hansen, C.M. (1967) *J. Paint Technol.*, **39**, 104–117.

28 Hansen, C.M. (1967) *J. Paint Technol.*, **39**, 505–510.
29 Hansen, C.M. and Skaarup, K. (1967) *J. Paint Technol.*, **39**, 511–514.
30 Small, P.A. (1953) *J. Appl. Chem.*, **3**, 71.
31 Hoy, K.L. (1970) *J. Paint Technol.*, **42**, 76–78.
32 van Krevelen, D.W. (1972) *Properties of Polymers: Correlations with Chemical Structure*, Elsevier, Amsterdam.
33 Coleman, M.M., Serman, C.J., Bhagwagar, D.E., and Painter, P.C. (1990) *Polymer*, **31**, 1187–1203.
34 Fredenslund, A., Gmehling, J., and Rasmussen, P. (1977) *Vapor–Liquid Equilibria Using UNIFAC*, Elsevier, Amsterdam.
35 Gmehling, J. (1998) *Fluid Phase Equilib.*, **144**, 37–47.
36 Derr, E.L. and Deal, C.H. (1969) *Inst. Chem. Eng. Symp. Ser.*, **3**, 40.
37 Klamt, A. and Schüürmann, G. (1993) *J. Chem. Soc., Perkin Trans. 2*, **5**, 799–805.
38 Jork, C., Kristen, C., Pierccini, D., Stark, A., Chiappe, C., Beste, Y.A., and Arlt, W. (2005) *J. Chem. Thermodyn.*, **37** (6), 537–558.
39 Klamt, A. (1995) *J. Phys. Chem.*, **99** (7), 2224–2235.
40 Klamt, A. (1998) COSMO and COSMO-RS, in *Encyclopedia of Computational Chemistry* (ed. P.v.R. Schleyer), John Wiley & Sons, Ltd, Chichester.
41 Klamt, A., Jonas, V., Bürger, T., and Lohrenz, J.C.W. (1998) *J. Chem. Phys. A*, **102** (26), 5074–5085.
42 Klamt, A. (2005) *COSMO-RS: From Quantum Chemistry to Fluid Phase Thermodynamics and Drug Design*, Elsevier, Amsterdam.
43 Paul, D.R. and Barlow, J.W. (1981) *Polym. Eng. Sci.*, **21**, 985–996.
44 Robeson, L.M. (2007) Chapter 2: Fundamentals of polymer blends, in *Polymer Blends*, Hanser, Munich.
45 Flory, P.J. (1962) *Principles of Polymer Chemistry*, Cornell University Press, Ithaca, NY.
46 Walker, B., Tamayo, A., Duong, D.T., Dang, X.-D., Kim, C., Granstrom, J., and Nguyen, T.-Q. (2011) *Adv. Energy Mater.*, **1** (2), 221–229.
47 Machui, F., Abbott, S., Waller, D., Koppe, M., and Brabec, C.J. (2011) *Macromol. Chem. Phys.*, **212**, 2159–2165.
48 Ruoff, R.S., Tse, D.S., Malhotra, R., and Lorents, D.C. (1993) *J. Phys. Chem.*, **97**, 3379–3383.
49 Yu, G., Gao, J., Hummelen, J.C., Wudl, F., and Heeger, A.J. (1995) *Science*, **270**, 1789–1791.
50 Kronholm, D. and Hummelen, J. (2007) *Mater. Matters*, **2**, 16–20.
51 Troshin, P., Hoppe, H., Renz, J., Egginger, M., Mayorova, J., Goryachev, A., Peregudov, A., Lyubovskaya, R., Gobsch, G., Sariciftci, N.S., and Razumov, V. (2009) *Adv. Funct. Mater.*, **19**, 779–788.
52 Hansen, C.M. and Smith, A.L. (2004) *Carbon*, **42**, 1591–1597.
53 Park, C.-D., Fleetham, T.A., Li, J., and Vogt, B.D. (2011) *Org. Electron.*, **12**, 1465–1470.
54 Lee, K.-G., Kim, J.Y., Park, S.H., Kim, S.H., Cho, S., and Heeger, A.J. (2007) *Adv. Mater.*, **19**, 2445–2449.
55 Ma, W.-L., Yang, C.-Y., Gong, X., Lee, K.-H., and Heeger, A.J. (2005) *Adv. Funct. Mater.*, **15**, 1617–1622.
56 Zhao, Y., Yuan, G., Roche, P., and Leclerc, M. (1995) *Polymer*, **36**, 2211–2214.
57 Mihailetchi, V.D., Xie, H.-X., de Boer, B., Popescu, L.M., Hummelen, J.C., and Blom, P.W.M. (2006) *Appl. Phys. Lett.*, **89**, 012107.
58 Chu, C.-W., Yang, H.-C., Hou, W.-J., Huang, J.-S., Li, G., and Yang, Y. (2008) *Appl. Phys. Lett.*, **92**, 103306.
59 Miller, E.W. and Miller, R.M. (1989) *Environmental Hazards: Air Pollution. A Reference Handbook*, ABC-Clio, Santa Barbara, CA.
60 David Cooper, C. and Alley, F.C. (1994) *Air Pollution Control: A Design Approach*, 2nd edn, Waveland Press, Inc., Long Grove, IL.
61 Dang, M.T., Hirsch, L., and Wantz, G. (2011) *Adv. Mater.*, **23**, 3597–3602.
62 Shaheen, S.E., Brabec, C.J., Sariciftci, N.S., Padinger, F., Fromherz, T., and Hummelen, J.C. (2001) *Appl. Phys. Lett.*, **78**, 841–843.

63 Hoppe, H., Niggemann, M., Winder, C., Kraut, J., Hiesgen, R., Hinsch, A., Meissner, D., and Sariciftci, N.S. (2004) *Adv. Funct. Mater.*, **14**, 1005–1011.

64 Hoppe, H. and Sariciftci, N.S. (2006) *J. Mater. Chem.*, **16**, 45–61.

65 Rispens, M.T., Meetsma, A., Rittberger, R., Brabec, C.J., Sariciftci, N.S., and Hummelen, J.C. (2003) *Chem. Commun.*, 2116–2118.

66 Ruderer, M., Guo, S., Meier, R., Chiang, H., Körstgens, V., Wiedersich, J., Perlich, J., Roth, S., and Müller-Buschbaum, P. (2011) *Adv. Funct. Mater.*, **21**, 3382–3391.

67 Yu, H. (2010) *Synth. Met.*, **160**, 2505–2509.

68 Kwong, C., Djurišić, A., Chui, P., Cheng, K., and Chan, W. (2004) *Chem. Phys. Lett.*, **384**, 372–375.

69 Park, S.H., Roy, A., Beaupré, S., Cho, S., Coates, N., Moon, J.S., Moses, D., Leclerc, M., Lee, K.-H., and Heeger, A.J. (2009) *Nat. Photon.*, **3**, 297–303.

70 Jaczewska, J., Budkowski, A., Bernasik, A., Moons, E., and Rysz, J. (2008) *Macromolecules*, **41**, 4802–4810.

71 Berson, S., De Bettignies, R., Bailly, S., and Guillerez, S. (2007) *Adv. Funct. Mater.*, **17**, 1377–1384.

72 Hoth, C.N., Choulis, S.A., Schilinsky, P., and Brabec, C.J. (2007) *Adv. Mater.*, **19**, 3973–3978.

73 Hoth, C.N., Schilinsky, P., Choulis, S.A., and Brabec, C.J. (2008) *Nano Lett.*, **8**, 2806–2813.

74 Kawano, K., Sakai, J., Yahiro, M., and Adachi, C. (2009) *Sol. Energy Mater. Sol. Cells*, **93**, 514–518.

75 Lange, A., Wegener, M., Boeffel, C., Fischer, B., Wedel, A., and Neher, D. (2010) *Sol. Energy Mater. Sol. Cells*, **94**, 1816–1821.

76 Chen, F.-C., Tseng, H.-C., and Ko, C.-J. (2008) *Appl. Phys. Lett.*, **92**, 103316.

77 Zhang, F.-L., Jespersen, K.G., Björström, C., Svensson, M., Andersson, M.R., Sundström, V., Magnusson, K., Moons, E., Yartsev, A., and Inganäs, O. (2006) *Adv. Funct. Mater.*, **16**, 667–674.

78 Wang, Z., Wang, E., Hou, L., Zhang, F., Andersson, M., and Inganäs, O. (2011) *J. Photon. Energy*, **1**, 011122.

79 Alem, S., Chu, T., Tse, S., Wakim, S., Lu, J., Movileanu, R., Tau, Y., Belanger, F., Desilets, D., Beaupre, S., Leclerc, M., Rodman, S., Waller, D., and Gaudiana, R. (2011) *Org. Electron.*, **12**, 1788–1793.

80 Park, Y.D., Lee, H.S., Chio, Y.J., Kwak, D., Cho, J.H., Lee, S., and Cho, K. (2009) *Adv. Funct. Mater.*, **19**, 1200–1206.

81 Moulé, A.J. and Meerholz, K. (2008) *Adv. Mater.*, **20**, 240–245.

82 Wang, W., Wu, H., Yang, C., Luo, C., Zhang, Y., Chen, J., and Cao, Y. (2007) *Appl. Phys. Lett.*, **90**, 183512.

83 Rogers, J., Schmidt, K., Toney, M., Kramer, E., and Bazan, G. (2011) *Adv. Mater.*, **23**, 2284–2288.

84 Lee, J.K., Ma, W.L., Brabec, C.J., Yuen, J., Moon, J.S., Kim, J.Y., Lee, K., Bazan, B.C., and Heeger, A.J. (2008) *J. Am. Chem. Soc.*, **130**, 3619–3623.

85 Peet, J., Soci, C., Coffin, R.C., Nguyen, T.Q., Mikhailovsky, A., Moses, D., and Bazan, G.C. (2006) *Appl. Phys. Lett.*, **89**, 252105.

86 Moet, D., Lenes, M., Morana, M., Azimi, H., Brabec, C., and Blom, P. (2010) *Appl. Phys. Lett.*, **96**, 213506.

87 Su, M.-S., Kuo, C.Y., Yuan, M.-C., Jeng, U.-S., Su, C.-J., and Wei, K.-H. (2011) *Adv. Mater.*, **23**, 3315–3319.

88 Agostinelli, T., Ferenczi, T.A.M., Pires, E., Foster, S., Maurano, A., Müller, C., Ballantyne, A., Hampton, M., Lilliu, S., Campoy-Quiles, M., Azimi, H., Morana, M., Bradley, D.D.C., Durrant, J., Macdonald, J.E., Stingelin, N., and Nelson, J. (2011) *J. Polym. Sci. B*, **49**, 717–724.

89 Liang, Y., Feng, D., Wu, Y., Tsai, S.-T., Li, G., Ray, C., and Yu, L. (2009) *J. Am. Chem. Soc.*, **131**, 7792–7799.

90 Chu, T.-Y., Tsang, S.-W., Zhou, J., Verly, P.G., Lu, J., Beaupré, S., Leclerc, M., and Tao, Y. (2011) *Sol. Energy Mater. Sol. Cells*. doi: 10.1016/j.solmat.2011.09.042.

91 Morana, M., Azimi, H., Dennler, G., Egelhaaf, H.-J., Scharber, M., Forberich, K., Hauch, J., Gaudiana, R., Waller, D., Zhu, H., Hingerl, K., van Bavel, S.S., Loos, J., and Brabec, C.J. (2010) *Adv. Funct. Mater.*, **20**, 1180–1188.

92 Baek, W.-H., Yang, H., Yoon, T.-S., Kang, C.J., Lee, H.H., and Kim, Y.-S. (2009) *Sol. Energy Mater. Sol. Cells*, **93**, 1263–1267.

93 Lide, D.R. (ed.) (2011–2012) *CRC Handbook of Chemistry and Physics*, 92nd edn (Internet version), CRC Press/Taylor & Francis, Boca Raton, FL.

94 Abbott, S., Hansen, C.M., and Yamamoto, H. (2008–2011) Hansen Solubility Parameters in Practice, (software) version 3.1.17, www.hansen-solubility.com (accessed October 2011).

95 Robeson, L.M. (2007) Chapter 5; Characterization of polymer blends, in *Polymer Blends*, Hanser, Munich.

96 Honda, S., Ohkita, H., Benten, H., and Ito, S. (2011) *Adv. Energy Mater.*, **1**, 588–598.

97 Campoy-Quiles, M., Ferenczi, T., Agostinelli, T., Etchegoin, P.G., Kim, Y., Anthopoulos, T.D., Stavrinou, P.N., Bradley, D.D.C., and Nelson, J. (2008) *Nat. Mater.*, **7**, 158–164.

98 Orim, A., Masuda, K., Honda, S., Benten, H., Ito, S., Ohkita, H., and Tsuji, H. (2010) *Appl. Phys. Lett.*, **96**, 043305.

99 Germack, D.S., Chan, C.K., Kline, R.J., Fischer, D.A., Gundlach, D.J., Toney, M.F., Richter, L.J., and DeLongchamp, D.M. (2010) *Macromolecules*, **43**, 3828–3836.

100 Jaczewska, J., Raptis, I., Budkowski, A., Goustouridis, D., Raczkowska, J., Sanopoulou, M., Pamula, E., Bernasik, A., and Rysz, J. (2007) *Synth. Met.*, **157**, 726–732.

101 Oh, J.Y., Jang, W.S., Lee, T.I., Myoung, J.-M., and Baik, H.K. (2011) *Appl. Phys. Lett.*, **98**, 023303.

102 Björström, C.M., Bernasik, A., Rysz, J., Budkowski, A., Nilsson, S., Svensson, M., Andersson, M.R., Magnosson, K.O., and Moons, E. (2005) *J. Phys.: Condens. Matter*, **17**, L529–L534.

103 Sumita, M., Sakakta, K., Asai, S., Miyasaka, K., and Nakagawa, H. (1991) *Polym. Bull.*, **25**, 265–271.

104 Wu, M. and Shaw, L. (2006) *J. Appl. Polym. Sci.*, **99**, 477–488.

105 Göldel, A., Kasaliwal, G., and Pätschke, P. (2009) *Macromol. Rapid Commun.*, **30**, 423–429.

106 Caudouin, A.-C., Devaux, J., and Bailly, C. (2010) *Polymer*, **51**, 1341–1354.

107 Ma, C.G., Zhang, M.Q., and Rong, M.Z. (2007) *J. Appl. Sci.*, **103**, 1578–1584.

108 Cheng, T.W., Keskkula, J., and Paul, D.R. (1992) *Polymer*, **33**, 1606–1619.

109 Morana, M., Wegscheider, M., Bonanni, A., Kopidakis, N., Shaheen, S., Scharber, M., Zhu, Z., Waller, D., Gaudiana, R., and Brabec, C. (2008) *Adv. Funct. Mater.*, **18**, 1757–1766.

110 Haugeneder, A., Neges, M., Kallinger, C., Spirkl, W., Lemmer, U., Feldmann, J., Scherf, U., Harth, E., Gügel, A., and Müllen, K. (1999) *Phys. Rev. B*, **59**, 15346–15351.

111 van Duren, J., Yang, X., Loos, J., Bulle-Lieuwma, C., Sieval, A., Hummelen, J.C., and Janssen, R.A.J. (2004) *Adv. Funct. Mater.*, **14**, 425–434.

112 Nismy, N.A., Jayawardena, K.D.G., Adikaari, A.A.D., and Silva, S.R.P. (2011) *Adv. Mater.*, **23**, 3796–3800.

113 Granström, M., Petritsch, K., Arias, A.C., Lux, A., Andersson, M.R., and Friend, R.H. (1998) *Nature*, **395**, 297–260.

114 Kim, J.S., Ho, P.K.H., Murphy, C.E., and Friend, R.H. (2004) *Macromolecules*, **37**, 2861–2871.

115 McNeill, C.R., Halls, J.J.M., Wilson, R., Whiting, G.L., Berkebile, S., Ramsey, M.G., Friend, R.H., and Greenham, N.C. (2008) *Adv. Funct. Mater.*, **18**, 2309–2321.

116 Goffri, S., Müller, C., Stingelin-Stutzmann, N., Breiby, D.W., Radano, C.P., Andreasen, J.W., Thompson, R., Janssen, R.A.J., Nielsen, M.M., Smith, P., and Sirringhaus, H. (2006) *Nat. Mater.*, **5**, 950–956.

117 Babel, A. and Jenekhe, S.A. (2004) *Macromolecules*, **37**, 9835–9840.

118 Brabec, C.J., Padinger, F., Sariciftci, N.S., and Hummelen, J.C. (1999) *J. Appl. Phys.*, **85**, 6866–6872.

119 Ferenczi, T., Müller, C., Bradley, D., Smith, P., Nelson, J., and Stingelin, N. (2011) *Adv. Mater.*, **23**, 4093–4097.

120 Müller, C., Goffri, S., Breiby, D.W., Andreasen, J.W., Chanzy, H.D., Janssen, R.A.J., Nielsen, M.M., Radano, C.P., Sirringhaus, H., Smith, P., and Stingelin-

Stutzmann, N. (2007) *Adv. Funct. Mater.*, **17**, 2674–2679.

121 Wolfer, P., Müller, C., Smith, P., Baklar, M.A., and Stingelin-Stutzmann, N. (2007) *Synth. Met.*, **157**, 827–833.

122 Kumar, A., Baklar, M.A., Scott, K., Kreouzis, T., and Stingelin-Stutzmann, N. (2009) *Adv. Mater.*, **21**, 4447–4451.

123 Asadi, K., Wondergem, H.J., Moghaddam, R.S., McNeill, C.R., Stingelin, N., Noheda, B., Blom, P.W.M., and de Leeuw, D.M. (2011) *Adv. Funct. Mater.*, **21**, 1887–1894.

124 Koppe, M., Scharber, M., Brabec, C., Duffy, W., Heeney, M., and McCulloch, I. (2007) *Adv. Funct. Mater.*, **17**, 1371–1376.

125 Mayer, A.C., Toney, M.F., Scully, S.R., Rivnay, J., Brabec, C.J., Scharber, M., Koppe, M., Heeney, M., McCulloch, I., and McGehee, M.D. (2009) *Adv. Funct. Mater.*, **19**, 1173–1179.

126 Collins, B., Gann, E., Guignard, L., He, X., McNeill, C., and Ade, H. (2010) *J. Phys. Chem. Lett.*, **1**, 3160–3166.

127 Kozub, D., Vakhshouri, K., Orme, L., Wang, C., Hexemer, A., and Gomez, E. (2011) *Macromolecules*, **44**, 5722–5726.

128 Ehrenstein, G.W., Riedel, G., and Trawiel, P. (2003) *Praxis der Thermischen Analyse von Kunststoffen 2*, Aufl. Carl Hanser Verlag, München.

129 Wunderlich, B. (1990) *Thermal Analysis*, Academic Press, San Diego, CA.

130 Müller, C., Ferenczi, T., Campoy-Quiles, M., Frost, J., Bradley, D., Smith, P., Stingelin-Stutzmann, N., and Nelson, J. (2008) *Adv. Mater.*, **20**, 3510–3515.

131 Zhao, J., Swinnen, A., Van Assche, G., Manca, J., Vanderzande, D., and Van Mele, B. (2009) *J. Phys. Chem. B*, **113**, 1587–1591.

132 Kim, J.Y. and Frisbie, C.D. (2008) *J. Phys. Chem. C*, **112**, 17726–17736.

133 Müller, C., Bergqvist, J., Vandewal, K., Tvingstedt, K., Anselmo, A., Magnusson, R., Alonso, M., Moons, E., Arwin, H., Campoy-Quilles, M., and Inganäs, O. (2011) *J. Mater. Chem.*, **21**, 10676–10684.

134 Zen, A., Saphiannikova, M., Neher, D., Grenzer, J., Grigorian, S., Pietsch, U., Asawapirom, U., Janietz, S., Scherf, U., Lieberwirth, I., and Wegner, G. (2006) *Macromolecules*, **39**, 2162–2171.

135 Brinkmann, M. and Rannou, P. (2007) *Adv. Funct. Mater.*, **17**, 101–108.

136 Huang, J.H., Chen, F.-C., Chen, C.-L., Huang, A.T., Hsiao, Y.-S., Teng, C.-M., Yen, F.-W., Chen, P., and Chu, C.W. (2011) *Org. Electron.*, **12**, 1755–1762.

137 Koppe, M., Egelhaaf, H.-J., Dennler, G., Scharber, M.C., Brabec, C.J., Schilinsky, P., and Hoth, C.N. (2010) *Adv. Funct. Mater.*, **20**, 338–346.

138 Li, N., Machui, F., Waller, D., Koppe, M., and Brabec, C.J. (2011) *Sol. Energy Mater. Sol. Cells*, **95**, 3465–3471.

2
Nanoscale Morphological Characterization for Semiconductive Polymer Blends
Joachim Loos

2.1
Introduction

During the past few years, a substantial part of my scientific activity was and still is focused on the overall research theme "Nanoscale Organization of Functional Polymer Systems," which comprises understanding and controlling of organization or assembly of functional polymer nanostructures. The main objective is tuning the morphology by physical methods at various length scales from (sub-)nanometer (intra- and intermolecular organization) up to hundreds of nanometers (e.g., phase separation and crystal superstructures) toward advanced performance of the corresponding functional systems and devices. Particularly, I am working on systems with applications in the research and development area of organic electronics with specific focus on polymer solar cells (PSCs).

Further, I am developing advanced microscopy methodologies. The thin-film nature of the photoactive layer of PSCs with typical thickness of about 100–200 nm and the need for local morphology information have led high-resolution microscopy techniques to become main investigation tools for morphology characterization. Transmission electron microscopy (TEM), scanning electron microscopy (SEM) – also in combination with preparative and imaging focused ion beam (FIB) – and scanning probe microscopy (SPM) – in particular atomic force microscopy (AFM) – have proven their versatility for detailed morphology characterization of the photoactive layer. The main difference between TEM on one hand and SPM and SEM on the other hand is that TEM provides mainly morphological information of the lateral organization of the thin-film samples by acquisition of projections through the whole film, whereas SPM and SEM probe the topography or phase demixing at the surface of such thin-film samples.

In this chapter, I will demonstrate how the utilization of advanced high-resolution microscopy techniques provides important information on the nanoscale organization of the photoactive layer of PSCs. Such information is mandatory to understand the influence of morphology on the overall properties of PSCs and to

Semiconducting Polymer Composites: Principles, Morphologies, Properties and Applications, First Edition.
Edited by Xiaoniu Yang.
© 2012 Wiley-VCH Verlag GmbH & Co. KGaA. Published 2012 by Wiley-VCH Verlag GmbH & Co. KGaA.

gain guidance when trying to tune organization and functionality toward ultimate performance of devices. Besides conventional TEM and AFM imaging modes for revealing morphological features of the samples under investigation, I will demonstrate how advanced microscopy methodologies can be applied for better understanding critical parameters determining the nanoscale organization and performance of functional polymer systems. In more detail, we have applied electron tomography (ET) to gain volume information, we have introduced scanning transmission electron microscopy (STEM), we have shown how conductive AFM (C-AFM) visualizes demixed acceptor and donor materials with high resolution and allows for local $I-V$ spectroscopy, and we have optimized FIB preparation of whole devices for further investigations by the above-mentioned techniques.

This chapter is composed mainly of work already published in various review articles, other book contributions, and original scientific publications.

2.2
The Importance of Morphology Control

In a typical polymer solar cell, light absorption creates strongly bound excitons (instead of free electrons and holes), which only can dissociate into free charges at a donor/acceptor interface by rapid electron transfer from the donor to acceptor or hole transfer to the donor in case of light absorbing acceptor materials. The need for exciton dissociation complicates the design of high-performance polymer solar cells since the light absorption depth is several times larger than the exciton diffusion range, which is limited to a few tens of nanometers at maximum (Figure 2.1). The challenge is to create such donor/acceptor organization in the photoactive layer that the interface area is maximized, with donor and acceptor materials forming nanoscale interpenetrating networks within the whole photoactive layer, and that continuous, preferably short, pathways for transport of charge carriers to the both electrodes are ensured; this is in its ideal case the so-called bulk heterojunction structure (Figure 2.2).

The following parameters have been identified as the most significant for their influence on the nanoscale morphology creation in the photoactive layers of bulk heterojunction PSCs: the chemical structure of donor and acceptor materials, which includes their molecular weights, molecular weight distributions, and ligand types, the solvent(s) used for processing, concentration in solution, the ratio between donor and acceptor, possible additives controlling phase separation during drying, and post-production treatments such as thermal annealing or exposure to solvent vapor (Figure 2.3) [2–4].

Since the photoactive layer is deposited from solution, the morphology determining features are of thermodynamic and kinetic nature, the latter mainly playing a role during the film formation process. Thermodynamic aspects are reflected in the chemical structure of the donor and acceptor compounds determining to a large extent their solubility in different solvents and the interaction (miscibility)

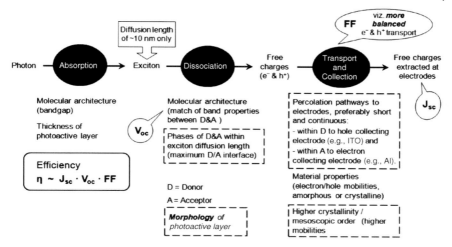

Figure 2.1 The key factors determining the power conversion efficiency (η) of bulk heterojunction PSCs, together with parameters of solar cell device performance: short-circuit current density J_{SC}, open-circuit voltage V_{OC}, and fill factor FF. All three basic processes, light absorption (characterized by efficiency η_A), exciton dissociation (η_{ED}), and transport and collection of charges (η_{CC}), should be efficient in order to get efficient PSCs. The efficiency determining factors are listed under each step; those dealing with photoactive layer morphology are shown in bold. (Reprinted with permission from Ref. [1]. Copyright 2010, Wiley-VCH Verlag GmbH.)

between these compounds taken in a certain ratio. The kinetic aspects have to do with duration of film formation, which is influenced, for example, by the solvent's boiling point, by solution viscosity, and so on, with the rate of (crystalline) ordering and thus accompanied reorganization, and with post-film formation treatments such as annealing that enables the diffusion and reorganization of one or both compounds in the blend, which ultimately leads to enhanced phase segregation.

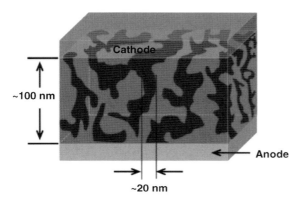

Figure 2.2 Schematic 3D representation of a bulk heterojunction (electron-donor and electron-acceptor constituents in different gray tones) with top and bottom electrodes. (Reprinted with permission from Ref. [2]. Copyright 2007, American Chemical Society.)

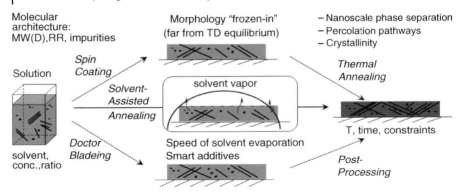

Figure 2.3 Compact summary of parameters determining morphology creation in a photoactive layer. (Reprinted with permission from Ref. [5]. Copyright 2012, Wiley-VCH Verlag GmbH)

Both thermodynamic and kinetic parameters show comparable significance in determining the morphology of the photoactive layer. Thermodynamics will, however, drive (and kinetics may limit) eventual morphological reorganization after films have been formed, and thus determine the long-term stability of the photoactive layer morphology and corresponding solar cell devices.

2.3
The Classic Blend: MDMO-PPV/PCBM as a Model for an Amorphous Donor System

The photoactive layer is composed of at least two components, the electron-donor and the electron-acceptor material. [6,6]-Phenyl-C_{61}-butyric acid methyl ester (PCBM) is by far the most widely utilized electron acceptor and intensive morphology studies have been performed on PCBM itself and polymer/PCBM systems [6, 7]. For instance, the crystallization behavior of PCBM when cast from solution, particularly the crystallization kinetics induced by different solvent evaporation dynamics, has been investigated [8–11]. It has been demonstrated that PCBM can be dissolved molecularly in the donor polymer [12] or that crystalline PCBM nanodomains develop depending on the processing conditions and the donor material used (Figure 2.4) [10]. In general, the dynamics of solvent evaporation during film formation and certain post-treatment procedures play a vital role for the crystallization behavior and morphology formation of PCBM from solution.

The influence of morphology formation on the performance of PSCs was first observed for systems where PCBM was blended with the electron-donor polymer poly[2-methoxy-5-(3′,7′-dimethyloctyloxy)-1,4-phenylene vinylene] (MDMO-PPV). A strong increase in power conversion efficiency was obtained by changing the solvent from toluene (0.9% efficiency) to chlorobenzene (2.5% efficiency) [13]. The better performance of MDMO-PPV/PCBM PSCs in case of using chlorobenzene (or o-dichlorobenzene) as a solvent rather than toluene was found to be due to smaller and thus more favorable scale of phase segregation (Figure 2.5). The higher

2.3 The Classic Blend: MDMO-PPV/PCBM as a Model for an Amorphous Donor System

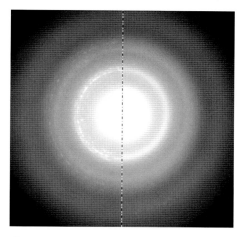

Figure 2.4 A comparison of electron diffraction pattern of superposed diffractions from larger PCBM crystals (left part) and diffraction rings from nanocrystalline PCBM (right part). (Reprinted with permission from Ref. [10]. Copyright 2010, Wiley-VCH Verlag GmbH.)

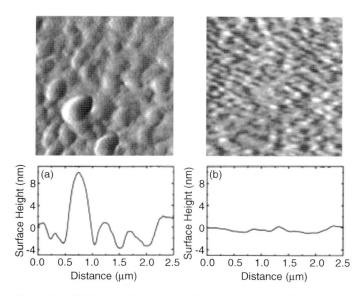

Figure 2.5 AFM images showing the surface morphology of MDMO-PPV/PCBM (1 : 4, w/w) blend films with a thickness of approximately 100 nm and the corresponding cross sections. (a) Film spin coated from a toluene solution. (b) Film spin coated from a chlorobenzene solution. The images show the first derivative of the actual surface heights. The cross sections of the true surface heights for the films were taken horizontally from the points indicated by the arrow. (Reprinted with permission from Ref. [13]. Copyright 2001, American Institute of Physics.)

solubility of PCBM in chlorobenzene results in smaller PCBM-rich domains in the MDMO-PPV-rich matrix that are formed during spin coating and subsequent solidification [8, 9, 13, 14].

Besides solubility, the evaporation rate of a solvent during film formation is also of importance for morphology formation. Even when a good solvent such as chlorobenzene is taken for MDMO-PPV/PCBM but its evaporation is slowed down, for example, by using lower spin speed during spin coating or by using drop casting instead of spin coating, coarse phase segregation is observed in the resulting films similar to faster spin coating from a less favorable solvent such as toluene [8]. Since a film has then a longer time to form and kinetic factors become less limiting, thermodynamically driven reorganization takes place already during the drying process so that large PCBM crystals are formed.

Besides the solvent used and the evaporation rate applied, the overall compound concentration and the ratio between two compounds in solution are important parameters controlling morphology formation, too. High compound concentrations induce large-scale phase segregation in MDMO-PPV/PCBM during film formation [15]. The maximum solubility of PCBM was determined to be roughly 1 wt% in toluene and 4.2 wt% in chlorobenzene (at room temperature), so that for concentrations above these critical concentrations aggregation of PCBM is anticipated already in the solvent and crystallization is enhanced even further during film formation [2].

For probing the morphological stability of polymer blends in general and the photoactive layer in particular, thermal annealing is a useful way. Apart from accelerating thermodynamically favored changes in the layer morphology, mild annealing also mimics practical conditions as solar cells can easily heat up during operation to temperatures of around 60 °C, which might be close to the glass transition temperature (T_g) of the polymer donor.

However, it has been shown that the dynamics of reorganization of a thin film upon thermal treatment not only is determined by the composition and organization of the film itself, but also depends on its local environment. Numerous experiments indicate that the mobility of polymer molecules in the vicinity of a surface or an interface is perturbed [16–18], and the extent to which they affect the mobility of polymer chains depends on the strength of their interactions with the surface/interface.

As discussed before, during spin coating of MDMO-PPV/PCBM dissolved in chlorobenzene, the high rate of solvent evaporation suppresses phase separation. Hence, the spin-coated films are probably not in their equilibrium state and likely there is a strong thermodynamic driving force to reorganize toward the stable equilibrium state. This process will be accelerated at elevated temperatures and is different for different environments, that is, different for freestanding, single-sided, or double-sided confinement; the last one reflects the situation as for whole devices [19].

Figure 2.6 shows bright-field TEM images and corresponding selected area electron diffraction (SAED) patterns of thin MDMO-PPV/PCBM film samples with 80 wt% PCBM. In the TEM images, the darker areas are attributed to PCBM-rich

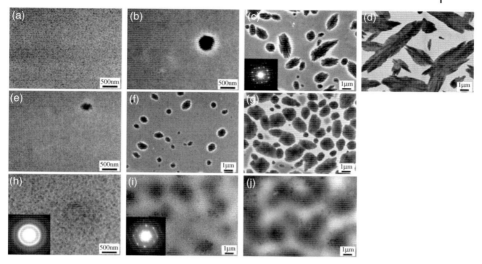

Figure 2.6 Bright-field TEM images demonstrating the formation of PCBM single crystals with time from 80 wt% MDMO-PPV/PCBM composite films upon annealing at 130 °C under different spatial confinements: (a) as spin-coated (fresh) sample; freestanding for (b) 10 min, (c) 20 min, and (d) 60 min; single-sided confined for (e) 10 min, (f) 20 min, and (g) 60 min; double-sided confined for (h) 20 min, (i) 38 min, and (j) 120 min. The insets are corresponding SAED patterns. For double-sided confined films, the metal cap was removed after annealing. (Reprinted with permission from Ref. [19]. Copyright 2004, American Chemical Society.)

domains because the electron scattering density of PCBM is higher than that of MDMO-PPV [20]. The initial film morphology consists of PCBM nanocrystals homogeneously distributed in the MDMO-PPV matrix (Figure 2.6a).

In case of freestanding film samples, PCBM clusters are formed upon annealing at 130 °C that can be identified in the TEM image as dark areas in a gray MDMO-PPV/PCBM matrix. Notably, the brighter areas surrounding the PCBM clusters reflect thinner regions of the film, being composed of almost pure MDMO-PPV, which is depleted from PCBM. We note that the dark PCBM clusters visualized in these images are single crystals, as evidenced from the corresponding SAED pattern (inset in Figure 2.6c). The PCBM single crystals develop gradually with annealing time and, particularly for these freestanding films, demonstrate a pronounced and highly elongated shape (Figure 2.6d). Notably, the aspect ratio of the single crystals, which is related to the anisotropy of the crystal growth in the lateral dimensions, increases with annealing time from about 1.5 (almost circular) at initial stage to 4 after 60 min thermal treatment.

For the case of single-sided confinement, a somewhat similar evolution of the morphology of thin MDMO-PPV/PCBM film can be followed. PCBM single crystals continuously grow during the annealing process (Figure 2.6e–g). However, when the thin-film samples are supported by a solid substrate (PEDOT/PSS), the PCBM crystals formed are smaller and less elongated as compared with freestanding film samples for the same annealing times. The aspect ratio of the crystals stays constant at approximately 1.5.

In contrast, the morphological evolution of samples in double-sided confinement follows a different route. For an annealing time of 20 min and after removing the top aluminum electrode, the morphology of the samples seems to be unchanged (Figure 2.6h). However, the appearance of the PCBM-rich clusters becomes more prominent, and the size of the clusters is increased from initially about 80 nm to about 120 nm. Corresponding SAED (Figure 2.6h, inset) analysis has revealed that the clusters are still composed of PCBM nanocrystals.

When the annealing time is increased, further reorganization of the PCBM-rich domains can be observed. Relatively dark regions emerge from the initially rather homogeneous film (Figure 2.6i and j), which seem to be PCBM crystals. Because the morphological appearance of these crystals is different when compared with the PCBM single crystals as seen for the cases of freestanding or single-sided confined films, additional SAED analysis was performed. The diffraction pattern (Figure 2.6i, inset) confirms that these dark regions are indeed PCBM single crystals possessing the same crystallographic structure as PCBM crystals formed for the other cases of confinement. Another feature of these crystals is that they have low contrast to the background and their contour is rather fuzzy. A possible reason may be that these PCBM crystals are fairly small in the direction perpendicular to the film plane. However, the lateral size is quite similar compared with those formed from films annealed at single-sided confinement, but they possess almost circular shapes and show no preferred growth direction. It should be noted that the PCBM crystals are not surrounded by a bright halo that corresponds to the PCBM-depleted regions.

For all the confinement conditions, large-scale crystallization of PCBM in the films is observed for the case of annealing temperatures above the glass temperature of bulk polymer. Elongated shapes of PCBM crystals are observed, preferably for the freestanding films. The crystal growth rate for this case is determined by the incorporation rate of PCBM molecules at the growth fronts instead of diffusion within the matrix. As more confinement is exerted to the composite film during annealing, the diffusion rate of PCBM molecules within the matrix is reduced and less elongated or even round PCBM crystals are obtained for single- and double-sided confined films, respectively.

Besides the influence of confinement on the morphology development, analyzing details of the reorganization process during annealing may help better understand the involved diffusion process of PCBM. For this reason, we have performed AFM investigations, which provide quantitative size and volume data of the growing PCBM crystals. Upon annealing, PCBM single crystals grow gradually with annealing time and stick out of the film plane (Figure 2.7). Notably, in these AFM topography images, the bright domains are PCBM single crystals (marked as "A" in Figure 2.7d), and the dark areas initially surrounding the PCBM crystals reflect thinner regions of the film, being composed of almost pure MDMO-PPV and are depleted from PCBM (marked as "B" in Figure 2.7d). In order to acquire exact growth kinetics for both the PCBM crystals and the depletion zones, volume

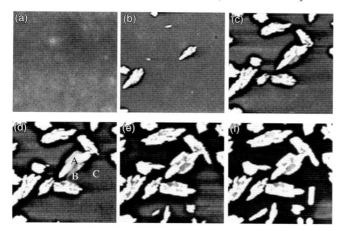

Figure 2.7 AFM topography images of MDMO-PPV/PCBM blend films (MDMO-PPV/PCBM = 1 : 4, w/w) *in situ* recorded upon annealing at 130 °C for (a) pristine film; (b) 12 min; (c) 22 min; (d) 27 min; (e) 38 min; (f) 73 min. Scan size: 15 × 15 µm^2; height range (from peak to valley): 200 nm. The letters A, B, and C marked in part (d) represent the region where the PCBM nucleates and the crystal grows (A), the depletion zone that is formed due to moving out of PCBM material toward the growing crystal (B), and the initial blend film still consisting of both MDMO-PPV and PCBM (C), respectively. (Reprinted with permission from Ref. [21]. Copyright 2004, American Chemical Society.)

quantification calculations were applied to topographic images from the composite film annealed for different times. Since the calculations are carried out based on quite large areas of the composite films, the results make statistical sense. More details on the procedures applied can be found in Ref. [21].

The principal reorganization scheme can be seen in Figure 2.8. At the very initial annealing time, as shown in Figure 2.8a, the volume amount of the film collapsed in the depletion area is smaller compared to that of the diffused PCBM inserted in the crystals. As annealing goes on, more and more PCBM is diffused toward the crystals, which ultimately causes a sudden collapse of large areas of the remaining MDMO-PPV matrix (Figure 2.8b). As annealing time moves on further, almost the whole MDMO-PPV matrix film has been collapsed; however, PCBM diffusion and crystal growth still continues (Figure 2.8c). Finally, diffusion rate of PCBM within the whole film decreases, reaching its equilibrium state, as shown in Figure 2.8d.

In summary, the prominent morphology evolution of thin MDMO-PPV/PCBM films at elevated temperature is a typical phenomenon observed at various annealing temperatures (even as low as 60 °C), with different PCBM ratios in the film and under various spatial confinements. This morphological change of the film is ascribed to the diffusion of PCBM molecules within the MDMO-PPV matrix even at temperatures below the glass transition temperature of the MDMO-PPV matrix and subsequent crystallization of PCBM molecules into large-scale crystals. However, for the high-performing polymer solar cell, the phase separation between electron-donor and electron-acceptor components should be controlled within a

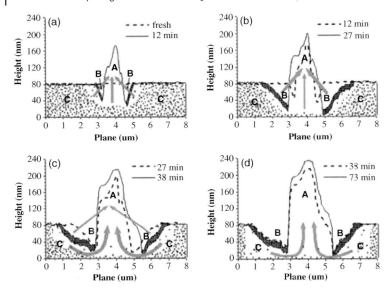

Figure 2.8 Schematic representations of the detailed morphology evolutions of thin MDMO-PPV/PCBM blend film upon thermal treatment. The dots in the profile represent PCBM molecules/nanocrystals and density of the dots represents the richness of PCBM; the diamond outlined regions represent the depletion zones after PCBM material moved out for crystal growth, in which the density of diamond outlines represents the richness of MDMO-PPV (sketches are similar to Figure 5 of Ref. [21]).

designed range to ensure a large interface for excitons to be dissociated efficiently. Large-scale phase separation enormously reduces the size of this interface area, which causes significantly decreased performance or even leads to failure of the device.

As mentioned above, MDMO-PPV/PCBM is a model for studying reorganization features of PCBM and an amorphous donor material, both being not in their equilibrium state after device preparation. The obtained information can be applied to similar but modern and high-performance systems for the photoactive layer reaching actually 10% efficiency (status 2012).

2.4
Intermezzo: Morphology Imaging with Scanning Transmission Electron Microscopy

As in the field of condensed matter sciences, polymer research has made extensive use of conventional transmission electron microscopy (CTEM) ever since its invention. Polymer research focuses on materials that mainly consist of carbon and other light elements. These materials are relatively weak electron scatterers and therefore give low mass thickness and phase contrast. Mass thickness contrast may be enhanced using an objective aperture, staining, and/or low acceleration voltages, whereas phase contrast is enhanced via defocusing. These methods, however, influence the structural organization of the specimen, introduce artifacts, limit the resolution of its projected image, and make the image difficult to interpret.

In contrast, STEM can be used in incoherent imaging condition to provide images that are easy to interpret due to the lack of phase contrast, the high signal-to-noise ratio, and the linearity of the signal intensity [22]. STEM accounts for mainly elastic scattering events consecutive to the interaction of the primary beam with the target material. The total scattering elastic cross section varies roughly as $Z^{3/2}$; therefore, local variation of elemental composition will generate relatively strong contrast in the STEM image, so-called Z-contrast [23]. As a consequence, STEM is rarely applied for the investigation of complex organic materials such as polymers, their blends or composites with carbon allotropic nanofillers (carbon black, graphene, nanotubes), or the vital class of biomaterials.

However, we have demonstrated that STEM imaging of various polymer systems – all purely carbon-based and unstained – by applying highly sensitive annular dark-field (ADF) detectors capable of single-electron counting may result in excellent contrast and allows for detailed morphological analysis with nanometer resolution [24, 25]. The detector collects with high efficiency elastically scattered electrons at large angle, which results in a dark-field image where bright contrast corresponds to the presence of scattering centers. With the average Z nearly constant for these systems, the signal intensity will vary linearly with the mass thickness—typically up to $\sigma t \sim 10^{-5}\,\text{g cm}^{-2}$ [26] – and contrast can be optimized by accurate tuning of convergence angle, camera length, and detector parameters.

In Figure 2.9, we compare conventional TEM with ADF-STEM images of MDMO-PPV/PCBM films. Because of its higher density, the PCBM appears bright and the surrounding MDMO-PPV matrix is dark when imaging in dark-field STEM conditions. For short camera lengths (i.e., large collection angles), good contrast between the PCBM domains and the matrix exists (Figure 2.9c). Contrast and appearance between the two components are somewhat similar with the TEM image acquired at low acceleration voltage; however, the PCBM domains show distinct contrast with the MDMO-PPV matrix. Applying a longer camera length (i.e., lower collection angles) but otherwise similar imaging conditions, the contrast between the PCBM domains and the MDMO-PPV matrix substantially increases (Figure 2.9d and e), probably because additional diffraction contrast of the PCBM nanocrystals contributes to the image formation. Most of the domains are connected with each other and form aggregates. Only few PCBM domains are isolated and both the PCBM and the MDMO-PPV form interpenetrating networks throughout the whole photoactive layer. Besides the dominant PCBM domains, tiny PCBM nanobridges are seen that cross the MDMO-PPV matrix and connect the domains with each other (Figure 2.9e). These nanobridges create additional interface in the photoactive film for effective exciton dissipation and a fine dispersed network for charge transport to the respective electrode. These unique morphological features adequately explain the high efficiency of corresponding devices.

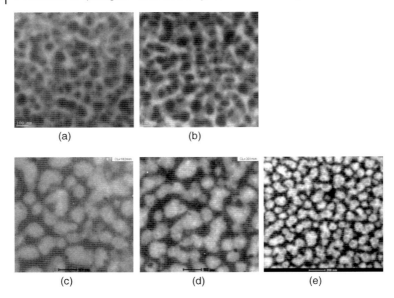

Figure 2.9 Bright-field TEM images of the PCBM/MDMO-PPV photoactive layer acquired at (a) 300 kV and (b) 80 kV acceleration voltage; ADF-STEM images of the same sample acquired with camera lengths of (c) 100 mm and (d) 300 mm; (e) contrast optimized ADF-STEM image showing the interconnected network of the PCBM domains. (Reprinted with permission from Ref. [25]. Copyright 2009, Cambridge University Press.)

2.5
Volume Characterization of the Photoactive Layer: Electron Tomography

The performance of PSCs strongly depends, as stated above, on the three-dimensional organization of the compounds within the photoactive layer. Donor and acceptor materials should form co-continuous networks with nanoscale phase separation to be able to effectively dissociate excitons into free electrons and holes, and to guarantee fast charge carrier transport from any place in the active layer to the electrodes. The most successful technique that provides local 3D morphology information with nanometer resolution in all three dimensions is electron tomography (ET), also referred to as transmission electron microtomography and 3D TEM [27–32].

In ET, a series of 2D projections is taken by TEM at different angles by tilting the specimen with respect to the electron beam. The tilt series thus obtained containing normally more than 100 images of the same specimen spot is then carefully aligned and used to reconstruct a 3D image of the specimen. The outcome of electron tomography can be used voxel by voxel (an abbreviation of "volume pixel") to study in detail the specimen's volume morphological organization.

In a first attempt to visualize the 3D organization of the system MDMO-PPV/PCBM having the optimum of 80 wt% of PCBM for highest performance of the device, we were able to obtain detailed 3D information about PCBM-rich domain sizes as well as connectivity of the PCBM domain network within the active

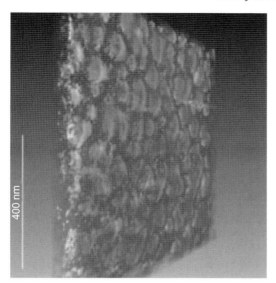

Figure 2.10 Snapshot from three-dimensional reconstruction of an electron tomography tilt series of the system MDMO-PPV/PCBM; the light gray clusters represent the PCBM-rich phase. (Reprinted with permission from Ref. [2]. Copyright 2007, American Chemical Society.)

layer (Figure 2.10). For the purpose of better contrast, we have tuned the morphology of the active layer such that PCBM-rich domains with sizes of 80–150 nm are formed after spin coating from chlorobenzene solution, which is slightly larger than that required for best performance but still small when compared with large-scale phase segregation for application of toluene as solvent. Besides size and shape, the reconstruction shows that the PCBM-rich domains are covered with MDMO-PPV, similar to the results from Sariciftci et al. [9].

Another example for successful volume morphology characterization by ET is the photoactive layer composed of the alternating copolymer containing a fluorene unit and a benzothiadiazole unit with two neighboring thiophene rings (PF10TBT) blended with PCBM [33]. The highest power conversion efficiencies were obtained for PSCs on the basis of PF10TBT characterized by a relatively high molecular weight of 188.6 kg mol^{-1} and a PCBM concentration of 80 wt%. The volume of the photoactive layer consists of interconnected polymer-rich strands forming a 3D network (Figure 2.11). The contrast originates from the density difference between the denser PCBM and the less dense polymer. The width of the polymer-rich strands, about several nanometers, is in the right range to benefit exciton dissociation at the interface between the polymer and PCBM. The existence of a polymer-rich 3D nanoscale network benefits hole transport within the whole volume of the active layer to the positive electrode, while adjacent PCBM-rich regions ensure efficient electron transport to the negative (metal) electrode. Indeed, higher efficiency of free charge formation and transport is reflected in higher J_{SC} and FF values of the devices made with the high molecular weight PF10TBT. Such beneficial

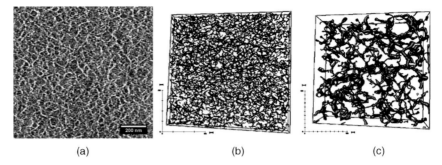

Figure 2.11 Resu t of electron tomography applied to the PF10TBT/PCBM films: (a) one slice taken out of the (x, y)-plane of the final 3D data set with bright-looking polymer strands and darker PCBM-rich regions; (b) a snapshot of the corresponding volume reconstruction showing the existence of a 3D nanoscale polymer network (volume dimensions are 1112 nm × 1090 nm × 75 nm; the dark polymer strands are made here thinner than they are in reality to facilitate visualization); (c) zoom-in of the same volume (with volume dimensions of 227 nm × 227 nm × 75 nm and the thickness of the polymer strands to scale). (Reprinted with permission from Ref. [34]. Copyright 2010, Wiley-VCH Verlag GmbH.)

morphology was not observed in PF10TBT/PCBM photoactive layers made using a low molecular weight polymer and gives rise to poorer solar cell devices.

Most intensively, we have studied by ET PCBM blended with poly(3-hexylthiophene) (P3HT). In contrast to MDMO-PPV and other amorphous polymer materials, P3HT can crystallize so that its crystallization behavior plays an additional role for morphology creation in the photoactive layer. After the pioneering work of Padinger *et al.* [35] and Waldauf and coworkers [36] in 2005, a series of studies was published dealing with the morphology-driven high performance of P3HT/PCBM-based PSCs [37–41]. In all studies, a remarkable increase in the performance is observed after annealing the devices, and power conversion efficiencies above 5% are reported [41].

Various reasons have been cited to account for morphology changes causing this efficiency improvement upon annealing, such as increased crystallinity of P3HT [42], favorable dimensions of (long and thin) P3HT crystals with width of about 20 nm, height of a few and length of hundreds of nanometers to several micrometers [43], suppressed formation of bulky PCBM clusters due to the presence of P3HT crystals [37, 40], improved light absorption of the P3HT/PCBM films as a result of morphological changes in P3HT [44], improved hole mobility, and hence more balanced hole and electron transport in P3HT/PCBM films [45–47]. This list, however long it may seem, is not complete as it does not include details on morphological organization throughout the volume of the photoactive layer, such as quality of percolating networks of nanocrystalline P3HT and PCBM and the exact scale of phase segregation.

In an attempt to reveal the three-dimensional organization in the photoactive layer of P3HT/PCBM PSCs, recently its volume morphology was analyzed in detail by means of ET and the critical morphology parameters contributing to the

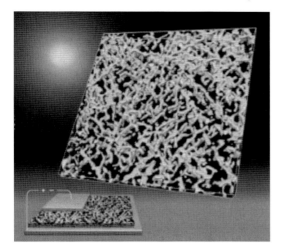

Figure 2.12 Volume representation obtained by electron tomography of P3HT nanorods (bright phase) in a P3HT/PCBM bulk heterojunction demonstrating the presence of a genuine 3D co-continuous network and concentration gradients within the thickness of the photoactive layer of both components. The experimental data are embedded in an artistic view on energy conversion by polymer solar cells. (Reprinted with permission from Ref. [49]; figure is part of the cover page. Copyright 2009, American Chemical Society.)

improved performance of P3HT/PCBM solar cell devices after thermal or solvent-assisted annealing were identified [48–50]. Figure 2.12 visualizes the final 3D morphology of the photoactive layer after annealing; only the P3HT phase is shown. A thermodynamically driven reorganization of the P3HT/PCBM morphology took place that resulted in highly crystalline and up to micrometers long P3HT nanorods composing a genuine 3D network and serving as physical barriers to PCBM diffusion, which suppressed a large-scale phase segregation of PCBM at any time of the photoactive layer preparation and post-treatment process. Moreover, recently a multiple-phase model for the organization of P3HT/PCBM is discussed, in which next to the pure crystalline phases of P3HT and PCBM amorphous P3HT and therein molecularly distributed PCBM are considered as an additional phase present at the interface when relating the properties of the photoactive layer with its organization.

The full potential of ET for volume morphology characterization of the photoactive layer has been demonstrated on the hybrid P3HT/ZnO system [51]. The concept of hybrid solar cells has been introduced and demonstrated by combining semiconducting polymers as donor with different inorganic materials, including CdSe [52, 53], TiO_2 [54, 55], and ZnO [56–58], as electron acceptor. Potential advantages of the inorganic semiconductors are a high dielectric constant, which facilitates the carrier generation process, high carrier mobility, and the thermal morphological stability of the photoactive layers.

To examine the morphology of the ZnO/P3HT photoactive layers, electron tomography was applied to freestanding films of different thicknesses obtained by spin coating on PEDOT:PSS. The snapshots of the reconstructed volume of these

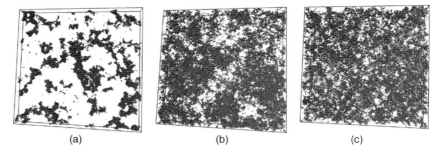

Figure 2.13 Reconstructed volumes of the P3HT/ZnO photoactive layers with thicknesses of (a) ∼60 nm, (b) ∼100 nm, and (c) ∼170 nm. The lateral size of the data sets is about 700 nm × 700 nm. The threshold was applied to the raw data such that ZnO appears gray and P3HT transparent. (Reprinted with permission from Ref. [34]. Copyright 2010, Wiley-VCH Verlag GmbH.)

films are shown in Figure 2.13. A threshold was applied to the raw data as obtained by ET in such a way that ZnO appears gray in the volume and P3HT looks transparent. Obviously, there is a large difference between the three films obtained, with finer phase-separated domains observed in thicker films. Especially, the thinnest film displays large domains for both the ZnO and the P3HT and these domains are substantially larger than the exciton diffusion length of about 5–20 nm. This is consistent with the poor device performance of thinner films.

In order to quantify the relevant morphological parameters, an extensive statistical analysis of the 3D data sets provided by ET was performed. The original 3D data of the bulk of the film were binarized to decide which voxels are ZnO and which are P3HT. The threshold for this binarization has a major impact on the final outcome, and hence error margins were estimated by applying two extremes for this threshold.

Based on the experimental 3D data, first the volume fraction of ZnO was determined. In the two thickest layers, the estimated ZnO volume fraction (about 21 vol.%) is close to the expected value of 19 vol.%, based on the ratio of diethylzinc and P3HT in the spin coating solution (for more information see Ref. [51]). The ZnO content in the thinnest layer is significantly lower at 13 vol.%. This is tentatively rationalized by a comparatively large fraction of the diethylzinc evaporating during spin coating with the higher spin speed applied for this thin layer. Next, spherical contact distances, defined as the distance from a certain voxel of one material to the nearest voxel of the other material, were determined. Because excitons are mostly generated inside P3HT, we focused on the distance distribution from P3HT to ZnO. Figure 2.14 shows the probability to find P3HT at a certain shortest distance to a ZnO domain. This analysis substantiates that coarser phase separation is present in thinner layers. Further, the efficiency of the charge carrier generation was then estimated by modeling the exciton diffusion through the P3HT phase. For this, the three-dimensional exciton diffusion equation was solved. As the result, an estimate of the fraction of excitons formed within P3HT reaching

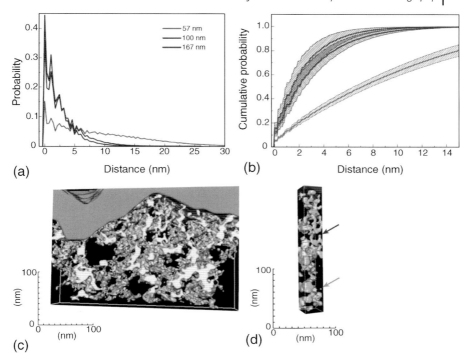

Figure 2.14 Statistical analysis of the 3D morphology. (a) Distribution of the probability to find a P3HT voxel at a certain distance from a ZnO domain for mixed P3HT/ZnO films of different thicknesses, calculated from the 3D data sets displayed in Figure 2.13. (b) Cumulative probability to have P3HT within a shortest distance to ZnO. The error margins indicated are obtained from the two most extreme thresholds possible for the binarization of the 3D data; grayscale code as in part (a). (c) Reconstructed volume of a cross section of the P3HT/ZnO device with gray ZnO domains in transparent P3HT matrix. (d) The part of this volume with the bright arrow indicating an isolated ZnO domain and the dark arrow indicating a ZnO domain connected to the top, but not through a strictly rising path. (Reprinted with permission from Ref. [51]. Copyright 2009, Nature Publishing Group.)

the interface with ZnO was obtained. Assuming that excitons efficiently dissociate into free charges at the interface with ZnO [59], the obtained numbers coincide with the efficiency of charge generation.

Besides charge carrier generation, carrier collection is also essential for solar cell operation. Efficient collection relies on continuous pathways for both carriers (Figure 2.14c and d). In view of the large volume excess of polymer in the blend, connectivity of this material will not be a limiting factor. The fraction of ZnO voxels that is interconnected via other ZnO voxels to the top of the investigated slab is quite high, at values well over 90% for all three layers, despite the low ZnO content. The connectivity is lower for thicker layers, likely because larger distances have to be crossed. Mere continuity of the ZnO phase may not be enough to effectively collect the charges. Within a continuous phase, pathways may exist that do not continue into the direction toward the collecting electrode. Due to the macroscopic electric field over the active layer of the device, charges may be trapped inside those cul-de-sacs and thus not collected (Figure 2.14d). Therefore, we also determined the fraction of ZnO connected to the top through a strictly rising path.

Looking at the combined effects of charge carrier generation and collection, we can conclude that the relatively poor performance of thin P3HT/ZnO solar cells is related to inefficient charge generation as a result of the low ZnO content, to the coarse phase separation, and to the exciton losses impaired by the electrodes. For thicker photoactive layers, charge generation is much more efficient, owing to a much more favorable phase separation. Thicker devices show superior efficiencies, but still the internal quantum efficiency (IQE) reaches only 50%. Since in thicker layers most excitons (around 80%) reach the P3HT/ZnO interface where they can dissociate into free charges, the IQE is most probably limited by inefficient charge transport. Electron transport may be limited by a low volume fraction of ZnO, whereas hole transport may be inefficient due to low hole mobilities in P3HT. In summary, careful quantification of high-resolution volume data obtained by electron tomography allows for determining characteristic morphology features within the photoactive layer and helps establishing structure–property relations in PSCs.

2.6
Measuring Nanoscale Electrical Properties: Conductive AFM

In general, performance measurements of PSCs are carried out on operational devices having at least the size of square millimeters. On the other hand, the characteristic length scale determining the functional behavior of the photoactive layer is on the order of 10 nm (exciton diffusion length) to 100 nm (layer thickness). Moreover, it is believed that the local organization of nanostructures dominantly controls the electrical behavior of devices. Thus, it is necessary to obtain electrical property data of nanostructures with nanometer resolution to be able to establish structure–property relations that link length scales from local nanostructures to large-scale devices. In this respect, a very useful analytical tool is scanning probe microscopy. In previous studies, scanning tunneling microscopy (STM) has been used for investigation of semiconducting polymers [60–62]. In particular, the current–voltage (I–$\bar{V}$) characteristics at the surface of PPV samples, widely used in polymeric photovoltaic devices, have been studied and modeled [61, 62]. However, in STM measurements variations of the topography and information on the electrical behavior are superimposed, and especially for electrically heterogeneous samples such as bulk heterojunctions, separation of electrical data from topography is difficult. Other SPM techniques probably better suited for analysis of the local functionality of polymer semiconductors are scanning near-field optical microscopy (SNOM) and atomic force microscopy. Near-field optical microscopy and spectroscopy have been used, for example, to study aggregation quenching in thin films of MEH-PPV [61]. The obtained results suggest that the size of aggregates in thin films must be smaller than the resolution limit of SNOM, roughly 50 nm. Further,

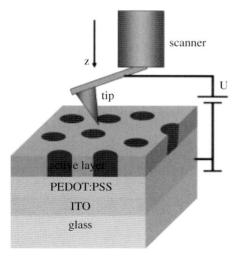

Figure 2.15 Scheme of the sample structure with segregated phases and the conductive AFM experimental setup including scanner and conductive tip.

SNOM has been applied to map topography and photocurrent of the active layer of some organic photovoltaic devices [63, 64]. However, the reported spatial resolution was only about 200 nm.

AFM equipped with a conductive probe, so-called conductive AFM [65, 66], can easily be applied to heterogeneous material systems and provides high spatial resolution (Figure 2.15). Because AFM uses the interaction force between probe and sample surface as feedback signal, both topography and conductivity of the sample can be mapped independently. Theoretically, the resolution of C-AFM is as small as the tip–sample contact area, which can be less than 20 nm. C-AFM is widely used for the characterization of electrical properties of organic semiconductors. For example, single crystals of sexithiophene have been studied [67], where the I–V characteristics of the samples were measured. Several electrical parameters such as grain resistivity and tip–sample barrier height were determined from these data. In another study, the hole transport in thin films of MEH-PPV was investigated and the spatial current distribution and I–V characteristics of the samples were discussed [68].

For conductive AFM measurements, the tip was kept in contact with the sample surface while the current through the tip was measured. In contrast to operating in intermittent contact (IC) mode, contact mode is characterized by a strong tip–sample interaction that can lead to destruction of the surface, especially in the case of soft polymer samples. Therefore, the load applied to the tip during C-AFM has to be small enough to reduce sample destruction and, at the same time, it must provide a reliable electric contact. We usually operated with a load of about 10–20 nN. The contact cantilevers used for C-AFM are suitable for operation in IC as well as in contact mode so that nondestructive testing of the sample surface could be performed before and after the C-AFM measurements. C-AFM measurements of the

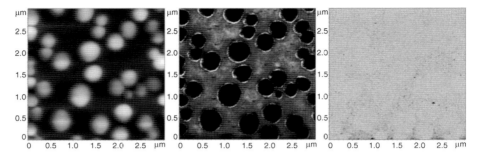

Figure 2.16 C-AFM image series acquired under ambient conditions in air of a thin PCBM/MDMO-PPV film spin coated from toluene solution: (a) topography image (gray scale represents 70 nm height variations), and current distribution images of the same area (b) for a positive bias at the tip of +2.3 V (gray values represent 1.2 nA current variation) and (c) for a negative bias at the tip of −10 V (gray values represent 250 pA current variation).

same sample area were done several times and resulted in completely reproducible data. Subsequent analysis of the surface performed in IC mode showed almost no destruction of the sample surface; only minor changes were detected from time to time.

Figure 2.16 shows a series of C-AFM images obtained under ambient conditions in air of a thin PCBM/MDMO-PPV film spin coated from toluene solution. All images were acquired with a gold-coated tip. For such preparation conditions, PCBM and MDMO-PPV segregate, and PCBM forms large nanocrystalline domains embedded in the MDMO-PPV matrix [3, 8]. The topography image (Figure 2.16a) shows that the PCBM domains (bright areas) have maximum diameters of about 500 nm. Phase segregation is responsible for the high roughness of the film: the PCBM domains stick out of the film plane few tens of nanometers, as already discussed before.

Figure 2.16b and c represents the current distribution image for bias voltages at the tip of +2.3 and −10 V, respectively, measured at the same sample area as the topography image. For positive bias, good contrast is obtained between the electron-donor and the electron-acceptor materials in the sample. From the corresponding energy level diagram, it follows that the difference between the HOMO level of MDMO-PPV and the Fermi levels of both electrodes, indium tin oxide (ITO)/poly(ethylenedioxythiophene):poly(styrene sulfonate) (PEDOT:PSS) and the Au tip, is rather small so that we expect ohmic contacts for hole injection and strong energy barriers for electrons [67, 69]. Therefore, a hole-only current through the MDMO-PPV is expected for both polarities of voltage in an ITO/PEDOT:PSS/MDMO-PPV/Au tip structure. On the other hand, we can conclude that areas of low current level correspond to the electron-acceptor PCBM, which is in accordance with the above-mentioned topographical observations. For negative bias, however, no differences between the two phases can be obtained, and the measured overall current level is below the noise level of our experimental setup.

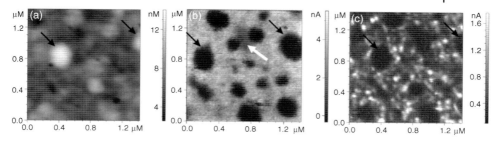

Figure 2.17 C-AFM images of the same area of a MDMO-PPV/PCNEPV active layer: (a) topography; (b) current distribution image with a positive bias at $U_{tip} = +8\,V$ (the white arrow in part (b) indicates a domain with reduced current); (c) current distribution image with a negative bias at $U_{tip} = -8\,V$; black arrows indicate same domains for reason of easy identification. (Reprinted with permission from Ref. [70]. Copyright 2006, Elsevier.)

In another study, the spatial distribution of electrical properties of realistic bulk heterojunctions has been performed by applying C-AFM with lateral resolution better than 20 nm [70]. For this study, we have blended MDMO-PPV with poly(oxa-1,4-phenylene-1,2-(1-cyanovinylene)-2,5-dialkoxy-1,4-phenylene-1,2-(2-cyanovinylene)-1,4-phenylene) (PCNEPV). Measurements of the electrical current distribution over the sample surface were performed with an Au-coated tip. A voltage was applied to the tip and the ITO front electrode was grounded. A topography image and the corresponding current distribution measured at +8 and −8 V on the tip are shown in Figure 2.17. All images were acquired subsequently so that some drift occurred. All pronounced domains in the topography image (Figure 2.17a) correlate with regions of minimal current in the C-AFM image (dark areas in Figure 2.17b).

For MDMO-PPV, the experimental setup with gold-coated AFM tip and ITO electrode represents again a hole-only device, as discussed above. The energy difference between the HOMO and LUMO of PCNEPV and the Fermi levels of both electrodes is about 1 eV, which means that a large barrier for electron injection exists in the structure ITO/PEDOT:PSS/PCNEPV/Au tip, too (some changes of barrier heights are possible when contact between metal electrodes and organic material occurs) [67, 69]. Because the hole mobility of an n-type polymer is typically smaller than that of a p-type polymer, a hole-only current through the MDMO-PPV is larger than that for PCNEPV in both bulk and contact-limited regimes. Therefore, we assume that the observed contrast in Figure 2.17b is due to a hole current, flowing through the MDMO-PPV-rich phase.

Figure 2.18 shows the typical I–V behavior of this system, and the corresponding C-AFM current distribution images for various biases when measured in inert atmosphere. The data were obtained by application of so-called I–V spectroscopy, which means that for a matrix of 128 × 128 points on the surface of the sample full I–V curves were measured for biases from −10 to +10 V. Three different types of I–V characteristics can be attained in the sample depending on the location of the measurement. For domains of the electron-acceptor compound PCNEPV, the current always is low, and the general contrast of the current distribution images

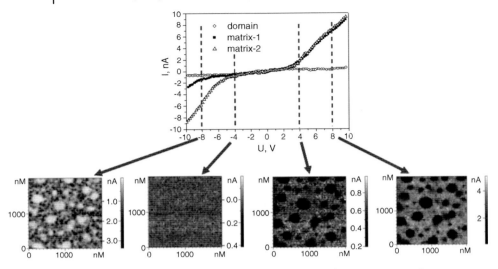

Figure 2.18 (top) Typical I–V curves as measured for each point of the I–V spectroscopy scan (128 × 128 pixels) demonstrating the heterogeneous I–V characteristics of the MDMO-PPV matrix; (bottom) for four biases the corresponding current distribution images are shown demonstrating the obvious contrast between PCNEPV and MDMO-PPV for high bias as well as contrast within the MDMO-PPV matrix with heterogeneities as small as few tens of nanometers.

depends only on the bias applied. In case of the electron-donor matrix compound MDMO-PPV, two different I–V characteristics can be obtained showing almost the same behavior for positive bias but varying significantly for negative bias. The current distribution images of Figure 2.18 demonstrate this behavior and provide additional information about lateral sizes of the MDMO-PPV heterogeneities. The point resolution of the images is about 15 nm. The log I–log V plot of the obtained data shows quadratic dependence of the current I on the voltage V measured on MDMO-PPV. This implies space–charge limited current that is in agreement with I–V measurements on complete devices [71, 72].

2.7
Current Progress and Outlook

In a number of studies, others and we have demonstrated that the performance of PSCs depends critically on the nanoscale organization and functionality of the photoactive layer, the interfaces with and type of the charge collecting electrodes, and the overall device architecture. For the last one, device performance can be improved, for example, by applying hole blocking layers [73], optical spacers to enhance light absorption in the layer of the same thickness [74, 75], and by using the tandem cell architecture [76–78], where two printable photovoltaic cells are added in series. In a tandem cell, it is possible to combine two, or more, thinner (more efficient) active layers and to use semiconductors with different bandgaps for more efficient light harvesting. Besides, since individual cells are added in series, the open-circuit voltage of a tandem cell is directly increased to the sum of the values of the corresponding individual cells.

In general, morphology investigation of whole devices becomes more and more important because only by this way we can achieve realistic information on the entire device; in particular, this is of importance for aging or lifetime studies. Such studies mainly assess the long-term performance of devices at elevated temperature but otherwise under inert conditions [79–81]. Commonly, during the aging process, all of the parameters that can be extracted from an I–V curve are recorded so that loss of performance can be identified. A similar approach is also applied for chemical degradation by exposing devices to oxygen, UV irradiation, and so on [82]. Such studies document the changes in the device performance, but give little insight into the microscopic mechanism or the chemical and physical processes involved. In particular, changes in the interaction of the photoactive layer with the electrode material and local organization at the interfaces (e.g., delamination) have barely been considered.

In recent years, we have demonstrated that focused ion beam is the tool of choice to prepare cross-sectional specimens out of full devices [51, 83, 84]. For example, Figure 2.14c represents an ET reconstruction of cross-sectional cut through a whole device; in the top part, the compact gray part represents the top aluminum electrode. Such specimens are suitable for further analysis of, for example, interface roughness, presence of local phase separation, compositional gradients, chemical architecture of the interface, and adhesion between the layers, to name but a few, after controlled aging experiments to identify critical parameters determining the lifetime of PSCs. Therefore, future volume organization studies certainly will focus on the analysis of whole devices including the electrodes. However, caution has to be taken that the powerful ion beam does not change the sample significantly during preparation.

The main advantages of scanning probe microscopy, on the other hand, are operation modes analyzing simultaneously the local morphology of the photoactive layer and measuring functional (optical and electrical) properties with nanometer resolution at the same spot. Such information is mandatory to better understand operation of PSCs at the nanoscale and to bring such devices to a new level of ultimate performance.

Acknowledgments

I would like to use this opportunity to thank Sasha Alexeev, Martijn Wienk, Jan Kroon, Sjoerd Veenstra, Volker Schmidt, Thijs Michels, and many others for helpful discussions. Particularly, I want to express my thank to René Janssen, who introduced me to the PSC theme, Xiaoniu Yang for his incredible hard and knowledgeable work on PSCs, which allowed us to set a mark in the area of morphological analysis of PSC, and Svetlana van Bavel and Erwan Sourty, who developed and applied ET and STEM for the first time on PSC systems. Further, I would like to thank the Dutch Polymer Institute (DPI) for support.

References

1 van Bavel, S., Veenstra, S., and Loos, J. (2010) *Macromol. Rapid Commun.*, **31**, 1835.
2 Yang, X. and Loos, J. (2007) *Macromolecules*, **40**, 1353.
3 Hoppe, H. and Sariciftci, N.S. (2006) *J. Mater. Chem.*, **16**, 45.
4 Thompson, B.C. and Frechet, J.M.J. (2008) *Angew. Chem., Int. Ed.*, **47**, 58.
5 Handbook of Nanoscopy, Editors: Gustaaf Van Tendeloo, Dirk Van Dyck, Stephen J. Pennycook, Wiley-VCH Verlag GmbH 2012
6 Hummelen, J.C., Knight, B.W., Lepeq, F., Wudl, F., Yao, J., and Wilkins, C.L. (1995) *J. Org. Chem.*, **60**, 532.
7 Wudl, F. (1992) *Acc. Chem. Res.*, **25**, 157.
8 Yang, X.N., van Duren, J.K.J., Janssen, R.A.J., Michels, M.A.J., and Loos, J. (2004) *Macromolecules*, **37**, 2151.
9 Hoppe, H., Niggemann, M., Winder, C., Kraut, J., Hiesgen, R., Hinsch, A., Meissner, D., and Sariciftci, N.S. (2004) *Adv. Funct. Mater.*, **14**, 1005.
10 Yang, X.N., van Duren, J.K.J., Rispens, M.T., Hummelen, J.C., Janssen, R.A.J., Michels, M.A.J., and Loos, J. (2004) *Adv. Mater.*, **16**, 802.
11 Yang, H.C., Shin, T.J., Yang, L., Cho, K., Ryu, C.Y., and Bao, Z.N. (2005) *Adv. Funct. Mater.*, **15**, 671.
12 Klimov, E., Li, W., Yang, X., Hoffmann, G.G., and Loos, J. (2006) *Macromolecules*, **39**, 4493.
13 Shaheen, S.E., Brabec, C.J., Sariciftci, N.S., Padinger, F., Fromherz, T., and Hummelen, J.C. (2001) *Appl. Phys. Lett.*, **78**, 841.
14 Rispens, M.T., Meetsma, A., Rittberger, R., Brabec, C.J., Sariciftci, N.S., and Hummelen, J.C. (2003) *Chem. Commun.*, **17**, 2116.
15 Merlo, J.A. and Frisbie, C.D. (2004) *J. Phys. Chem. B*, **108**, 19169.
16 Forrest, J.A., DalnokiVeress, K., and Dutcher, J.R. (1997) *Phys. Rev. E*, **56**, 5705.
17 Frank, B., Gast, A.P., Russell, T.P., Brown, H.R., and Hawker, C.J. (1996) *Macromolecules*, **29**, 6531.
18 Reiter, G. (1993) *Europhys. Lett.*, **23**, 579.
19 Yang, X.N., Alexeev, A., Michels, M.A.J., and Loos, J. (2005) *Macromolecules*, **38**, 4289.
20 Bulle-Lieuwma, C.W.T., van Gennip, W.J.H., van Duren, J.K.J., Jonkheijm, P., Janssen, R.A.J., and Niemantsverdriet, J.W. (2003) *Appl. Surf. Sci.*, **203**, 547.
21 Zhong, H.F., Yang, X.N., de With, B., and Loos, J. (2006) *Macromolecules*, **39**, 218.
22 Nellist, P.D. and Pennycook, S.J. (1998) *J. Microsc. (Oxford)*, **190**, 159.
23 Howie, A., Marks, L.D., and Pennycook, S.J. (1982) *Ultramicroscopy*, **8**, 163.
24 Loos, J., Sourty, E., Lu, K.B., de With, G., and van Bavel, S. (2009) *Macromolecules*, **42**, 2581.
25 Sourty, E., van Bavel, S., Lu, K.B., Guerra, R., Bar, G., and Loos, J. (2009) *Microsc. Microanal.*, **15**, 251.
26 Colliex, C. and Mory, C. (1994) *Biol. Cell*, **80**, 175.
27 Weyland, M. (2002) *Top. Catal.*, **21**, 175.
28 Weyland, M. and Midgley, P.A. (2004) *Mater. Today*, **7**, 32.
29 Jinnai, H. and Spontal, R.J. (2009) *Polymer*, **50**, 1067.
30 Mobus, G. and Inkson, B.J. (2007) *Mater. Today*, **10**, 18.
31 Cormack, A.M. (1963) *J. Appl. Phys.*, **34**, 2722.
32 Radermacher, M. (1980) Ph.D. thesis, Department of Physics, University of Munich, Munich, Germany.
33 Slooff, L.H., Veenstra, S.C., Kroon, J.M., Moet, D.J.D., Sweelssen, J., and Koetse, M.M. (2007) *Appl. Phys. Lett.*, **90**, 143506.
34 van Bavel, S.S. and Loos, J. (2010) *Adv. Funct. Mater.*, **20**, 3217.
35 Padinger, F., Rittberger, R.S., and Sariciftci, N.S. (2003) *Adv. Funct. Mater.*, **13**, 85.
36 Schilinsky, P., Waldauf, C., Hauch, J., and Brabec, C.J. (2004) *Thin Solid Films*, **451**, 105.

37 Yang, X.N., Loos, J., Veenstra, S.C., Verhees, W.J.H., Wienk, M.M., Kroon, J.M., Michels, M.A.J., and Janssen, R.A.J. (2005) *Nano Lett.*, **5**, 579.

38 Al-Ibrahim, M., Ambacher, O., Sensfuss, S., and Gobsch, G. (2005) *Appl. Phys. Lett.*, **86**, 201120.

39 Reyes-Reyes, M., Kim, K., and Carroll, D.L. (2005) *Appl. Phys. Lett.*, **87**, 083506.

40 Ma, W.L., Yang, C.Y., Gong, X., Lee, K., and Heeger, A.J. (2005) *Adv. Funct. Mater.*, **15**, 1617.

41 Reyes-Reyes, M., Kim, K., Dewald, J., Lopez-Sandoval, R., Avadhanula, A., Curran, S., and Carroll, D.L. (2005) *Org. Lett.*, **7**, 5749.

42 Zhao, Y., Yuan, G.X., Roche, P., and Leclerc, M. (1995) *Polymer*, **36**, 2211.

43 Ihn, K.J., Moulton, J., and Smith, P. (1993) *J. Polym. Sci. B*, **31**, 735.

44 Chirvase, D., Parisi, J., Hummelen, J.C., and Dyakonov, V. (2004) *Nanotechnology*, **15**, 1317.

45 Mihailetchi, V.D., Xie, H.X., de Boer, B., Koster, L.J.A., and Blom, P.W.M. (2006) *Adv. Funct. Mater.*, **16**, 699.

46 Savenije, T.J., Kroeze, J.E., Yang, X.N., and Loos, J. (2005) *Adv. Funct. Mater.*, **15**, 1260.

47 Koster, L.J.A., Mihailetchi, V.D., Lenes, M., and Blom, P.W.M. (2008) *Organic Photovoltaics*, Wiley-VCH Verlag GmbH, Weinheim.

48 van Bavel, S., Sourty, E., de With, G., Veenstra, S., and Loos, J. (2009) *J. Mater. Chem.*, **19**, 5388.

49 van Bavel, S.S., Sourty, E., de With, G., and Loos, J. (2009) *Nano Lett.*, **9**, 507.

50 van Bavel, S., Sourty, E., de With, G., Frolic, K., and Loos, J. (2009) *Macromolecules*, **42**, 7396.

51 Oosterhout, S.D., Wienk, M.M., van Bavel, S.S., Thiedmann, R., Koster, L.J.A., Gilot, J., Loos, J., Schmidt, V., and Janssen, R.A.J. (2009) *Nat. Mater.*, **8**, 818.

52 Huynh, W.U., Dittmer, J.J., and Alivisatos, A.P. (2002) *Science*, **295**, 2425.

53 Wang, P., Abrusci, A., Wong, H.M.P., Svensson, M., Andersson, M.R., and Greenham, N.C. (2006) *Nano Lett.*, **6**, 1789.

54 Kwong, C.Y., Djurisic, A.B., Chui, P.C., Cheng, K.W., and Chan, W.K. (2004) *Chem. Phys. Lett.*, **384**, 372.

55 Kuo, C.Y., Tang, W.C., Gau, C., Guo, T.F., and Jeng, D.Z. (2008) *Appl. Phys. Lett.*, **93**, 033307.

56 Beek, W.J.E., Wienk, M.M., and Janssen, R.A.J. (2004) *Adv. Mater.*, **16**, 1009.

57 Beek, W.J.E., Wienk, M.M., and Janssen, R.A.J. (2006) *Adv. Funct. Mater.*, **16**, 1112.

58 Olson, D.C., Lee, Y.J., White, M.S., Kopidakis, N., Shaheen, S.E., Ginley, D.S., Voigt, J.A., and Hsu, J.W.P. (2007) *J. Phys. Chem. C*, **111**, 16640.

59 Ravirajan, P., Peiro, A.M., Nazeeruddin, M.K., Graetzel, M., Bradley, D.D.C., Durrant, J.R., and Nelson, J. (2006) *J. Phys. Chem. B*, **110**, 7635.

60 Alvarado, S.F., Riess, W., Seidler, P.F., and Strohriegl, P. (1997) *Phys. Rev. B*, **56**, 1269.

61 Rinaldi, R., Cingolani, R., Jones, K.M., Baski, A.A., Morkoc, H., Di Carlo, A., Widany, J., Della Sala, E., and Lugli, P. (2001) *Phys. Rev. B*, **63**, 075311.

62 Kemerink, M., Alvarado, S.F., Muller, P., Koenraad, P.M., Salemink, H.W.M., Wolter, J.H., and Janssen, R.A.J. (2004) *Phys. Rev. B*, **70**, 045202.

63 Huser, T. and Yan, M. (2001) *Synth. Met.*, **116**, 333.

64 McNeill, C.R., Frohne, H., Holdsworth, J.L., Furst, J.E., King, B.V., and Dastoor, P.C. (2004) *Nano Lett.*, **4**, 219.

65 Shafai, C., Thomson, D.J., Simardnormandin, M., Mattiussi, G., and Scanlon, P.J. (1994) *Appl. Phys. Lett.*, **64**, 342.

66 Dewolf, P., Snauwaert, J., Clarysse, T., Vandervorst, W., and Hellemans, L. (1995) *Appl. Phys. Lett.*, **66**, 1530.

67 Kelley, T.W. and Frisbie, C.D. (2000) *J. Vac. Sci. Technol. B*, **18**, 632.

68 Lin, H.N., Lin, H.L., Wang, S.S., Yu, L.S., Perng, G.Y., Chen, S.A., and Chen, S.H. (2002) *Appl. Phys. Lett.*, **81**, 2572.

69 Koch, N., Elschner, A., Schwartz, J., and Kahn, A. (2003) *Appl. Phys. Lett.*, **82**, 2281.

70 Alexeev, A., Loos, J., and Koetse, M.M. (2006) *Ultramicroscopy*, **106**, 191.

71 Blom, P.W.M., deJong, M.J.M., and Vleggaar, J.J.M. (1996) *Appl. Phys. Lett.*, **68**, 3308.

72 Tanase, C., Blom, P.W.M., and de Leeuw, D.M. (2004) *Phys. Rev. B*, **70**, 193202.

73 Hayakawa, A., Yoshikawa, O., Fujieda, T., Uehara, K., and Yoshikawaa, S. (2007) *Appl. Phys. Lett.*, **90**, 163517.

74 Hansel, H., Zettl, H., Krausch, G., Kisselev, R., Thelakkat, M., and Schmidt, H.W. (2003) *Adv. Mater.*, **15**, 2056.

75 Gilot, J., Barbu, I., Wienk, M.M., and Janssen, R.A.J. (2007) *Appl. Phys. Lett.*, **91**, 113520.

76 Kim, J.Y., Lee, K., Coates, N.E., Moses, D., Nguyen, T.Q., Dante, M., and Heeger, A.J. (2007) *Science*, **317**, 222.

77 Gilot, J., Wienk, M.M., and Janssen, R.A.J. (2007) *Appl. Phys. Lett.*, **90**, 143512.

78 Hadipour, A., de Boer, B., Wildeman, J., Kooistra, F.B., Hummelen, J.C., Turbiez, M.G.R., Wienk, M.M., Janssen, R.A.J., and Blom, P.W.M. (2006) *Adv. Funct. Mater.*, **16**, 1897.

79 De Bettignies, R., Leroy, J., Firon, M., and Sentein, C. (2006) *Synth. Met.*, **156**, 510.

80 Al-Ibrahim, M., Roth, H.K., Zhokhavets, U., Gobsch, G., and Sensfuss, S. (2005) *Sol. Energy Mater. Sol. Cells*, **85**, 13.

81 Conings, B., Bertho, S., Vandewal, K., Senes, A., D'Haen, J., Manca, J., and Janssen, R.A.J. (2010) *Appl. Phys. Lett.*, **96**, 163301.

82 Jorgensen, M., Norrman, K., and Krebs, F.C. (2008) *Sol. Energy Mater. Sol. Cells*, **92**, 686.

83 Loos, J., van Duren, J.K.J., Morrissey, F., and Janssen, R.A.J. (2002) *Polymer*, **43**, 7493.

84 van Duren, J.K.J., Loos, J., Morrissey, F., Leewis, C.M., Kivits, K.P.H., van Ijzendoorn, L.J., Rispens, M.T., Hummelen, J.C., and Janssen, R.A.J. (2002) *Adv. Funct. Mater.*, **12**, 665.

3
Energy Level Alignment at Semiconductive Polymer Interfaces: Correlating Electronic Energy Levels and Electrical Conductivity

Nobuo Ueno

3.1
Introduction

It is well known that performance of organic devices, such as organic light-emitting diodes (OLEDs), organic solar cells (OSCs), and organic field-effect transistors (OFETs), is dominated by energy level alignment (ELA) at the interface. The electronic structure, particularly near the top of valence bands and the bottom of conductions bands, plays a key role in realizing devices with higher performance. As there are many literatures that present the energy level alignment at organic-based interfaces for various organic systems, we discuss mainly science, experimental background, and some selected results that are necessary for understanding electronic properties of the organic semiconductors and related interface phenomena.

This chapter describes fundamental aspects of the electronic structure of organic semiconductors (small molecules and polymers), and their interfaces, and the method to bridge the electronic structure and electrical property more directly using ultraviolet photoemission spectroscopy (UPS). Penning ionization electron spectroscopy (PIES), which is the most surface-sensitive method, is also introduced for study of electronic states at outermost surfaces of solids, which are responsible for charge exchange through the interfaces between different materials when they get contact to form a hybrid system.

3.2
General View of Electronic Structure of Organic Solids

3.2.1
Introduction to Correlating Electronic Structure and Electrical Conductivity

Here we introduce why we need electronic structure studies in relation to charge transport. Figure 3.1 illustrates the energy level alignment between organic semiconductor and electrodes for hole transport in an OFET.

Semiconducting Polymer Composites: Principles, Morphologies, Properties and Applications, First Edition.
Edited by Xiaoniu Yang.
© 2012 Wiley-VCH Verlag GmbH & Co. KGaA. Published 2012 by Wiley-VCH Verlag GmbH & Co. KGaA.

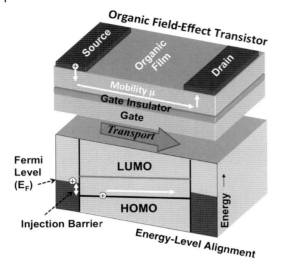

Figure 3.1 Schematic illustration for charge injection barrier (energy level alignment) and charge transport in an organic field-effect transistor using hole transport. The upper part depicts a typical structure of OFET and the lower part illustrates simplified energy levels with a source–drain potential that corresponds to the Fermi level difference between the source and the drain electrodes.

The electrical conductivity (σ) is given by

$$\sigma = nq\mu, \tag{3.1}$$

where n is the carrier concentration, q is the charge of the carrier concerned, and μ is the charge carrier mobility. This relation simply indicates that we must increase n and μ, if we need larger conductivity and electrical current. To obtain sufficient current in organic films, we need to inject charge carriers effectively from electrodes to increase n. As the carrier injection is dominated by the charge injection barrier height that is the energy difference between the Fermi level (E_F) and the highest occupied molecular orbital (HOMO) state (for hole) or the lowest unoccupied molecular orbital (LUMO) state (for electron) of the organic film, a large number of studies using UPS and inverse photoemission spectroscopy (IPES) have been carried out to study the energy level alignment at organic/metal interfaces [1–6]. These experiments have provided quantitative information on the position of E_F in the HOMO–LUMO bandgap of an organic layer at organic/metal or organic/organic heterojunctions, and motivated studies on the origin of the Fermi level pinning to give models such as the charge neutrality level and/or the induced density of interface states [7–15] and the integer charge transfer states (ICT model) due to polaron [16–18] in the HOMO–LUMO gap near the interfaces. However, the mechanism of the organic-related ELA has not yet been fully understood especially for weakly interacting interfaces and is still the subject of the argument.

To increase μ, we need to know principal origin of μ of organic material concerned, namely, coherent band conduction or hopping conduction. The band mobility, which is derived by coherent carrier motion (wave-like behavior) and is generally much larger than the hopping mobility, is dominated by the energy band dispersion [19, 20], whereas the hopping mobility by the charge reorganization energy that is related to electron–phonon coupling (terminology in solid-state physics) [20–25]. Unfortunately, however, quantum-mechanical/chemical information on the charge mobility has been elusive in experimental field due to difficulty in realizing high energy resolution UPS measurements on organic thin films [20, 26]. Such fundamental understanding based on experimental evidence is critical if we are going to improve charge carrier mobility from quantum-chemical molecular design and material engineering.

In this way, electronic structure at interfaces and in the bulk is of critical importance to unravel electrical conduction and charge exchange-related phenomena in organic devices [20].

3.2.2
Evolution of Electronic Structure from Single Molecule to Molecular Solid

As polymer solids in general consist of assembly of long polymeric molecules with weak intermolecular interaction, we first overview evolution of electronic structure from single molecule to molecular solid with a small molecule and discuss the energy bands typically appearing in the solid due to the intermolecular interaction. Evolution of electronic structure of polymer will be considered in Section 3.2.3 for a polymer chain.

Figure 3.2a shows schematically the electronic structure of a polyatomic molecule, where the molecule is made of three atoms. The potential well is formed by

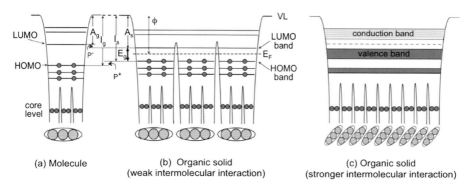

Figure 3.2 Evolution of electronic structure, from single molecule (a) to solid (b and c). When intermolecular electronic interaction is weak, the width of energy bands is very narrow (b). With increase in the intermolecular interaction, the bandwidth becomes larger (c). VL: vacuum level; E_F: Fermi level; A_g: electron affinity of gas phase; A_s: electron affinity of solid; I_g: ionization energy of gas phase; I_s: ionization energy of solid; P^-: polarization energy for negative ion in solid; P^+: polarization energy for positive ion in solid; E_g: bandgap energy; ϕ: work function.

the Coulomb potential of each atomic nucleus. The effective potential well of the molecule for an electron is formed by the atomic nuclei as well as other electrons. However, we here neglect contribution of other electrons for simplicity. The wells of the nuclei are merged in the upper part to form a broad well, where various molecular orbitals (MOs) exist and produce discrete energy levels that are different from atomic energy levels. Each MO level (energy level) is occupied by two electrons with spin up and down, respectively, in the ground state. The uppermost part of the potential well is the vacuum level (VL), at which an electron that exists outside the molecule stops moving because its kinetic energy is zero. The electron excited above the VL can escape from the molecule to vacuum. Electrons at deeper levels are localized in the atomic potential well (core levels), and thus have the feature of those in atomic orbitals because of very high potential barrier between the atoms. The upper energy levels, MO levels, involve interatomic interaction to form MOs delocalized over the molecule.

The energy separations from the HOMO or LUMO level to the VL are defined as the gas-phase ionization energy (I_g) or the electron affinity (A_g) of the molecule, respectively. When molecules assemble to form an organic solid, the electronic structure becomes like Figure 3.2b. Since the molecules interact only by the weak van der Waals interaction in many organic solids, wavefunctions of the occupied valence states (or valence bands) and the lower unoccupied states (conduction bands) are mainly localized in each molecule, yielding narrow intermolecular energy band of the bandwidth $<\sim 0.4\,eV$ [20]. Thus, the electronic structure of an organic solid approximately preserves that of a molecule, and the validity of usual band theory is often limited in discussing charge transport in an organic solid [20], which means that such an organic solid often shows two faces, in some cases face of single molecule and in other cases face of solid state. The former situation in the electronic structure of the solid allows us to simply write the band structure such as the HOMO and the LUMO levels by using "line" due to very narrow bandwidth as in Figure 3.2b. When intermolecular interaction becomes larger, both occupied (valence) and unoccupied (conduction) bands become wider because of larger overlapping of relevant MOs of adjacent molecules (Figure 3.2c). This also means that the HOMO does not necessarily show the widest band in occupied valence bands, since the bandwidth is related to spatial spread of the MO as well as the intermolecular distance.

From a molecule to a solid, the energy levels change from Figure 3.2a to b, where I_g and A_g change to the ionization energy (I_s) and the electron affinity (A_s) of the solid, respectively. When a hole (electron) is introduced into the HOMO (LUMO) of the solid, the electronic polarization of the molecules near the charge (ionized molecule) stabilizes the energy of the system by screening the charge, leading to a decrease in I (for HOMO) and an increase in A (for LUMO) compared to those in the gas phase (Figure 3.2a and b). The difference between I_g (A_g) and I_s (A_s) is related to the difference between the potential energy of an electron in the Coulomb potential of the cation in vacuum, $U = e^2/4\pi\varepsilon_0 r$, and that in the solid, $U = e^2/4\pi\varepsilon r$, as shown in Figure 3.3, where ε_0 and ε are the permittivity of vacuum and the solid, respectively. As the polarization effect may be different for the hole

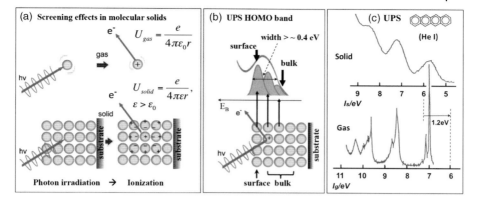

Figure 3.3 Screening effects in an organic semiconductor film and in a gas-phase molecule (a), and origin of the bandwidth of a UPS feature for a molecular thin film (b). The potential (U) of the photogenerated hole (positive ion) acting on the photoelectron is described in panel (a). Panel (b) illustrates a historical model showing that UPS bandwidth is determined by superposition of photoelectrons from surface molecules (with low kinetic energy/higher binding energy) and bulk molecules (with higher kinetic energy/lower binding energy), where screening effects depend on the number of molecules surrounding the ion. Panel (c) show comparison of gas and thin-film UPS on naphtacene (spectra are taken from Ref. [29]), where the ionization energy of gas-phase spectrum is shifted by ∼1.2 eV, polarization energy (P^+), to align the HOMO positions.

(cation) and the electron (anion) because (i) a molecular solid is not continuum medium and (ii) spatial distributions of the HOMO and the LUMO are different, there are two polarization energies P^+ and P^- for the hole and the electron, respectively [27–29]; thus, we write

$$I_s = I_g - P^+, \quad A_s = A_g + P^-, \quad |P^+| \neq |P^-|. \tag{3.2}$$

UPS of organic thin films gives I_s that is smaller by P^+ than I_g, and IPES provides A_s larger by P^- than A_g. The polarization energy is also called the relaxation energy.

Although we usually do not write a hole explicitly in the occupied MO level in the energy level scheme but two electrons, the scheme always means that the occupied MO level corresponds to that of one-hole state and the unoccupied MO level to one-electron state by assuming Koopmans' theorem [30]. The valence band is thus called the hole band and the conduction band is called the electron band. For ionization, Koopmans' theorem gives

$$E_N - E_{N-1} = \varepsilon_i, \tag{3.3}$$

where E_N and E_{N-1} are total electron energies of N- and (N − 1)-electron systems in closed-shell Hartree–Fock approximation, respectively, and ε_i is the energy of ith MO level from which an electron is ejected. Therefore, the MO energy of a molecule in the ground state corresponds to the ionization energy of corresponding MO level. Based on such understanding, we use the energy level diagram in Figure 3.1 for UPS and IPES, and also for discussion of hole and electron conduction in organic solids. We note here that Koopmans' theorem works less for systems with stronger electron correlation and for more localized valence states.

3.2.3
Evolution of Electronic Structure from Single Atom to Polymer Chain

With increase in the number of atoms in a molecule with quasi-one-dimensional structure, we finally obtain a polymer chain with infinite chain length. Figure 3.4 depicts schematically the evolution of the electronic structure of the polymer chain. When the number of atoms is small, one has a small molecule (oligomer) that is the element of the molecular solid discussed in Section 3.2.2. The electronic structure of such a molecule is discussed by molecular orbital theory and specified by discrete occupied and unoccupied molecular orbital states. With increase in the number of atoms, each atomic energy level splits into many levels depending on the number of interacting atoms, and then finally yields energy band in a very long molecular chain. The energy band is specified by the momentum $P = \hbar k$; thus, we describe $E = E(k)$. Here $k = 2\pi/\lambda$, where λ is the wavelength of electron concerned. The conduction and valence bands are separated by bandgap, where there is in principle no electronic state, but there exist actually very small densities of states that are usually produced by structural imperfection and impurities and act as charge trapping centers.

When these chains come together to form a polymer solid with weak interchain electronic coupling, the valence and conduction bands are stabilized to have smaller IP and larger A_g by the screening effect as in the case of the molecular solid in Section 3.2.2.

We can relate energy levels described by molecular orbitals and the energy bands by comparing a large oligomer (finite chain length) and corresponding polymer with infinite chain length, since the electronic structures of these finite

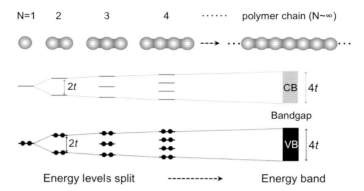

Figure 3.4 Schematic evolution of electronic structure from single atom to polymer chain. Conduction band (CB: unoccupied band)) and valence band (VB: occupied band) are shown. In tight binding approximation, the energy band width is given by $4t$ (t: transfer integral describing interatomic electronic interaction).

and infinite chains become similar with increase in the number of atoms in the oligomer. We can obtain the dispersion relation, $E = E(k)$, by assuming a tight binding model for a one-dimensional array with nearest neighbor interaction. In this case, the energy with a certain value of k (thus λ) can be correlated to one of such split levels (specified by molecular orbitals) [20, 31]. That is, using linear combination of atomic orbitals (LCAO), the Bloch wavefunction of the one-dimensional infinite system ($\Psi_B(k) = \sum_j e^{ik \cdot ja} \phi(x-ja)$) is correlated to the molecular orbital ($\Psi_{MO}(x) = \sum_{j=0}^{j=N-1} C_j \phi(x-ja)$) for N interacting atoms with the distance a, where ϕ is the responsible atomic orbital. For the N interacting atoms, there are N energy levels described by different MOs, each of which has different set of MO coefficients, C_j, and corresponds to a Bloch state with specific value of k. For example, the Bloch wavefunction with $k = 0$ ($\lambda = \infty$) corresponds to the MO with $C_0 = C_1 = C_2 = \cdots = C_{N-1}$. Therefore, a MO with many nodes corresponds to a Bloch state with larger k.

Since carbon atom has the electronic structure of $1s^2 2s^2 2p^2$, carbon atoms form four nearest neighbor bonds. In σ-bond polymers, the C atoms are sp^3 hybridized, as in polyethylene (PE), and each C atom has four σ-bonds. In such polymers, the electronic structure of the chain that comprises the backbone of the macromolecule consists of only σ-bonds. Large bandgap (E_g^σ) appears in σ-bonded polymers, which renders these polymer materials electrically insulating. In polyethylene, for example, which consists of a unit cell (repeating unit) of $-(CH_2-CH_2)-$, the bandgap is 8–9 eV [31, 32]. On the other hand, in a pendant group polymer such as polystyrene, *benzene* is σ-bonded to the backbone unit as $-(CHC_6H_5-CH_2)-$; thus, π-electronic MOs are well separated with each other, *localized* on the side phenyl groups, and their HOMO levels exist in the large bandgap of the backbone. The valence electronic structure of polystyrene may be approximated by those of model compound ($CH_3-CHC_6H_5-CH_3$) of polymer unit and the electronic properties of polystyrene are dominated by the phenyl group, since the HOMO of polystyrene comes from the phenyl group [33].

In conjugated polymers, however, there exists a continuous network of adjacent unsaturated carbon atoms, that is, carbon atoms in the sp^2 (or sp) hybridized state. Each of these sp^2 C atoms has three σ-bonds, and a remaining C $2p_z$ atomic orbital, which exhibits π-state overlapping with the corresponding C $2p_z$ atomic orbitals of the nearest neighbor carbons. This leads to the formation of π-conjugated states delocalized along the polymer chain. In a system with one-dimensional periodicity, these π-states form energy bands, with a bandgap (E_g^π), accounting for optical absorption at lower photon energies. The essential properties of the delocalized π-electron system are as follows:

i) $E_g^\pi < E_g^\sigma$, leading to low-energy electronic excitations and thus semiconductive property.
ii) The polymer chains can be easily oxidized or reduced by charge transfer from dopant species.
iii) Carrier mobility is large enough to result in high electrical conductivities in the doped state.

When polymer chains are assembled by weak interchain interaction to form a polymer solid, the electronic structure is modified as described in Section 3.2.2.

3.2.4
Polaron

As UPS experiment introduces a hole to a molecular solid, the polarization that is faster than the ionization time contributes to the P^+ (see Section 3.2.2). This is generally contribution of very rapid electron rearrangement after photoionization. If we consider slower ionization process such as charge hopping, polarization associated with deformation of geometrical structure of the molecular ion (change in the atomic positions within the ion that is related to molecular vibration (local phonon in solid-state physics terminology)) and of surrounding molecules (crystal phonon) contributes to P^+. In this way, there are three polarization contributions: (i) electronic polarization (fast, and called electronic polaron from quasiparticle picture), (ii) intramolecular geometrical polarization (slower than (i) and called small polaron), and (iii) intermolecular geometrical polarization (deformation of the crystal structure near the ion, large polaron, expected to be slower than (ii)). Thus, P^+ in UPS may involve mainly electronic polarization effects (i), and contribution of (ii) and (iii) has been considered to be too slow to be detected effectively by conventional UPS. Accordingly, UPS measures mainly a positive electronic polaron, while the hopping mobility in devices involves also effects of the small and large polarons (positive/negative polaron for hole/electron) depending on timescale of the hopping as well as the electronic polaron [24, 34]. Therefore, information on these polarons, especially polaron binding energy (E_{pol}: stabilization energy by polarization) and the timescale, is important in discussion of photoionization and charge transport.

A charge can be introduced to a molecular solid by impurity doping or contact to another solid. In this case, we generally consider fully relaxed state for polaron including (i), (ii), and (iii). Figure 3.5 illustrates small and large polarons with both electronic and geometrical polarization after hole injection to organic systems. The polarons are self-localized, which leads to stabilization of the energy level in the sense that the presence of electronic charge leads to local changes in the molecular structure and the lattice structure.

In polymer solids, for example, if a charge is introduced into a polymer chain, polaron appears to form new electronic state as depicted in Figure 3.6a, which may correspond to molecular (small) polaron since major structural change is inside the chain. If second charge is injected, we have bipolaron (Figure 3.6b). As in the case of molecular solid of small molecules, there must be various polarons depending on the timescales of polarizations. Therefore, readers must be very careful again in using term "polaron." Furthermore, it is not easy to distinguish directly the polaron state in alkali-doped polymer solid from the electronic states due to chemical reaction of the metal atom and the polymer with UPS, although the electronic states appearing in the bandgap upon the doping have in general been ascribed to the polaron and/or bipolaron [35, 36].

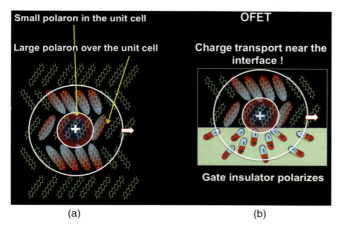

Figure 3.5 Schematic of polaron (positive) and polarization effects after hole injection. (a) The hole is in the bulk. (b) The hole is near the interface between the gate insulator and organic semiconductor in OFET (see Figure 3.1). When the area of polarization is in (beyond) the unit cell, we use the term of small (large) polaron. If intramolecular polarization dominates the polarization, it would be better to call very small polaron, since in general the unit cell involves two molecules.

3.2.5
Energy Level Alignment at the Interface

The energy level alignment means relationship between energy levels at the interface of two solids after their contact, which plays a key role in the charge injection and band bending phenomena. A typical ELA is shown in Figure 3.7 for metal–n-type inorganic and metal–n-type organic semiconductor interfaces. Upon the contact, thermal equilibrium concept strictly requires that the Fermi levels of the interacting two solids must align to have "single" Fermi level throughout the system. If

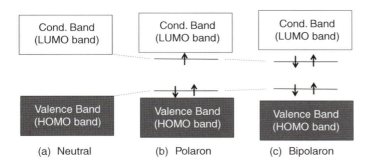

Figure 3.6 Scheme of energy level of negative polaron and bipolaron in conjugated polymers after full relaxation of the polymer structure (the self-localized states). (a) A neutral polymer. (b) Polaron state formed upon the addition of an extra electron. (c) Bipolaron state formed upon the addition of a second electron, which corresponds to the combination of two polarons.

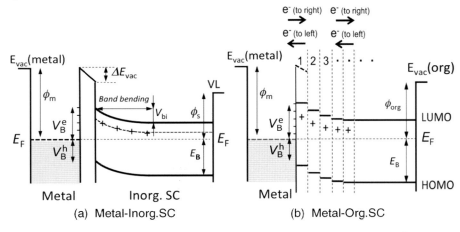

Figure 3.7 Energy level alignment at metal–semiconductor interface. (a) n-type inorganic semiconductor assuming continuum medium model for the semiconductor, where the interface dipole effect (ΔE_{vac}) is shown. (b) Organic semiconductor with interfaces between molecular layers, where "the interface dipole" is not shown since the organic first layer (1) may act as "the interface dipole layer."

the work functions of the two solids are different, electrons in one solid flow to the other solid until they have a common Fermi level, where thermal equilibrium of the system is achieved. This means that flow of electrons from the left material to the right material through the interface is the same as the flow from the right to the left. If one considers virtual interfaces in the organic semiconductor, the right going and the left going flows are the same (Figure 3.7b). The number of thermally excited mobile electrons (holes) in conduction (valence) band is specified by the Fermi–Dirac distribution function

$$f(E) = \frac{1}{e^{[E-E_F(T)]/k_B T} + 1}, \tag{3.4}$$

where k_B is the Boltzmann constant, T is the absolute temperature, E is the electron energy, and E_F is the Fermi level. E_F is located at the center (or very close to the center) of the bandgap for intrinsic semiconductors, while it is positioned closer to the bottom of conduction band (LUMO) for n-doped semiconductors and to the top of valence band (HOMO) for p-doped semiconductors. Thus, the ELA is related to where the Fermi level is located in the HOMO–LUMO gap in the bulk as well as at the interface (see Figure 3.7).

A static potential inside the film changes energy levels. Thus, both the electric dipole at the interface (interface dipole) and ionized impurities induce changes in energy levels (see Figure 3.7). The energy level changes produced by the interface dipole and by ionization of the impurities may not be distinguished by experiments for organic semiconductor molecules, since the molecular size is very large (particularly for standing orientation of the molecules).

3.3 Experimental Methods

3.3.1 Ultraviolet Photoelectron Spectroscopy

Photoelectron spectroscopy is also known as photoemission spectroscopy, where the kinetic energy of photoelectrons at the irradiation of high-energy monochromatic light is analyzed. Photoelectron spectroscopy using vacuum ultraviolet light is thus called ultraviolet photoelectron spectroscopy and probes valence electronic states with much higher energy resolution than X-ray photoelectron spectroscopy (XPS). The binding energy of occupied electronic states of materials can be directly measured with respect to the vacuum level and the Fermi level (E_F) using energy conservation role. He gas discharge lamp is conventionally used as the excitation light source of UPS, as it mainly emits He I (21.22 eV) or He II (40.82 eV) depending on the discharge conditions. UPS is a well-established and the most widely used technique in characterizing valence electronic structure at interfaces, including metal–molecule interfaces and solids. A schematic UPS process is shown in Figure 3.8, where the energy and momentum conservation rules

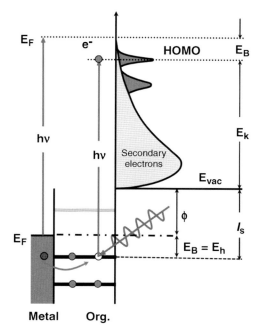

Figure 3.8 Electronic structure probed by ultraviolet photoemission spectroscopy and important parameters in discussing organic devices. Here, binding energy (E_B) refers to the Fermi level (E_F). E_k, ϕ, and E_h are the kinetic energy of the photoelectron, work function, and hole injection barrier, respectively.

are used in analyzing observed spectra. As the photogenerated hole exists during the photoionization of a molecular solid, the spectrum reflects the one-hole state. Therefore, UPS feature of HOMO is directly related to the hole transport state.

Thus, for example, the binding energy (E_B) from E_F can be obtained by the energy conservation rule as

$$E_B = h\nu - E_k - \phi, \tag{3.5}$$

where E_k and ϕ are the kinetic energy of photoelectron and the work function of molecular film, respectively. The vacuum level energy (E_{vac}), at which $E_k = 0$, is identified by the cutoff position of the secondary electron. E_F is measured by UPS of the conductive/metal substrate. In some cases, the binding energy (E_B^{vac}) is measured from E_{vac} as

$$E_B^{vac} = E_B + \phi = h\nu - E_k. \tag{3.6}$$

When a photon creates a photoelectron, a photogenerated hole exists during the traveling and detection of the photoelectron. If the photogenerated hole is not eliminated by the electron transfer from the conductive substrate before the second photon ionizes the dielectric sample (many of organics), we observe charging effects upon the second ionization. Thus, one needs to use thin films for the UPS measurement of electrical insulators (most of organic solids).

In order to obtain the energy band dispersion from UPS experiments, we need to use the momentum conservation role as well as the energy conservation role upon photoelectron emission. A three-step model is generally adopted for the photoelectron spectroscopy process, which consists of an optical dipole excitation in the solid, followed by transport to the surface and emission to the vacuum [37, 38]. General assumptions are as follows: (i) both the energy and momentum of the electrons are conserved during the optical transition, (ii) the momentum component parallel to the surface is conserved while the electron escapes through the surface, and (iii) the final continuum state in the solid is a parabolic free-electron-like band in a constant inner potential V_0,

$$E = \frac{\hbar^2 k^2}{2m^*} + V_0, \tag{3.7}$$

where m^* is the effective mass of the photoexcited electron in the final state (conduction band) and k is the electron wave vector (see Figure 3.9). V_0 represents the effective potential step to be crossed by the photoexcited electron to leave the surface. The kinetic energy E_{kin} and the wave vector K of observed photoelectron are described by following relation, with the surface normal ($K_\perp$) and parallel ($K_{//}$) components of K, respectively,

$$E_{kin} = \frac{\hbar^2 K^2}{2m_0} = \frac{\hbar^2 (K_\perp^2 + K_{//}^2)}{2m_0}, \tag{3.8}$$

$$K_\perp = K \cos\theta, \quad K_{//} = K \sin\theta, \tag{3.9}$$

where m_0 is the free electron mass and θ is the photoelectron emission angle from surface normal (see Figure 3.9b). The surface normal and parallel components of

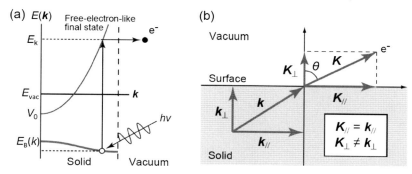

Figure 3.9 Energy and momentum conservation rules for measurements of valence band dispersion $E_B(k)$ (a) and the momentum conservation upon photoelectron escape to vacuum (b).

the wave vector k of the photoexcited electron in the solid, respectively, can be expressed as

$$k_\perp = \sqrt{\frac{2m^*}{\hbar^2}}\sqrt{E_{kin}\cos^2\theta + V_0}, \qquad (3.10)$$

$$k_{//} = \sqrt{\frac{2m_0}{\hbar^2}}\sqrt{E_{kin}}\sin\theta = 0.51 A^{-1}\sqrt{E_{kin}\,(\text{eV})}\sin\theta. \qquad (3.11)$$

The binding energy (E_B^{vac}) from E_{vac} of the electron in the initial state is written as Eq. (3.6). The free electron mass, m_0, is often assumed for m^* in the computation of $k_\perp$. The band dispersion along either $k_{//}$ or $k_\perp$ in the organic film may be investigated by changing the electron emission angle θ or E_{kin} of the photoelectrons via tuning of the energy of the incident photons, $h\nu$. As understood from Eqs. (3.10) and (3.11), however, the inner potential V_0 must be determined in obtaining $k_\perp$, while $k_{//}$ can be determined without V_0. Thus, the simplest way to obtain band dispersion is to measure angle-resolved UPS (ARUPS) of a single crystal specimen as a function of θ in order to tune $k_{//}$ (method I). A principal difficulty in measuring $k_{//}$ for organic semiconductors and insulators is that electrical conductivity is very low and thus charging of the specimen upon photoemission hinders the use of single-crystal organic samples in ARUPS. Thus, one needs to use oriented thin films on conductive substrates to realize the measurement. Organic thin films, being free from charging, are divided into two groups: (i) uniaxially oriented thin films where direction of the periodic structure or molecular stacking direction is along the surface normal, and (ii) oriented thin films where direction of the periodic structure (one-dimensional molecules) or molecular stacking direction is along the surface. For the former, in order to tune $k_\perp$ and probe the electronic band existing along the periodic direction perpendicular to the substrate, experimental setup is chosen such that the electrons are collected normal to the surface ($\theta = 0°$), that is, $k_{//}$ is zero, while varying $h\nu$ of the incident photons (method II). The latter is similar to the measurements of the single crystal with method I.

The use of actual single crystal in UPS measurements requires elimination of the charging effects by using high-quality single crystal with less charge trapping centers as well as using photoconduction to reduce the number of the trapped holes. These experiments are very difficult and still in challenging stage.

3.3.2
Penning Ionization Electron Spectroscopy

Penning ionization electron spectroscopy is also called (i) metastable atom electron spectroscopy (MAES), (ii) metastable deexcitation spectroscopy (MDS), (iii) metastable quenching spectroscopy (MQS), and (iv) metastable impact electron spectroscopy (MIES), where atom at metastable state is used instead of photon to ionize the target material. When a slow, long-lived, electronically excited metastable atom hits a solid surface, most of its excitation energy is used to eject electrons from the surface. Unlike photons used for UPS, metastable atoms do not penetrate into the bulk of the solid. PIES, therefore, excites the outermost surface layer selectively [39].

The metastable He (2^3S ($1s^\uparrow 2s^\uparrow$)) (He*) is generally used in PIES. As He* is at the triplet state, decay from triplet He* to the singlet ground state is forbidden by the spin selection rule. When third electron interacts with He*, He* decays to the ground state by the following process on an insulator surface, such as organic solids, where penning ionization (PI) (also called Auger deexcitation (AD)) takes place. As schematically shown in Figure 3.10, an electron in the valence state of the surface fills the He* 1s hole and the 2s electron is emitted to the continuum state simultaneously. The PI process yields a single-hole spectrum, as in the case of UPS. Hence, a valence state exposed outside the surface interacts with He* more effectively than that localized on the surface as depicted in Figure 3.11.

This phenomenon offers following unique possibility: (i) qualitative but direct information of molecular orientation can be obtained easily by comparing

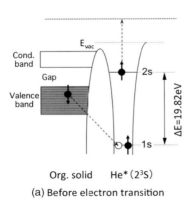
(a) Before electron transition

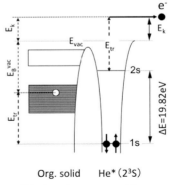
(b) After electron transition

Figure 3.10 Penning ionization of a bandgap solid by metastable He (He*) impact. The energy released by the transition of a valence electron to the 1s state (E_{tr}) excites the 2s electron of He* to the state above the vacuum level. Thus, the kinetic energy of emitted electron E_k is given by $E_k = \Delta E - E_B^{vac}$ if we neglect energy level changes in He* near the surface.

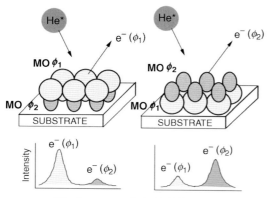

Figure 3.11 Direct detection of the molecular orientation using ultrahigh surface sensitivity of PIES. When molecules are oriented as in the left scheme, electrons in molecular orbital states of ϕ_1 are mainly excited by He* to give stronger band in PIES spectra, while ϕ_2 electrons yield stronger band when the orientation is reversed as in the right scheme.

intensities of corresponding features in PIES and UPS, (ii) formation of a complete monolayer on a substrate can be uniquely confirmed by measuring the deposition amount dependence of PIES, and (iii) the *size* of wavefunction at the surface and interface can be studied, because electrons at the outermost surface can be selectively detected. Due to ultrahigh surface sensitivity, PIES is also useful to study the outermost surface reaction [40, 41].

Figure 3.12 shows an example that demonstrates the ultrahigh surface sensitivity of PIES, where orientation, aggregation, and monolayer formation of titanyl phthalocyanine (OTiPc, also abbreviated as OTiPc) molecule on graphite surface are detected [42].

3.4
Valence Electronic Structure of Organic Semiconductors: Small Molecules

3.4.1
Energy Band Dispersion and Band Transport Mobility

If the intermolecular band dispersion ($E_B = E_B(k)$) is measured with angle-resolved UPS, the effective mass of hole (m_h^*) is obtained experimentally as [20]

$$m_h^* = \hbar^2 \left[\frac{d^2 E_B(k)}{dk^2} \right]^{-1}. \tag{3.12}$$

In the case that the band dispersion is described by a tight binding model,

$$E_B(k) = E_0 - 2t \cos(ak), \tag{3.13}$$

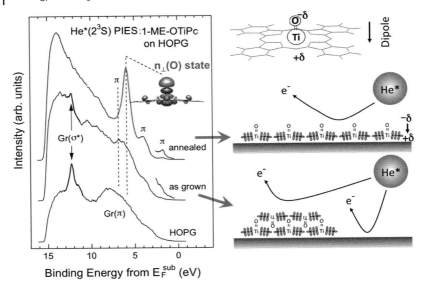

Figure 3.12 PIES spectra of monolayer-equivalent (ME) OTiPc film on HOPG surface. As-grown film consists of islands with the paired OTiPc's, where π-states of the Pc part exist at the outermost surfaces of islands. After annealing at ~150 °C, the molecules spread to form a monolayer with orientation shown in the figure, which results in the intense peak from $n_\perp(O)$ electrons and disappearance of the σ^* conduction peak from the graphite surface. The spectra are adapted from Ref. [42].

m_h^* becomes

$$m_h^* = \hbar^2 \left[\frac{d^2 E_B(k)}{dk^2}\right]^{-1} = \frac{\hbar^2}{2ta^2}, \quad (3.14)$$

where E_0 is the energy of the band center, t is the transfer integral that specifies the intermolecular interaction, and a is the lattice constant for relevant direction. Here for the tight binding dispersion, the cosine curve is approximated by a parabola near the top of the band. In a broadband model (bandwidth $(W) \gg k_B T$), the band transport mobility (drift mobility) of a hole (μ_h) can be estimated from the uncertainty principle [19, 20],

$$\tau \geq \frac{\hbar}{W}, \quad \tau \geq \frac{\hbar}{k_B T}, \quad (3.15)$$

$$\mu_h = \frac{e\tau}{m_h^*} \geq \frac{e\hbar}{m_h^* W} \cong 20 \frac{m_0}{m_h^*} \cdot \frac{300}{T}, \quad (3.16)$$

where τ is the relaxation time of the hole due to scattering and T is the temperature.

The first experimental determination of intermolecular band dispersion was reported by Hasegawa et al. [43] for an oriented thin film of bis(1,2,5-thiadiazolo)-p-quinobis(1,3-dithiole) (BTQBT). They used an oriented multilayer (30 Å) grown

3.4 Valence Electronic Structure of Organic Semiconductors: Small Molecules

on a cleaved MoS$_2$ single-crystal surface with the molecular planes nearly parallel to the surface as confirmed by quantitative analysis of the photoelectron angular distribution [44]. So far, some other dispersion measurements are available for oriented thin films (see review in Ref. [20]). Here we discuss results on pentacene and rubrene, both of which are important organic semiconductors for organic devices.

In Figure 3.13, the results of pentacene [45] are shown as an example. For pentacene (Pn), the values of m_h^* in the $\Gamma-X_{Pn}$ and $\Gamma-Y_{Pn}$ directions at 300 K are determined to be $3.02m_0$ and $1.86m_0$, respectively, by applying Eq. (3.14) to the measured dispersion relations. This result demonstrates the presence of the anisotropy of the hole mobility in pentacene crystals also at room temperature. Furthermore, by comparing with the other experimental $E(k)$ relations in pentacene films [46–48], it was confirmed that the band structure of pentacene films is very sensitive to the minor difference in the film structure, especially molecular tilt angle, because of bumpy special distribution of MO of the molecule. This was also

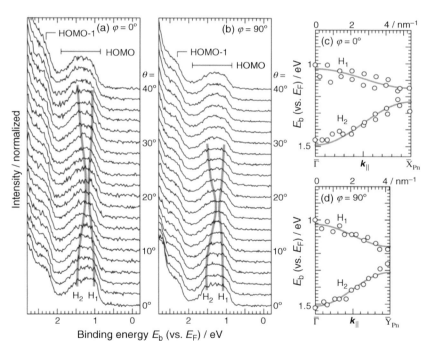

Figure 3.13 Photoelectron takeoff angle (θ) dependence of the ARUPS spectra for the highly ordered upright-standing pentacene multilayer film on Cu(110) measured at $\varphi =$ (a) 0° and (b) 90° and the experimental HOMO band dispersion. The incidence photon energy is 20 eV and the sample temperature is 300 K. E_b is the binding energy relative to the Fermi level (E_F) of the substrate. (c and d) $E(k_{//})$ relation for the highly ordered upright-standing pentacene multilayer film on Cu(110) at sample azimuthal angle (φ) = (c) 0° and (d) 90°, where φ is measured from the [110] direction. The figure is adapted with permission from Ref. [45]. Copyright (2008) Wiley.

shown with band structure calculation [49] and is an important characteristic of molecular solids.

The HOMO band dispersion in pentacene suggests that there is a density of states (DOS) structure and therefore the DOS may be observed even for polycrystalline films. Such evidence was observed by Fukagawa et al. [50] prior to the band dispersion measurements and accelerated the ARUPS experiments. The experimental two DOS features in the HOMO band of polycrystalline pentacene are shown in Figure 3.14 [50]. Actually, the maximum difference in binding energy between the top of higher E_B band and the bottom of lower E_B band in Figure 3.13, 460–500 meV, is in good agreement with the largest energy separation between two components of the HOMO [45, 48, 51] (see also Figure 3.13). The two DOS

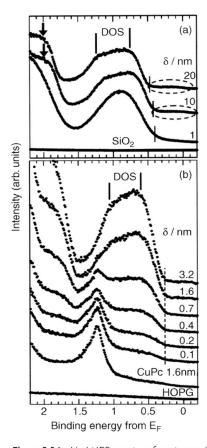

Figure 3.14 He I UPS spectra of pentacene/ SiO_2/Si(100) (a) and pentacene/CuPc/HOPG (b) as a function of the deposition amount of pentacene (δ) in the HOMO region. In (b), the underlying CuPc film consists of flat-lying CuPc. All of the spectra were measured at 295 K. Bars indicate DOS structure of the pentacene HOMO band. δ denotes the film thickness. The figure is adapted with permission from Ref. [50]. Copyright (2006) by the American Physical Society.

components are observed even for polycrystalline monolayer when it is prepared on CuPc(monolayer (ML))/GeS(001) and CuPc(ML)/HOPG (Figure 3.14b) [51], where the surfaces of GeS(001) and HOPG (highly oriented pyrolytic graphite) are passivated by flat-lying CuPc(ML). Note that one can obtain a clear evidence of the intermolecular band dispersion even for the polycrystalline film of monolayer range when the molecular packing structure in each grain is sufficiently good. If such DOS splitting is not observed, one should understand that the molecular packing in the film is not sufficient, thus yielding a very low hole mobility in pentacene OFET.

Recently, band dispersion measurements of a single crystal were realized by Machida et al. for rubrene that showed the largest mobility in organic semiconductors so far [52]. In this experiments, the second light (laser) illuminated the single crystal to eliminate the charging of the crystal during ARUPS measurements. The results are shown in Figure 3.15. The HOMO band dispersion width was found to be 0.4 eV along the Γ–Y direction (b-direction of the crystal), whereas it is very small along the Γ–X direction (a-direction). The dispersion along the Γ–Y direction gives $m_h^* = 0.65(\pm 0.1)m_0$ using Eq. (3.14). In the case of rubrene single crystal, reliable μ_h is known, so the lifetime (τ_h) or mean free path (l_h) of the HOMO hole could also be obtained as $\tau_h = 15$ fs and $l_h = 2.1$ nm or longer by combining experimental m_h^*, μ_h, and following relations:

$$l_h = \bar{v}\tau, \quad \bar{v} = \sqrt{\frac{3k_B T}{m_h^*}}, \tag{3.17}$$

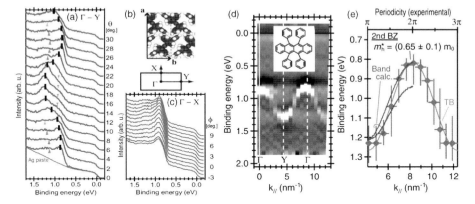

Figure 3.15 ARUPS spectra of a rubrene single crystal and band dispersion. (a) The spectra along the Γ–Y direction (b-direction of the crystal). The upward and downward triangles indicate the high- and low-E_B shoulders, respectively. The bottom curve represents a spectrum of the silver paste. (b) Schematic of the crystalline a–b plane ($a = 1.44$ nm, $b = 0.72$ nm [53]) and corresponding reciprocal lattice. (c) ARUPS spectra along the Γ–X direction. (d) Second derivative of the ARUPS spectra mapped on the E–$k_{//}$ plane. (e) E–$k_{//}$ diagram of the main peaks along the Γ–Y direction in the second BZ. Theoretical band dispersions [54] and a fitting curve obtained by the tight binding approximation are also shown. The figure is adapted with permission from Ref. [52]. Copyright (2006) by the American Physical Society.

where $\bar{v}$ is the speed of the holes and is assumed to follow the Maxwell–Boltzmann distribution. This value of l_h is three times larger than the lattice constant, suggesting that the hole transport behavior in the rubrene single crystal may be dominated by coherent band transport. Similar results were obtained by Ding et al. [55].

3.4.2
Electron–Phonon Coupling and Hopping Mobility

3.4.2.1 Fundamental Aspects on Charge Hopping

A charge traveling through intermolecular interfaces couples strongly with phonons, and the hopping charge mobility is dominated by so-called electron–phonon coupling. Here phonons involve both delocalized lattice phonons and localized phonons at each molecule in a molecular solid. The latter corresponds well to molecular vibrations of a free ionized molecule; thus, in many cases we use electron–vibration coupling as the keyword when we consider the coupling with localized phonons. For hole transport, it is necessary to consider hole–vibration coupling, which in principle can be measured by UPS as vibration-related shake-up satellites of a spectral peak (HOMO for hole conduction in organic devices).

The first reliable experimental evidence of HOMO hole–vibration coupling in organic semiconductor film was reported in 2002 [56]. As the HOMO band in UPS spectrum involves information on the coupling between the conduction hole and vibrations of the molecular ion, UPS measurements can offer key information that is necessary to unravel the fundamental mechanism in the carrier transport properties of organic devices.

Charge hopping dominates the mobility in molecular systems with very narrow bandwidths and even in systems with larger bandwidths if the mean free path of the traveling charge is very small (the order of the intermolecular distance). Charge transfer processes and carrier dynamics of organic molecular solids have been studied in various fields and detailed theoretical descriptions can be found in several reviews [22–24, 57].

Since the electronic state concerned is well localized, charge transport occurs via hopping from one molecule to the next. Moreover, changes in the hopping rate due to variation of relative intermolecular geometry need to be considered for static disorder systems, which are often found in actual organic solids. When an electron or hole is injected into a molecular solid, electron–phonon coupling leads to localization due to polaron binding energy and transport occurs via localized charge hopping, which is closely related to Marcus electron transfer theory [21]. The semiclassical Marcus hopping model for self-exchange charge transfer has been widely studied. Here, we present one semiclassical approach to estimate hopping mobility. The molecular parameters, which have been used in theoretical models, can experimentally be obtained by using high energy resolution UPS measurements.

The hopping mobility (μ) in the high-temperature regime can be approximated from the electron transfer rates by considering the Einstein relation for diffusive motion [20, 23],

$$\mu = \frac{ea^2}{k_B T} k_{CT}, \quad \text{where} \quad k_{CT} = \frac{2\pi}{\hbar} t^2 \sqrt{\frac{1}{4\pi\lambda k_B T}} \exp\left[-\frac{\lambda}{4k_B T}\right], \tag{3.18}$$

where a is the intermolecular distance, k_{CT} is the charge transfer/hopping probability per unit time, λ denotes the reorganization energy induced by the charge transfer, and t corresponds to the intermolecular transfer integral that describes the strength of the electronic interaction between adjacent molecules as in Eq. (3.13). Thus, there are two major parameters that determine the charge hopping mobility: (i) the electronic coupling (transfer integral t) between adjacent molecules, which needs to be large, and (ii) the reorganization energy λ, which needs to be small for obtaining efficient hopping mobility. This requirement is also necessary for obtaining higher band mobility. For hole transport, t can be experimentally obtained from the HOMO band dispersion of a molecular stacking system or from the splitting of the HOMO level of a dimer molecule. λ corresponds to the sum of the geometry relaxation energies ($\lambda_{rel}^{(1)}$ and $\lambda_{rel}^{(2)}$) in Figure 3.16. The contribution of each vibration mode to relaxation energy $\lambda_{rel}^{(2)}$ can be determined by the intensities of vibration satellites in high-resolution UPS. The satellite intensities are described by the values of the Huang–Rhys factors, S, which in the harmonic approximation are related to $\lambda_{rel}^{(2)}$ by

$$\lambda_{rel}^{(2)} = \sum_i S_i h \nu_i. \tag{3.19}$$

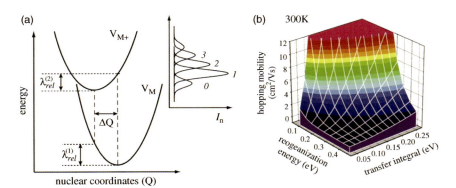

Figure 3.16 (a) Typical adiabatic energy surfaces of neutral (V_M) and ionized (V_{M^+}) states and two relaxation energies $\lambda_{rel}^{(1)}$ and $\lambda_{rel}^{(2)}$ at ionization ($M \rightarrow M^+$) and neutralization ($M^+ \rightarrow M$) processes. The UPS intensity of vibration satellites I_n on photoionization is shown on the right. $\lambda_{rel}^{(2)}$ is obtained by measuring I_n. (b) Typical simulation results of charge hopping mobility as a function of transfer integral (t) and reorganization energy ($\lambda = 2\lambda_{rel}^{(2)}$) using Eq. (3.18) at 300 K and $a = 0.32$ nm. The figure is adapted with permission from Ref. [26]. Copyright (2009) Elsevier.

There are two λ_{rel} components, $\lambda_{rel}^{(1)}$ and $\lambda_{rel}^{(2)}$, corresponding to going to the ionized state and returning to the neutral state (Figure 3.16a). If $\lambda_{rel}^{(1)}$ is not too different from $\lambda_{rel}^{(2)}$, λ can be written as

$$\lambda = \lambda_{rel}^{(1)} + \lambda_{rel}^{(2)} \approx 2\lambda_{rel}^{(2)}. \tag{3.20}$$

One thus obtains λ using the following relation:

$$\lambda \approx 2\lambda_{rel}^{(2)} = 2\sum_i S_i h\nu_i. \tag{3.21}$$

When the neutral state is in the vibrational ground state, the intensities of the vibrational progression resemble a Poisson distribution

$$I_n = \frac{S^n}{n!} e^{-S}, \tag{3.22}$$

where I_n is the intensity of the nth vibrational satellite for ith vibration mode of $h\nu_i$. These relations mean that λ can be experimentally estimated by measuring $h\nu_i$ and corresponding I_n with high-resolution UPS. Although Eq. (3.18) is too simplified for calculating μ using experimental λ, no good theoretical description is available so far to obtain μ from the analyses of the vibration satellites in the thin-film UPS. Furthermore, one should be careful in considering the hopping mobility obtained by above-described approach, since (i) the validity of the Einstein relationship has been debated for nonequilibrium conditions, for example, in an actual organic field-effect transistor, and (ii) the above-described hopping mobility corresponds to that in an ideal crystalline region, where there is no energy distribution/spread of the HOMO. For nonideal crystals, which have various bandgap states/trap states due to crystal imperfectness and domain boundaries, we need to consider distributions of the electronic energy levels, the intermolecular distance that depends on relative molecular orientation of relevant adjacent molecules, and t, and thus need progress of theoretical study on these contributions as well as on the effects of temperature and low-energy (crystal) phonons. It is true, however, that μ in actual devices of poor molecular packing should be smaller than that estimated with the above-described method.

As the polaron binding energy (E_{pol}) is defined as stabilization energy when the hole (electron) is localized on a single lattice site, E_{pol} is directly related to the relaxation energy and thus the reorganization energy [22–24]. For intramolecular relaxation (for a very small positive polaron) in the limit of weak intermolecular electronic interaction ($t \sim 0$), E_{pol} may be obtained by using $\lambda_{rel}^{(2)}$ that is determined from the UPS vibration satellites,

$$E_{pol} = \lambda_{rel}^{(2)} \cong \frac{\lambda}{2}. \tag{3.23}$$

This relation means that the high-resolution UPS gives E_{pol} even for molecular systems with a large difference between $\lambda_{rel}^{(1)}$ and $\lambda_{rel}^{(2)}$.

3.4.2.2 Reorganization Energy and Small Polaron Binding Energy

The bandwidth of UPS features contains information on electron/hole–phonon/vibration interaction, electron (hole) lifetime, and electron-defect scattering provided that final state band structure (dispersion) effects are not taken into account. It is necessary to obtain high-resolution UPS spectra to observe vibration satellites for the HOMO band. There are few examples of small molecules on metal surfaces where vibrational features can be resolved in a valence photoemission spectrum and where structural heterogeneity can be eliminated from molecule–metal interfaces to allow the hole–vibration coupling to be quantitatively determined. For organic molecules, however, this measurement has been believed to be nearly impossible.

The first result is shown in Figure 3.17, which was observed at room temperature (295 K) for a CuPc submonolayer on HOPG [56] with the molecular plane parallel to the surface. The HOMO (labeled A) in panel (a) appears as a sharp peak at 1.2 eV below E_F. Panel (b) shows an enlarged view of the HOMO peak after the contribution from the HOPG substrate has been subtracted. The peak is asymmetric and can be decomposed into three vibrational features ($v = 0, 1, 2$) with an energy separation of 170 meV (this value has been improved slightly in later work [26, 51]). This gives a full-width at half-maximum (FWHM) of 172 meV for the main component ($v = 0$). Considering an instrument resolution (ΔE) of 80 meV, the intrinsic linewidth was obtained as 150 meV, which corresponds to a lifetime of 2.2 fs within the assumption of using an uncertainty relation [56]. This value can be an upper limit for the lifetime of the hole generated by the photoemission process. The transient hole was considered to be filled by electron transfer from the HOPG

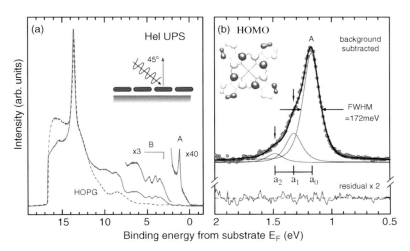

Figure 3.17 (a) He I UPS of CuPc (0.2 nm) on a HOPG substrate (solid curve) and clean HOPG (dashed curve) at sample temperature of 295 K at $\theta = 0°$. (b) Expanded spectrum of top band (A) (HOMO) region after subtraction of background. Circles represent observed spectrum after subtraction of background. Three components (a_0–a_2) of pseudo-Voigt profiles used in curve fitting are indicated by thin solid curves. Solid curve that overlaps observed spectrum is convolution of three fitting curves. Residual of curve fitting is also shown at bottom of (b). Orbital pattern of HOMO of CuPc is also illustrated. The figure is adapted with permission from Ref. [56]. Copyright (2002) Elsevier.

substrate; thus, this lifetime of the HOMO hole may correspond to the electron transfer rate to the HOMO.

Figure 3.18 shows the angle-integrated UPS spectra ($\theta = 0$–$60°$) of a pentacene (ML)/HOPG system compared with the convoluted curves of 18 A_g vibration modes [26, 58]. The energy scale is relative to the 0–0 transition peak. Here, the thin-film spectra have also been integrated for the azimuthal angle around the surface normal by the azimuthal disorder of the single-crystal domains in the HOPG surface. The convolutions were carried out using Voigt functions, in which the intensity of the vibration satellite is given by 0–0, 0–1, and 0–2 transitions as represented by the Poisson distribution. The convoluted curve with $W_G = 5$ meV and $W_L = 65$ meV for Voigt functions and with the vibration intensities given by the gas-phase S factor (S_{gas}) and the gas-phase vibration energies (hv_{gas}), which were used to analyze the gas-phase spectrum [59], is in excellent agreement with the gas-phase spectrum to yield a λ_{gas} of 97 meV. However, there was marked disagreement between this convoluted curve and the 49 K spectrum both in the satellite

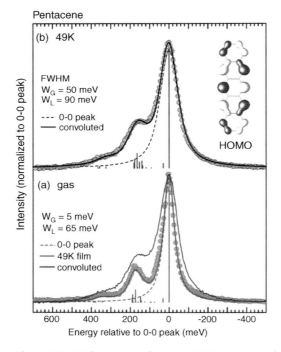

Figure 3.18 High energy resolution He I UPS spectra of gaseous (a) (circles) and monolayer film (b) (circles) for pentacene, compared with convoluted curves of 18 A_g vibrational modes (solid curves). Energy is relative to 0–0 transition peak (dashed curve). (a) Convolution curve obtained by Voigt functions ($W_G = 5$ meV and $W_L = 65$ meV) with S_{gas} and hv_{gas}, compared with gas-phase and 49 K film spectra (thin curve). Vertical bars indicate 0–0, 0–1, and 0–2 transition intensities. (b) Convolution curve obtained by Voigt functions ($W_G = 50$ meV and $W_L = 90$ meV), assuming values of $1.2S_{gas}$ and $0.95hv_{gas}$. Gas spectrum is adapted from Ref. [59]. The figure is adapted with permission from Ref. [26]. Copyright (2009) Elsevier.

intensities and in the linewidth. Better agreement between the 49 K and convoluted spectrum is obtained for $W_G = 50$ meV, $W_L = 90$ meV, and $S_{film} = 1.2 S_{gas}$ for all A_g vibrational modes, in which all $h\nu_{gas}$'s are contracted by 0.95, which is obtained from direct measurements of the vibration energy for the satellite peak (158 meV for 49 K film and 167 meV for gas-phase UPS). Therefore, the λ for the pentacene/HOPG (λ_{film}) was thus obtained from Eq. (3.21) as $\lambda_{film} = 109$ meV at 49 K. λ_{film} is larger than λ_{gas}, indicating that the hole mobility at the interface and in the bulk of oligoacene crystal is smaller than that expected from the gas-phase results. Using Eq. (3.18), the hopping mobility in a crystalline pentacene can be estimated as $1 \sim 2$ cm^2 V^{-1} s^{-1} from $\lambda_{film} = 0.10$–0.11 eV obtained from monolayer film, $t = 0.04$–0.06 eV from band dispersion measurements [45, 48], $a = 0.32$ nm, and $T = 300$ K. Note that the hopping mobility estimated in this way may give the upper limit, since (i) Eq. (3.18) does not reflect crystal phonons (intermolecular effects) and (ii) λ_{film} and t depend on the direction of hopping in the crystal.

Figure 3.19a shows high energy resolution UPS of HOMO band of rubrene monolayer on HOPG, which was measured at 35 K, after subtracting contribution of photoelectron from the substrate and integrating electron emission angular dependence of the satellite intensities [60]. Peak T$_1$ (P$_1$) is assigned to the 0–0 transition of twisted (planar) rubrene and the other peaks labeled T$_i$ (P$_i$) to its vibrational progressions. This assignment of peaks thus gives an $h\nu_{tw}$ of 134 meV and an $h\nu_{pl}$ of 127 meV for the twisted and planar conformations, respectively. The

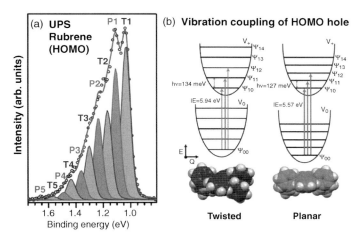

Figure 3.19 (a) The HOMO hole–vibration coupling of a monolayer rubrene on HOPG measured by high energy resolution UPS (after the background subtraction and integration for photoelectron emission angle (θ)). Fit to the angle-integrated HOMO peak. Dots correspond to measured data. T$_1$–T$_2$ and P$_1$–P$_5$ are the hole–phonon (vibration) coupling satellites of twisted conformation and planar conformation molecules, respectively. (b) Schematic of adiabatic energy surfaces of neutral (V$_0$) and ionized (V$^+$) rubrene in the twisted (left) and planar (right) conformation. E denotes energy and Q the configurational coordinate. The ionization energy (IE) is defined by the difference of the 0–0 transition peak position and the vacuum level. The figure is adapted with permission from Ref. [60]. Copyright (2012) Wiley.

Huang–Rhys factor S is 0.58 for the twisted conformation and 0.62 for the planar conformation; thus, the small polaron binding energies can be determined to be 78 and 79 meV for the twisted and the planar conformation, respectively, which gives charge reorganization energies of $\lambda_{tw} = 155$ meV and $\lambda_{pl} = 157$ meV, which are slightly (however, not significantly) smaller than the calculated value of 159 meV for an isolated molecule [54].

The details of this coupling (e.g., the vibrational energies) are slightly different between the twisted and planer conformations. However, the relevant parameters λ and E_{pol} are essentially the same and could be averaged to 156 and 78.5 meV, respectively, as values for condensed rubrene. With the averaged value of λ and the transfer integral t, which has been determined by room-temperature (295 K) ARUPS measurements of a rubrene single crystal along the direction of largest dispersion to $t = 0.11$ eV [52], the hopping probability k_{ET} can be calculated (Eq. (3.18)) to be 1.1×10^{14} s^{-1} at 295 K and 3.6×10^{9} s^{-1} at 35 K. That is, using the intermolecular distance ($a = 0.72$ nm) from the single-crystal structure [19], an upper limit of the rubrene hopping mobility can be determined to be 22.7 cm^2 V^{-1} s^{-1} at 295 K and 6.2×10^{-3} cm^2 V^{-1} s^{-1} at 35 K [60]. The room-temperature value is an outstanding high value for hopping mobility of an organic solid. However, it is only around half of the highest measured mobility values of rubrene single crystals of up to 40 cm^2 V^{-1} s^{-1} [61], thus demonstrating again band transport in rubrene single crystals (see section 3.4.1). Note again that Eq. (3.18) is valid only within the high-temperature approximation ($t \ll kT$) and the strong coupling limit ($\lambda \gg t$).

3.5
Valence Electronic Structure of Polymers

3.5.1
Quasi-One-Dimensional Band Dispersion Along Polymer Chains

3.5.1.1 σ-Bond Polymer
Experimental results on the band dispersion in σ-bond polymers are very limited due to difficulty in preparing thin films with oriented chains [20, 31, 32, 62]. Here, we introduce the band dispersion of quasi-one-dimensional polymer polyethylene. Early work on the band structure study was carried out on systems with alkyl chains and was aimed at understanding the electronic structure of polyethylene, in particular, the possible existence of one-dimensional band structure in thin films where molecular chains assemble via weak interchain interactions. There is renewed interest in the band dispersions as they determine carrier transport properties in nanoscale molecular electronics [63].

The first successful experiment was carried out in 1983 by Seki et al. at HASYLAB, DESY, and published in 1984 [64], where the authors used hexatriacontane (n-CH$_3$(CH$_2$)$_{32}$CH$_3$) vacuum deposited on metal surfaces that are not "clean" to obtain standing molecular orientations. As the alkyl chains orient with their long axes perpendicular to the surface, the intramolecular band dispersion can be determined from $h\nu$ dependence of ARUPS spectra measured at $\theta = 0$ (normal emission). The second successful measurement was carried out by Ueno *et al.* [65] also at HASYLAB using Langmuir–Blodgett (LB) films of Cd arachidate $\{Cd^{2+}[CH_3(CH_2)_{18}COO^-]_2\}$, in which alkyl chains are oriented with their long axes perpendicular to the surface, as verified by the *in situ* observation of a large angular dependence of photoelectron intensity. These two studies including determination of V_0 were fully reported in Ref. [31] with comparison to various theoretical calculations. Among the theoretical band dispersions, which were largely different from each other at these era, they found that the results with *ab initio* method [66] showed the best agreement with experimental findings. Due to limitation of available photon energies in these experiments, they could not observe dispersion over the whole Brillouin zone. The measurement of band dispersion over the whole Brillouin zone was later realized by the same groups using newly constructed BL8B2 line at UVSOR, Institute for Molecular Science [67–69]. Here, the results on evaporated thin films of pentatriacontan-18-one $[CH_3(CH_2)_{16}CO(CH_2)_{16}CH_3]$ are introduced. The molecules are in the standing molecular orientation. The results, shown in Figure 3.20, demonstrate that (i) the top of the valence band at the C point consists of a band of B_1 (B_{2g} C point) symmetry that is derived by C 2p atomic orbitals directed along the carbon chain, and (ii) the chemical disorder due to the C=O group in the alkyl chain does not affect the band structure so much and the molecule shows dispersion similar to that in an ideal polyethylene chain. The experimental results demonstrated that *only eight repeating units of $(CH_2)_2$ are sufficient* to give the band dispersion. This finding suggests that it is possible to measure intermolecular band dispersion using very thin films consisting of small stack of molecules that are easily charged upon photoionization. The intermolecular dispersion related to small number of repeating units can be thus discussed as described in Section 3.2.3. From these experiments, we could fully understand the electronic structure of the quasi-one-dimensional polyethylene chain, and furthermore learned that molecular orientation can be controlled by choosing substrates, for example, by using metal surfaces passivated/covered by organic "contamination."

3.5.1.2 π-Conjugated Polymer Chain

Preparation of a thin film of single crystal or of well-oriented chains is critically necessary in experimental band dispersion measurements using ARUPS. A method for the preparation of thin single-crystal films of polydiacetylene has been developed by Berrehar *et al.* [70, 71]. Typical film areas are a fraction of a square centimeter. Such samples allow charging to be avoided, while keeping perfect chain order. Salaneck *et al.* [72] measured the electronic band structure of thin single crystals of poly{5,7-dodecadiyene-1,2-diol bis[((4-butoxycarbonyl)methyl)urethane]} (4-BCMU polydiacetylene) using ARUPS with polarized light from synchrotron radiation. The spectra are collected as a function of θ in the K–b plane, namely, as a function of $K_\parallel$ along the polymer chain, where K is the photoelectron momentum vector and b is the vector parallel to the chain (see Figure 3.21). A plot of the dispersion relations in units of the width in momentum of the first Brillouin zone, π/a, is

3 Energy Level Alignment at Semiconductive Polymer Interfaces

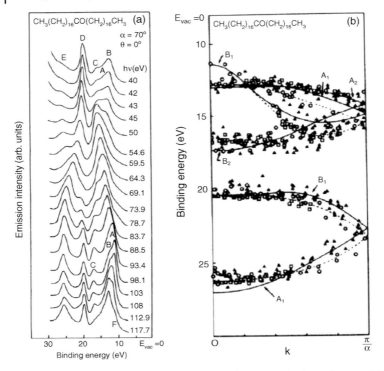

Figure 3.20 (a) Photon energy ($h\nu$) dependence of the normal emission spectra of vertically oriented pentatriacontabe-18-one with photon incidence angle $\alpha = 70°$. (b) Comparison of experimental energy band dispersions for pentatriacontan-18-one, hexatriacontane, and Langmuir–Blodgett films of cadmium arachidate in reduced zone scheme. Open circles: pentatriacontan-18-one {$CH_3(CH_2)_{16}CO$ $(CH_2)_{16}CH_3$} [69]; open triangles: hexatriacontane [n-$CH_3(CH_2)_{34}CH_3$] [31, 64]; closed triangles: hexatriacontane[67]; open squares: Langmuir–Blodgett films of cadmium arachidate ([$CH_3(CH_2)_{18}COO^-]_2Cd^{2+}$) [31, 65]. The theoretical result by Karpfen [66] is shown by a solid curve after contraction of the energy scale by 0.8 and shift for the best fit with the experimental results. The dotted curves indicate the experimentally deduced dispersion curves for the A_1 and B_1 bands. The figure is adapted with permission from Ref. [69]. Copyright (1990) by the American Physical Society.

shown in Figure 3.21. Here, a is the dimension of the unit cell of 4-BCMUPDA in the b-direction, 4.88 ± 0.1 Å.

3.5.2
Pendant Group Polymers: Is the Surface of Solution-Cast Film Clean on Molecular/Atomic Scale?

A great advantage of using polymers is easy preparation of thin films from their solutions. However, when they are not prepared in ultrahigh vacuum, surfaces of solids are in general covered by contaminants and therefore not clean from surface science. This hindered use of electron spectroscopies for valence electronic

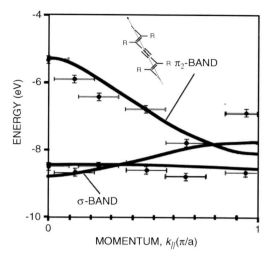

Figure 3.21 Experimental band dispersion of upper valence bands obtained from the ARUPS results of 4-BCMUPDA (chemical structure shown in the figure) in comparison with portion of the VEH band structure (solid curves). The ordinate corresponds to the binding energy from the vacuum level. The figure is adapted with permission from Ref. [72]. Copyright (1994) Elsevier.

structure of polymers. We here demonstrate that surfaces of polymer and other organic films prepared by solution processes and/or in ambient air are extremely clean, which is impossible to be realized for metals and most inorganic solids. We show two examples of polymers that demonstrate such evidence.

Figure 3.22 shows comparison of UPS and PIES spectra for thin films of poly(2-vinylnaphthalene) (PvNp) and poly(9-vinylcarbazole) (PvCz) prepared by spin casting on Au-coated H–Si(100) substrates [33]. For both polymers, the energy positions of spectral peaks in the PIES and UPS spectra are in good agreement with those of the calculated DOS of model compounds of PvNp (CH_3–$CHC_{10}H_7$–CH_3) and PvCz (CH_3–$CHNC_{12}H_8$–CH_3). Very clear appearance of spectral features in PIES spectra, as in the UPS, demonstrates that the outermost surfaces of the PvNp and PvCz thin films are not contaminated but extremely clean at atomic/molecular level. These results allow us to apply surface-sensitive techniques to investigate electronic structure of polymer thin films and other organic thin films that are prepared by solution processes in ambient air.

3.5.3
P3HT: Electronic Structure and Control of π-Electron Density Distribution at the Surface for Realizing a Functional Interface

Solution-processed regioregular conjugated poly(3-hexylthiophene) (P3HT) thin films can easily lead to well-organized conformation leading to highly oriented polymer films, and such films in general have complex microstructures with ordered

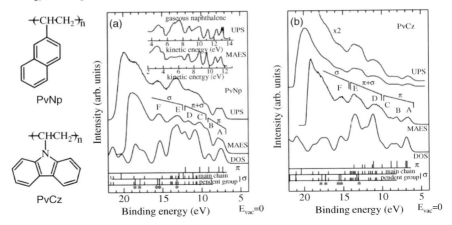

Figure 3.22 He*(2^3S) PIES (indicated as MAES) and He I UPS spectra of PvNp (a) and PvCz (b). The vertical bars show the molecular orbital energies by *ab initio* MO calculation (6–31G with diffuse function) for the model compound of the polymers (see the text). The energy levels of π- and σ-orbitals (at pendant group and at main chain) are separately shown, where orbitals involving contributions from both pendant group and main chain are represented by the shorter bars. σ-orbitals at the C—H bonds of the pendant group are marked by an asterisk. The DOS was obtained by a broadening with Gaussian function (width = 0.8 eV). He*(2^3S) PIES (indicated with MAES) and He I UPS spectra of gaseous naphthalene are compared in panel (a). The figure is adapted with permission from Ref. [33]. Copyright (2001) Elsevier.

microcrystalline domains embedded in an amorphous matrix [73, 74]. Due to its unique characteristics with rigid backbone and solubilizing flexible side chains, P3HT usually serves as a prototypical semiconducting polymer to be employed in the application and properties characterization. The long-range order and π–π interchain stacking in the P3HT thin films play an important role in the charge transport process and the device performance, which have been characterized comprehensively [73, 75, 76]. The structural characteristics are therefore well understood in the bulk and thin films, but few studies were reported focusing on the details of the surface layers. Delongchamp et al. [77] have pointed out that molecular orientation and hole mobility of P3HT thin films can be varied by spin coating speed. Molecular orientation is investigated by near-edge X-ray absorption fine structure spectroscopy (NEXAFS) that probes more bulk of the film than UPS and PIES. The bulk behavior, however, does not necessarily extend to the surface due to the different solvent removal process between the bulk and the surface [78]. Therefore, it is essential to determine the conformation and stacking at surfaces, especially at the outermost surfaces that are most adjacent to the overlayer molecules, because the spatial electron density distribution at these regions might be inconsistent with the bulk region due to the different chain conformation resulting from variation of coating process, and furthermore, it is of particular significance to understand the charge transfer mechanism with an overlayer molecule due to the π–π electron wavefunction overlap.

The spectroscopic tools using soft X-ray-based photons as probes can detect the electronic structures of about top ~2 nm or even thicker layer inside the polymer

3.5 Valence Electronic Structure of Polymers

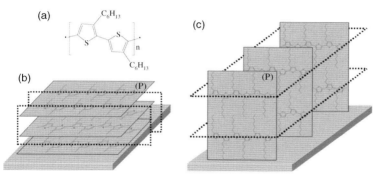

Figure 3.23 The chemical structure of P3HT (a) and the schematic diagrams of P3HT thin-film conformations: (b) face-on and (c) edge-on. Thiophene ring planes are parallel to the planes (P) shown in (b) and (c); the dashed planes perpendicular to planes (P) correspond to those for lamellar structure. When molecules are deposited on the film (b), π electrons in P3HT interact with the overlayer molecule, while they cannot on the film (c).

surface and those using ultraviolet photons as probes like UPS can observe the electronic density distribution of about top ∼1 nm inside the surface. Contrarily, PIES using He* as probes can observe the outermost surface electrons selectively as described in Section 3.3.2. Unlike the ordinary surface-sensitive spectroscopic techniques using photons or electrons, PIES can thus be adopted for probing the electronic states exposed outside the polymer surface to understand the chain conformation and spatial distribution of π-electron density at outermost surface.

We describe hereafter that there is clear evidence showing that the chain conformation and therefore the spatial distribution of the electron density at surface and outermost surface can be controlled by employing different coating parameters/methods [79, 80]. The schematic diagrams of P3HT thin-film conformation are given in Figure 3.23: (b) face-on conformation and (c) edge-on conformation. It is clearly evident that for the face-on conformation π-electrons in the polymer backbone are exposed to the outside and interact with overlayer molecules, while for the edge-on conformation they are shielded by side alkyl chains and cannot interact with the overlayer molecules.

To understand π-electronic states contributed from the polymer backbone, the surface electronic structure of evaporated thin films of polythiophene (PT) without side chains was investigated by UPS and PIES, as shown in Figure 3.24 [79]. Because there is no effect brought by side pendant groups on the electronic states of PT, the observed features appearing on the spectra are definitely from the polymer backbone. Figure 3.24 gives the π-electronic states region of (1) UPS spectra (upper curve), (2) PIES spectra (medium curve) for thin films of PT, and (3) calculated density of valence states (bottom curve, calculated for 12 repeated thiophene ring units). Due to many molecular orbitals in close vicinity to each other in the polymer chains as well as solid-state broadening (Section 3.2.2) and vibration coupling of the UPS features, the individual molecular orbital cannot be resolved in the spectra. Therefore, several molecular orbitals usually contribute to a single UPS

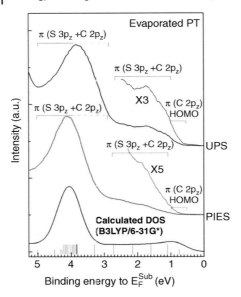

Figure 3.24 UPS spectra (upper curve), PIES spectra (medium curve) for thin film of PT without annealing, and calculated density of valence states (bottom curve, calculated by a DFT method with B3LYP/6–31G* basis set on PT with 12 repeated thiophene ring units; the longer bars indicate that the electronic states include the contribution from S $3p_z$ atomic orbitals) [79]. The binding energy was measured from the Fermi level of the substrate (E_F^{Sub}). The figure is adapted with permission from Ref. [79]. Copyright (2007) American Chemical Society.

feature in the measured spectra. The observed features of π-electronic states can be assigned with the help of the calculations in the low binding energy region from 0 to 5 eV. The features located around ~0.5 to ~1 eV may be assigned to the strongly delocalized, all-C $2p_z$-derived π-electronic states, and the features appearing at ~1–2.5 eV should be contributed mainly from π-states of C $2p_z$ atomic orbitals (AOs) and partly from the S $3p_z$ AOs. In addition, the peaks located around ~4 eV in UPS and PIES may include localized π-states with strong S $3p_z$ contribution and weak contribution of the C $2p_z$ AOs. Because PIES is more surface sensitive than UPS, the peak position difference in the spectra between UPS and PIES may be due to the difference of AO distribution, C $2p_z$ and S $3p_z$, at the outermost surface. Usually, S $3p_z$ AOs spread more widely than C $2p_z$ AOs and give more contribution to the electronic wavefunction distributed at the outermost surface, which can interact more easily with the metastable He* atoms in PIES. Therefore, the peak in PIES that is located at a relatively higher binding energy position (~4.1 eV) than that in UPS (~3.8 eV) should come from π-states with more contribution from S $3p_z$ AOs, and the peak in UPS should include more contribution from π-states of C $2p_z$ AOs.

Surface electronic structures of spin-coated and dip-coated P3HT thin films have also been observed by UPS and PIES spectra of P3HT thin films prepared by different coating conditions. P3HT powder was dissolved into chloroform and the spin casting was performed for 60 s with speed of 3000 rpm (high speed) and 400 rpm (low speed) for the solution concentration of 2.5 and 0.5 mg ml^{-1}, respectively. The dip casting was also carried out by dipping the substrate into the solution (0.5 mg ml^{-1}) for 60 s and then drying slowly in a chloroform vapor environment. The thickness is about 5–10 nm controlled by the solution concentration and spin coating process parameters, which is thick enough to cover the substrate and thin enough to minimize the charging artifacts during UPS and PIES measurements. The solution-processed films were annealed at 453 K under UHV for 1 h to further remove the contamination/solvent and improve the crystallinity. By comparing with the NEXAFS that has much deeper probing depth, we can get the clear picture of electronic states distribution from inside to outside. Overlap of π-electronic wavefunction within the lamellae planes should be responsible for the high intra-layer mobility in P3HT films. On the other hand, π-electronic states distributed outside the polymer surface provide the possibility of forming overlap with the π-electronic wavefunction of an overlayer molecular material, consequently achieving efficient charge transfer in the related heterojunction structure used for polymer devices.

Figure 3.25a shows NEXAFS spectra of P3HT thin films prepared with different coating conditions. The transition peaks situated at 287.8 eV originate from transitions to the superimposed C–H/C–S σ* orbitals, while the broader peak located at 292.9 eV may be associated with transitions to C–C σ* orbitals. The peak due to

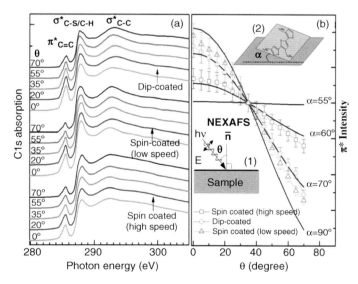

Figure 3.25 (a) Normalized C K-edge NEXAFS spectra as a function of incident X-ray angle of annealed P3HT thin films. (b) C=C π* intensity variation as a function of the incident angle (θ) (computed curves labeled with different tilt angle α; experimental data for the high-speed spin-coated sample with closed squares, the low-speed spin-coated sample with open triangles, and the dip-coated samples with open circles) [79]. Inset (1) gives the measurement geometry of NEXAFS and inset (2) shows the definition of the tilt angle α of the polymer backbone. The figure is adapted with permission from Ref. [79]. Copyright (2007) American Chemical Society.

C=C π^* state exhibits a strong systematic variation with the incident angle (θ) (as depicted in Figure 3.25b) that can determine the average conjugated plane orientation. Calculated results for the π^* intensity as a function of the incident angle (θ) are also shown in Figure 3.25b using the computation procedure as reported previously [80, 81]. Comparing the experimental π^* intensities with the computed results, the average molecular orientation of polymer backbone can be estimated. The average tilt angle (α) of the plane of polymer backbone with respect to the substrate surface is 62°, 70°, and 72° for samples spin coated with high spin speed and low spin speed, and dip-coated samples, respectively, which match with the computed curves very well. Therefore, although all the samples have tilted conformation, dip-coated and low-speed spin-coated samples have greater edge-on characters than the high-speed spin-coated sample, whereas the high-speed spin-coated sample has more face-on characters. The slower solvent removal rate that results in materials closer to their equilibrium structure should be responsible for the edge-on conformation in dip-coated and low-speed spin-coated samples [77].

Figure 3.26a gives the UPS spectra for (A) annealed dip-coated, (B) annealed low-speed spin-coated, (C) annealed high-speed spin-coated, and (D) non-annealed high-speed spin-coated P3HT thin films and (e) calculated density of valence states (calculated for 10 repeated thiophene ring units). The left panel shows the spectra of the complete valence region, and the right one shows the HOMO spectral region. The binding energy was measured from the Fermi level of the substrate (E_F^{Sub}). The observed features in the spectra can be assigned correspondingly by the calculated results. The stronger peak located at about 3–4 eV corresponds to the localized π-states contributed from S $3p_z$ and C $2p_z$ AOs. The observed π-states whose features are located around ~0.5–1 eV in UPS spectra are due to the strongly delocalized nature of HOMO state of C $2p_z$ AOs, and the feature around ~1 to ~2.5 eV comes from contribution of C $2p_z$ and partly contribution of S $3p_z$ AOs. No clear discernible features appear for all the samples between 5 and 12 eV, which are related to the σ-states from the main chain and the hexyl side groups due to the complexity of the polymer structure. The spectra are comparable to the theoretical calculation in the low binding energy region from 0 to 5 eV. Also, comparing the UPS spectra of PT (Figure 3.24) and P3HT thin film in the low binding energy (0–5 eV), the π-electronic structure is found to be conserved approximately after attaching the side hexyl chain onto the polymer backbone. In addition, by comparing spectra C and D of the high-speed spin-coated thin films, the observed electronic structure does not change much except for the stronger intensity for the annealed thin films due to the improved crystallinity and ordered structure after annealing [80].

As aforementioned in the NEXAFS results, the high-speed spin-coated sample has more face-on characters and low-speed spin-coated and dip-coated samples have more edge-on characters. The different conformation should be responsible for the difference in the UPS spectra of different samples. It can be seen clearly, from Figure 3.26a, that the intensity of peaks related to π-states is stronger in the spectrum for the high-speed spin-coated sample than that for low-speed spin-coated and dip-coated samples. This suggests that π-electronic wavefunction is

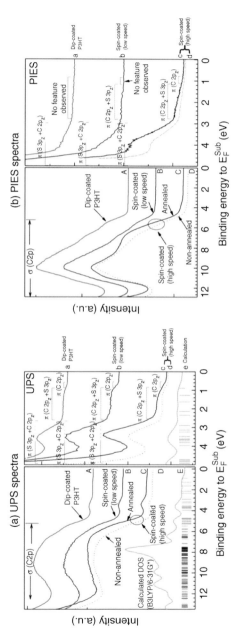

Figure 3.26 (a) UPS spectra for (A) annealed dip-coated, (B) annealed low-speed spin-coated, (C) annealed high-speed spin-coated, and (D) non-annealed high-speed spin-coated P3HT thin films, and (E) calculated density of valence states (calculated by a DFT method with B3LYP/6-31G* basis set on P3HT with 10 repeated thiophene ring units; the longer bars indicate that the electronic states include the contribution from S $3p_z$ atomic orbitals). (b) PIES spectra for (A) annealed dip-coated, (B) annealed low-speed spin-coated, (C) annealed high-speed spin-coated, and (D) non-annealed high-speed spin-coated P3HT thin films. The left panel in (a) and (b) shows the spectra of the complete valence region, while the right one shows the HOMO spectral region. The figure is adapted with permission from Ref. [79]. Copyright (2007) American Chemical Society.

exposed more outside the surface for the sample with face-on conformation and can be detected with less influence of the side chains, and correspondingly generate more photoelectrons for higher spectral intensity. For the high-speed spin coating process, there is too little time to achieve equilibrium structure in the thin film due to rapid evaporation of the solvent and therefore the polymer chains at surface regions may be assembled with more face-on conformation before any significant crystallization or molecular orientation reorganization occurs [77]. However, the π-states are shielded more by the side chains for the samples with edge-on conformation. Although the probing depth allows UPS to detect the electronic states of the polymer backbone, the corresponding intensity in the UPS spectra becomes weaker due to the shielding by side chains. On the basis of this assumption, we may conclude that the dip-coated sample most likely has more edge-on confirmation than the low-speed spin-coated sample in the top one or two monolayers near the surface because the intensity of peaks related to the π-states is much weaker in the spectrum for the dip-coated sample than that for low-speed spin-coated sample. This should be due to the slower solvent removal rate of dip-coated sample at not only the bulk region but also the surface region, which enables the polymer chain to have sufficient time for achieving the equilibrium edge-on conformation [77]. Considering that the PIES has much greater surface sensitivity, it is adopted to prove the viewpoint mentioned in the UPS results.

PIES spectra for (A) annealed dip-coated, (B) annealed low-speed spin-coated, (C) annealed high-speed spin-coated, and (D) non-annealed high-speed spin-coated P3HT thin films are given in Figure 3.26b. The left panel shows the spectra of the complete valence region, and the right one shows the HOMO spectral region. The binding energy was measured from the Fermi level of the substrate (E_F^{Sub}). For high-speed spin-coated sample, the π-states (~1 to ~2.5 eV and ~3 to ~4 eV) contributed by C $2p_z$ and S $3p_z$ AOs are shown clearly in the spectra, indicating that the π-electronic wavefunction from the polymer backbone is distributed outside the polymer surface for the high-speed spin-coated sample and can effectively interact with metastable He* atoms. Correspondingly, the HOMO molecular orbital from the polymer backbone also can be detected by metastable He* atoms showing a clear feature in the range of the binding energy of ~0.5 to ~1 eV, which is believed to be facing the vacuum at the polymer surface. However, there are less obvious features for low-speed spin-coated and dip-coated samples in the π-states region (0–4 eV), indicating that the electronic states from the polymer backbone are more shielded by the side alkyl chain than the high-speed sample and cannot interact with the metastable atoms. The very weak shoulders around ~3–4 eV in the spectra of both low-speed spin-coated and dip-coated samples and those around ~1–2.5 eV in the spectra of low-speed spin-coated thin film can be observed because of the wider spreading nature of S $3p_z$ AO. Due to the great surface sensitivity of PIES, we believe that high-speed spin-coated samples adopt more face-on orientation in the surface region different from the tilted conformation inside the film, whereas the low-speed spin-coated and dip-coated samples adopt tilted or more edge-on conformation at both bulk and surface regions. The intensity of σ-states related to the hexyl side groups in high-speed spin-coated thin film is also stronger than that in another two samples, which is shown as an obvious shoulder in the range of 6–8 eV in the spectrum, exhibiting that the σ-orbitals (i.e., the side hexyl chains) are preferentially oriented parallel to the surface. It is noted that the π-electronic structure of high-speed spin-coated P3HT is approximately similar to that of PT in the low binding energy region (0–5 eV). The difference in the higher binding energy region (larger than 5 eV) comes from the effect of the σ-electronic states brought by the attached hexyl chains. There is not much difference observed for low-speed spin-coated sample and dip-coated sample in PIES spectra, because the metastable He* atoms can only interact with the electronic states of side hexyl chains extending outside the surface in both samples. The extended side hexyl chains prevent the He* atoms interacting with the polymer main chain and therefore there is no clear feature observed in the low binding energy region (0–3 eV), indicating that the electron density contributed from polymer backbone is buried by the side chains.

Comparison of spectra C (UPS) and D (IPES) in Figure 3.26a and b shows that the annealed thin films give stronger intensities in the observed features due to the improved crystallinity and ordered structure after annealing, as discussed previously [80]. The π-electronic wavefunction from the polymer backbone can be detected by the most surface-sensitive PIES for high-speed spin-coated P3HT thin films before and after annealing, indicating that the outermost surface layer adopts a similar face-on conformation for these two thin films. Therefore, it assures that annealing-induced increment of molecular tilt angle (α) does not occur in the outermost surface region that differs from the bulk region in the thin films.

In passing the dependence of the molecular orientation on film thickness was also reported for polystyrene film prepared from solution [82].

3.6
Role of the Interface Dipole Layer: Its Impact on the Energy Level Alignment

Mechanism of the energy level alignment has not yet been clarified. It may be related to the density of gap states (DOGS), since DOGS dominates the position of the Fermi level in the HOMO–LUMO gap. Electrical measurements have indicated that charge trapping states exist in the bandgap even for nondoped organic single crystals [83, 84]. The gap states act as hopping conduction sites and control the Fermi level if the density is not too small. Density of these gap states is very low and cannot be detected by conventional UPS because of its insufficient sensitivity.

Fukagawa et al. [85] observed the VL shift as well as the E_F shift for a well-defined interface system, pentacene/ClAlPc(ML)/HOPG, where two-dimensional dipole layer of ClAlPc (ClAl phthalocyanine) [86] is inserted between pentacene and graphite (HOPG). As shown in Figure 3.27a, upper part of the HOMO band of pentacene (standing orientation) is located above the E_F of the ClAlPc(ML)/HOPG substrate before contact, and moves below the E_F and the corresponding shift of the VL appears due to transfer of 0.01 electrons per pentacene molecule to the HOPG

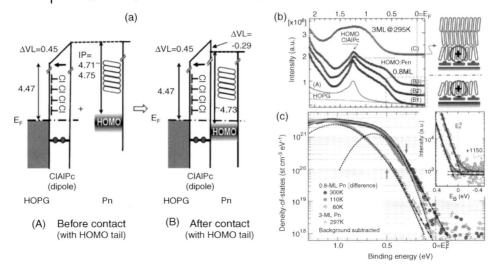

Figure 3.27 Impact of charge injection on the energy level alignment. (a) ELA at pentacene and ClAlPc (dipole layer)/graphite before (A) and after (B) contact (holes are injected to Pn layer) (experimental data in Ref. [85]), where ΔVL is vacuum level shift and energy is given in eV. (b) UPS spectra of HOMO region of the system in (a)-(B), where spectra of HOPG, ClAlPc/HOPG (A), Pn (0.8 ML)/ClAlPc/HOPG (B1 at 300 K, B2 at 110 K, and B3 at 60 K), and Pn (3 ML)/ClAlPc/HOPG at 295 K(C) are shown. (c) Density of states near the HOMO and the bandgap states (on log scale) at 300, 110, and 60 K obtained from ultrahigh-sensitivity He I$_{\alpha\alpha}$ UPS spectra. The results of Pn (3 ML)/ClAlPc/HOPG at 297 K are also plotted for comparison. The arrows correspond to boundaries of the two distributions of the gap states (Gaussian tail and exponential tail). The exponential tail reaches the Fermi level. The inset shows the spectra of the Pn (0.8 ML) at 60 K and Pn (3 ML) at 297 K on log scale after subtraction of background photoelectrons. Parts (b) and (c) are adapted with permission from Ref. [87] and shown after slight modification. Copyright (2009) American Institute of Physics.

through the ClAlPc dipole layer after contact [85]. Sueyoshi *et al.* detected DOGS in the pentacene on the VL-increased HOPG system (pentacene/ClAlPc(0.8ML)/HOPG) with ultrahigh-sensitivity UPS measurements as shown in Figure 3.27c [87]. They demonstrated that (i) there are two types of DOGS between the pentacene HOMO and the E_F, with Gaussian energy distribution near the HOMO and exponential energy distribution near the E_F, (ii) the exponential DOGS reaches the E_F for 0.8 ML pentacene (interface pentacene), and the DOGS near the E_F is lower in a thicker pentacene film (3 ML) than in the interface pentacene to result in a higher E_B for the 3 ML HOMO (Figure 3.27b), and (iii) the DOGS determines the E_F (namely, HOMO position from E_F). They attributed the DOGS to imperfect molecular packing at the interface, and pointed out a possibility that spontaneous disorder appears near the weakly interacting organic interface. Note that fully relaxed positive polaron must exist in the pentacene overlayer in the pentacene/ClAlPc/HOPG system, since holes exist in the pentacene overlayer after the contact [85, 87]. It is thus evident that a part of the DOGS may be related to the fully relaxed positive polaron in the pentacene overlayer.

Ultralow-density DOGS reaching to the E_F was also directly observed recently in a CuPc film after exposure to 1 atm N_2 gas [88] and at organic–organic interfaces [89].

3.7
Future Prospects

It is well known that the energy level alignment at organic interfaces is extremely important for organic devices. However, its mechanism is still far from full understanding. The issue is related to nature of molecular solids where molecules with low structure symmetry assemble by weak intermolecular interaction. Moreover, motion of charges is affected by phonons/molecular vibrations in molecular solids. We need more precise knowledge on the electronic structure at surfaces, interfaces, and bulk of molecular solids, including low-density electronic states in the bandgap, role of spatial distribution of the wavefunctions, and electron–phonon coupling to unravel electronic functions of organic-based devices. Precise information on these is indispensable to understand quantum-chemical origin of the electronic functions, namely, difference between organic semiconductors and their inorganic counterparts. In this chapter, we described fundamental aspects on the above key parameters as well as on electronic structure of molecular solids and polymers. Experimental studies on the key parameters were recently realized on small-molecule organic semiconductors, but corresponding studies on polymers have been left to be challenged.

Acknowledgments

The author is deeply grateful to the late Prof. K. Seki of Nagoya University for collaboration work for more than 30 years. He also thanks Prof. S. Kera and Prof. K. Okudaira, Dr. H. Yamane, Dr. S. Duhm, and many Ph.D. students of Chiba University, and Dr. X.T. Hao of the University of Melbourne for their contribution.

References

1 Ishii, H., Sugiyama, K., Ito, E., and Seki, K. (1999) *Adv. Mater.*, **11**, 605.
2 Salaneck, W.R., Seki, K., Kahn, A., and Pireaux, J.-J. (eds) (2002) *Conjugated Polymers and Molecular Interfaces: Science and Technology for Photonic and Optoelectronic Applications*, Marcel Dekker, New York.
3 Koch, N. (2007) *ChemPhysChem*, **8**, 1438.
4 Hwang, J., Wan, A., and Kahn, A. (2009) *Mater. Sci. Eng. R*, **64**, 1.
5 Gao, Y. (2010) *Mater. Sci. Eng. R*, **68**, 39.
6 Heimel, G., Salzmann, I., Duhm, S., and Koch, N. (2011) *Chem. Mater.*, **23**, 359.
7 Heine, V. (1965) *Phys. Rev.*, **138**, A1689.
8 Tejedor, C., Flores, F., and Louis, E. (1977) *J. Phys. C*, **10**, 2163.

9. Flores, F. and Tejedor, C. (1987) *J. Phys. C*, **20**, 145.
10. Mönch, W. (1994) *Surf. Sci.*, **299**, 928.
11. Vázquez, H., Oszwaldowski, R., Pou, P., Ortega, J., Pérez, R., Flores, F., and Kahn, A. (2004) *Europhys. Lett.*, **65**, 802.
12. Vázquez, H., Gao, W., Flores, F., and Kahn, A. (2005) *Phys. Rev. B*, **71**, 041306 (R).
13. Kahn, A., Zhao, W., Gao, W., Vázquez, H., and Flores, F. (2006) *Chem. Phys.*, **325**, 129.
14. Vázquez, H., Dappe, Y.J., Ortega, J., and Flores, F. (2007) *J. Chem. Phys.*, **126**, 144703.
15. Vázquez, H., Flores, F., and Kahn, A. (2007) *Org. Electron.*, **8**, 241.
16. Braun, S. and Salaneck, W.R. (2007) *Chem. Phys. Lett.*, **438**, 259.
17. Braun, S., Salaneck, W.R., and Fahlman, M. (2009) *Adv. Mater.*, **21**, 1.
18. Braun, S., Liu, X., Salaneck, W.R., and Fahlman, M. (2010) *Org. Electron.*, **11**, 212.
19. Meier, H. (1974) Chapter 10, in *Organic Semiconductors*, vol. **2** (ed. H.F. Ebel), Verlag Chemie, Weinheim.
20. Ueno, N. and Kera, S. (2008) *Prog. Surf. Sci.*, **83**, 490.
21. Marcus, R.A. (1993) *Rev. Mod. Phys.*, **65**, 599.
22. Silinsh, E.A. and Čápek, V. (1994) *Organic Molecular Crystals: Interaction, Localization and Transport Phenomena*, Springer, Berlin.
23. Brédas, J.-L., Beljonne, D., Coropceanu, V., and Cornil, J. (2004) *Chem. Rev.*, **104**, 4971.
24. Coropceanu, V., Cornil, J., da Silva Filho, D.A., Oliver, Y., Silbey, R., and Brédas, J.-L. (2007) *Chem. Rev.*, **107**, 926; see also Erratum in *Chem. Rev.* **107** (2007) 2165.
25. Kera, S., Yamane, H., Sakuragi, I., Okudaira, K.K., and Ueno, N. (2002) *Chem. Phys. Lett.*, **364**, 93.
26. Kera, S., Yamane, H., and Ueno, N. (2009) *Prog. Surf. Sci.*, **84**, 135.
27. Gutmann, F. and Lyons, L.E. (1967) *Organic Semiconductors*, John Wiley & Sons, Inc., New York.
28. Seki, K. (1989) *Mol. Cryst. Liq. Cryst.*, **171**, 255.
29. Sato, N., Seki, K., and Inokuchi, H. (1981) *J. Chem. Soc., Faraday Trans. 2*, **77**, 1621.
30. Koopmans, T. (1933) *Physica*, **1**, 104.
31. Seki, K., Ueno, N., Karlsson, U.O., Engelhardt, R., and Koch, E.E. (1986) *Chem. Phys.*, **105**, 247.
32. Ueno, N., Sugita, K., Seki, K., and Inokuchi, H. (1986) *Phys. Rev. B*, **34**, 6386.
33. Okudaira, K.K., Kera, S., Setoyama, H., Morikawa, E., and Ueno, N. (2001) *J. Electron Spectrosc. Relat. Phenom.*, **121**, 225.
34. Ueno, N., Kera, S., and Kanai, K. (2012) Chapter 7, in *The Molecule–Metal Interface* (eds N. Koch, N. Ueno, and A.T.S. Wee), John Wiley & Sons, Inc., in press.
35. Miyamae, T., Ueno, N., Hasegawa, S., Saito, Y., Yamamoto, T., and Seki, K. (1999) *J. Chem. Phys.*, **110**, 2552.
36. Salaneck, W.R., Friend, R.H., and Bredas, J.L. (1999) *Phys. Rep.*, **319**, 231.
37. Berglund, C.N. and Spicer, W.E. (1964) *Phys. Rev. A*, **136**, 1030.
38. Feibelman, P.J. and Eastman, D.E. (1974) *Phys. Rev. B*, **10**, 4932.
39. Harada, Y., Masuda, S., and Ozaki, H. (1997) *Chem. Rev.*, **97**, 1897.
40. Kera, S., Setoyama, H., Onoue, M., Okudaira, K.K., Harada, Y., and Ueno, N. (2001) *Phys. Rev. B*, **63**, 115204.
41. Setoyama, H., Kera, S., Okudaira, K.K., Hara, M., Harada, Y., and Ueno, N. (2003) *Jpn. J. Appl. Phys.*, **42**, 597.
42. Kera, S., Okudaira, K.K., Harada, Y., and Ueno, N. (2001) *Jpn. J. Appl. Phys.*, **40** (Part 1, No. 2A), 783.
43. Hasegawa, S., Mori, T., Imaeda, K., Tanaka, S., Yamashita, Y., Inokuchi, H., Fujimoto, H., Seki, K., and Ueno, N. (1994) *J. Chem. Phys.*, **100**, 6969.
44. Hasegawa, S., Tanaka, S., Yamashita, Y., Inokuchi, H., Fujimoto, H., Kamiya, K., Seki, K., and Ueno, N. (1993) *Phys. Rev. B*, **48**, 2596.
45. Yamane, H., Kawabe, E., Yoshimura, D., Sumii, R., Kanai, K., Ouchi, Y., Ueno, N., and Seki, K. (2008) *Phys. Status Solidi (b)*, **245**, 793.
46. Koch, N., Vollmer, A., Salzmann, I., Nickel, B., Weiss, H., and Rabe, J.P. (2006) *Phys. Rev. Lett.*, **96**, 156803.
47. Annese, E., Viol, C.E., Zhou, B., Fujii, J., Vobornik, I., Baldacchini, C., Betti, M.G., and Rossi, G. (2007) *Surf. Sci.*, **601**, 4242.
48. Kakuta, H., Hirahara, T., Matsuda, I., Nagao, T., Hasegawa, S., Ueno, N., and

Sakamoto, K. (2007) *Phys. Rev. Lett.*, **98**, 247601.
49 Yoshida, H. and Sato, N. (2008) *Phys. Rev. B*, **77**, 235205.
50 Fukagawa, H., Yamane, H., Kataoka, T., Kera, S., Nakamura, M., Kudo, K., and Ueno, N. (2006) *Phys. Rev. B*, **73**, 245310.
51 Ueno, N., Kera, S., Sakamoto, K., and Okudaira, K.K. (2008) *Appl. Phys. A*, **92**, 495.
52 Machida, S., Nakayama, Y., Duhm, S., Xin, Q., Funakoshi, A., Ogawa, N., Kera, S., Ueno, N., and Ishii, H. (2010) *Phys. Rev. Lett.*, **104**, 156401.
53 Chapman, B.D., Checco, A., Pindak, R., Siegrist, T., and Kloc, C. (2006) *J. Cryst. Growth*, **290**, 479.
54 Filho, D.A.d.S., Kim, E.-G., and Brédas, J.-L. (2005) *Adv. Mater.*, **17**, 1072.
55 Ding, H., Reese, C., Mäkinen, A.J., Bao, Z., and Gao, Y. (2010) *Appl. Phys. Lett.*, **196**, 222106.
56 Kera, S., Yamane, H., Sakuragi, I., Okudaira, K.K., and Ueno, N. (2002) *Chem. Phys. Lett.*, **364**, 93.
57 Zhu, X.-Y. (2004) *Surf. Sci. Rep.*, **56**, 1.
58 Yamane, H., Nagamatsu, S., Fukagawa, H., Kera, S., Friedlein, R., Okudaira, K.K., and Ueno, N. (2005) *Phys. Rev. B*, **72**, 153412.
59 Malagoli, M., Coropceanu, V., da Silva Filho, D.A., and Brédas, J.-L. (2004) *J. Chem. Phys.*, **120**, 7490.
60 Duhm, S., Xin, Q., Hosoumi, S., Fukagawa, H., Sato, K., Ueno, N., and Kera, S. (2012) *Adv. Mater.*, **24**, 901.
61 Takeya, J., Yamagishi, M., Tominari, Y., Hirahara, R., Nakazawa, Y., Nishikawa, T., Kawase, T., Shimoda, T., and Ogawa, S. (2007) *Appl. Phys. Lett.*, **90**, 102120.
62 Miyamae, T., Hasegawa, S., Yoshimura, D., Ishii, H., Ueno, N., and Seki, K. (2000) *J. Chem. Phys.*, **112**, 3333.
63 Shpaisman, H., Seitz, O., Yaffe, O., Roodenko, K., Scheres, L., Zuilhof, H., Chabal, Y.J., Sueyoshi, T., Kera, S., Ueno, N., Vilan, A., and Cahen, D. (2012) *Chem. Sci.*, **3**, 851.
64 Seki, K., Karlsson, U., Engelhardt, R., and Koch, E.E. (1984) *Chem. Phys. Lett.*, **103**, 343.
65 Ueno, N., Gaedeke, W., Koch, E.E., Engelhardt, R., Dudde, R., Laxhuber, L., and Moewald, H. (1985) *J. Mol. Electron.*, **1**, 19–23.
66 Karpfen, A. (1981) *J. Chem. Phys.*, **75**, 238.
67 Fujimoto, H., Mori, T., Inokuchi, H., Ueno, N., Sugita, K., and Seki, K. (1987) *Chem. Phys. Lett.*, **141**, 485.
68 Ueno, N., Fujimoto, H., Sato, N., Seki, K., and Inokuchi, H. (1990) *Phys. Scr.*, **41**, 18.
69 Ueno, N., Seki, K., Sato, N., Fujimoto, H., Kuramochi, T., Sugita, K., and Inokuchi, H. (1990) *Phys. Rev. B*, **41**, 1176.
70 Berrehar, J., Lapersonne-Meyer, C., and Schott, M. (1986) *Appl. Phys. Lett.*, **48**, 630.
71 Berrehar, J., Hassenforder, P., Lapersonne-Meyer, C., and Schott, M. (1990) *Thin Solid Films*, **190**, 181.
72 Salaneck, W.R., Fahlman, M., Lapersonne-Meyer, C., Fave, J.-L., Schott, M., Lögdlund, M., and Brédas, J.-L. (1994) *Synth. Met.*, **67**, 309.
73 Sirringhaus, H., Brown, P.J., Friend, R.H., Nielsen, M.M., Bechgarrd, K., Langeveld-Voss, B.M.W., Spiering, A.J., Jassen, R.A.J., Meijer, E.W., Herwig, P., and de Leeuw, D.M. (1999) *Nature*, **401**, 685.
74 Yang, C.Y., Soci, C., Moses, D., and Heeger, A.J. (2005) *Synth. Met.*, **155**, 639.
75 Derue, G., Coppée, S., Gabriele, S., Surin, M., Geskin, V., Monteverde, F., Leclère, P., Lazzzroni, R., and Damman, P. (2005) *J. Am. Chem. Soc.*, **127**, 8018.
76 Cascio, A.J., Lyon, J.E., Beerbom, M.M., Schlaf, R., Zhu, Y., and Jenekhe, S.A. (2006) *Appl. Phys. Lett.*, **88**, 062104.
77 Delongchamp, D.M., Vogel, B.M., Jung, Y., Gurau, M.C., Richter, C.A., Kirillov, O.A., Obrzut, J., Fisher, D.A., Sambasivan, S., Richter, L.J., and Lin, E.K. (2005) *Chem. Mater.*, **17**, 5610.
78 Kline, R.J., McGehee, M.D., and Toney, M.F. (2006) *Nat. Mater.*, **5**, 222.
79 Hao, X.T., Hosokai, T., Mitsuo, N., Kera, S., Okudaira, K.K., Mase, K., and Ueno, N. (2007) *J. Phys. Chem. B*, **111**, 10365.
80 Hao, X.T., Hosokai, T., Mitsuo, N., Kera, S., Okudaira, K.K., Mase, K., and Ueno, N. (2006) *Appl. Phys. Lett.*, **89**, 182113.
81 Patnaik, A., Okudaira, K.K., Setoyama, H., Mase, K., and Ueno, N. (2005) *J. Chem. Phys.*, **122**, 154703.

82 Ueno, N., Azuma, Y., Tsutsui, M., Okudaira, K.K., and Harada, Y. (1998) *Jpn. J. Appl. Phys.*, **37**, 4979.
83 Lang, D.V., Chi, X., Siegrist, T., Sergent, A.M., and Ramirez, A.P. (2004) *Phys. Rev. Lett.*, **93**, 086802.
84 Kalb, W.L., Haas, S., Krellner, C., Mathis, T., and Batlogg, B. (2010) *Phys. Rev. B*, **81**, 155315.
85 Fukagawa, H., Kera, S., Kataoka, T., Hosoumi, S., Watanabe, Y., Kudo, K., and Ueno, N. (2007) *Adv. Mater.*, **19**, 665.
86 See for example Fukagawa, H., Hosoumi, S., Yamane, H., Kera, S., and Ueno, N. (2011) *Phys. Rev. B*, **83**, 085304, and references therein.
87 Sueyoshi, T., Fukagawa, H., Ono, M., Kera, S., and Ueno, N. (2009) *Appl. Phys. Lett.*, **95**, 183303.
88 Sueyoshi, T., Kakuta, H., Ono, M., Sakamoto, K., Kera, S., and Ueno, N. (2010) *Appl. Phys. Lett.*, **96**, 093303.
89 Mao, H.Y., Bussolotti, F., Qi, D.-C., Wang, R., Kera, S., Ueno, N., Wee, A.T.S., and Chen, W. (2011) *Org. Electron.*, **12**, 534.

4
Energy and Charge Transfer
Ralf Mauer, Ian A. Howard, and Frédéric Laquai

4.1
Introduction

Organic light-emitting diodes and organic photovoltaic devices are both emerging technologies that make use of composite systems of conjugated molecules and polymers. In order for these devices to function efficiently, the flow of charge and energy through the materials must be carefully engineered. This chapter reviews the fundamental aspects of charge and energy transfer in conjugated polymers and donor–acceptor blends. It begins by quickly introducing the semiconducting electronic structure of conjugated polymers and then turns to examine the physics of energy transfer. Due to the significant Coulomb interaction caused by the generally low dielectric constants of organic materials, plus the substantial electron–electron correlation and geometric relaxation effects [1], the primary photoexcitations in conjugated polymers are strongly bound Frenkel-type excitons. Depending on the spin composition of the state, the exciton can be either a singlet or a triplet, which completely alters the nature of the energy transfer process. In either case, the excitons do not extend over the entire polymer chain but are localized over a few repeat units, so motion both within a single chain and within a composite film depends on hopping. After reviewing the mechanism of both singlet and triplet energy transfer, along with examples of their importance in device structures, this chapter presents the theory of charge transfer. Charge carriers are also localized in conjugated polymers; in general, they are delocalized to an even lesser extent than excitons. This is due to the strong electron–phonon coupling in conjugated polymers, which leads to significant structural relaxation of the polymer backbone around the charge. Charge transfer is therefore also described by hopping models. Of particular importance to photovoltaic devices made of conjugated polymers is the quenching of excitons at a donor–acceptor interface due to separation of the electron and hole onto the two different materials. Hence, the last section of this chapter presents a detailed review of this important process. In general, this chapter will stress how different the physics of energy and charge transfer are in conjugated polymer composites and inorganic semiconductors. Device models developed for inorganic materials based on notions such as weak exciton binding energy and band-like

Semiconducting Polymer Composites: Principles, Morphologies, Properties and Applications, First Edition.
Edited by Xiaoniu Yang.
© 2012 Wiley-VCH Verlag GmbH & Co. KGaA. Published 2012 by Wiley-VCH Verlag GmbH & Co. KGaA.

transport are therefore not clearly directly applicable to conjugated polymer-based devices, and can only be used with extreme discretion and caution. As the understanding of energy and charge transfer progresses alongside the development of new conjugated polymer materials, a broad spectrum of new applications and engineering opportunities is likely to be opened.

4.2 Energy Transfer

4.2.1 Electronic Structure and Excited States of Conjugated Polymers

In nonconjugated polymers such as polyethylene, each carbon atom forms a bond with four other atoms. In this case, the valence shell orbitals of the carbon atom ($2s^2 2p^2$) hybridize to form four equivalent sp^3 bonding orbitals. Thus, the valence electrons are strongly bound in σ-bonds holding the adjoining atoms in a tetrahedral structure. In conjugated polymers, the primary bonding between carbon atoms also occurs due to these σ-bonds, but alternate carbon atoms also have a π-bond formed between them. π-bonds occur between carbon atoms that are each connected to only three other atoms. In this case, only two of the three p orbitals hybridize with the s orbital forming sp^2 hybrid orbitals. The p_z atomic orbital is unchanged in this case, and the overlap of the p_z orbitals between adjacent carbon atoms forms an additional bond, a π-bond, between the two atoms. The alternating π-bonds increase the rigidity of the chain, and also the π-electrons dictate the optoelectronic properties of the conjugated polymers. Various quantum-mechanical models have been developed to describe the electronic properties of conjugated polymers. These range in sophistication from Hückel theory (which assumes a static geometry and no electron–electron interactions) to the combination of the Su–Schrieffer–Heeger and the Pariser–Parr–Pople models [2]. The combination of these latter models takes into account interacting electrons and mobile nuclei to allow a quite complete picture of the energy levels and transition rates of a material to be developed. A good review of these models was published by Barford in 2005 [2]. For the purposes of this chapter, in order to investigate energy and charge transfer in conjugated polymers, a simplified qualitative picture of the electronic structure is sufficient. We simply consider that fairly dense energy bands in the conjugated polymer have developed from splitting of the highest occupied molecular orbital (HOMO) and lowest unoccupied molecular orbital (LUMO) of its monomer as monomers are joined into longer oligomers and then finally into a long polymer backbone. This process is schematically illustrated in Figure 4.1.

4.2.1.1 Excitons: The Nature of Excited States in Conjugated Polymers

Transitions between symmetry-allowed electronic states give rise to optical absorption. Simplistically, the absorption of a photon promotes an electron from the HOMO to the LUMO of the polymer, creating a singlet excited state with the energy

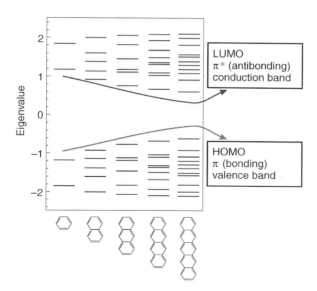

Figure 4.1 As the conjugated π system is extended, the initially degenerate HOMO and LUMO levels split into regions of energy with many states. In the limit of a polymer, these states essentially form continuous bands. (Reprinted from Ref. [3] with permission from Macmillan Publishers Ltd. Copyright 2006.)

of the polymer bandgap. However, the electron and hole are attracted to each other by Coulomb forces that bind them together into a quasiparticle called an exciton. The nature of the exciton depends on its binding energy. In three-dimensional inorganic semiconductors with a high dielectric constant, the exciton binding energy is small (<0.1 eV) compared to the thermal energy kT at room temperature and the electron–hole separation is large indicative of a Mott–Wannier-type exciton. In molecular crystals, the dimensionality and dielectric constant mean that the electron and hole are more strongly coupled (binding energy ∼1 eV) and Frenkel-type excitons are created. In conjugated polymers (and semiconductor nanoparticles that constrain the exciton extent due to the limited physical extent of the particle), the exciton binding energy is in an intermediate regime of roughly 0.5 eV. This means that the optical bandgap is roughly 0.5 eV lower in energy than the HOMO–LUMO difference. This relatively large binding energy means that splitting excitons into free charges is a challenge in a pristine polymer, and leads to the need to use partially demixed blends of two conjugated materials (an electron-donating one and an electron-accepting one) in an organic solar cell. In these blends, charge transfer excitons and/or free charge carriers are the product of the dissociation of the primary Frenkel-type excitons generated after photoexcitation. Per definition, CT excitons describe a state in which the electron and hole occupy neighboring sites, with the positive charge carrier being located on the donor and the negative charge carrier on the acceptor. The qualitative characteristics of the various exciton categories are shown in Figure 4.2.

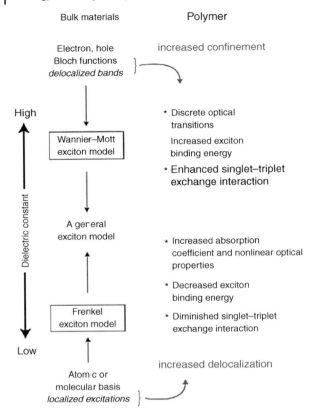

Figure 4.2 The properties of polymer excitons compared to Mott–Wannier and Frenkel excitons. (Reprinted from Ref. [3] with permission from Macmillan Publishers Ltd. Copyright 2006.)

In conjugated polymers, the exciton is delocalized along the polymer backbone. The delocalization of the exciton can be deduced from the redshift of the energy of the singlet state with increasing number of monomers that form an oligomer. This effect is well known and, for instance, has been characterized for ladder-type oligomers and oligothiophenes [4]. Typically, a dependence of $E(n) = E_P + \text{const} \times 1/n$ is used to extrapolate the singlet state energies of the oligomers to the polymer limit [5]. However, the singlet state energy of the polymer exceeds the energy determined from the simple extrapolation of $n \to \infty$. This points to a limitation of the spatial extension of the exciton along the polymer chain, typically interpreted as a finite conjugation length. The latter can be approximated from the extrapolated $E(n) = E_P + \text{const} \times 1/n$ relation, but recent theories suggest that the quantitative assessment is problematic, since a deviation from the $1/n$ law at large chain lengths is expected [6]. A comprehensive overview of experimental data and quantum-chemical calculations has recently been published by Gierschner et al. [7] and the deviation from the $1/n$ dependence for a set of step-ladder-type oligomers has recently been shown by Laquai et al. [8].

While in conjugated homopolymers the singlet exciton can be delocalized over several monomer units depending on the distortion of the backbone, the singlet exciton may experience a stronger localization in copolymers such as alternating donor–acceptor-type polymers. Figure 4.3a shows theoretical data from the literature that illustrate the size and nature of an exciton on the conjugated copolymer poly(9,9-dioctyluorene-*alt*-benzothiadiazole) (F8BT). It shows that the electron and hole in the exciton are often (48% probability) separated, with the one residing on a BT subunit while the other is on the fluorene unit, or (40% probability) both on the same BT unit. In either case, the electron–hole separation is small, but the excitation is delocalized over the three repeat units of the chain, as shown in the figure. Higher lying excitons often have wavefunctions in which the electron–hole separation is increased, and as a side note it is an active area of investigation as to which extent this may play a role in the charge separation reactions critical for solar cell performance. Figure 4.3b shows how the morphology of a polymer chain affects the conjugation length, and therefore imposes limits on the size of excitons at different locations. Kinks and twists in the polymer backbone break the conjugation and can constrain the exciton size. They also introduce energetic disorder and, along with variations in the local electric field and dielectric constant, introduce significant inhomogeneous broadening of the density of states (DOS). This high degree of inhomogeneous energetic disorder plays a significant role in the photophysics of polymers in the solid state. In fact, only in very well-ordered systems at low temperature the exciton can delocalize over the entire polymer chain. A prominent example is the observation of macroscopic exciton coherence in single chains of polydiacetylene obtained by light-induced solid-state polymerization in the crystal of its monomer [9]. Here, the crystalline matrix separates the individual polymer chains from each other and the solid-state photopolymerization leads to stretched and defect-free polymer chains. It was shown that in these systems the exciton is effectively delocalized over the entire polymer chain of several micrometer lengths [9].

As opposed to singlet excitons that are readily created by optical absorption, triplet excitons in conjugated polymers are formed either by intersystem crossing of a singlet exciton primarily generated by photoexcitation or through the recombination of polarons after charge injection from the electrodes, for instance, in an organic light-emitting diode. The exchange energy in conjugated polymers is relatively high, typically around 0.7 eV [11]. Values for the triplet energy levels for a wide range of polymers have been obtained by phosphorescence, transfer from triplet sensitizers, and photoacoustic calorimetry [11, 12]. This large exchange energy is further evidence for tightly bound excitons. The singlet–triplet splitting can be slightly increased by torsional distortion of the backbone, which further reduces the delocalization length of the exciton. Intersystem crossing can occur by spin–orbit coupling, hyperfine interaction, or spin–lattice relaxation. The timescale for spin lattice relaxation in PPV was recently measured to be approximately 30 µs [13]. This is much longer than the nanosecond lifetime of the singlet exciton, and so this mechanism of triplet

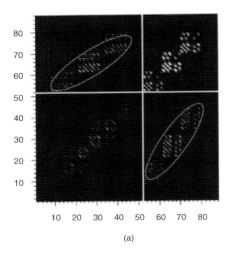

(a)

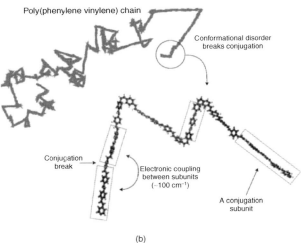

(b)

Figure 4.3 Panel (a) shows the exciton size in the polymer F8BT. The contour plot shows the probability (brighter denotes greater) that the electron and hole are located on the ith and jth atoms. Thus, the diagonal elements represent cases where the electron and hole are very close to each other and the off-diagonal elements represent cases where the electron and hole are further separated. The circled region of the molecular structure shows the area over which the exciton is delocalized. Panel (b) shows how the bending and twisting of the polymer backbone limits conjugation length. (Reprinted with permission from Ref. [10]. Copyright 2003, American Institute of Physics.)

generation can be neglected. The hyperfine interaction involves the mixing of the singlet and triplet states by the coupling of the electronic and nuclear spins. It can be significant when the electron and hole are separated and the exchange energy is low, but in the case of a confined exciton state it will be negligible. This leaves spin–orbit coupling, which has been used to describe the intersystem crossing in the excitonic state with good qualitative agreement with experiment if not a precise quantitative match [14]. In conjugated polymer composites, triplet excitons can also arise from intersystem crossing in a charge transfer state or from charge recombination [15]. In comparison to singlet excitons, triplets are localized to a significantly shorter part of the polymer chain and thus their energy depends less strongly on the number of repeat units [5]. In fact, van Dijken *et al.* have shown that the triplet state does not extend to more than four adjacent phenyl rings by studying the phosphorescence spectra of various fluorene and carbazole-type copolymers [16]. Hence, it is not straightforward to manipulate the energy of the triplet state, for instance, to decrease the singlet–triplet gap in conjugated polymers. This would require breaking the conjugation along the polymer chain after a few repeat units only, for instance, by introducing *meta* instead of *para* connections between adjacent phenyl rings. For a more detailed overview, the interested reader is referred to a recent review on triplet states in organic semiconductors by Köhler and Bässler [17].

4.2.2
Excited State Dynamics in Conjugated Polymers

4.2.2.1 Role of Disorder in Energy Transfer

Due to the variety of local conformations and dielectric environments of polymer chain segments in pristine polymer films and composite blends, the density of states of all excited species is significantly inhomogeneously broadened. Theoretically, the effect of this inhomogeneous broadening is explored in the Gaussian disorder model (GDM) [18], and/or through the introduction of an exponential tail to the density of states [19]. This disorder in site energy influences the energy transfer in a conjugated polymer. The rate of energy transfer can become slower over the lifetime of the excited state due to relaxation into the tail states of the inhomogeneously broadened DOS, which is assumed to be of Gaussian shape according to

$$\varrho(\varepsilon) = \sqrt{(2\pi\sigma^2)} \exp\left(-\frac{\varepsilon^2}{2\sigma^2}\right),$$

with σ the variance of the Gaussian distribution and ε the site energy with respect to the center of the DOS [20]. Photoexcitation will create singlet excitons randomly distributed within the density of states. As these excitations hop from site to site, they will preferentially move to sites of lower energy until they reach their dynamic thermal equilibrium. The elementary step of the incoherent transport process is hopping from site ε_i to ε_j, which can be mathematically

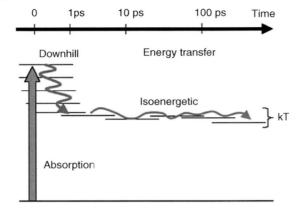

Figure 4.4 Illustration of singlet exciton transfer in a polymer film with an inhomogeneously broadened density of states. The exciton hops quickly on the timescale of a few picoseconds until it reaches a quasi-equilibrium level in the bottom section of the density of states. At this point, the hopping slows down, due to the increased average distance to energetically accessible acceptor sites. (Reprinted with permission from Ref. [23]. Copyright 2007, American Chemical Society.)

described by Miller–Abrahams jump rates:

$$\nu_{ij} = \nu_0 \exp(-2\alpha R_{ij}) \times \begin{cases} \exp\left(-\dfrac{\varepsilon_j - \varepsilon_i}{kT}\right), & \varepsilon_j > \varepsilon_i, \\ 1, & \varepsilon_j < \varepsilon_i, \end{cases}$$

with ν_{ij} the hopping frequency, ν_0 the attempt-to-escape frequency, R the distance between the two sites, and α the inverse wavefunction localization radius [21]. After relaxation is finished, the excited states populate the occupied density of states (ODOS). It has been shown that the center of the ODOS is located at $-\sigma^2/kT$ with respect to the center of the DOS [22].

A simplified scheme of the relaxation is shown in Figure 4.4. This general picture of the time dependence of the rate of energy transfer is applicable to other states if they are also created at a random position in the density of states. In the case of both triplets and charges, exothermic energy transfer is much faster than thermally activated hopping, and again leads to excited states preferentially residing within the tail of the energetically broadened density of states. Thus, singlet and triplet excitons as well as polarons (as discussed later) can have a time-dependent diffusion constant/mobility.

4.2.2.2 Singlet Exciton Energy Transfer

Excitons in conjugated polymers can move by transfer from their current location (the "donor" site) to an acceptor location on the same polymer chain or on another nearby polymer backbone. This motion of energy plays important roles in the photophysics of the material, and is critical to the function of optoelectronic devices such as solar cells. Singlet motion is usually considered in terms of Förster resonant energy transfer (FRET), where it is assumed that the coupling is incoherent and that the donor and acceptor sites can be approximated by point dipoles. The rate of transfer can be written as $k_{\text{ET}} = k_d (R_0/R)^6$, where k_d is the rate of decay of the excited donor in the absence of an acceptor, R is the separation of the donor and acceptor, and R_0 is the Förster radius defined as $R_0^6 = (9 \ln 10/128 \pi^5 N_a)(\kappa^2 \Phi_D/n^4) J$. Here N_a is the Avogadro constant, κ is an orientational factor taking into account the relative direction of the two dipoles, n is the effective refractive index of the medium, Φ_D is the photoluminescence quantum efficiency for the donor in the absence of an acceptor, and J is the spectral overlap of the donor emission and acceptor absorption. A good review of the applicability of these equations in various situations is provided by Braslavsky et al. [24]. One major problem with applying this equation to motion of excitons in conjugated polymers is that the point dipole assumption is valid only if the size of the chromophores (in case of conjugated polymers the size of the excitons) is small compared to the distance that they transfer. In reality, this is not the case, and so this approach breaks down. In order to more accurately describe the motion, the total dipole of the chromophores can be broken down into a sum of smaller dipoles, and the coupling of all of these smaller dipoles can be evaluated to give the net transfer rate [25]. It is assumed that each of the k units of the acceptor and each of the l units of the donor have discrete wavefunctions and their overlaps are pairwise computed and then summed. In this line dipole approach, the rate of energy transfer can be calculated as $k_{\text{ET-LD}} = k_d (9 \ln 10/128 \pi^5 N_a)(\Phi_D/n^4) J [\sum_{l,k} \psi_k \kappa_{kl} R_{kl}^{-3} \psi_l]^2$, where $\psi_{k,l}$ are the wavefunctions of the kth and lth subsegments, respectively, κ_{kl} is the orientation factor, and R_{kl} is the separation between the kth and lth subsegments. This is schematically illustrated in Figure 4.5. Using Hückel theory of a linear chain, $\psi_{k,l}$ can be approximated as $\psi_k = [\sin(\pi k/(l_n + 1))]/\sum_n \sin(\pi n/(l_n + 1))$, where l_n is the number of segments that each dipole is split into. When the separation between the donor and acceptor sites is similar or smaller than the size of the singlet exciton, as is normally the case in composite polymer films, this line dipole approach better describes the rate of exciton transfer than the standard Förster approach [25]. Although interchain exciton transfer in polymer films can be well treated by this approach, the assumption that intrachain singlet motion is incoherent has recently been challenged. Evidence from two-dimensional photon echo spectroscopy and two-time anisotropy decay measurements suggests that exciton transport along a polymer chain can occur in the intermediate coupling regime, wherein the exciton "hops" but quantum phase information is retained [26].

4.2.2.3 Triplet Exciton Dynamics

While singlet exciton motion in conjugated organic molecules can be described by Förster-type resonance energy transfer as pointed out above, triplet energy transfer occurs usually by a correlated two-electron exchange process termed Dexter-type energy transfer. A simplified scheme of Dexter-type energy transfer is presented in Figure 4.6.

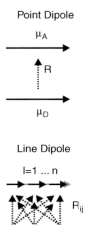

Figure 4.5 Schematic illustration of the standard point dipole approach to calculating the Förster rate, and the line dipole method. The line dipole method divides each chromophore into several dipoles whose distance from each other and wavefunction overlap are individually calculated and then used to calculate the overall transfer rate. This approach allows energy transfer rates to be calculated on the small length scales similar to the size of the singlet exciton.

The main reason for the vastly different energy transfer mechanisms is that triplet energy transfer via the Förster-type process would require a spin-flip by intersystem crossing for each individual hop of the triplet exciton, which is rather unlikely due to the low intersystem crossing rate of purely hydrocarbon-based molecules. Hence, the triplet exciton is transferred from the donor molecule to the acceptor by simultaneous transfer of an electron from the LUMO of the donor to the LUMO of the acceptor and vice versa from the HOMO of the acceptor to the HOMO of the donor. This process effectively circumvents the necessity for a spin-flip during the energy transfer process. However, Förster-type triplet energy transfer may be

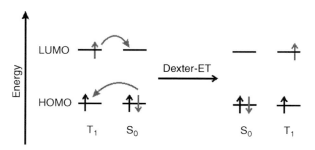

Figure 4.6 Simplified representation of Dexter-type triplet energy transfer as a correlated two-electron exchange process.

effective between molecules that exhibit a significant oscillator strength of the singlet to triplet transition as observed for heavy metal-containing complexes, where the presence of the heavy metal atom facilitates intersystem crossing due to enhanced spin–orbit coupling [27].

Dexter-type triplet energy transfer intrinsically requires a significant orbital overlap of the donor and acceptor molecules and thus exhibits a pronounced distance dependence, which is expressed in an exponential dependence of the transfer rate on the molecular distance between donor and acceptor:

$$k_{ET}(R) = \frac{1}{\tau_0} \exp\left[\gamma\left(1 - \frac{R_{AD}}{R_0}\right)\right],$$

with τ_0 the phosphorescence lifetime of the donor, R_0 the critical radius where the rate of energy transfer equals the total rate of radiative and nonradiative deactivation, and $\gamma = 2R_0/L$, where L denotes the effective average Bohr radius of typically 0.1–0.2 nm. Thus, Dexter-type ET is mostly operative on short distances similar to the van der Waals distance, that is, between 0.5 and 1.0 nm, unlike FRET that allows energy transfer over distances up to tens of nanometers. A straightforward consequence is that Dexter-type ET is significantly slower than FRET; however, the longer lifetime of triplet states can compensate the lower diffusivity and lead to similar or even enhanced exciton diffusion lengths. The importance of molecular interaction for the efficiency of Dexter-type ET has been demonstrated in polymer chains [28]. It was shown that triplets transfer much faster in the direction of the polymer chain, that is, by intramolecular energy transfer, than perpendicular to the chain, that is, by intermolecular energy transfer, since the typical interchain separation in conjugated polymers is about 0.4 nm, which hampers interchain transfer and thus leads to the anisotropic energy transfer process.

As explained above, triplet transfer can be understood by a quantum-mechanical Dexter-type energy transfer process, that is, by a correlated two-electron exchange process. This implies that triplet transfer may be treated analogous to a charge transfer process, the difference being that charge transfer involves a single-electron transfer only, while two electrons need to be exchanged for the triplet transfer. Thus, the diffusivity of charges is expected to be higher than that of triplets, which in fact has been confirmed in experiments on thin amorphous films of CBP, for instance. Here, the triplet diffusivity was determined to be $1.4 \times 10^{-8}\,\text{cm}^2\,\text{s}^{-1}$, while a charge carrier diffusivity of $3 \times 10^{-5}\,\text{cm}^2\,\text{s}^{-1}$ has been calculated [29].

A prominent and widely applied model of electron transfer in organic molecules is Marcus theory. It is thoroughly introduced in Section 4.3.1.1 and hence we concentrate here on its applicability to triplet energy transfer supported by recently reported experimental studies of the temperature dependence of the triplet diffusivity and of the effect of disorder on triplet transfer in organic molecules. We would also like to refer the interested reader to a recent overview on experimental and theoretical work concerning a unified description of triplet energy transfer by Köhler and Bässler [30].

In classical Marcus theory (see below), the parameters that determine the rate of an electron transfer process between a donor and an acceptor are the reorganization energy λ, the electronic coupling element J, and the free energy G^* of the electron transfer process. Electron transfer between isoenergetic molecules (i.e., $G^* = 0$), thus requires according to the classical Marcus theory an activation energy of $\lambda/4$ for the electron transfer process. This activation energy has to be provided by thermal fluctuations of the system. While Marcus theory has found widespread application in the field of chemistry and biology, physicists are more familiar with Holstein's small polaron theory [31, 32]. In fact, Marcus theory is mathematically equivalent to the high-temperature limit of Holstein's theory. However, Holstein derived his model for electron transfer in molecular crystals, where the charge distorts the surrounding crystal lattice and thus the motion of the charge is coupled to the motion of the lattice distortion, resulting in the formation of a quasiparticle, namely a polaron. According to Holstein, polaron motion is a temperature-activated process in the high-temperature regime, while in the low-temperature regime, that is, if the thermal energy is small compared to vibrational energies, tunneling of the charge prevails. The latter is, however, impossible to observe in disordered and amorphous materials such as many conjugated organic molecules, where local fluctuations of the electrostatic potential cause a distribution of the energy levels, typically termed density of states. Since the fluctuations of the electrostatic potential are entirely statistical, the distribution of energy levels is often considered to be of Gaussian shape and the Gaussian disorder model (GDM) has found widespread application to describe the photophysical properties of disordered organic materials. In this case of a distribution of energy levels, the tunneling process, which does not require any energy in perfect molecular crystals according to the low-temperature regime of the Holstein model, now requires some activation energy that can be provided by a single phonon. This is taken into account in the well-known Miller–Abrahams expression for the hopping rate (see above). In the following, we will now review some recent experimental results that allow to gain better insight into the microscopic mechanism of triplet diffusion and into the dynamics of triplet states.

Triplet states are a ubiquitous phenomenon in organic materials. However, there direct observation by optical spectroscopic techniques is not a simple task. The main reasons are (i) the low intersystem crossing efficiency in most organic molecules impeding formation as well as radiative decay of triplet states, (ii) their susceptibility to quenching by impurities and by molecular oxygen, and (iii) their high nonradiative decay rates due to coupling of the triplet state to vibronic sublevels of the singlet ground state. Furthermore, some experimental issues can prevent the observation of triplet states, for instance, their emission can easily be superimposed and thus covered by the strong fluorescence of organic molecules and they often emit at comparably low energy, where most detectors do not have sufficient sensitivity. These experimental issues have to some extent been overcome by the development of gated intensified charge-coupled device (iCCD) detectors and the development of high-sensitivity and cooled NIR cameras. Hence, the past decade has witnessed the development of a decent understanding of the energetics of the triplet state, but due to the above-mentioned shortcomings the motion of triplets was not well understood and only recently a better picture of the mechanism of triplet diffusion could be

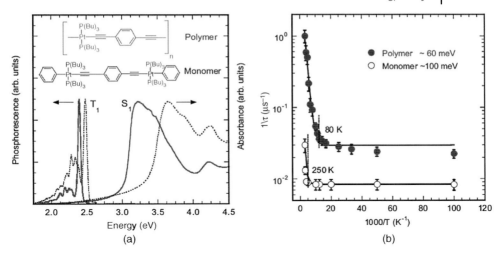

Figure 4.7 (a) Normalized phosphorescence spectra at 10 K and absorption spectra at 300 K of the Pt-containing polymer (solid line) and its monomer (dotted line). The inset shows the chemical structures of the polymer and monomer. (b) Arrhenius plot of the phosphorescence decay rate against the inverse temperature. The black curves correspond to the equation $1/\tau = a \times \sqrt{1/T} \exp[-E_a/k_B T] + b$ with an activation energy E_a of 60 meV for the polymer and 100 meV for the monomer. The dotted lines indicate the transition temperatures 80 and 250 K. (Reproduced with permission from Ref. [28]. Copyright 2008, American Physical Society.)

developed. In fact, temperature-dependent experiments have previously let to largely varying conclusions with temperature dependence of phosphorescence and delayed fluorescence suggesting hopping-like motion and temperature independence pointing to a band-like transport. The interpretation of experimental results is complicated by the presence of traps and disorder, especially if amorphous organic materials and conjugated polymers are considered. However, recently Sudha Devi *et al.* presented an experimental study on the temperature dependence of triplet energy transfer in weakly disordered organometallic small molecules and conjugated polymers [28]. Their model system consists of a platinum-containing monomer and its corresponding polymer (see Figure 4.7).

The presence of the heavy metal atom facilitates intersystem crossing and thus allows a straightforward detection of the phosphorescence spectrum, intensity and lifetime of the triplet states, and their temperature dependence. Furthermore, the disorder of the materials is very low compared to typical organic semiconductors as proven by the narrow linewidth of the phosphorescence spectra. Hence, donor and acceptor site energies could be treated as isoenergetic, which simplifies the data analysis and interpretation and allows disentangling polaronic contributions from effects caused by the presence of energetic disorder. The main findings of the authors are (i) a strongly temperature-dependent phosphorescence intensity in the polymer, but little temperature dependence in the monomer up to 200 K, and (ii) a similar dependence of the phosphorescence lifetime on temperature as observed for the phosphorescence intensity. Both effects were attributed to a thermally activated diffusion of triplet states to quenching sites. An Arrhenius plot of the decay rate allowed to determine the transition temperature, above which a strong temperature dependence of the phosphorescence decay is observed. This transition temperature was found to be much higher for the monomer (250 K) compared to the polymer (80 K). Furthermore, the activation energy of the triplet energy transfer could be estimated to 100 meV for the former and 60 meV for the latter (see Figure 4.7). Interestingly, the authors observed that below the transition temperature the decay rate of the triplets becomes constant for the monomer and only weakly temperature dependent for the polymer. The results were interpreted by assuming a thermally activated diffusion of triplets along the polymer backbone, which leads to trapping or annihilation and hence reduces the phosphorescence lifetime and intensity. This process occurs more efficiently in the polymer than in the monomer indicating faster triplet diffusion in polymer films, which is a result of less geometric distortion along the polymer chain and strong intrachain electronic coupling. Furthermore, the temperature dependence and the agreement of the experimentally determined and modeled activation energies strongly indicate that in the high-temperature regime triplet transfer can indeed be successfully described by Marcus theory in these systems.

The absence of any temperature dependence of the decay rate in films of the monomer was interpreted as immobilization of triplets at low temperatures. However, a weakly thermally activated diffusion with an activation energy of 5 meV was observed for the polymer in the low-temperature regime indicative of triplet migration at temperatures below the transition temperature. Further evidence was provided from the observation of bimolecular triplet annihilation at high pump fluence supporting triplet diffusion even at low temperatures. Thus, it appears that triplet motion is controlled by tunneling in the low-temperature regime, with the residual energetic disorder causing the weak temperature dependence. The authors pointed out that their data are entirely consistent with the Holstein small polaron model, where charge transport occurs by tunneling in the low-temperature limit and by phonon-assisted hopping in the high-temperature limit, consistent with Marcus theory. In subsequent work reported by some of the same authors, a polaron theory taking into account the influence of disorder was reported and the impact of disorder on polaronic transport in a series of poly(p-phenylene)-type polymers and oligomers was investigated experimentally [33, 34]. By using time-resolved photoluminescence spectroscopy and a Franck–Condon analysis of the obtained data, the authors could assess the polaronic and the disorder contribution to triplet exciton transport. Combining their experimental data with their polaron theory, the authors came to the conclusion that indeed Marcus-type jump rates with dominantly polaronic activation energies control triplet motion. This is often in contrast to charge transport, where the effect of energetic disorder is more pronounced.

Having discussed the triplet generation and the microscopic mechanism of triplet motion in organic molecules, we will now turn to have a closer look at the dynamics of triplet states in ordered and disordered systems. In general, the dynamics of triplet states can be described by the following rate equation:

$$\frac{d[T]}{dt} = G_T - \beta_0[T] - \gamma_{TTA}[T]^2,$$

with G_T the generation rate of triplet states, k_T the sum of all monoexponential decay channels, and γ the bimolecular annihilation coefficient.

At low triplet densities, the bimolecular term can be neglected and thus the intensity of the phosphorescence, which is given by $I(t)_{Ph} = k_r[T](t)]$, follows:

$$I_{Ph}(t) = k_r[T_0]e^{-\beta_0 t},$$

while at high triplet densities the bimolecular term determines the triplet dynamics and thus the phosphorescence intensity:

$$I_{Ph}(t) = \frac{k_r[T_0]}{1 + \gamma_{TTA}[T_0]t}.$$

The dynamics of the delayed fluorescence (DF) are given by

$$I_{DF}(t) = \frac{1}{2}f\gamma_{TTA}[T(t)]^2,$$

with f the fraction of triplet encounters that lead to annihilation and the prefactor 1/2 accounting for the fact that two triplets are consumed during annihilation.

Inserting the equations derived for the time dependence of the triplet dynamics at low and high triplet densities results in

$$I_{DF}(t) = \frac{1}{2}f\gamma_{TTA}[T_0]^2 e^{-2\beta_0 t},$$

$$I_{DF}(t) = \frac{1}{2}f\gamma_{TTA}\frac{[T_0]}{(1 + \gamma_{TTA}[T_0]t)^2}.$$

A straightforward result is that in the limit of low triplet densities the DF decays exponentially with half the lifetime of the triplet. In the limit of high triplet densities, the DF approaches a t^{-2} power law at long delay times while at short times the decay is independent of the triplet density.

However, these time dependences of the phosphorescence and DF can only be expected in well-ordered molecular crystals. In fact, early experiments on triplet states used the DF decay in anthracene crystals to obtain the triplet lifetime and diffusivity [35, 36]. However, a general problem is that impurities can act as traps or quenching sites for triplets, thus reducing their lifetime and making the intrinsic lifetime of the triplets difficult to determine. The main reason is the long lifetime of the triplet state that allows for many sites to be visited during the triplet diffusion.

In disordered systems such as amorphous films of small organic molecules and polymers, the excited states are typically generated at an arbitrary position within the DOS and they are prone to relaxation toward the tail states of the DOS, which are lower in energy. The time dependence of the relaxation process in the DOS exhibits a logarithmic decay law. The straightforward consequence of the energetic relaxation is that the bimolecular annihilation "constant" is no longer a constant but becomes a function of time. Consequently, the DF reflects the time dependence of γ_{TTA} at short times and at long times, when γ_{TTA} approaches a constant value again, the DF approaches the expected t^{-2} power law. This has been theoretically shown by Monte Carlo simulations [37] and experimentally by the Bässler group on polyfluorene thin films [38]. In their experiments, the authors probed the delayed fluorescence and phosphorescence dynamics by time-resolved optical spectroscopy over a wide dynamic range. It was shown that (i) the DF intensity follows the squared phosphorescence intensity pointing to TTA as the origin of DF, (ii) at low excitation density the phosphorescence decays exponentially with a lifetime of 1.1 s, while the DF decays with half the lifetime of the phosphorescence, and (iii) at high excitation densities the intensity of the DF follows a power law with an exponent between 1.0 and 1.3 and it approaches the theoretically expected t^{-2} law at long delay times.

Very recently, the generation of delayed fluorescence by triplet–triplet annihilation regained considerable attention, apart from its general usefulness for studies of triplet motion and dynamics in molecular donor–acceptor systems, in which both molecules, that is, donor and acceptor, have the same molecular structure. In fact, if a donor molecule with high intersystem crossing efficiency (hereafter called sensitizer) is combined with an acceptor with low intersystem crossing efficiency, whose triplet energy is lower than that of the donor, the triplet population of the acceptor can be efficiently sensitized. Furthermore, if a system is chosen in which the singlet state of the acceptor is higher than that of the sensitizer, then the latter can be selectively excited at energies lower than the ground state absorption of the acceptor and successive triplet transfer from the sensitizer to the acceptor followed by triplet–triplet annihilation among the acceptor molecules can cause sensitized delayed fluorescence from the acceptor molecules (compare Figure 4.8).

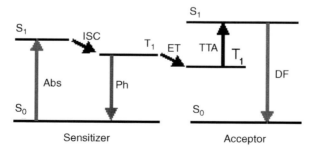

Figure 4.8 Simplified energy level scheme of a sensitizer–acceptor system used for photon energy upconversion (sensitized delayed fluorescence). The first step is absorption (Abs) of a photon by the sensitizer followed by efficient intersystem crossing (ISC) in the sensitizer molecules. The triplet is then transferred by Dexter-type energy transfer (ET) to the lower energy triplet state of the acceptor. Here triplet–triplet annihilation (TTA) of two triplet states occurs and leads to delayed fluorescence (DF). Note that for clarity only the most relevant processes are shown.

This effect may also be termed photon energy upconversion as the delayed fluorescence occurs at higher energy than used for the excitation of the sensitizer [39]. The phenomenon occurs also in systems in which the triplet level of the acceptor is higher than that of the sensitizer and different mechanisms have been suggested depending on the individual system [40, 41]. However, the efficiency of the upconversion is highest, if the triplet level of the acceptor is lower than that of the sensitizer due to the efficient downhill energy Dexter-type energy transfer. A prominent example of such an upconversion system is a sensitizer–acceptor pair consisting of a metallized porphyrin and an anthracene derivative, for instance, platinum octaethylporphyrin (PtOEP) and diphenylanthracene (DPA) as reported by Baluschev *et al.* [42] The upconversion works best in carefully degassed solutions, since the triplets are highly susceptible to quenching by molecular oxygen. Another issue limiting the efficiency is the position of the singlet state energy of the acceptor with respect to the sensitizer that allows for long-range Förster-type singlet energy transfer to be operative and thus can lead to quenching of the delayed fluorescence due to singlet energy transfer to the sensitizer. In fact, this process may severely limit the efficiency of such upconversion systems in the solid state as already suggested by Laquai *et al.* [43]. However, this drawback can be partially circumvented by using high-viscosity low molecular weight polystyrene as a matrix for the upconversion molecules as demonstrated by the Castellano group [44]. Meanwhile this effect has been demonstrated for several combinations of sensitizers and acceptors and some applications such as upconversion displays and upconverters for dye-sensitized solar cells have been demonstrated [45]. Recent studies also show that the theoretically expected limit of the upconversion process, which is set by the spin statistics of the triplet–triplet annihilation, can actually be outperformed in real-world systems [46]. However, significantly more work is required to increase the efficiency of the upconversion process in order to put it into application.

4.2.3
Energy Transfer: Relevance to Device Performance

Energy transfer plays important roles in both organic light-emitting diodes and solar cells. In organic solar cells, the exciton created by photon absorption must reach an interface between the two materials creating the composite in order for it to be split into separated charges. Therefore, to efficiently split excitons in organic solar cells the average domain size of two materials in the composite must be commensurate with the average diffusion length of an exciton within its lifetime. Several methods have been employed to measure the exciton diffusion length in conjugated polymers. A recent study presents a robust technique that involves investigating the photoluminescence decay of thin polymer films of varying thickness deposited on an exciton-quenching and non-exciton-quenching substrate [47]. From these measurements, a diffusion coefficient in the direction perpendicular to the substrate of $1.8 \times 10^{-3} \, \text{cm}^2 \, \text{s}^{-1}$ was found, which, along with measurements of the exciton lifetime, led to an estimation for the diffusion length of an exciton of 8.5 ± 0.7 nm. Another method of measuring the exciton diffusion length based on pulsed laser techniques is to observe the rate of exciton–exciton annihilation after high densities of excitons are created by a short duration but intense laser pulse. By varying the intensity of the laser excitation, the mean initial separation between excitons can be varied and by studying how the rate of annihilation varies with this initial separation estimates for the exciton diffusion length can be obtained [18]. The roughly 10 nm length scale of the exciton diffusion in conjugated polymers imposes the condition that the components of the composite solar cell be intimately intermixed. A detrimental consequence of the high interfacial surface area that is a corollary of this requirement is that recombination of free charges becomes more likely. It would be desirable to increase the diffusion length of the exciton, thereby allowing composites with larger domain sizes and smaller internal interfacial areas to be used. One approach that has been suggested is to efficiently convert singlet excitons into longer lived triplet excitons (with lifetimes of microseconds as compared to nanoseconds). This could be achieved by utilizing materials in which singlet fission is efficient, such as pentacene [48], and two triplet excitons can be spontaneously generated from a single singlet exciton. However, although the lifetime of the triplet exciton is much longer than the lifetime of the singlet exciton, the diffusion coefficient of the triplet exciton is also much smaller than that of the singlet exciton meaning that it also moves much more slowly. For example, the triplet diffusion coefficient in a polyfluorene:polythiophene copolymer was measured to be 1.6×10^{-6} cm^2 s^{-1} [19], which is roughly three orders of magnitude slower than that measured for the singlet exciton in polythiophene. Therefore, although relaxation of the morphological constraint imposed on organic solar cells may possibly be achieved by efficient conversion of singlet excitons to triplet excitons, this is by no means certain. In any case, the transfer of energy from the point at which the photon is absorbed to an interfacial site at which the energy can be separated into opposite charge carriers is of fundamental importance to the efficiency of composite organic photovoltaics, and increasing the distance over which this energy can be transferred is an open challenge in the field.

In organic light-emitting diodes, singlet and triplet excitons are created at the interface between carrier transport layers. Color tuning can be achieved by introducing various chromophores at this interface that accept energy from the excitons and re-emit it at longer wavelengths. This funneling of energy to different chromophores can lead to white organic light-emitting diodes with high efficiencies [49]. The concept of using energy transfer to enhance the performance of an organic composite optoelectronic device has also been applied in the field of lasing. In order to avoid many of the quenching mechanisms that preclude lasing when high densities of charges and excitons necessary to approach population inversion are present in the composite layer, a bilayer structure has been used wherein a conventional inorganic diode is used to transport charges and create excitons that are then transferred to an organic layer in which population inversion and lasing can be achieved [50].

4.3
Charge Transfer in Polymer/Fullerene Composites

4.3.1
Theoretical Background

In this section, we try to provide a comprehensive framework for understanding the basic principles underlying the transition from an initially excited molecule to spatially separated charge carriers in organic semiconductor composites. We do this by briefly reviewing the theoretical background of the charge transfer process at an interface that leads to the formation of a so-called charge transfer state and the subsequent dissociation of this state into separated charges.

4.3.1.1 Theory of Charge Transfer

Charge transfer is a transition process of a charge carrier moving between localized sites that are separated by an energy barrier. This transfer is accompanied by a structural change of the conformation of the involved molecule(s) as well as of the surrounding medium, which in turn reshapes the energy landscape of the charge carrier. Besides developing a deeper understanding of the involved processes, the aim of a theory describing charge transfer is to predict the rate constant of the transfer process, that is, the inverse of the average time it takes for a charge carrier to be transferred. A full many-particle quantum-mechanical analysis of the problem, as first performed by Jortner [51], is beyond the scope of this chapter. Instead, we will rather concentrate on the more intuitive classical approach by Marcus [52], and refer the interested reader to more detailed literature [53].

Reorganization-Assisted Transfer In a simplified quasi-one-particle picture, the transfer of an electron from an excited donor site D^* to an acceptor site A in the form $D^*A \rightarrow D^+A^-$ involves the movement of an electron in the potential energy landscape created by the atomic nuclei of the donor and the acceptor site. This potential energy landscape essentially creates a donor and an acceptor quantum well with width and depth depending on the momentary configuration of the atomic nuclei. A sketch of such quantum wells along with the corresponding electronic energy levels is shown in Figure 4.9a. Assuming no exchange of energy between the electron and its environment, the electronic energy levels of the donor and the acceptor have to be equal in order to satisfy the conservation of energy during the transfer. Consequently, all nuclei involved in the transfer process have to reorganize by statistical fluctuations about their equilibrium positions in such a way that the energy levels in the donor and the acceptor quantum well match. This can also be illustrated with the help of so-called potential energy surfaces. In Figure 4.9b, these potential energy surfaces before (labeled D^*A) and after charge transfer (labeled D^+A^-) are represented by parabolas plotted against a generalized reaction coordinate accounting for all degrees of freedom of the nuclear configurations. Vibrations, rotations, and similar fluctuations of the nuclei cause the system to oscillate along the potential energy surface D^*A, as indicated by

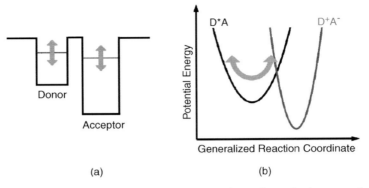

Figure 4.9 Quantum well and potential energy surfaces relevant for charge transfer. Arrows indicate fluctuations about the generalized reaction coordinate leading to changes in the energy levels of the quantum wells, which corresponds to oscillations along the potential energy surface.

arrows in Figure 4.9. When the system reaches the point where the energies of the states before and after charge separation are identical (intersection of the parabolas), charge transfer can occur.

Marcus Theory In classical Marcus theory, the molecular fluctuations along a potential energy surface are described by harmonic oscillations about an equilibrium position (minimum of a parabola). The parameters used for describing the problem are defined in Figure 4.10. Both potential surfaces are assumed to be parabolas with identical curvature c. The equilibrium configurations of the initial

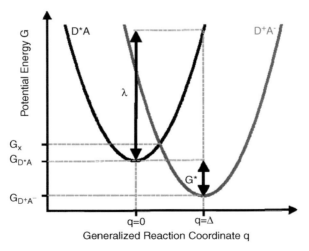

Figure 4.10 Potential energy surfaces of the initial (D*A) and the charge transferred (D$^+$A$^-$) state approximated by parabolas.

($D^*A(q) = cq^2 + G_{D^*A}$) and of the charge separated ($D^+A^-(q) = c(q-\Delta)^2 + G_{D^+A^-}$) states are denoted with $q=0$ and $q=\Delta$ and they are energetically displaced by the value of $G^* = G_{D^*A} - G_{D^+A^-}$. G_B^* is the energy barrier the system needs to overcome before charge transfer can occur and G_x is the intersection of the parabolas. The reorganization energy $\lambda = D^*A(\Delta) - D^*A(0) = c\Delta^2$ is defined as the amount of energy that would be needed to change the molecular configuration of the initial state to that of the equilibrium position Δ of the charge transferred state. It can be divided into the so-called inner sphere and outer sphere contributions. In this context, inner sphere describes the contribution of the molecule(s) directly involved in the transfer process, while outer sphere refers to all molecules in the closer vicinity of the directly involved molecule(s). All energies mentioned here are Gibbs free energies to account for entropic effects.

From geometrical considerations follows that the energy barrier is given by $G_B = (\lambda + G^*)^2/4\lambda$. While the reorganization of the system directly contributes as a barrier for charge transfer, the influence of the energetic displacement G^* of the equilibrium configurations is a bit more complicated. In the regime $0 > G^* \geq -\lambda$, it acts as a driving force for charge transfer by decreasing the energy barrier. However, if the absolute value of G^* exceeds λ, $G^* < -\lambda$, then it has the opposite effect and increases the barrier. The prediction of this so-called Marcus inverted region is one of the main achievements of Marcus theory. In thermodynamic equilibrium, the charge transfer rate of the system can be determined by Boltzmann statistics as

$$k_{CT} = A \exp\left\{-\frac{G_B}{k_B T}\right\} = A \exp\left\{-\frac{(G^* - \lambda)^2}{4\lambda k_B T}\right\}, \tag{4.1}$$

with Boltzmann's constant k, at temperature T, and with the prefactor A that depends on the specific type of transfer reaction (e.g., intra- or intermolecular transfer).

So far we have only considered the energetic requirements that are necessary for the system to reach the configuration at which charge transfer can happen. However, the actual probability for charge transfer to occur in this configuration is not necessarily unity and depends on the strength of the coupling between the initial and the charge transferred state. Landau and Zener independently developed a model to estimate the influence of the coupling on the transition probability and consequently on the charge transfer rate. In short, each time the system reaches the transfer configuration the transition probability P_{CT} is a product of the probability for the charge to be transferred and to stay in the charge transferred state. It is given by $P_{CT} = e^{-\gamma}(1 - e^{-\gamma})$, with γ the so-called Massey parameter, which depends on the electronic coupling and the total free energy G of the system. The transition rate for a given energy of the system then calculates as the transition probability times the transfer attempt frequency $\nu = 2G/h$ (it is twice the oscillation frequency G/h, because the system crosses the transfer configuration twice during each oscillation), with Planck's constant h:

$$k(\nu) = \nu e^{-\gamma}(1 - e^{-\gamma}). \tag{4.2}$$

Assuming thermodynamic equilibrium and Boltzmann statistics, this rate can be used to determine an average transfer rate at a given temperature T:

$$k_{CT} = \frac{\int_{-\infty}^{2G_+/h} k(\nu) e^{-h\nu/kT} \, d\nu}{\int_{-\infty}^{\infty} e^{-h\nu/kT} \, d\nu}. \tag{4.3}$$

If the transition probabilities and thus also the transfer rates are high, that is, for strong coupling of the initial and the charge transferred state, the transfer process is called adiabatic. The weakly coupling regime is called nonadiabatic charge transfer. The fundamental principles influencing the strength of the coupling will be discussed in the next chapter about a semiclassical extension of Marcus theory.

Quantum-Mechanical Considerations Classical Marcus theory is adequate for analyzing the dependence of transfer rates on the energetic offset between initial and charge transferred states as well as on the reorganization energy for systems at room temperature and with moderate transfer barriers. At low temperatures or very high barriers, direct tunneling of the charge through the barrier dominates over the classical pathway requiring reorganization by thermally activated fluctuations. In such cases, quantum-mechanical effects need to be considered for a full description.

In such a case, the rate constant of electron transfer is given by Fermi's golden rule:

$$k = \frac{2\pi}{\hbar} |V|^2 \varrho(G^*), \tag{4.4}$$

where $|V|^2$ is the coupling strength between initial and final states and $\varrho(G^*)$ is the Frank–Condon weighted density of states. In the high-temperature limit, this density of states can be approximated by

$$\varrho(G^*) = \frac{1}{\sqrt{4\pi \lambda k_B T}} \exp\left\{ -\frac{(G^* - \lambda)^2}{4\lambda k_B T} \right\}, \tag{4.5}$$

which, when inserted into Eq. (4.4), reproduces the classical rate constant from Eq. (4.1).

Diffusion-Limited Charge Transfer In some cases, not every site in a material is suitable for charge transfer so that an additional transport step is preceding the transfer process. If this transport step is slow compared to the actual transfer time, then the overall rate of the transfer process is reduced. For three-dimensional isotropic diffusion being the transport mechanism, an estimation of the reduced transfer rate can be obtained. Assuming an excitation on average has to travel a distance x between the site where it was created and a site suitable for charge transfer, the average rate k_D for the excitation to reach a transfer site is given by

$$k_D = \frac{D}{x^2}, \tag{4.6}$$

with diffusion constant D. In the limit of very fast transfer and very slow diffusion, the total charge transfer rate k_{CT} can be approximated by the diffusion rate $k_{CT} \approx k_D$.

4.3.1.2 Theory of Field and Temperature Dependence of Charge Separation

Onsager–Braun Theory In polymer:fullerene composites, dissociation of the exciton via electron transfer from the polymer to the fullerene does not necessarily lead to the formation of free charge carriers. As the dielectric constant of organic semiconductors is generally low (~3), the electron on the fullerene and the hole on the polymer still feel their mutual coulombic attraction, leading to the formation of a geminately bound pair of charge carriers in a charge transfer state. However, as the distance between electron and hole is increased after the transfer, the binding energy of such a geminate pair is reduced compared to that of an exciton. This reduction potentially makes it susceptible to dissociation under the influence of an external electric field or thermal activation. A commonly used theory for such electric field- and temperature-assisted dissociation of charge transfer states was first developed by Onsager in the 1930s and later refined by Braun [54] and Tachiya. The pathways for generation, dissociation, and recombination of CT states (the last one is also called geminate recombination) according to this model are sketched in Figure 4.11.

The relaxed charge transfer state is assumed to be metastable with a field- and temperature-dependent lifetime $\tau(F, T) = [k_{CT \to SSC}(F, T) + k_{CT \to GS}]^{-1}$. The probability for a charge transfer state to dissociate into free charges $P(F, T)$ then

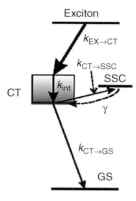

Figure 4.11 Schematic representation of the Onsager–Braun model including generation of a charge transfer (CT) state from an exciton followed by vibrational relaxation, dissociation to and reformation from spatially separated charges (SSC), and recombination to the ground state (GS). (Adapted with permission from Ref. [55]. Copyright 2010, American Chemical Society.)

follows from the product of the dissociation rate and its lifetime:

$$P(F, T) = \frac{k_{CT \to SSC}(F, T)}{k_{CT \to SSC}(F, T) + k_{CT \to GS}},$$

where the dissociation rate is given by

$$k_{CT \to SSC}(F, T) = \nu \exp[-\Delta E/kT]\left(1 + b + \frac{b^2}{3} + \frac{b^3}{18} + \cdots\right),$$

with the separation attempt frequency

$$\nu = \frac{3\mu e}{4\pi\varepsilon\varepsilon_0 a^3}$$

and the effective field parameter

$$b = \frac{e^3 F}{8\pi\varepsilon\varepsilon_0 (kT)^2}.$$

In this context, the dissociation barrier ΔE is given only by the Coulomb binding energy at separation distance a between the charge carriers and does not consider that the energetic environment of the electron in the fullerene is much different from that in the polymer. The value of this dissociation barrier is typically found to be 0.1–0.5 eV corresponding to a separation of the charge carriers in the charge transfer state of roughly 1–4 nm [56–60]. Also the explicit morphological structure of the interface is not taken into account by Onsager–Braun theory and the interface is assumed to be perpendicular to the applied electric field, which is generally not the case in randomly distributed polymer:fullerene composites. In fact, the only way to consider any difference in the materials is to use appropriate values for the spatially averaged sum of electron and hole mobilities μ and the dielectric constant ε. Following along this line, an increase in the dielectric constant should cause a decrease in the charge transfer binding energy. However, in a compositional study, Vandewal et al. [61] found that if increasing the average dielectric constant by increasing the amount of PCBM in a polymer:PCBM blend, the charge transfer state is stabilized by a higher binding energy. This implies that other factors have to be taken into account in order to explain all aspects of charge transfer binding energies.

Even though Onsager–Braun theory being a limited effective medium approximation, it can still give some useful qualitative insight into the separation mechanism and especially into its dependence on electric field, charge transfer distance, and temperature. Figure 4.12 shows the field dependence of the dissociation probability at various charge separation distances and temperatures for a representative set of parameters. For fields below 10^8 V m^{-1} (corresponding to an applied voltage of 10 V at a sample thickness of 100 nm), both the charge separation distance after the charge transfer step (a) and the temperature (b) have a strong influence on the dissociation probability. According to this model, charge separation distances of more than 1 nm, high charge carrier

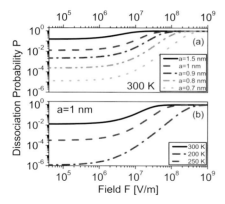

Figure 4.12 Dissociation probability P as a function of electric field F at various charge separation distances a and room temperature (a) and at various temperatures T and fixed charge separation distance of $a = 1$ nm (b). The parameters used for the calculation are $\mu = 10^{-4}\,\mathrm{m^2\,V^{-1}\,s^{-1}}$, $\varepsilon = 3$, and $k_{\mathrm{CT}\to\mathrm{GS}} = 10^8\,\mathrm{s^{-1}}$.

mobilities, and high temperatures are prerequisites for an efficient dissociation of charge transfer states at moderate electric fields.

4.3.2 The Role of Charge Transfer States for Charge Separation

In this section, we will discuss experimental evidence for the role of charge transfer states for charge separation. Much of the material presented herein has also been reviewed excellently by Clarke [62] and Deibel [63].

The charge transfer process in polymer:fullerene composites, that is, the separation of an exciton at an interface, has been found to be ultrafast, happening on a sub-100 fs timescale, and very efficient with only a small percentage of fullerene needed for almost quantitative dissociation of the exciton [64–69]. In material systems with a high degree of separation between the phases of the polymer and the fullerene, the rate of the transfer process is usually limited by the diffusion of excitons to the interface within less than 100 ps after their formation [55, 70, 71]. However, as already indicated above, charge transfer does not necessarily lead to the creation of spatially separated charges but can also result in bound charge transfer states at the interface.

Such charge transfer states can be uniquely identified by their photophysical properties. When two organic semiconductors are blended together and charge transfer states at the interface are formed, sometimes additional features in both absorption and emission spectra are observed that cannot be assigned to any of the individual components. As charge transfer states are generally lower in energy than singlet excitons, these features show a bathochromic shift relative to the transitions of the unblended materials. Electroluminescence measurements are particularly well suited for analyzing the emission from charge transfer states, because here the charge transfer states are formed directly by the recombination of injected free

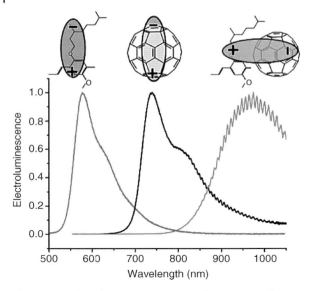

Figure 4.13 Electroluminescence spectra of MDMO-PPV (first curve), PCBM (second curve), and the blend of these two materials (third curve). (Reprinted with permission from Ref. [72]. Copyright 2009, American Chemical Society.)

charge carriers so that there is no emission from higher lying singlet excitons that is superimposed on the charge transfer emission [72, 73]. An example for this is shown in Figure 4.13 by the means of electroluminescence spectra of MDMO-PPV (first curve), PCBM (second curve), and a blend of these two materials (third curve). In this case, it is very obvious that the emission spectrum of the blend is not a superposition of the emission of the pristine materials, but rather originates from a charge transfer state lower in energy. Similar studies have provided evidence for the existence of charge transfer states in numerous polymers blended with PCBM, as well as in polymer:polymer composites. Only in a few cases, for example, in polymers with very low highest molecular orbitals, where the energy levels of triplet excitons can be below that of the charge transfer state and can thus form the lowest excited state, no indication for the occupation of charge transfer states at the interface could be found.

A study of the photoluminescence quenching of a series of polythiophenes blended with PCBM performed by Ohkita et al. [64] indicated that even though the charge transfer process itself was equally efficient for all polythiophenes of the series blended with PCBM, the yield of spatially separated charges differed by two orders of magnitude, depending on the ionization potentials of the polymers. This suggests that the occupation of charge transfer states is indeed a step involved in charge separation. Further evidence for this assumption came from observations of the correlation between the reduction of the photoluminescence intensity of charge transfer states upon the application of an electric field and a simultaneous increase of the yield of spatially separated charge carriers by Offermans et al. [58].

There are several factors that influence the ratio of recombination through charge transfer states and spatial separation of charge carriers. In the following sections, we will discuss the role of excess energy, morphology, charge carrier mobility, electric field, and temperature on the suppression of geminate recombination and gain in the yield of spatially separated charge carriers.

4.3.2.1 Parameters Influencing the Separation of Charge Transfer States

Excess Energy In many cases, the energy difference between the relaxed exciton, that is, the initial state of the charge transfer process, and the relaxed charge transfer state is much larger (typically $0.5\,\text{eV} < G^* < 1\,\text{eV}$) [64] than the reorganization energy of the transfer process. In this situation, the rate of the charge transfer between these two relaxed states is reduced, because the transition occurs in the Marcus inverted regime (see above). However, because of the strong electron–phonon coupling in organic semiconductors, the charge transfer state features a quasicontinuum of vibrationally excited states. The transition rate from the exciton to each of these levels is governed by the ratio of the respective energetic offset and the reorganization energy, with the highest transition rate to the level for which this ratio approaches unity. Consequently, the charge transfer will mainly occur between the relaxed exciton and a "hot" vibrationally excited charge transfer state. These hot states have a larger electron hole distance than the relaxed state and are thus more likely to split into free charge carriers. In fact, Morteani *et al.* [74] found that the electron–hole distance in relaxed charge transfer states is so small that they do virtually not decompose into free spatially separated charges any more, but only recombine geminately to the ground state. The generation of spatially separated charges was consequently assigned to the dissociation of the hot charge transfer state in kinetic competition with its vibrational relaxation.

Ohkita *et al.* [64] studied the influence of this excess energy with respect to the relaxed charge transfer state on the yield of spatially separated charge carriers with a series of polythiophenes with differing singlet exciton energy levels (Figure 4.14). They found that increasing the excess energy from 0.6 to 1 eV resulted in an increase of the yield of spatially separated charge carriers by two orders of magnitude, strongly indicating that excess energy is beneficial for the separation of charge transfer states. However, in a similar study, Clarke *et al.* [75] reported a very high yield of free charge carriers for a low bandgap polymer:fullerene composite with an excess energy of less than 0.2 eV, demonstrating that high excess energies are not a necessary prerequisite for efficient charge separation.

Morphology Morphology can potentially have a huge influence on the separation of charge transfer states in various ways and consequently on a broad range of length scales, reaching from the orientation of the molecules at the interface to the formation of interpenetrating phase-segregated networks of the blended materials. Unfortunately, collecting nonambiguous and reliable structural information on the involved very short length scales is not a straightforward task. This is why most of the studies dealing with the influence of morphology on charge separation

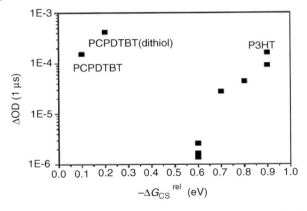

Figure 4.14 Yield of spatially separated charge carriers ($\Delta OD(1\,\mu s)$) as a function of excess energy G_{cs}^{rel} (eV). (Reprinted with permission from Ref. [75]. Copyright 2009, Royal Chemical Society.)

concentrate on the larger end of this scale, that is, the influence of the extent and purity of phase segregation along with the size of phase-segregated domains. To this end, a variety of methods for controlling the morphology have been exploited experimentally. Among these are variations of the sample preparation conditions, for example, use of different solvents [68] or annealing conditions [66, 76], slight modifications in the molecular structure such as a replacement of side chains [55, 70], or alterations of the composition of two blended materials [27]. All of these approaches have in common that they involve rather subtle changes within a series of samples under study in order to keep secondary effects that could influence charge separation, such as excess energy, as minimal as possible. Following the majority of publications on this topic, we will mainly concentrate on the nanoscale phase segregation within the blend films in this chapter and will only give a brief outlook toward smaller length scales.

The optimal size for nanoscale phase segregation in terms of charge separation is determined by two processes. On the one hand, there is an upper limit given by the exciton diffusion length (see above). Once the phase domains are larger than this diffusion length, not all excitons will reach the interface to perform charge transfer but instead will recombine. On the other hand, the charge carriers have to able to move far enough from the interface to be considered spatially separated. In this context, often the Coulomb capture radius is quoted as an estimation of the distance at which the coulombic attraction of electron and hole is on the order of the thermal energy kT and thus negligible:

$$r_c = \frac{e^2}{4\pi\varepsilon\varepsilon_0}\frac{1}{kT}.$$

For an organic semiconductor blend with an average dielectric constant of 3, this Coulomb capture radius is 19 nm at room temperature, which is somewhat larger than the typical exciton diffusion length of 10 nm. Of course for this estimation it should be considered that an ideal nanomorphology does probably not exclusively consist of spherical domains, but rather of elongated structures that satisfy efficient exciton quenching along the smaller axis and at the same time allow for a sufficient separation of the charge carriers along the longer axis.

The effect of the compositional dependence of a fluorine copolymer:PCBM blend was studied by Veldman et al. [77]. They used atomic force microscopy (AFM) and transmission electron microscopy (TEM) on blend films of a broad range of mixing ratios and found that while at low concentrations the PCBM is finely dispersed in the polymer, at very high concentrations the PCBM tends to aggregate and forms crystalline domains that are up to 100 nm large. Simultaneously, they performed time-resolved photoluminescence measurements and found that the emission from the charge transfer state, which decayed on a few nanoseconds timescale in the near-infrared wavelength region, strongly depended on this variation in morphology. First of all, the initial amplitude of the emission was reduced by the coarsening of the phase segregation. This can have three possible explanations. The simplest would be that the optical density of the films was reduced due to the larger amount of PCBM in the blend, which unfortunately was not quoted in their experimental details. Assuming the absorption was not strongly influenced by the composition, this means that either excitons were quenched less efficiently by the higher amount of PCBM or excitons were quenched as efficiently but the relaxed (emissive) charge transfer state was not populated as efficiently, meaning that at higher PCBM concentrations it might be more favorable for hot charge transfer states to dissociate than to relax compared to those at lower PCBM concentrations. Thus, the initial amplitude of the emission does not give an unambiguous conclusion as all of these effects can occur simultaneously. However, the lifetime of the emission is reduced at higher PCBM concentrations, which means that the charge transfer state is depopulated much faster. Assuming that the recombination of the charge transfer state is not accelerated by the addition of PCBM, this strongly suggests that the charge transfer states are separated into free charge carriers more efficiently if the phase segregation between the polymer and the PCBM phases is increased. They also employed photoinduced absorption (PIA) experiments and measured a higher yield of long-lived spatially separated charge carriers at higher PCBM concentrations, supporting these findings. Quist et al. [78] performed time-dependent microwave photoconductivity experiments and TEM on blends of MDMO-PPV and PCBM for a range of compositions prepared from both chlorobenzene and toluene solutions. They observed the same trend in phase segregation and formation of nanocrystallites with higher PCBM content. In addition, the amount of the segregation was found to be different for the two solvents used, with a stronger segregation for the toluene-cast films. Again, they found that the yield of spatially separated charges strongly correlates with the domain size.

Guo [70] and Howard [55] investigated the kinetics of exciton quenching and the formation of charge transfer states and spatially separated charges in blends of poly (3-hexylthiophene) (P3HT) and PCBM, the "fruit fly" of organic semiconductor composites [79], with transient absorption spectroscopy. They observed that for P3HT with a random distribution of head-to-tail and head-to-head couplings within the polymer chains (i.e., regiorandom) excitons were quenched ultrafast within 100 fs and that the yield of spatially separated charges from the resulting charge transfer states was very low. From this they concluded that regiorandom P3HT and PCBM do not phase segregate even at modestly high PCBM concentrations (of 50%). The results for P3HT with an almost exclusively head-to-tail coupled chain (i.e., regioregular) blended with PCBM were quite different. Quenching of excitons was much slower, occurring on a 10 ps timescale, and led to a significantly higher separation probability of the charge transfer states. This was attributed to a much more pronounced phase segregation. In pure P3HT domains, excitons have to diffuse to the interface before charge transfer can occur, essentially reducing the charge transfer rate to the diffusion-limited case (see above). As in the studies by Veldman and Quist, the larger phase segregation was found to be beneficial for the dissociation of the charge transfer state. Annealing in P3HT:PCBM was found to have a similar effect [66, 76] and the simultaneous recovery of the photoluminescence [55, 70], indicating an incomplete exciton quenching, further supported the conclusion that larger pure phase domains lead to a more facile dissociation of charge transfer state.

Monte Carlo simulations by Groves et al. [80] suggested that in a polymer:polymer bilayer charges can be considered effectively free when they are separated from each other by about 4 nm, which is clearly shorter than would be expected from the Coulomb capture radius. Nonetheless, in a blend a further increase of the phase segregation to domain sizes of up to 16 nm still increased the yield of spatially separated charges. This strongly supports the experimental findings reviewed above.

Holcombe et al. [81] directly controlled the separation distance between electrons and holes at the interface of a series of polythiophene:acceptor blends and correlated this distance with the binding energy of the charge transfer state and its dissociation probability. Not surprisingly, a larger distance between the donor and acceptor molecules resulted in a reduced charge transfer state binding energy and a higher dissociation probability, showing that the precise interfacial structure strongly influences the formation of spatially separated charges (Figure 4.15). In a similar manner, Müller et al. [71] assigned an ultrafast (<0.5 ps) depolarization of the hole in a MDMO-PPV:PCBM blend to an intermolecular delocalization in the well-ordered MDMO-PPV phase that by increasing the distance between electron

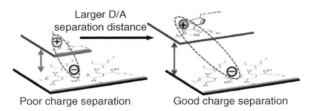

Figure 4.15 Schematic of the influence of the distance between donor and acceptor on the charge separation yield. (Reprinted with permission from Ref. [81]. Copyright 2011, American Chemical Society.)

and hole at the interface destabilized the charge transfer state and led to a more efficient generation of spatially separated charges. Deibel took this idea one step further and utilized Monte Carlo simulations to show that not only the order in one of the phases of the blend but also the order within single polymer chains at the interface, that is, their conjugation length, can considerably enhance the charge separation yield [82].

Further, theoretical calculations and experiments suggest that certain molecular arrangements can lead to the formation of dipoles at the interface that can reduce the coulombic attraction between an electron and a hole in a charge transfer state, thus increasing the dissociation probability [83–85].

Electric Field From calculating the influence of the electric field on the dissociation of charge transfer states with the Onsager–Braun model (see Section 4.3.1.2), we expect that electric fields exceeding 10^6 V m^{-1} have a pronounced influence on the formation of spatially separated charge carriers. Fortunately, in many organic semiconductor composites this can be directly examined by field-dependent time-resolved photoluminescence spectroscopy of the emission from the charge transfer state [58, 74, 77]. As typical film thicknesses of such composites are on the order of 100 nm, voltages of up to 100 V are more than sufficient to generate the required electric field strengths for these experiments.

Veldman *et al.* [77] investigated the electric field-induced quenching of the emission from charge transfer states in a fluorine copolymer:PCBM blend (Figure 4.16). There are three especially remarkable aspects of the results of these measurements. First of all, the lifetime of the unquenched charge transfer state is roughly 2.5 ns, which corresponds to a depopulation rate of the charge transfer state of about 4×10^8 s^{-1}. Second, the depopulation is strongly accelerated by the application of an electric field, increasing the total rate to above 1×10^9 s^{-1} at an electric field of

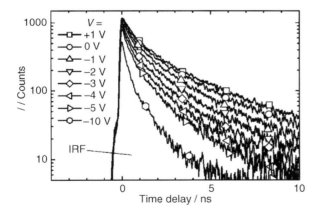

Figure 4.16 Field-dependent time-resolved photoluminescence measurement of a copolymer:PCBM blend film of 220 nm thickness. The voltages given in the legend of the figure correspond to electric field strengths of $F = 0$–4.5×10^7 V m^{-1}. (Adapted with permission from Ref. [77]. Copyright 2008, American Chemical Society.)

$F = 4.5 \times 10^7 \, \text{V m}^{-1}$, strongly suggesting that charge transfer states can indeed be split by the application of an electric field. Furthermore, by fitting the photoluminescence transients to a biexponential decay, they found that the sum of the initial amplitudes is independent of the electric field. This indicates that the formation of the CT states is not influenced by the electric field, which was supported by the observation that the emission from singlet excitons in the polymer was not quenched simultaneously. It should be noted that the choice of a biexponential fit was merely based on empirical findings and not on a solid consideration of the rate equation for the depopulation of the charge transfer state. Thus, the true physical meaning of the parameters extracted from fitting is somewhat unclear. A biexponential decay does suggest that the charge transfer state is depopulated through exactly two decay channels. Applying Onsager–Braun theory, these channels could be assigned to recombination to the ground state and dissociation to spatially separated charge carriers. However, Veldman et al. observed that both rate constants are increased by the application of an electric field, whereas Onsager–Braun theory assumes that the recombination to the ground state is independent of the electric field. One possible explanation of this deviation could be that in a polymer:fullerene composite the orientation of the interface with respect to the electric field is distributed randomly throughout the entire film. Consequently, there are some orientations of the charge carriers in the charge transfer state that have a component parallel to the external field and some that have an antiparallel component. While the former will be more likely to dissociate into spatially separated charge carriers, the latter could show an enhanced likelihood for recombination, so that both rate constants would be increased. Such an effect is not considered by Onsager–Braun theory that only takes a homogeneous medium into account.

Kern et al. [86] also used field-induced photoluminescence quenching of the emission from charge transfer states to determine the binding energy of the charge transfer state according to the Onsager–Braun model. They observed that in this particular material system fields in excess of 10 V had to be applied in order to dissociate a significant fraction of the charge transfer states. Estimations of the binding energy yielded a value of 0.2 eV that corresponds to an average separation distance between the charge carriers of 2 nm.

It is further interesting to note that results from transient absorption spectroscopy suggest that in certain material systems such as P3HT:PCBM a very high yield of spatially separated charges can be found even in the absence of an external electric field, so that the electric field can only have a minor influence on the overall charge separation yield [55, 70, 87].

Other Factors While excess energy, morphology, and electric field are the most frequently studied parameters influencing the dissociation of charge transfer states, there are still a number of other factors that have to be considered. Among these are the temperature and charge carrier mobilities as already mentioned in the introduction of Onsager–Braun theory. But also the details of the charge transfer process such as electronic coupling or reorganization energies or effects of energetic disorder are of importance.

In P3HT:PCBM blends, Mauer *et al.* [88] reported that the high yield of spatially separated charges was not influenced by reducing the temperature down to as low as 80 K. While this seems surprising considering that from Onsager–Braun theory one would expect an exponential dependence on the temperature, it can be argued that the thermal activation by $\sim$25 meV can be neglected compared to the excess energy of 1 eV that results from the charge transfer process. Furthermore, the mobility of holes and electrons is reduced by about two orders of magnitude when decreasing the temperature to 80 K. This suggests that in this case the charge carrier mobilities also do not play a crucial role for the dissociation of the charge transfer states. This finding was supported by observations by Ohkita *et al.* who found no correlation between the hole mobility and dissociation yield when studying polymers with hole mobilities varying over four orders of magnitude [64].

Density functional theory calculations of the localization of the involved molecular wavefunctions predicted a bridging state across the interface of P3HT and fullerene molecules that promotes an adiabatic charge transfer process [89]. In addition, this bridging state was found to have a considerable overlap with LUMO states in the fullerene, thus increasing the separation probability of the charge transfer state.

4.3.3
Charge Transfer: Relevance to Device Performance

The power conversion efficiency of solar cells made of organic semiconductor composites results from a cascade of steps reaching from the absorption of photons to the extraction of spatially separated charge carriers. It is determined by the ability of a material to generate as many extractable charges per absorbed photon as possible without losing energy during the conversion process. As discussed above, not only the migration of excitons to the interface but also the charge transfer process and the resulting charge transfer state play a crucial role in this context.

The dissociation behavior of charge transfer states is of huge interest because it directly influences the amount of spatially separated charge carriers available for extraction. Consequently, Onsager–Braun theory has frequently been used to simulate current–voltage characteristics of solar cell devices under solar illumination [90–93]. However, as discussed above this model has its limitations in adequately predicting experimental results on the influence of parameters such as temperature or charge carrier mobility on dissociation yields. Such simulations use the charge transfer dissociation rate as free parameter to fit calculations to experimental current–voltage characteristics. This usually results in dissociation rates that are incompatible with those directly determined by transient absorption or emission spectroscopy, being several orders of magnitude slower [92]. Furthermore, Onsager–Braun theory is used to model the field dependence of the photocurrent in solar cell devices despite photoluminescence experiments suggesting that in many material systems electric fields much higher than those typically present in solar cells under normal working conditions are necessary in order to achieve a reasonable impact on device performance [94]. Nonetheless, charge transfer states have a strong influence on the power conversion efficiency. For example, in confined morphologies when charge carriers cannot dissociate due to geometrical limitations, the recombination of charge carriers occupying charge transfer states will be the main loss mechanism.

Besides the dissociation yield itself, the energy loss during the charge transfer and dissociation process limits the power conversion efficiency. As the charge transfer state is the energetically lowest excited state involved in photocurrent generation (assuming that the formation of triplet excitons can be neglected), it determines the minimal energy a charge carrier can possess when being extracted at an electrode. This essentially means that the energy difference between an absorbed photon and the charge transfer state is lost to thermal relaxation during charge transfer and dissociation. Consequently, the energy of the charge transfer state should be adapted to the optical bandgap of the semiconductor composite as closely as possible to optimize the power conversion efficiency. Fortunately, as reported above despite its strong influence such an excess energy is not a necessary prerequisite for efficient charge transfer and dissociation so that a trade-off between the conservation of energy and the yield of dissociated charge carriers can be avoided.

Concluding these findings, a material for an ideal solar cell would quantitatively dissociate all excitons into spatially separated charge carriers without needing any energetic driving force that involves the dissipation of a part of the absorbed energy. Modern donor–acceptor copolymers blended with PCBM seem to be on a good way to reach this goal [95].

References

1 Bredas, J.L. and Silbey, R. (2009) *Science*, **323**, 348.
2 Barford, W. (2005) *Electronic and Optical Properties of Conjugated Polymers*, Clarendon Press, Oxford.
3 Scholes, G.D. and Rumbles, G. (2006) *Nat. Mater.*, **5**, 683.
4 Wasserberg, D., Marsal, P., Meskers, S.C.J., Janssen, R.A.J., and Beljonne, D. (2005) *J. Phys. Chem. B*, **109**, 4410.
5 Hertel, D., Setayesh, S., Nothofer, H.G., Scherf, U., Mullen, K., and Bässler, H. (2001) *Adv. Mater.*, **13**, 65.
6 Jansson, E., Jha, P.C., and Aren, H. (2007) *Chem. Phys.*, **336**, 91.
7 Gierschner, J., Cornil, J., and Egelhaaf, H.J. (2007) *Adv. Mater.*, **19**, 173.
8 Laquai, F., Mishra, A.K., Ribas, M.R., Petrozza, A., Jacob, J., Akcelrud, L., Mullen, K., Friend, R.H., and Wegner, G. (2007) *Adv. Funct. Mater.*, **17**, 3231.
9 Dubin, F., Melet, R., Barisien, T., Grousson, R., Legrand, L., Schott, M., and Voliotis, V. (2006) *Nat. Phys.*, **2**, 32.
10 Cornil, J. et al. (2003) *J. Chem. Phys.*, **118**, 6615.
11 Köhler, A. and Beljonne, D. (2004) *Adv. Funct. Mater.*, **14**, 11.
12 Burrows, H.D., de Melo, J.S., Serpa, C., Arnaut, L.G., Monkman, A.P., Hamblett, I., and Navaratnam, S. (2001) *J. Chem. Phys.*, **115**, 9601.
13 Yang, C.G., Ehrenfreund, E., and Vardeny, Z.V. (2007) *Phys. Rev. Lett.*, **99**, 157401.
14 Beljonne, D., Shuai, Z., Pourtois, G., and Bredas, J.L. (2001) *J. Phys. Chem. A*, **105**, 3899.
15 Westenhoff, S., Howard, I.A., Hodgkiss, J.M., Kirov, K.R., Bronstein, H.A., Williams, C.K., Greenham, N.C., and Friend, R.H. (2008) *J. Am. Chem. Soc.*, **130**, 13653.

16 van Dijken, A., Bastiaansen, J., Kiggen, N.M.M., Langeveld, B.M.W., Rothe, C., Monkman, A., Bach, I., Stossel, P., and Brunner, K. (2004) *J. Am. Chem. Soc.*, **126**, 7718.
17 Köhler, A. and Bässler, H. (2009) *Mater. Sci. Eng. R*, **66**, 71.
18 Borsenberger, P.M., Pautmeier, L., and Bässler, H. (1991) *J. Chem. Phys.*, **94**, 5447.
19 Nelson, J. (2003) *Phys. Rev. B*, **67**, 155209.
20 Kador, L. (1991) *J. Chem. Phys.*, **95**, 5574.
21 Miller, A. and Abrahams, E. (1960) *Phys. Rev.*, **120**, 745.
22 Bässler, H. (1993) *Phys. Status Solidi (b)*, **175**, 15.
23 Scheblykin, I.G., Yartsev, A., Pullerits, T., Gulbinas, V., and Sundstroem, V. (2007) *J. Phys. Chem. B*, **111**, 6303.
24 Braslavsky, S.E., Fron, E., Rodriguez, H.B., Roman, E.S., Scholes, G.D., Schweitzer, G., Valeur, B., and Wirz, J. (2008) *Photochem. Photobiol. Sci.*, **7**, 1444.
25 Westenhoff, S., Daniel, C., Friend, R.H., Silva, C., Sundström, V., and Yartsev, A. (2005) *J. Chem. Phys.*, **122**, 094903.
26 Collini, E. and Scholes, G.D. (2009) *Science*, **323**, 369.
27 Wasserberg, D., Meskers, S.C.J., and Janssen, R.A.J. (2007) *J. Phys. Chem. A*, **111**, 1381.
28 Sudha Devi, L., Al-Suti, M.K., Dosche, C., Khan, M.S., Friend, R.H., and Köhler, A. (2008) *Phys. Rev. B*, **78**, 045210.
29 Giebink, N.C., Sun, Y., and Forrest, S.R. (2006) *Org. Electron.*, **7**, 375.
30 Köhler, A. and Bässler, H. (2011) *J. Mater. Chem.*, **21**, 4003.
31 Holstein, T. (1959) *Ann. Phys.*, **8**, 325.
32 Holstein, T. (1959) *Ann. Phys.*, **8**, 343.
33 Fishchuk, I.I., Kadashchuk, A., Devi, L.S., Heremans, P., Baessler, H., and Koehler, A. (2008) *Phys. Rev. B*, **78**, 045211.
34 Hoffmann, S.T., Scheler, E., Koenen, J.-M., Forster, M., Scherf, U., Strohriegl, P., Baessler, H., and Koehler, A. (2010) *Phys. Rev. B*, **81**, 165208.
35 Avakian, P. and Merrifield, R.E. (1968) *Mol. Cryst.*, **5**, 37.
36 Ern, V., Avakian, P., and Merrifield, R.E. (1966) *Phys. Rev.*, **148**, 862.
37 Scheidler, M., Cleve, B., Bässler, H., and Thomas, P. (1994) *Chem. Phys. Lett.*, **225**, 431.
38 Hertel, D., Bässler, H., Guentner, R., and Scherf, U. (2001) *J. Chem. Phys.*, **115**, 10007.
39 Islangulov, R.R., Kozlov, D.V., and Castellano, F.N. (2005) *Chem. Commun.*, **30**, 3776.
40 Baluschev, S., Yakutkin, V., Wegner, G., Minch, B., Miteva, T., Nelles, G., and Yasuda, A. (2007) *J. Appl. Phys.*, **90**, 181103.
41 Keivanidis, P.E., Laquai, F., Robertson, J.W.F., Baluschev, S., Jacob, J., Muellen, K., and Wegner, G. (2011) *J. Phys. Chem. Lett.*, **2**, 1893.
42 Baluschev, S., Miteva, T., Yakutkin, V., Nelles, G., Yasuda, A., and Wegner, G. (2006) *Phys. Rev. Lett.*, **97**, 143903.
43 Laquai, F., Wegner, G., Im, C., Busing, A., and Heun, S. (2005) *J. Chem. Phys.*, **123**, 074902.
44 Islangulov, R.R., Lott, J., Weder, C., and Castellano, F.N. (2007) *J. Am. Chem. Soc.*, **129**, 12652.
45 Miteva, T., Yakutkin, V., Nelles, G., and Baluschev, S. (2008) *New J. Phys.*, **10**, 103002.
46 Cheng, Y.Y., Fuckel, B., Khoury, T., Clady, R., Tayebjee, M.J.Y., Ekins-Daukes, N.J., Crossley, M.J., and Schmidt, T.W. (2010) *J. Phys. Chem. Lett.*, **1**, 1795.
47 Shaw, P.E., Ruseckas, A., and Samuel, I.D.W. (2008) *Adv. Mater.*, **20**, 3516.
48 Rao, A., Wilson, M.W.B., Hodgkiss, J.M., Albert-Seifried, S., Bässler, H., and Friend, R.H. (2010) *J. Am. Chem. Soc.*, **132**, 12698.
49 Reineke, S., Lindner, F., Schwartz, G., Seidler, N., Walzer, K., Luessem, B., and Leo, K. (2009) *Nature*, **459**, 234.
50 Vasdekis, A.E., Moore, S.A., Ruseckas, A., Krauss, T.F., Samuel, I.D.W., and Turnbull, G.A. (2007) *Appl. Phys. Lett.*, **91**, 051124.
51 Jortner, J. (1976) *J. Chem. Phys.*, **64**, 4860.
52 Marcus, R.A. (1956) *J. Chem. Phys.*, **24**, 966.
53 May, V. and Kühn, O. (2004) *Charge and Energy Transfer Dynamics in Molecular Systems*, 1st edn, Wiley-VCH Verlag GmbH, Weinheim.
54 Braun, C.L. (1984) *J. Chem. Phys.*, **80**, 4157.
55 Howard, I.A., Mauer, R., Meister, M., and Laquai, F. (2010) *J. Am. Chem. Soc.*, **132**, 14866.

56 Blom, P.W.M., Mihailetchi, V.D., Koster, L.J.A., and Markov, D.E. (2007) *Adv. Mater.*, **19**, 1551.

57 Hallermann, M., Haneder, S., and Como, E.D. (2008) *Appl. Phys. Lett.*, **93**, 053307.

58 Offermans, T., van Hal, P.A., Meskers, S.C.J., Koetse, M.M., and Janssen, R.A.J. (2005) *Phys. Rev. B*, **72**, 045213.

59 Rand, B.P., Burk, D.P., and Forrest, S.R. (2007) *Phys. Rev. B*, **75**, 115327.

60 Veldman, D., Meskers, S.C.J., and Janssen, R.A.J. (2009) *Adv. Funct. Mater.*, **19**, 1939.

61 Vandewal, K., Gadisa, A., Oosterbaan, W.D., Bertho, S., Banishoeib, F., Van Severen, I., Lutsen, L., Cleij, T.J., Vanderzande, D., and Manca, J.V. (2008) *Adv. Funct. Mater.*, **18**, 2064.

62 Clarke, T.M. and Durrant, J.R. (2010) *Chem. Rev.*, **110**, 6736.

63 Deibel, C. and Dyakonov, V. (2010) *Rep. Prog. Phys.*, **73**, 096401.

64 Ohkita, H., Cook, S., Astuti, Y., Duffy, W., Tierney, S., Zhang, W., Heeney, M., McCulloch, I., Nelson, J., Bradley, D.D.C., and Durrant, J.R. (2008) *J. Am. Chem. Soc.*, **130**, 3030.

65 Brabec, C.J., Zerza, G., Cerullo, G., De Silvestri, S., Luzzati, S., Hummelen, J.C., and Sariciftci, S. (2001) *Chem. Phys. Lett.*, **340**, 232.

66 Hwang, I.-W., Moses, D., and Heeger, A.J. (2008) *J. Phys. Chem. C*, **112**, 4350.

67 Kraabel, B., Lee, C.H., McBranch, D., Moses, D., Sariciftci, N.S., and Heeger, A.J. (1993) *Chem. Phys. Lett.*, **213**, 389.

68 Zhang, F., Jespersen, K.G., Björström, C., Svensson, M., Andersson, M.R., Sundström, V., Magnusson, K., Moons, E., Yartsev, A., and Inganäs, O. (2006) *Adv. Funct. Mater.*, **16**, 667.

69 Bakulin, A.A., Hummelen, J.C., Pshenichnikov, M.S., and van Loosdrecht, P.H.M. (2010) *Adv. Funct. Mater.*, **20**, 1653.

70 Guo, J., Ohkita, H., Benten, H., and Ito, S. (2010) *J. Am. Chem. Soc.*, **132**, 6154.

71 Müller, J.G., Lupton, J.M., Feldmann, J., Lemmer, U., Scharber, M.C., Sariciftci, N.S., Brabec, C.J., and Scherf, U. (2005) *Phys. Rev. B*, **72**, 195208.

72 Tvingstedt, K., Vandewal, K., Gadisa, A., Zhang, F., Manca, J., and Inganäs, O. (2009) *J. Am. Chem. Soc.*, **131**, 11819.

73 Tvingstedt, K., Vandewal, K., Zhang, F., and Inganas, O. (2010) *J. Phys. Chem. C*, **114**, 21824.

74 Morteani, A.C., Sreearunothai, P., Herz, L.M., Friend, R.H., and Silva, C. (2004) *Phys. Rev. Lett.*, **92**, 247402.

75 Clarke, T., Ballantyne, A., Jamieson, F., Brabec, C., Nelson, J., and Durrant, J. (2009) *Chem. Commun.*, **1**, 89.

76 Savenije, T.J., Kroeze, J.E., Yang, X., and Loos, J. (2005) *Adv. Funct. Mater.*, **15**, 1260.

77 Veldman, D., Ipek, O., Meskers, S.C.J., Sweelssen, J., Koetse, M.M., Veenstra, S.C., Kroon, J.M., Bavel, S.S.V., Loos, J., and Janssen, R.A.J. (2008) *J. Am. Chem. Soc.*, **130**, 7721.

78 Quist, P.A.C., Martens, T., Manca, J.V., Savenije, T.J., and Siebbeles, L.D.A. (2006) *Sol. Energy Mater. Sol. Cells*, **90**, 362.

79 Kamm, V., Battagliarin, G., Howard, I.A., Pisula, W., Mavrinskiy, A., Li, C., Müllen, K., and Laquai, F. (2011) *Adv. Energy Mater.*, **1**, 297.

80 Groves, C., Marsh, R.A., and Greenham, N.C. (2008) *J. Chem. Phys.*, **129**, 114903.

81 Holcombe, T.W., Norton, J.E., Rivnay, J., Woo, C.H., Goris, L., Piliego, C., Griffini, G., Sellinger, A., Bredas, J.L., Salleo, A., and Frechet, J.M.J. (2011) *J. Am. Chem. Soc.*, **133**, 12106.

82 Deibel, C. (2009) *Phys. Status Solidi (a)*, **206**, 2731.

83 Arkhipov, V.I., Heremans, P., and Bässler, H. (2003) *Appl. Phys. Lett.*, **82**, 4605.

84 Verlaak, S. and Heremans, P. (2007) *Phys. Rev. B*, **75**, 115127.

85 Sreearunothai, P., Morteani, A.C., Avilov, I., Cornil, J., Beljonne, D., Friend, R.H., Phillips, R.T., Silva, C., and Herz, L.M. (2006) *Phys. Rev. Lett.*, **96**, 117403.

86 Kern, J., Schwab, S., Deibel, C., and Dyakonov, V. (2011) *Phys. Status Solidi (RRL)*, **5**, 364.

87 Kniepert, J., Schubert, M., Blakesley, J.C., and Neher, D. (2011) *J. Phys. Chem. Lett.*, **2**, 700.

88 Mauer, R., Howard, I.A., and Laquai, F. (2010) *J. Phys. Chem. Lett.*, **1**, 3500.

89 Kanai, Y. and Grossman, J.C. (2007) *Nano Lett.*, **7**, 1967.
90 Buxton, G.A. and Clarke, N. (2006) *Phys. Rev. B*, **74**, 085207.
91 Deibel, C., Wagenpfahl, A., and Dyakonov, V. (2008) *Phys. Status Solidi (RRL)*, **2**, 175.
92 Koster, L.J.A., Smits, E.C.P., Mihailetchi, V.D., and Blom, P.W.M. (2005) *Phys. Rev. B*, **72**, 085205.
93 Mihailetchi, V.D., Koster, L.J.A., Hummelen, J.C., and Blom, P.W.M. (2004) *Phys. Rev. Lett.*, **93**, 216601.
94 Häusermann, R. (2009) *J. Appl. Phys.*, **106**, 104507.
95 Etzold, F., Howard, I.A., Mauer, R., Meister, M., Kim, T.-D., Lee, K.-S., Baek, N.S., and Laquai, F. (2011) *J. Am. Chem. Soc.*, **133**, 9469.

5
Percolation Theory and Its Application in Electrically Conducting Materials
Isaac Balberg

5.1
Introduction

Percolation theory is concerned with the connectivity in inhomogeneous systems. In particular, the connectivity requirement of a flow through such systems is intimately associated with their percolation properties. In this chapter, we will concentrate on the electrical conductivity in composite materials that are made of a disordered mixture of insulating and conducting phases. However, even the latter subject is a very wide field of research that is much beyond the scope of this chapter. Correspondingly, we limit ourselves here to the direct current (dc) properties of these systems in light of percolation theory. In fact, we limit ourselves even further, by considering only the dependence of the electrical conductivity, σ, on the partial volume content of the conducting phase, x, in those composites. This $\sigma(x)$ dependence is the generally accepted fundamental characterization property of electrically conducting composites and it provides the very basic physics of their electrical behavior. Since the conducting phase in the composite is usually made of individual particles, we divide the numerous composites into two basic types: those in which the increase of x brings about the coalescence of the conducting particles and those where the particles keep their individual character for any x. The conspicuous examples of the first type are the granular metals or granular semiconductors, at relatively high x values, while the conspicuous examples of the second type are carbon or metal particles that are embedded in a polymer matrix.

In this chapter, the emphasis will be then on trying to understand the $\sigma(x)$ dependence as observed in those composites in view of the recent extensions of classical lattice percolation theory to systems in the continuum. The basic characteristic of $\sigma(x)$ that is common to all composite materials is a sharp rise in σ as x increases. This rise is followed by a monotonic moderate rise in the $\sigma(x)$ dependence. In order to get the desired understanding of this behavior, the first part of this chapter will provide the background to percolation theory [1–4] and the tools required for the discussion of the $\sigma(x)$ dependence, while the other part will be devoted to a comparison of the experimental observations of ours and others to the expectations that follow percolation theory. In particular, we will try to understand the multiple sharp

Semiconducting Polymer Composites: Principles, Morphologies, Properties and Applications, First Edition.
Edited by Xiaoniu Yang.
© 2012 Wiley-VCH Verlag GmbH & Co. KGaA. Published 2012 by Wiley-VCH Verlag GmbH & Co. KGaA.

rises at some "critical" x values and the particular mathematical forms of the $\sigma(x)$ dependence. The relation between those values and forms, and the structure of the composites, will be then the main topic of this chapter. On the other hand, no broader description of the electrical conductivity and other electrical properties as well as the practical applications [5] of these composites will be attempted in this chapter. We will not include in this chapter the details of the fabrication of the systems studied, the experimental procedure used in the measurements, and the numerical methods used in various studies, but some references to those will be given. Also, in view of the limited scope of this chapter, we will not critically review specific works since our focus will be on the basic features of the published results and the overall physical principles and trends that can be interpreted in light of percolation theory and its extensions. In passing we note that there are previous excellent reviews on percolation theory in general [1–4] and on the electrical properties of composites in particular [5–10]. Correspondingly, the emphasis here will be then on the new understanding of the $\sigma(x)$ dependences that we developed recently [11] and its application to particular composite systems that became of great interest in recent years.

The structure of this chapter is as follows. In Section 5.2, we provide the main concepts of percolation theory and in particular those relevant to the $\sigma(x)$ behavior. This will be done by using the well-understood framework of lattices. Following the fact that the composite materials of concern are disordered systems in the three-dimensional (3D) continuum, we present in Section 5.3 the extensions of lattice percolation theory to ensemble of conducting objects that are embedded in the insulating continuum. Since the interpretation of the results obtained in many composites requires also the consideration of the tunneling between pairs of particles, the introduction of that mechanism within the framework of percolation theory is elaborated on in Section 5.4. Having the above background, we turn in Section 5.5 to a brief description of the structures of the two major types of composite materials and the expected implication of these structures concerning the qualitative features of their $\sigma(x)$ dependences. In Section 5.6, we provide an overview of the $\sigma(x)$ dependence in composite materials and interpret them in light of the background given in Sections 2–4. Then, in Section 5.7, we summarize our understanding of the $\sigma(x)$ dependences in composites and hint at the still open questions.

5.2
Lattice Percolation

To introduce the basic concepts of percolation theory [1–3], let us consider the very simple model of a two-dimensional square lattice as illustrated in Figure 5.1 [11, 12]. In this system, we assume a probability, p, that a lattice site is occupied and we denote the occupied sites by closed circles. In the present lattice case, two sites are considered connected only if they are nearest neighbors and if they are both occupied. We define a (connected) cluster as an ensemble of occupied sites such that each of them is connected to at least another occupied site of the ensemble. In

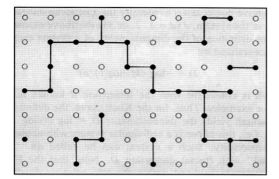

Figure 5.1 A two-dimensional illustration of a portion of an infinite lattice. The open circles represent unoccupied lattice sites and the closed circles represent occupied sites. The segments connecting two nearest neighbor occupied sites represent a local connection, or a bond. In this illustration, there are finite clusters of connected sites (of size 1, 2, and 4) and a "spanning" cluster that connects two opposite edges of the system. In the infinite size system, the latter is known as the "infinite" or the "percolation" cluster. (after Ref. [11])

Figure 5.1, we see a cluster that connects opposite edges of the system. The latter is called the spanning cluster. In percolation theory, we are concerned mainly with "infinite" systems and thus we call the spanning cluster the percolation or the infinite cluster. The illustration given in Figure 5.1 reflects here a finite (small) portion of the infinite system but it has to be remembered that quantitative characterizations of the various properties are considered meaningful only at the "infinite" system limit.

From Figure 5.1 it is easy to appreciate that the numbers and sizes of the clusters will grow with the site occupation probability, p [1]. The central consequence, however, is that there will be a value of p that is known in the literature as the critical probability, p_c, such that only for $p \geq p_c$ there is a spanning cluster. This p_c is known as the *percolation threshold* of the lattice [1–3]. The behavior of the various geometrical or physical properties around p_c as a function of $p - p_c$ is known as the *critical behavior*. The origin of that name followed the rigorous proof that the behavior of various properties of systems such as that of Figure 5.1 can be mapped onto a phase transition [13]. Correspondingly, these properties are well described by power laws of the proximity to the threshold $p - p_c$. For example, the average "geometrical size" of the finite clusters, i.e., their effective (e.g., gyration) radius, λ, that defines then the length scale of the "percolating" system is given by [1]

$$\lambda \propto |p - p_c|^{-\nu}. \tag{5.1}$$

The critical exponent ν depends only on the dimensionality D, and that feature is known as universality. The well-known values of ν in percolation theory are 1.33 for $D = 2$ and 0.88 for $D = 3$.

Turning to the dynamic properties associated with the system's connectivity, in general [1, 12, 13], and the electrical [1, 14] or other flows [4, 12], in particular, we consider here only the most studied property, that is, the global electrical conductance of the system that is also the main property of interest in composite materials [6–11]. Let us consider then the system shown in Figure 5.1 from its possible electrical conductivity point of view. We assume that the "bonds" (between two connected sites) in the system are resistor bars so that there can be electrical conduction through the system only if there is a spanning cluster, such as the one in the figure. We call the ensemble of bonds that conduct current the *backbone* of the percolation cluster, while the other bonds of the percolation cluster are known as "dead ends." If we have two sites that are connected in a lattice volume smaller than λ^D only by a single bond, we refer to the corresponding bond as a "singly connected bond." The parts of the backbone that are not singly connected bonds are known as its "blobs". The backbone of the percolation cluster in the infinite sample can be envisioned as consisting of a topologically equivalent square or a cubic network of "links" of length λ that intersect at nodes. In this *links–nodes–blobs* (LNB) model [1, 6], the average resistance of a link is R_λ. If the sample's size is L, we have on the average a series of L/λ links that connect opposite edges of the sample and there are parallel $(L/\lambda)^{D-1}$ such series. Hence, the resistance of the whole backbone network, R_L, will be given by [1, 6]

$$R_L = R_\lambda (\lambda/L)^{D-2}. \tag{5.2}$$

We are left then "only" with the need to estimate the value of R_λ. The lower bound of R_λ is simply $R_\lambda = r_0 L_1$, where r_0 is the value of the single bond resistor and L_1 is the number of singly connected bonds within a link. A simple argument can show that $L_1 \propto -(p-p_c)^{-1}$ [6]. If we include the blob resistors, we expect that $R_\lambda > r_0 L_1$. For our present purpose, it is enough, however, to note that $R_\lambda \propto (p-p_c)^{-\zeta}$, where ζ, which represents the diminishing number of blobs as the percolation threshold is approached [15], has the values $\zeta \approx 1.3$ for $D=2$ and $\zeta \approx 1.1$ for $D=3$ [3, 15]. All the above results can be summarized now by the critical behavior of the global specific conductivity of a percolation lattice ($\sigma \propto 1/R_L$):

$$\sigma = \sigma_{0p}(p-p_c)^t, \tag{5.3}$$

where σ_{0p} is some characteristic conductivity of the system and t is the critical exponent of the conductivity that is given, in view of the above, by

$$t = (D-2)\nu + \zeta. \tag{5.4}$$

Following the above, the corresponding universal values of t, t_{un}, are ≈ 1.3 for $D=2$ and ≈ 2 for $D=3$.

The question that is expected to be relevant to systems (such as the composites) in the continuum is: What happens if the resistors in the system do not all have the same r_0 value? Since the L_1 resistors in the LNB model are connected in series, the link's resistance R_λ can be represented by $\langle r_0 \rangle L_1$, where $\langle r_0 \rangle$ is the average of the individual resistors in the system and thus in the link. This simple average has a meaning only if we assume (as we will do throughout) that there is no correlation between the position of the resistors in the system and their values. Now we note that each group of p_c resistors that is chosen randomly provides a percolation spanning cluster and this applies in particular to the p_c resistors of the lowest values in

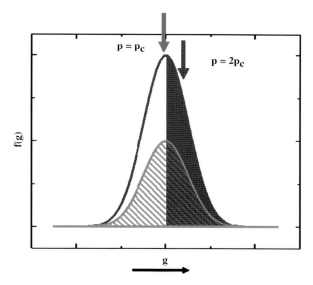

Figure 5.2 An illustration of a Gaussian distribution of the local conductances, g, in a lattice for which the requirement for percolation is that the probability p for a site occupation is p_c. The average value of g that is needed for the fulfillment of this condition is indicated by the first arrow, while if $p > p_c$ this average will increase, as indicated by the second arrow.

the system. Obviously, this cluster will provide the lowest possible resistance of the system. To get a feeling for the effect of the distribution on $\langle r_0 \rangle$, let us compare the scenario where $p = p_c$ with the scenario where $p = 2p_c$ in the case where the values of the local conductances $g \equiv (1/r_0)$ have a Gaussian distribution. As indicated in Figure 5.2, the average conductance in the first scenario is at the center of the distribution while the effective average conductance in the second scenario will be higher. In other words, the average resistor ($\langle r_0 \rangle = \langle 1/g \rangle$) that will control the global conductance will be smaller the higher the value of p. If the distribution of the values of the conductances in the system, $f(g)$, has a strong enough divergence toward $g \to 0$, the value of the average resistor may also diverge as $p \to p_c$. In that case, this average can take the form $\langle r_0 \rangle \propto (p - p_c)^{-u}$ and thus the effective critical exponent will become $t = t_{un} + u$. We refer then to the critical behavior of the conductivity as a nonuniversal one [14, 16]. For example, for $f(g) = [(1 - \alpha)]g^{-\alpha}$, it can be easily shown [4, 14, 16] that (since $f(g)$ is normalizable for $\alpha < 1$) for the range $1 > \alpha > 0$ we will get

$$t = t_{un} + \alpha/(1 - \alpha). \tag{5.5}$$

In our above simple square lattice, we have considered only nearest neighbors and the bonding between them. Let us consider now the case where we also include occupied second nearest neighbors [11]. In that case, under the same site occupation probability p, where "farther" (second nearest) neighbors are also considered, the percolation onset will be achieved at a p for which no percolation existed in the nearest neighbors-only case. Let us define now the critical p_c values for the two

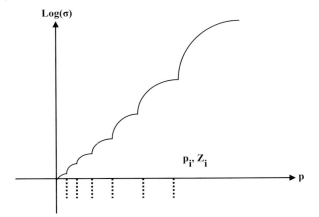

Figure 5.3 A qualitative presentation of the expected staircase dependence of the conductivity on p when the intersite conduction is assumed to be by tunneling. The dotted segments indicate the percolation thresholds p_i when "all" the corresponding near neighbors (up to the ith one), Z_i, are considered.

possible scenarios, by p_{1c} for the nearest neighbors-only case and p_{2c} ($<p_{1c}$) for the case that includes also the next nearest neighbors.

Turning to the implications of the above geometrical percolation behavior on the electrical conductance of the system, let us assume that larger local bond resistors in the network are associated with the next nearest neighbor. We will have then that the global conductance beyond that achieved by both the near and the next nearest neighbors at p_{2c} will exhibit a jump at $p = p_{1c}$ since the system will become connected then (in practice) only by the (dominant p_{1c}) higher conductances of the nearest neighbor resistors. In general, this effect will become more pronounced the stronger the decrease of the conductance g values of the farther neighbors. In particular, when the latter decrease exponentially (as is the case of tunneling, see Section 5.4), a "staircase" of the $\sigma(p)$ dependence such as the one shown in Figure 5.3 is expected. While in each stair we expect a corresponding universal behavior, in cases where we have a "blurring" of the position of the lattice points we may find deviation from universality. The wider the distribution, the more emphasized the nonuniversal-like behavior is expected. In passing we note also that by "virtually" changing the ratio between the number of the first nearest neighbors z_{NN} and the number of next nearest neighbor sites z_{NNN}, we will change the second (p_{2c}) percolation threshold so that if z_{NNN} is made somehow to decrease, p_{2c} will increase. This is a significant extension of the above model to cases where for the same p we change ("externally") the relative concentration of the "potentially participating" bonds in the system.

In our above discussion, the lattice was assumed to be made of mathematical dots and segments. In principle, as shown in Figure 5.4a, we can occupy the lattice sites by conducting spheres with a volume v each, such that two nearest neighbor sites can be considered bonded geometrically and electrically (at the point where

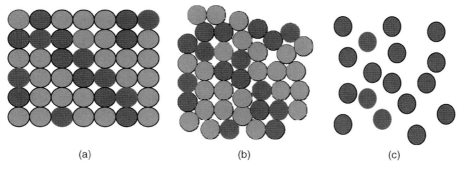

Figure 5.4 A small portion of a square lattice that is comprised of "conducting" (dark gray) and insulating (light gray) circles is shown in panel (a). The local connectivity is defined here by the "single point" contact of two dark gray circles, and the "global connectivity" is determined by the spanning cluster of touching dark gray circles. The same ensemble of circles is shown in panel (b) but in the configuration of the random close packing of circles. It is seen there that the cluster structure and the global connectivity of the dark gray circles are very similar to those in the lattice configuration. A system of conducting dark gray circles, where there are no "insulating" light gray spheres, is shown in panel (c). Such a system with the same amount of "conducting" circles, generated by a random implantation of circles, does not seem to have a local and thus a global connectivity. It is important to note that in panel (c) the "particles" (the dark gray circles) are uniformly dispersed while in panel (b) they form isolated "blobs" (or isolated compact clusters). The competition between the dispersion and the blob formation in real continuum systems influences the critical concentration needed for the onset of percolation. (after Ref. [11])

there is a contact between them) if they are both occupied. In that case (Figure 5.4a), we replaced the site occupation probability by the total fractional occupied volume $x = \nu N = pf$, where N is the concentration of spheres in a unit cube and f is the fill coefficient of the lattice. Correspondingly, this lattice-occupied-by-spheres model of Scher and Zalen (S&Z) [2, 17] will enable to replace Eq. (5.3) by

$$\sigma \propto (x - x_c)^t. \tag{5.6}$$

In approaching the discussion on the continuum systems of composites, we illustrate in Figure 5.4 two steps that are associated with shifting from the lattice to the continuum. In the first, which leads to the configuration of Figure 5.4b, we simply take the conducting (dark gray) and insulating (light gray) spheres of Figure 5.4a and rearrange them randomly (say, by dropping spheres, with the desired probability, "gravitationally" from the top [18]). The very important finding of S&Z was that x_c in that case is the same as in lattices, thus concluding that the connectivity in the two cases appears to be practically the same [2, 17]. However, if we consider only randomly dispersed (dark gray) spheres in a continuous insulating matrix, we get the configuration that is shown in Figure 5.4c. In that case, it is clear that global connectivity will be obtained here for a much higher content of dark gray spheres than in the cases of the mixed spheres [14]. This is a rather important point (and generally an overlooked one [19]) since the usual continuum models simply assume the uniform dispersion case (see Section 5.3). The corresponding realization will be of prime importance when we consider the electrical properties of composites in Section 5.6.

5.3
Continuum Percolation

In the previous section, we have briefly reviewed the basics of the lattice models of percolation where the systems have sites (or bonds) that are occupied or empty. In continuum percolation, the systems are composed of objects (or members) that are *randomly placed in space*. These objects may be of *various sizes and shapes*. If the latter are nonisotropic, one also considers the *distribution of their orientations*. Correspondingly, the values of the physical parameters that determine the bonding between two objects may vary from bond to bond, depending, say, on the local geometry and/or properties of the bond.

While the use of the fractional volume x as a parameter that is equivalent in a way to p in lattices was an important step in launching a theory for continuum percolation, the x_c values of S&Z were misused later [19]. For example, the systems described by Figure 5.4a and b are by no means general or prototypes of "real" continuum system [11]. In order to account, however, for the situation of Figure 5.4c, let us consider the case of permeable spheres of volume τ for which it can be easily shown [20] that the critical fractional volume of space that has to be occupied by randomly dispersed permeable spheres is given by $\phi_c = 1 - \exp(-B_c/2^D)$, where B_c is the critical concentration of bonds per sphere, which is given then by $N_c \tau 2^D$, where N_c is the critical concentration for the onset of percolation (defined by an infinite cluster of partially overlapping spheres). It can be easily appreciated then that as $N \to N_c$ or $B \to B_c$, we have [14, 21]

$$\phi - \phi_c \propto B - B_c \propto N - N_c. \tag{5.7}$$

In the above example of permeable spheres, we can define then an "excluded volume" of a sphere (which is actually the "interaction volume") as $V_{ex} = \tau 2^D$ of those spheres and thus replace the above equation for ϕ_c by $\phi_c = 1 - \exp(-B_c \tau/V_{ex})$. The system of two permeable spheres and their mutual excluded volume is illustrated in Figure 5.5a. Following that we can generalize the case of spheres to cases of other permeable object shapes by simply writing it for any permeable object [21, 22] in which τ is the volume of the object and V_{ex} is its excluded volume. Hence, since the values of B_c and τ are well known, once we can calculate V_{ex} we would have the critical volume content ϕ_c. A conspicuous example for a nonspherical object is the volume of "soft" capped cylinder of length L and radius b for which $\tau = (4\pi/3)b^3 + \pi b^2 L$ and [21]

$$\langle V_{ex} \rangle = (32\pi/3)b^3 + 8b^2 L + 4bL^2 \langle \sin \theta \rangle, \tag{5.8}$$

where $\langle \sin \theta \rangle$ is the average over the orientation distribution of the cylinders [23]. The "nontrivial" message of the latter result is that $\langle V_{ex} \rangle$ is not proportional to τ and that in the large aspect ratio, L/b, limit we find that $\tau/\langle V_{ex} \rangle \propto b/L$, while for the

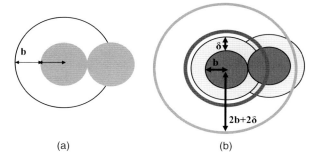

Figure 5.5 The area (volume) of adjacent permeable circles (spheres) of radius b and the corresponding attached excluded circles of radius $2b$ (a). The area of circles of a hard core radius b and soft shell δ, their corresponding total excluded area, a circle of radius $2(b+\delta)$, and the excluded area of the hard core, a circle of radius $2b$ (b). The net excluded of this hard core–soft shell circle has an area of $4\pi\delta(2b+\delta)$.

small aspect ratio of 1 ($L = 0$) we recover the known $\tau/V_{ex} = 1/8$ result for spheres. Let us assume now (in view of the topological invariance) that these cylinders are actually topologically "stretched spheres" so that the value of B_c is the same as that of the system of permeable spheres. Correspondingly, we find that the critical capped cylinders number N_c is given by $B_c = N_c \langle V_{ex} \rangle$.

Another object that is very helpful for the description of various systems is the disc with a radius b and thickness $š$ [24, 25] that has a "soft" object volume of $\tau = \pi b^2 š$, while its excluded volume (in the isotropic case) is given by $V_{ex} = \pi^2 b^3$. It is apparent then that for this disc the much smaller than 1 aspect ratio ($š/b$) yields that $\tau/V_{ex} \propto š/b$ explaining the very small ϕ_c values observed in sedimentary rocks, in systems of geological cracks [25], and in cellular composites [26].

In order to approach our interest in composite materials, we also examine the above results for systems that have a hard core and a soft shell with a width δ, such that two particles are considered connected if there is some overlap between their soft shells, but there is not any overlap between their hard cores [27–29]. Such a system of spheres is illustrated in Figure 5.5b. In passing we note that this type of system represents very well systems of microemulsions [27, 28]. In the figure, we illustrate the excluded volume of hard cores, the excluded volume of the whole objects, and the net excluded volume of the system (i.e., the excluded volume of the whole object minus that of the hard cores, which is ring-like in 2D and spherical shell-like in 3D). Using then the above $B_c = \langle V_{ex} \rangle N_c$ and $x_c = v N_c$ relations, we get that the total critical volume of the hard core spheres x_c is given by

$$x_c [8(3b^2\delta + 3b\delta^2 + \delta^3)] = B_c b^3, \tag{5.9}$$

From this result we have that in the dilute limit ($\delta \gg b$), $\delta/b \approx (B_c/x_c)^{1/3}/2$, while for the other extreme, dense, case ($b \gg \delta$) we have $\delta/b \approx B_c/(24 x_c)$.

Following the above reasoning, let us turn to discs of radius b and thickness $š$ such that $b \gg š$. In that case (to first order), the total "potential" volume for overlap is $\pi^2(b+\delta)^3$ but then (as we saw above) the excluded volume due to the hard cores is $\pi^2(b)^3$ and thus the net possible overlap volume is $\pi^2(3b^2\delta + 3b\delta^2 + \delta^3)$, while the volume of the disc is $v = \pi b^2 t$. We can continue now as in the case of spheres finding that $\pi^2(3b^2\delta + 3b\delta^2 + \delta^3)N_c = B_c$. In the $\delta \gg b$ limit, we have then the dilute particle dispersion case that yields $\delta^3/b^2 t \approx B_c/(\pi x_c)$. For other extreme (i.e., $b \gg \delta$) that is more interesting (for the tunneling problem considerations in Section 5.4), we have $x_c \approx (\check{s}/\delta)(B_c/3\pi)$, that is,

$$\delta/t \approx B_c/(3\pi x_c), \tag{5.10}$$

which is somewhat surprising being (to first order) independent of t/b.

Similar considerations apply to the capped cylinder that we considered above. This system is interesting of course in the $L \gg b$ limit and as such the net excluded volume of this cylinder is $\pi\delta L^2$ while the hard core volume of the cylinder is $\pi L b^2$. This yields that $x_c \approx B_c(b/\delta)(b/L)$, that is,

$$(\delta/b) \approx B_c(b/L)/(x_c). \tag{5.11}$$

As to be discussed in Section 5.4, this result yields quite a general prediction for composites of elongated particles that are embedded in a polymer.

5.4
Percolation Behavior When the Interparticle Conduction Is by Tunneling

Most probably, the best known approach to the conduction by tunneling in random systems is the one used in the theories of hopping [6, 30]. However, as discussed below, this approach does not predict a percolation-like behavior as given by Eq. (5.6) with which we were concerned above. One starts the consideration of the hopping model by recalling the exponential decrease of the interparticle conductance g with the distance r, so that

$$g = g_0 \exp[-2(r/\xi)], \tag{5.12}$$

where ξ is the localization (or the wavefunction decay) length of the electron out of the particle to which it belongs. In the model of nearest neighbor hopping, that is, where all sites are at the same energy or the system is at a high enough temperature, one assumes a random uniform distribution of sites (or particles of nearly zero volume) with a density N [6]. The onset of a globally connected network, of the smallest possible resistors in the system with a concentration of N particles, requires that (on the average) the smallest r to be considered is r_c, which is defined (see Section 5.3) by [6, 11, 31]

$$(32\pi/3)(r_c^3 - b^3)N = B_c. \tag{5.13}$$

We define then the average critical local conductance by $g_c = g(r_c)$ in Eq. (5.12). For simplicity, let us consider the case that is commonly discussed in the literature [6], that is, the $b/r_c \to 0$ limit. One also assumes then that $r_c \gg \xi$ so that, practically, for all the interparticle distances $r > r_c$, $g = 0$ (blocked conductance), while for all $r < r_c$ distances one assumes that $g = \infty$ (shorts). One can show, as expected, that the global conductivity σ is simply proportional to g_c [32, 33]. Hence, following Eqs. (5.12) and (5.13), and using $B_c = 2.8$ [20, 31] for the dilute (or small b/r_c) case, the dependence of σ on N will be given by [6]

$$\sigma \propto g_c \propto \exp[-1.73(2/\xi)(N^{-1/3})]. \tag{5.14}$$

We have then that the hopping model, while frequently called "the percolation model" [6], is essentially a *percolation threshold model* that is not associated a priori with the critical behavior of the conductivity as in percolation theory (Eq. (5.6)). In passing we further note that in the limiting case of a high particle density (i.e., $r_c - 2b \ll 2b$) but still assuming a uniform N (see Section 5.3), the combination of Eqs. (5.12) and (5.13) yields that $\log(\sigma)$ varies as N^{-1} rather than the above $\log(\sigma) \propto N^{-1/3}$ result of the dilute N limit [11] that is given by Eq. (5.14). This yields that the conductance of the system, as a function of x, will be given by [11]

$$\sigma(x) = \sigma_{0h}\exp[-Ax^{-\beta}], \tag{5.15}$$

where the value of A is easily derived for composites [11] from the basic hopping theory (as was the case for Eq. (5.14) [6]). We note here then that generally $0 < \beta < 1$. Since the $\beta = 1/3$ case is the more commonly considered dependence, we will relate mostly to this dependence in the comparisons that we will make with the experimental data in Section 5.6.

Let us consider now the problem of a system with interparticle tunneling from the percolation point of view and assume that while the conduction between particles is by tunneling (that in principle has an infinite range) there is still "somehow" a well-defined threshold. This assumption (which is in contrast with the above hopping model) will be discussed further below. The next step is then the consideration of the critical behavior that will result from the distribution of the g values, by following the average local resistance in the system. Using now the simple relation between the local conductance distribution function $f(g) = h(r)(dr/dg)$ and the functions $g(r)$ (Eq. (5.12)) and considering an $h(r)$ distribution of the form $[1/2(\tilde{a} - b)]$ $\exp[-(r - 2b)/(2\tilde{a} - 2b)]$ [14, 34, 35], where $\tilde{a}$ is given by $\tilde{a} = aB_c^{2/3}$ and $(4\pi/3)a^3N = 1$ [4], we get that $f(g)$ has the form $g^{-\alpha}$, where $\alpha = 1 - \xi/4(\tilde{a} - b)$. Hence, from Eq. (5.11) and the fact that $0 < \alpha < 1$, we have

$$t - t_{un} = 4(\tilde{a} - b)/\xi - 1. \tag{5.16}$$

While a more accurate expression has been given for $t - t_{un}$ [36], we will consider here this more simple transparent equation in what follows. The fact that both Eqs. (5.14) and (5.16) depend on the interparticle distances suggests that when σ varies with N, the dominant effect in the case of tunneling, when there is an appropriate interparticle distance distribution, will be the variation of the local conductances rather than the connectivity of the system.

In the above description of the $\sigma(x)$ behavior, when tunneling is the mechanism of interparticle conduction we got then two results. One is the hopping result as given by Eq. (5.15) and the other is the percolation result given by Eq. (5.6), but with a t value that is given by Eq. (5.16). Since the physics is essentially the same for hopping and percolation in the large t case, we would expect that *the two behaviors will merge* in the $(2\tilde{a} - 2b)/\xi \gg 1$ limit. Indeed, the above two presentations turn to yield fits of similar quality in numerical and experimental results (Balberg, I., unpublished.).

In the above discussion, we overlooked the meaning of x_c in Eq. (5.6). However, the observed x_c's in simulations and experiments may or may not have some meaning [37]. The conspicuous case where a finite x_c (>0) has meaning is in a system where we have a staircase behavior and the available data are associated with a given single stair (see Section 5.2). In that case, x_c represents the critical concentration that is needed for the onset of percolation by the ensemble of given equal (or nearly equal) value resistors [11, 35, 38]. In the tunneling problem when more than a single stair is revealed, these are the dominant (or representative) local conductances of the corresponding stairs. One can easily see that if a few stairs are present but are not clearly separated and we try to fit the entire data to a single simple Eq. (5.6) like dependence, we will get a larger t and a smaller x_c, the larger the number of stairs that are included. In that case, x_c does not have a well-defined ("intrinsic") meaning.

5.5
The Structure of Composite Materials

The subject of this chapter is the electrical conduction in inhomogeneous media with an emphasis on the connectivity aspect of the conduction in solid systems. Basically, we consider media that are made of two phases, a conducting one and an insulating one. In particular, we are concerned with the dependence of the conductivity σ on the content of the conducting phase x. For the purpose of this chapter, we call such systems composite materials and we limit ourselves to the two most abundant types of such composites. In the first type of system two particles coalesce when they touch, while in the other type they cannot be considered touching and thus the conduction between them takes place by tunneling. The prototype of the first type is granular metals that are made of metal grains (in which the typical conductivity is on the order of $10^5 \, \Omega^{-1} \, cm^{-1}$ and a dielectric matrix (typically glass or alumina) where the conductivity is smaller than $10^{-8} \, \Omega^{-1} \, cm^{-1}$ [39]). The other prototype of composites is that of carbon black (CB) particles (with a typical conductivity of $10^2 \, \Omega^{-1} \, cm^{-1}$) that are embedded in a polymer (with a typical conductivity of $10^{-16} \, \Omega^{-1} \, cm^{-1}$ [40]).

Starting with granular materials we show in Figure 5.6 high-resolution transmission electron micrographs (HRTEMs) of a semiconductor–insulator composite where the semiconductor grains are silicon nanocrystallites that are embedded in glass [41]. For the low volume content of the conducting phase of the silicon, x, the particles are isolated spheres (Figure 5.6a). The increase of x yields the touching between some of them and the corresponding distortion of their shapes. For the much higher Si contents, almost all the particles have touching neighbors and a large cluster of such neighbors forms. In this semiconductor–glass composite, as in its metallic counterparts [39], the conducting particles change their

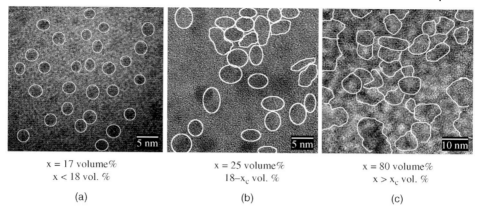

Figure 5.6 HRTEM images of our cosputtered Si-SiO$_2$ composites for three Si phase contents: (a) 17 vol.%, (b) 25 vol.%, and (c) 80 vol.%. The crystallites, detected by the Si lattice fringes, are marked by their borders. Note that for $x <$ 18 vol.% the crystallites are geometrically isolated, nearly perfect, spheres and for $x =$ 80 vol.% they take the shapes needed for their high-density accommodation. Also indicated in the figures are the x regimes for which each image is typical. (after Ref. [41])

concentration, size, and shape with the increase of x. From the electrical conductivity point of view, it was found that two touching particles (marked by their white perimeters in Figure 5.6) in both the granular metals [42] and the granular semiconductors [41] behave, for all practical purposes, as a single electrical unit. We see then that in general, with increasing x, the system goes from that of a mixture of individually dispersed spherical nanoparticles to that of single isolated particles and clusters of touching particles (Figure 5.6b). Then, as seen in Figure 5.6c for high x values, long chains of touching particles form. Considering the fact that the HRTEM image is two dimensional but the investigated system is three dimensional, it is apparent that for the high x regime there is an infinite cluster of touching particles that we call below the percolation–coalescence cluster. For such a system, we expect that there will be a sharp rise of the conductivity at the x value at which such an infinite cluster forms. In principle then, the system can be discussed in light of continuum percolation. On the other hand, for the smaller $x < x_c$ values, the conduction can be assumed to take place in a system such as the one shown in Figure 5.6a. This is of course provided only by tunneling. Hence, this system (of $x < x_c$) can be discussed, as done in Section 5.6, in light of the analysis of Section 5.4.

The prototypes of systems where the conducting particles that are embedded in the composite do not coalesce and there is an insulating phase of variable thickness between each two of them such that their shape and size are independent of x [41] are the composites of CB conducting particles that are embedded in a polymer matrix. In Figure 5.7, we illustrate then some typical shapes of CB particles and give their typical sizes recalling that the conductivity of the "bulk" CB material is on the order of $10^2 \, \Omega^{-1} \, cm^{-1}$ and that of the polymer many orders of magnitude

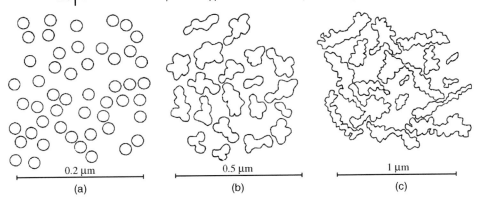

Figure 5.7 Schematic descriptions of the carbon blacks that are dispersed in the corresponding polymer composites: (a) low structure black, (b) intermediate structure black, and (c) high structure black. The scales are, respectively, typical of N990, Raven-430, and XC-72 blacks. (after Ref. [34])

smaller [40]. For our purposes, we will ignore the microstructure of the individual CB particles and look at them as capped cylinders with a variable (1–10) aspect ratio. In passing we add that similar considerations apply to composites of metal particles that are embedded in polymers; that is, there is no coalescence of the particles and some polymer layer always exists between adjacent particles. On the other hand, in contrast with the CB particles the latter particles have a conductivity that approaches the much higher bulk value of metals rather than the conductivity value of graphite.

In view of the recent developments in the study of other carbon particle–polymer composites, let us mention three conspicuous types of them. The first is the carbon nanotube (CNT)–polymer composites for which an excellent review is available [43]. These nanotubes have typical (simple graphite-based cylinder-like) length of up to 100 μm while their diameter is in the range of 2–50 nm. Another system that is made of elongated particles is the carbon nanofibers (CNFs) [44], which is similar but does not has as extreme aspect ratio as the CNTs; that is, it has similar length but a diameter of 50–200 nm. The other type of carbon–polymer-based composites involves the extreme, very small ($\ll 1$) aspect ratio graphene particles. These can be described as discs with a radius on the order of 1 μm and a thickness of a couple of nm [45, 46]. In addition, there are various other graphite sheet–polymer composites that have extremely low ($\ll 1$) aspect ratios, and those are reminiscent of the graphene–polymer composites [47, 48]. From the percolation point of view, the similar property of the above three types (prolate and oblate) of particles is that in the composites their percolation thresholds are strongly dependent on the very extreme, high or low, aspect ratios as to be expected from the considerations given in Section 5.3.

In this chapter, in general, and in this and the following section, in particular, we do not consider properties of the composites other than their $\sigma(x)$ dependences. Also we do not describe their fabrication procedures and their commercial applications for which there is ample literature. Following, however, the basic structures of these composites, we do describe, in the next section, the experimentally observed σ(x) dependences in them as well as our trial to understand these dependences in light of percolation theory.

5.6 The Observations and Interpretations of the σ(x) Dependence in Composite Materials

In this section, we describe the basic features of the $\sigma(x)$ dependences found experimentally in conductor–insulator composites and we try to explain them in view of the expectations that were outlined in Sections 5.3 and 5.4. Considering the fact that the two basic features of the $\sigma(x)$ dependence are the abrupt "jumps" of $\sigma(x)$ at some x_c values and the particular mathematical $\sigma(x)$ dependence for $x > x_c$, we start with the typical x values of the sharp rise in the conductivity that is commonly called (though as we saw above not always well defined) the percolation threshold, x_c. The other part of the discussion will be concerned with the fundamental $\sigma(x)$ dependence and its relation to the predictions of percolation and hopping theories. Since in the literature people quantify the x and x_c values in terms of vol.% and fractional volume contents (i.e., vol.%/100), we will in this section use either of these quantities as required by the context of the discussion. In particular, in the comparison of experimental data with the relations (5.6) and (5.15), we will assume that x stands for the fractional volume content. Also, we note in passing that while the volume is the relevant physical parameter, in many cases the wt% is the much easier to determine parameter. Correspondingly, we will use here the vol.% noting that the difference in the value of vol.% and that of the wt% does not alter the discussion.

5.6.1 The Percolation Threshold

Starting with granular metals and concentrating on the coalescence–percolation transition, it is found that, as shown in Figure 5.8, in granular metals and granular semiconductors the coalescence-induced x_c values are in the range of 30–50 vol.% [35, 39, 41]. The basic reason that these x_c values are considerably larger than those of S&Z of, say, 16 vol.% is well understood by the fact that contacts between adjacent particles will be established in those composites only for higher metallic contents than those "automatically" provided in the S&Z model (see Figure 5.4a and b) and thus a continuous "contacted" network will be established only upon approaching the close random packing of 67% [2]. The latter explanation [11, 19] is well confirmed in Figure 5.6 where it is shown that a continuous network in a "granular semiconductor" is established only beyond $x = 25$ vol.%, which is higher than the 16 vol.% found for the S&Z model.

While annealing the granular metals brings about the formation of larger particles (or "aggregates" (Balberg, I., unpublished.), [39]) that amounts to "shifting"

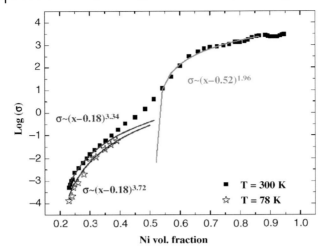

Figure 5.8 The measured $\sigma(x)$ dependence of a Ni-SiO$_2$ granular metal composite. The curves represent best fits of the data to the coalescence–percolation and the tunneling–percolation regimes, at two temperatures. The critical concentrations and exponents that were derived are given in the figure. (after Ref. [35])

from Figure 5.6c to Figure 5.6b, the stirring of the melt of which the composite is made in CNTs is found to "decompose" the aggregates and disperse the individual particles [43]. Hence, the effect of stirring is as shifting in the opposite direction in Figure 5.4. Still both effects (while seemingly opposite) yield an increase in the value of x_c. At first, this is surprising, but then, in light of the finding for microemulsion-like systems [27, 28, 49], we note that, for the same x, aggregation of particles will initially reduce the percolation threshold (by providing the "touching" necessary for global connectivity) but too much of aggregation, at the expense of dispersion, will yield large isolated aggregates and thus their separation. The latter effect will increase then the percolation threshold [44]. In fact, an initial consideration of that problem was given previously [50] but has not been followed. The important point is, however, that in real composites the determination of x_c from Eqs. (5.9)–(5.11) is indeed limited to cases of a uniform dispersion and thus can serve in practice only as an ideal system guideline.

In carbon black–polymer composites, the value of x_c varies from about 40 vol.% in the case of spherical CB (e.g., N990) particles [51] to about 5 vol.% (e.g., for XC-72 [51] and Ketjenblack [52]) following the increase of the aspect ratio of the particles to the order of 10 [34]. The latter result is in excellent agreement with the excluded volume theory that predicts such a dependence of x_c on the aspect ratio (see Eq. (5.11)) if we identify the shell thickness δ with a critical average distance, δ_c, that the tunneling electrons should cross [29]. This trend is continued when we consider the x_c values for CNTs [43], where x_c values on the order of 1 vol.% are observed for aspects ratios on the order of 10^2, and 0.1 vol.% for aspect ratios on the order of 10^3–10^4.

It turns out that by taking the most commonly accepted value of ξ ($\approx$1–10 nm) [53] we get that in the cases of spherical [34, 51] and cylindrical ("high structured") CBs [34, 54], CNFs [50, 55, 56], and CNTs [43], the x_c values are quite consistent with the actual b and L sizes (see Section 5.3). On the other hand, for the graphene–polymer [45, 46, 57] and graphite sheet–polymer [47, 48, 57, 58] composites, there is no such agreement. In particular for graphene, where $b \approx 3$ nm, the predicted x_c values for the above ξ values should have been on the order of 10 vol.% rather than the 1 vol.%, or even lower, values that are actually observed. This discrepancy with the simple prediction of the excluded volume expectation (Eq. (5.10)) has not been explained thus far. However, based on our above realization of the role of dispersion, we suggest the following plausible explanation for that, rather unexpected, observation. In the excluded volume theory that we developed in Section 5.3 for hard core disks embedded in the continuous insulating matrix, we made an assumption that seems to be inappropriate for graphene and graphite sheets, that is, that these sheets are rigid. Removing that assumption amounts to the practical elimination of the hard core exclusion that we used in Section 5.3, and thus we can approximate the system by totally permeable disks for which $x_c \propto v/V_{ex} \propto t/b$ as in the soft core case (rather than $x_c \propto t/\delta \approx t/\xi$ as in the rigid, hard core, case). Due to the fact that the sheets are flexible, the only effect that remains from the actual hard core is that of providing "automatic" contacts once a "would be overlap" (permeable-like contact) is established.

So far we have considered the cases where only one stair in the $\sigma(x)$ dependence is revealed and we associated the x_c values either with coalescing (or nearly coalescing) percolation behavior or with the domination of the nearest neighbor tunneling. However, the above considerations apply to the cases of more than a single percolation stair. In particular, the highest x_c, when stairs are observed (Figure 5.8), is associated with coalescence (or with tunneling) while all the lower x_c values are associated with tunneling as explained in Section 5.4. The presence of all stairs is in principle a result of a deviation from the uniform or a simple monotonic dispersion of the particles.

5.6.2
The Critical Behavior of σ(x)

The other feature that emerges from the experimental data is the functional dependence of the observed $\sigma(x)$ that is manifested in particular by the conductivity exponent t. Having the framework given in Sections 5.2–5.4, we can understand the various observed values of t as follows. In the cases where the onset of percolation is associated with the coalescence of particles (as in granular metals [42]) or coalescence-like (as in semiconductor–dielectric composites [41] shown in Figure 5.6), the values of t [35, 39, 40, 42, 59] are, within the experimental uncertainty, very close to the classical critical values of $1.7 < t < 2$ that were derived from calculations or computations [1, 3, 15]. This observation is well understood by following two considerations. Either, as suggested above, we approach the S&Z picture, or we can virtually divide the continuous phase network into small elements (say, spheres or cubes) that exactly touch each other [60]. Hence, close to x_c the system is reminiscent of the S&Z case apart from the nonuniform dispersion of the particles.

When the system is composed of nontouching particles so that the conduction is by interparticle tunneling, the general observations disclose t values that are equal to or larger than the universal value [35, 51]. In particular, larger than universal t values were observed in granular metals for x below the above grain coalescing transition [35], and in CB–polymer composites in which the particles have a spherical shape [34, 35, 51]. An example of that behavior, for Ni-SiO$_2$ granular metal, is shown in Figure 5.8. It is seen (as demonstrated in more detail elsewhere [59]) that for the high x (>50 vol.%) regime the data are well fitted by the universal value while for lower x values [35] the data provide higher than universal t values and are temperature dependent. In passing one should note that in examining $\sigma(x)$ data (such as in the $x < 0.5$ range in Figure 5.8) and trying to evaluate it by a dependence of the type $\sigma = \sigma_0(x - x_c)^t$, for a given σ_0 and x_c, the larger the value of t (>1) the "slower" the increase of $\sigma(x)$ above the corresponding x_c. This is a consequence of the (maybe somewhat counterintuitive) power law dependence, which for $(x - x_c) < 1$ values plays the opposite role than for a $\sigma = \sigma_0(y)^t$ dependence when $y > 1$.

Turning to carbon black composites, one finds that in contrast with the as high as $t = 10$ values when the composite consists of spherical CB particles, for composites of CB particles of large aspect ratios [34, 51], for CNTs [43] (and CNFs [56]), and for graphene [45, 57, 61, 62], the t values are very close to the universal one. Qualitatively, these observations can be explained as follows. In the case of spherical particles, there is a distribution of the nearest neighbor distances that yield a diverging distribution of the conductances so that a tunneling-induced nonuniversal t value is obtained [34, 51] (see Section 5.4). However, the possible "friction" between particles upon the melt mixing may increase the tendency to weaken the divergence of the "theoretical" $f(g)$ distribution. This is because for particles with aspect ratios that deviate considerably from unity the strong "interaction" (the large excluded volume) between the particles and the possible friction between them (see Figure 5.7) are such that the particles approach each other so that their interparticle distance (at the points of nearest approach [46]) is limited only by the thinnest possible polymer layer between them [63–67]. This "forced" type of contacts is expected to yield that the junctions between adjacent particles are, in practice, all the same and thus the conductances that matter are of nearly the same value [34]. This conclusion is strongly supported by the extensive review on nanotube polymer composites in which it was concluded that the t exponents derived in some hundred works are "in the range from 1.3 to 4 peaked around $t = 2$" [43]. While the $t \approx t_{un}$ values are well understood from that and the above discussion in Section 5.4, lower than the 3D universal t ($1.7 < t < 2$) values are not well accounted for. However, values in the $1.7 > t > 1.3$ range can be explained simply by assuming that the system studied is effectively a 2D network [46, 57] that may result (say, by a shear effect) during the melt mixing of composites that are made of nonisotropic particles.

Our attribution of the nonuniversal behavior to tunneling and the relation between tunneling and hopping suggest that the critical behavior and hopping are closely related. Mathematically, this can be seen by a comparison of Eq. (5.15) and the application of Eq. (5.16) in Eq. (5.6) (for $t > t_{un}$). Physically, both hopping and percolation behaviors are controlled by the (average) critical resistor and as such account for the same process. Hence, as we pointed out recently [37], one should take precaution when making conclusions solely on the basis of the quality of each of these fits. The above merger of the predictions of the two theories is the key to the understanding of the temperature-activated or variable range hopping-like behavior in granular metals (for x_c below the coalescence transition) in terms of percolation theory (see Eq. (5.13)), as follows (Balberg, I., unpublished.). With increasing T, the system will approach nearest neighbor hopping and thus r_c will decrease. Hence, the increase of T should manifest itself by a decrease in t (i.e., the decrease in r_c or effective $\tilde{a}$) and an increase in x_c. Indeed, by a new analysis of the raw data of Ref. [68] on the W-Al$_2$O$_3$ composite, we confirmed the above expectations (Balberg, I., unpublished.).

From relation (5.6) and the "jumps" mentioned for lattices, we should have expected to see a staircase such as in Figure 5.3 with two or more stairs in composite materials. The a priori expected two transitions are that of the coalescence (when it takes place) and that of the tunneling, as in Figure 5.8. Support for this expectation of a staircase, beyond these two, was provided recently in our experimental results on Ag-Al$_2$O$_3$ granular metal (Balberg, I., unpublished.) as well as by simulations and by calculations that utilized effective medium approximation (EMA) and computer simulation of a S&Z-like system [69]. In fact, the first transition (highest x_c value) in all these works is a coalescence–touching transition or a tunneling transition while the following ones are all associated with tunneling to farther and farther neighbors. In contrast, while there were very many experimental $\sigma(x)$ results that indicate stairs in various carbon particle-based composites, their presence was usually ignored and the analyses, when applied, have tried to fit the whole set of available data to a single $\sigma(x)$ stair such as that given by Eq. (5.6) (see, for example, Refs. [43, 70–72]). On the other hand, no results have shown a systematic or an ordered behavior as expected from the staircase model described here in Section 5.2 or its S&Z modification [69].

Following these considerations, we have prepared and studied a "dense series" of carbon black–polymer composites in which nearly spherical conducting particles (N990) with a diameter of 250 nm are embedded in a polyethylene matrix, expecting that this series can yield the possibility of examining the predictions of our staircase model in such composites. The properties of this "low structure" carbon black [51, 73] and their polymer composites, as well as our method of the composites' fabrication in the form of strips (a few cm long, a few mm wide, and 0.25 mm thick) with silver paste coplanar contacts and our four-probe resistivity measurements, have been described previously [74, 75]. The rather detailed $\sigma(x)$ dependence that we found in the above-described measurements on the spherical N990 CB–polymer composite is shown in Figure 5.9 (Ravid G. and Balberg I., unpublished). The presence of four stairs is very apparent from these data. However, for relating the above stairs to our model (as described in Section 5.2), we had to confirm that the $\sigma(x)$ dependence yields a close to the universal, $t \approx 2$, value for each stair. Indeed, we found that this expectation is quasi-quantitatively fulfilled as can be seen by the curves in the figure. We note of course that even for that "dense series" of samples we have for each stair only three experimental points per stair and thus the fits cannot be considered as

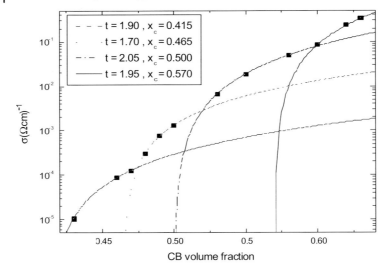

Figure 5.9 The CB volume fraction dependence of the electrical conductivity in a CB–polymer composite in which the CB particles are essentially spherical. The curves fitted to the data and the corresponding parameters, for each of the four distinct stairs in the staircase, are given in the legend. (after Ravid G. and Balberg I., unpublished)

quantitatively satisfactory. However, the fact that we can fit the universal t value to all four stairs that are shown (with a quality that is well satisfied by the eye) is a strong indication for the presence of stairs with that basic feature. The more serious difficulty with accepting our data as a proof for the staircase model in those CB systems is the possibility that the data deviation from a smooth $\sigma(x)$ dependence (that yields an overall nonuniversal behavior) is due to fluctuations in the data taking process. To eliminate this possibility, we have confirmed the presence of the same stairs in the $\sigma(x)$ dependence that we found for the same set of samples but by a completely different method of microwave transmission through the sample [54]. Hence, we concluded that for the series of samples that we used, there is for each x regime a dominant resistor and the stairs represent a hierarchy of dominant resistors. With that confidence in the data we returned to the results shown in Figure 5.9 checking the other feature that is predicted for our stairs model of Section 5.2, that is, that the jumps should be of the order suggested by tunneling through interparticle separations as expected from RDF-based calculations [69] (since no corresponding experimentally determined RDF data are available for such samples in general and our samples in particular). Inserting the corresponding interneighbor distances revealed that these distances can account for the relatively small jumps in Figure 5.9 only if we assume that they are of the order of a few nm rather than the expected tens of nm for the N990 particles. We explain then the observations as follows. It is well known that the carbon or metallic particles interact with the polymer [64, 67] and thus a polymer layer with a thickness of a few nm forms around the particles [64]. Hence, it is not impossible that "quantized," nm thick, polymer layers ("anion-like") form around a CB particle and we have a peaked RDF of the polymer layers around each particle. Correspondingly, the variation of the critical resistor is at least an order of magnitude smaller than expected from the RDF model of Section 5.2 and this fits reasonably well the experimental data. Another alternative is that these stairs are associated with a dispersion effect such as the one that we suggest below for non spherical carbon particles.

While we have assumed above that for elongated particles we will have that, due to the effective interaction (the large excluded volume), the first stair of "nearly coalescing" particles will dominate the conductivity, one finds that in many cases there is more than a single stair and the jumps in the corresponding dilute system cannot be associated with tunneling stairs as above (see, for example, Refs. [43, 70, 71]). We suggest then that in those cases the stairs are due to a dispersion phenomenon. We base the above explanation on the computer simulation of Hu *et al.* [76] who found that in the case where compact aggregates (of low resistance) form upon the implantation of elongated particles, there are jumps in the $\sigma(x)$ dependence. They have further shown that these jumps disappear as the tendency to form aggregates is weakened (in analogy with lattices when the ratio between z_{NN} and z_{NNN} in Figure 5.4 increases). This suggestion would further explain why the stairs found in the corresponding composites are not as regular as expected from the tunneling model. Experimentally, further support to the dispersion model is provided by the significant variation of the percolation threshold due to different polymer matrices and/or the effect of spinning during the melt mixing of the composite [43].

The conclusion from all that is that when a t much larger than t_{un}, or a series of explicit jumps, is observed, we have a hierarchy of resistors, the dominant one of each is associated with a specific x regime. In the first case, that is, if no distinguishable stairs are observed, the dependence represents a monotonic or a dilute-like dispersion of the particles (i.e., a hopping-like system). In contrast, if only a single stair with a universal t is observed, the contributions of resistors other than the single-value dominant ones are negligible. This explains numerous data in the literature where at first sight the data cannot be accounted for by the classical behavior of Sections 5.2–5.5. However, from the basic physics point of view, the most important conclusion is the meaning of x_c that can be summarized as follows. There is a hierarchy of resistors in the system and each x_c is associated with a particular resistance value (when the resistors are listed from small to large) that is necessary for a global connectedness of the system. This meaningful x_c values are relatively easy to identify in cases of sharp jumps between two stairs and a close to universal behavior just above x_c. We have also demonstrated the existence of such a hierarchy by local measurements on granular metals in Ref. [38].

5.7
Summary and Conclusions

Percolation theory is about connectedness in systems. This theory was originally formulated for discrete lattice systems and its extension to systems in the continuum involves essentially the mapping of the latter on the former by recalling that the important parameter that governs the connectivity in both types of systems is the number of bonds per mathematical or real object. When the electrical properties of the system are evaluated, the bonds are considered as resistors. As such, they have values and thus the connectivity, from the transport point of view, may be considered "weak" or "strong" (say, according to the value of the corresponding conductance of a resistor bond). In particular, in the continuum where the local environment of a particle is not the same for different particles, the "value" of the bond, that is, the value of the attached resistor, may depend on the structural and/or the physical properties of this environment. A relatively simple case is the one where the particles merge upon touching so that the bonds can be considered to have a single value. The more complicated case is the one where the particles are not touching so that the interparticle transport is by tunneling, and thus, the resistors in the system are determined by the distance between the particles. In the latter case, a priori, all particles are connected and thus the main premise of percolation theory, that is, the percolation threshold x_c, is missing. On the other hand, due to the strong dependence of the bond resistors on distance, the system can also be described by hopping theory. Obviously, the hopping and percolation pictures would merge as the x_c value approaches 0, which is essentially the "threshold" for the hopping conduction. However, if some cutoff in the tunneling range is imposed, a largest participating resistor is determined and a percolation threshold is well defined so that the system becomes then a bona fide percolation system. In that case, the distribution of resistor values yields a nonuniversal behavior of the global conductivity that is manifested by larger than the universal conductivity exponent, t_{un}.

When one turns to real systems such as composites of conducting particles embedded in an insulating matrix, all the above is valid but the nonuniform dispersions of the particles yield much richer behaviors. Upon sample preparation, the components of the system and the fabrication conditions yield the formation of specific configurations. In particular, the formation of compact particle aggregates at the expense of a uniform dispersion of the particles or vice versa may bring about stairs in the $\sigma(x)$ dependence. A more complicated variation of the percolation threshold is expected to take place if the result of the competition between the dispersion and the aggregation depends on x. In a composite of spherical particles, the conduction between which is by tunneling, the stairs represent the peaks in the radial distribution function of the interparticle distances, while for ensembles of "interacting" anisotropic particles, the stairs represent the statistics of the aggregate connecting bonds. In all cases, the resistance of the sample will be basically determined by the (relatively) high-resistance bond resistors (that connect the low-resistance bond resistors) that are required for the onset of connectivity.

Overlooking the above background, the experimentally observed $\sigma(x)$ may be misinterpreted. Two conspicuous examples of those should be mentioned. First, the presence of stairs associated with a particular dispersion was overlooked (by considering the entire sets of data points) and the data in the literature were analyzed to yield x_c and t values that are not actually representative of the more detailed (or more genuine percolation-like) description of the system. Second, the trials to decide between the hopping behavior and the percolation behavior, just by the fits of the experimental $\sigma(x)$ to the corresponding theoretical dependencies, are not amenable since, as we saw here, for a smooth radial distribution of the interparticle distances the two should merge. We also note that the limited range and the limited accuracy of the data, in particular for analysis according to the hopping model, are quite generally not enough to decide conclusively between the two pictures.

Following the above considerations, let us turn to suggest systematic guidelines in analyzing the $\sigma(x)$ data when one tries to evaluate the conducting network and the transport mechanism in composite materials. One should start by considering additional data, such as structural and other corresponding electrical data, in order to conclude whether stairs may or may not exist in the system. If the $\sigma(x)$ data could be separated to a few distinguishable stairs, each stair should be fitted to Eq. (5.6) with a universal t, or a value quite close to it (e.g., in 3D, up to $t=4$). Once the interpretation of the stairs (e.g., the tunneling model or the aggregation–dispersion model) seems justified, one can learn on the interparticle conduction mechanism from the magnitude of the "jumps" between the stairs. If no more than a single stair is evident, the value of t would suggest a coalescence (a universal t) or a tunneling–hopping-like (a nonuniversal t) behavior. In the latter case, if t is large (say, larger than 4), one should suspect that the data correspond to a few "smeared" stairs or a smooth distribution of the interparticle distances. On the other hand, a single stair with a close to universal t can be a result of particle coalescence, but this is of course appropriate also for cases where close to a single value tunneling resistors constitute the system (such as in cases where the particles were "pushed" against each other during the melt mixing of the composite). In such cases, the many orders of magnitude differences between the values of those resistors and the other resistors in the system (e.g., for particles that did not "interact" during the mixing) is the reason for the observation of only a single stair. From the experimental point of view, we note that, in principle, if one would have an extremely sensitive large-resistance determination tool, one should be able to detect subsequent stairs, in all composite systems.

Finally, regarding the x_c values we note that these were usually taken as an experimentally given parameter and very few attempts were made to account for their particular values in specific composites. Following the discussion in this chapter, we conclude that these are very sensitive to the dispersion of the particles, which is determined by their "interaction" during the fabrication of the composites. In particular, we note that the simple theories of continuum percolation were proper only under the assumption of a uniform dispersion of the particles and as such can serve only as indicators, or as giving bounds, for the x_c values. Hence, for the determination of the x_c values theoretically, or for the evaluation of these values experimentally, one needs to provide a description of the dispersion in a quantitative way. Attempts in this direction have begun only recently.

Acknowledgments

This work was supported in part by the Israel Science Foundation (ISF), in part by the Israeli Ministry of Science and Technology, and in part by the Enrique Berman Chair of Solar Energy Research at the Hebrew University. The author would like to thank D. Azulay, Y. Goldstein, C. Grimaldi, M.B. Heaney, J. Jedrzejewski, O. Millo, G. Ravid, E. Savir, I. Wagner, and M. Wartenberg for their valuable contribution to the work reported here.

References

1 Stauffer, D. and Aharony, A. (1994) *Introduction to Percolation Theory*, Taylor & Francis, London.
2 Zallen, R. (1983) *The Physics of Amorphous Solids*, John Wiley & Sons, Inc., New York.
3 Bonde, A. and Havlin, S. (eds) (1991) *Fractals and Disordered Systems*, Springer, Berlin.
4 Hunt, A. and Ewing, R. (2009) *Percolation Theory for Flow in Porous Media*, Springer, Berlin.
5 Chung, D.D.L. (2004) *J. Mater. Sci.*, **39**, 2645.
6 Shklovskii, B.I. and Efros, A.L. (1984) *Electronic Properties of Doped Semiconductors*, Springer, Berlin.
7 Sahimi, M. (2003) *Heterogeneous Materials. I. Linear Transport and Optical Properties*, Springer, New York.
8 Bergman, D.J. and Stroud, D. (1992) *Solid State Phys.*, **46**, 147.
9 Nan, C.W. (1993) *Prog. Mater. Sci.*, **37**, 1.
10 Torquato, S. (2009) *Random Materials: Microscopic and Macroscopic Properties*, Springer, New York.
11 Balberg, I. (2009) *J. Phys. D*, **42**, 064003.
12 Berkowitz, B. and Balberg, I. (1993) *Water Resour. Res.*, **29**, 775.
13 (a) Kasteleyn, P. and Fortuin, C.M., (1969) *J. Phys. Soc. Jpn. Suppl.*, **26**, 11. (b) Fortuin, C.M. and Kastelyan, P. (1972) *Physica*, **57**, 536.
14 Balberg, I. (2009) *The Springer Encyclopedia of Complexity*, vol. **2**, Springer, Berlin, p. 1443.
15 Weight, D.C., Bergman, D.J., and Kantor, Y. (1985) *Phys. Rev. B*, **33**, 396.
16 Kogut, P.M. and Straley, J. (1979) *J. Phys. C*, **12**, 2151.
17 Scher, H. and Zallen, R. (1970) *J. Chem. Phys.*, **53**, 3759.
18 Powell, M.J. (1979) *Phys. Rev. B*, **20**, 4194.
19 Balberg, I. and Binenbaum, N. (1987) *Phys. Rev. B*, **35**, 8749.
20 Shante, V.K.S. and Kirkpatrick, S. (1971) *Adv. Phys.*, **20**, 325.
21 Balberg, I., Anderson, C.H., Alexander, S., and Wagner, N. (1984) *Phys. Rev. B*, **30**, 3933.
22 Wagner, N., Balberg, I., and Klein, D. (2006) *Phys. Rev. E*, **74**, 011127.
23 Neda, Z., Florian, R., and Brechet, Y. (1999) *Phys. Rev. E*, **59**, 3717.
24 (a) Charlaix. E., Guyon, E., and Rivier, N., (1984) *Solid State Commun.*, **50**, 999. (b) Charlaix, E. (1986) *J. Phys. A*, **19**, L533.
25 Balberg, I. (1986) *Phys. Rev. B*, **33**, 3618.
26 Chiteme, C., McLachlan, D.S., and Balberg, I. (2003) *Phys. Rev. B*, **67**, 024207.
27 Grest, G.S., Webman, I., Safran, S.A., and Bug, A.L.R. (1986) *Phys. Rev. A*, **33**, 2842.
28 Drory, A., Balberg, I., and Berkowitz, B. (1994) *Phys. Rev. E*, **49**, R949; *Phys. Rev. E*, **52**, 4482 (1995)
29 Ambrosetti, G., Grimaldi, C., Balberg, I., Maeder, T., Danani A., and Ryser, P. (2010) *Phys. Rev. B*, **81**, 155434.
30 Mott, N.F. and Davis, E.A. (1979) *Electron Processes in Non-Crystalline Materials*, Clarendon Press, Oxford.
31 Balberg, I. and Binenbaum, N. (1987) *Phys. Rev. A*, **35**, 5174.
32 Tye, S. and Halperin, B.I. (1989) *Phys. Rev. B*, **39**, 877.
33 Berman, D., Orr, B.G., Jaeger, H.M., and Goldman, A.M. (1986) *Phys. Rev. B*, **33**, 4301.
34 Balberg, I. (1987) *Phys. Rev. Lett.*, **59**, 1305.

35 Balberg, I., Azolay, D., Toker, D., and Millo, O. (2004) *Int. J. Mod. Phys. B*, **18**, 2091.
36 Johner, N., Grimaldi, C., Balberg, I., and Ryser, P. (2008) *Phys. Rev. B*, **77**, 174204.
37 Balberg, I., Millo, O., Azulay, D., Grimaldi, C., and Ambrosetti, G. (2011) *Phys. Rev. Lett.*, **106**, 079701.
38 Toker, D., Azulay, D., Shimoni, N., Balberg, I., and Millo, O. (2003) *Phys. Rev. B*, **68**, 041403(R).
39 Abeles, B., Sheng, P., Coutts, M.D., and Arie, Y. (1975) *Adv. Phys.*, **24**, 407.
40 Sichel, E.K. (ed.) (1982) *Carbon Black–Polymer Composites*, Marcel Dekker, New York.
41 (a) Antonova, I.V., Gulyaev, M., Savir, E., Jedrzejewski, J., and Balberg, I., (2008) *Phys. Rev. B*, **77**, 125318. (b) Balberg, I., Savir, E., and Jedrzejewski, J. (2011) *Phys. Rev. B*, **83**, 035318.
42 Peng, D.L., Konno, T.J., Wakoh, K., Hihara, T., and Sumiyama, K. (2001) *Appl. Phys. Lett.*, **78**, 1535.
43 Bauhofer, W. and Kovacs, J.Z. (2000) *Compos. Sci. Technol.*, **69**, 1486.
44 Al-Saleh, M.H. and Sundararaj, U. (2009) *Carbon*, **47**, 2.
45 Stankovich, S., Dikin, D.A., Dommett, G.H.B., Kohlhaas, M., Zimmy, E., Stach, E.A., Piner, R.D., Nguyen, S.T., and Ruoff, R.S. (2006) *Nature*, **440**, 282.
46 Pang, H., Che, T., Zhang, G., Zeng, B., and Li, Z.-M. (2010) *Mater. Lett.*, **64**, 2226.
47 Panwar, V., Kang, B., Park, J.-O., Park, S., and Mehra, R.M. (2009) *Eur. Poly. J.*, **45**, 1777.
48 Weng, W., Chen., G., Wu, D., Chen, X., Lu, J., and Wang, P. (2004) *Polym. Sci. B*, **42**, 2844.
49 Bug, A.L.R., Safran, S.A., Grest, G.S., and Webman, I. (1985) *Phys. Rev. Lett.*, **55**, 1896.
50 Ezquerra, T.A., Connor, M.T., Roy, S., Kulescza, M., Fernandes-Nascimeto, J., and Balta-Calleja, F.J. (2001) *Compos. Sci. Technol.*, **61**, 903.
51 Rubin, Z., Sunshine, S.A., Heaney, M.B., Bloom, I., and Balberg, I. (1999) *Phys. Rev. B*, **59**, 12196.
52 Balberg, I. and Bozowski, S. (1982) *Solid State Commun.*, **44**, 551.
53 Zweifel, Y., Plummer, C.J.G., and Kausch, H.-H. (1998) *Polym. Bull.*, **40**, 259.
54 Connor, M.T., Roy, S., Esqueza, T.A., and Balta Calleja, F.J. (1998) *Phys. Rev. B*, **57**, 2286.
55 Carmona, F., Prudhon, P., and Barreau, F. (1984) *Sold State Commun.*, **51**, 255.
56 Gordejev, S.A., Macedo, F.J., Ferreira, J.A., van Hattum, F.W.J., and Bernardo, C.A. (2000) *Physica B*, **279**, 33.
57 Fan, Y., Wang, L., Li, J., Li, J., Sun, S., Chen, F., Chen, L., and Jiang, W. (2010) *Carbon*, **48**, 1743.
58 Chen, G., Wu, C., Weng, W., Wu, D., and Yan, W. (2003) *Polym. Commun.*, **44**, 1781.
59 Balberg, I., Wagner, N., Goldstein, Y., and Weisz, S.Z. (1990) *Mater. Res. Soc. Symp. Proc.*, **195**, 233.
60 McCarthy, J.F. (1987) *Phys. Rev. Lett.*, **58**, 2242.
61 Chen, G.-H., Wu, D.J., Weng, W.G., He, B., and Yan, W.L. (2001) *Polym. Int.*, **50**, 980.
62 Kalaitzidou, K., Fukusjma, H., and Drzal, L.T. (2007) *Compos. Sci Technol.*, **67**, 2045.
63 Brown, D., Marcadon, V., Mele, P., and Allerola, N.D. (2008) *Macromolecules*, **41**, 1499.
64 Picu, R.C. and Rakshit, A. (2007) *J. Chem. Phys.*, **126**, 144909.
65 Moczo, J. and Pukanszky, B. (2008) *J. Ind. Eng. Chem.*, **14**, 535.
66 Smuckler, J.H. and Finnerty, P.M. (1974) *Adv. Chem. Ser.*, **134**, 171.
67 Feng, J.Y., Li, J.X., and Chan, C.M. (2002) *J. Appl. Polym. Sci.*, **85**, 358.
68 Abeles, B., Pinch, H.L., and Gittleman, J.I. (1975) *Phys. Rev. Lett.*, **36**, 257.
69 Ambrosetti, G., Balberg, I., and Grimaldi, C. (2010) *Phys. Rev. B*, **82**, 134201.
70 Dang, Z.-M., Wang, L., Yin, Y., Zhang, Q., and Lei, Q.-Q. (2007) *Adv. Mater.*, **19**, 852.
71 Yuen, S.-M., Ma, C.-C.M., Wu, H.-H., Kuan, H.-C., Chen, W.-J., Liao, S.-H., and Hsu, C.W. (2007) *J. of Appl. Poly. Science* **103**, 1272.
72 Shekhar, S., Prasad, V., and Subramanyam, S.V. (2006) *Carbon*, **44**, 334.
73 Samarzija-Jovanovic, S., Jovanovic, V., Markovic, G., and Marinovic-Cinovic, M. (2009) *J. Therm. Anal. Calorim.*, **98**, 275.
74 Heaney, M.B. (1995) *Phys. Rev. B*, **52**, 1.
75 Heaney, M.B. (1995) *Phys. Rev. B*, **52**, 12477.
76 Hu, N., Masuda, Z., Yan, C., Yamamoto, G., Fukunaga, H., and Hashida, T. (2008) *Nanotechnology*, **19**, 215701.

6
Processing Technologies of Semiconducting Polymer Composite Thin Films for Photovoltaic Cell Applications

Hui Joon Park and L. Jay Guo

6.1
Introduction

Cost-effective and highly efficient renewable energy is becoming ever more important due to the rising energy price and the serious issue of global warming from burning the fossil fuels. Among these, solar cells, which utilize a nonexhaustible and green solar energy, are being developed as a promising solution. Especially, polymer photovoltaic (PV) cells [1–3] offer a promising alternative to inorganic solar cells, which have been the main stream of solar cell researches, due to their low cost, easy fabrication, and compatibility with flexible substrates over a large area. Since their first report [4], the power conversion efficiency (PCE) of polymer PVs has steadily increased and now reached up to 4–8% [5–9]. The recent great success of polymer PVs as an emerging technology is mostly due to the development of advanced structures of semiconducting polymers in photoactive layer, efficiently generating photocharges, and the effective fabrication processes to achieve those nanostructures.

In the early 1990s, PCEs in single-component intrinsic semiconductor-type polymer PVs were still limited to less than 0.1% [10–13] (Figure 6.1a). This was mainly due to the weak driving force that is inefficient to dissociate the relatively high binding energy of photogenerated excitons of polymer semiconducting materials, ranging between 0.1 and 1 eV [14–18], and secondly due to the low carrier mobilities after exciton dissociation in the photoactive layer. A key advance to solve the problem was to introduce a second electron-acceptor material that has an energy offset with respect to the electron-donor photoactive layer to facilitate the charge separation [1, 4, 19]. Excitons generated near the heterojunction can diffuse to the interface between two semiconductors and undergo forward electron or hole transfer (Figure 6.1b). This charge transfer lead to the spatial separation of the electrons and holes, thereby preventing direct recombination and allowing the transport of electrons to one electrode and holes to the other.

The external quantum efficiency (EQE) of polymer PVs based on exciton dissociation at a donor–acceptor interface is $\eta_{EQE} = \eta_A \times \eta_{ED} \times \eta_{CC}$. Here, η_A is the absorption efficiency. The exciton diffusion efficiency, η_{ED}, is the fraction of photogenerated

Semiconducting Polymer Composites: Principles, Morphologies, Properties and Applications, First Edition.
Edited by Xiaoniu Yang.
© 2012 Wiley-VCH Verlag GmbH & Co. KGaA. Published 2012 by Wiley-VCH Verlag GmbH & Co. KGaA.

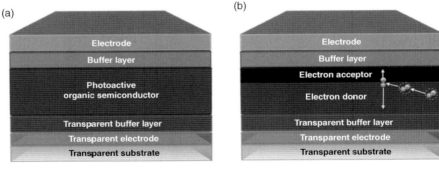

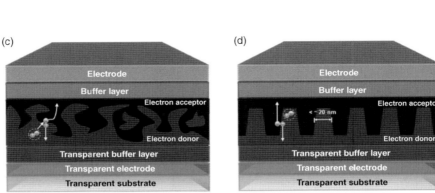

Figure 6.1 Four device architectures of conjugated polymer-based photovoltaic cells: (a) single-layer polymer PV cell; (b) bilayer polymer PV cell; (c) disordered bulk heterojunction; (d) ordered heterojunction.

excitons that reaches a donor–acceptor interface before recombining. The carrier collection efficiency, η_{CC}, is the probability that a free carrier, generated at a donor–acceptor interface by dissociation of an exciton, reaches its corresponding electrode. Typically, in bilayer donor–acceptor organic PVs, η_{CC} is approaching almost 100%. Because there are essentially no minority free carriers in the undoped semiconductors, there is little chance of carrier recombination once the charges move away from the interface, despite the long transit times to the electrodes. However, since the exciton diffusion length, measured to be 4–20 nm [19–23], is typically an order of magnitude smaller than the thickness to fully absorb the light, a large fraction of the photogenerated excitons remains unused for photocurrent generation [1], limiting η_{EQE} and hence the PCE of this type of planar junction cell.

This limitation was overcome by the concept of the bulk heterojunction (BHJ) composite (Figure 6.1c), where the donor and acceptor materials are intimately blended throughout the bulk [24–26]. In this way, excitons do not need to travel long distances to reach the donor–acceptor interface, and charge separation can take place throughout the whole depth of the photoactive layer. Thus, the active zone for photocurrent generation extends throughout the volume maximizing

η_{ED}. However, due to the randomly distributed blend morphologies inducing the poor free carrier pathways, η_{CC} can be significantly decreased; therefore, it is one of the most important tasks to optimize free charge carrier pathways while preserving the domain size within the exciton diffusion length for high-efficiency BHJ polymer PV cells.

As another efficient structure, the interdigitated networks of an electron-donor and an electron-acceptor material have been highlighted as the most ideal structure to achieve high-efficiency polymer PVs, as shown in Figure 6.1d. It is desirable to limit the size of each domain to within the exciton diffusion length. By making such controlled nanoscale morphology between the donor and the acceptor, and having the donor–acceptor interface vertically oriented to the cathode and the anode, the excitons can be fully dissociated to electrons and holes, and can be efficiently transported to the electrodes before recombination, maximizing both η_{ED} and η_{CC}. One of the major challenges to realize this type of structure is the difficulty to access those periodic pillar and hole structures of tens of nanometer pitch by using the semiconducting polymer and fullerene derivate as a donor and acceptor pair.

In this chapter, we address the advanced processing technologies to develop the efficient nanostructures of semiconducting polymers in photoactive layer. First, a new process inducing superior bulk heterojunction morphology will be introduced. Compared with conventional annealing-based methods, the optimized BHJ morphology showing well-organized charge transport pathways with high crystallinity is achieved. Furthermore, this approach is scalable to a high-speed roll-to-roll process while preserving the high device performances. The second part of the chapter will describe our recent effort to fabricate ideal polymer PV structures with vertically interdigitated donor–acceptor structures and having domain sizes comparable to the exciton diffusion length of organic semiconductor. Nanopillar- and nanohole-type nanoimprint lithography (NIL) stamps with sub-20 nm feature size are fabricated from a self-assembled block copolymer template, and NIL-based nanopatterns are made in organic semiconductor.

6.2
Optimization of Bulk Heterojunction Composite Nanostructures

In order to find ways to optimize bulk heterojunction composite morphology of photoactive layer, a number of works, based on poly(3-hexylthiophene) (P3HT): [6,6]-phenyl-C_{61}-butyric acid methyl ester (PCBM) as a model system, have been reported. Currently, thermal annealing (TA) and solvent-assisted annealing (SAA) treatments after spin casting the blend film are widely accepted as general approaches to control the blend morphology for high-efficiency polymer solar cells, because well-organized interpenetrating networks composed of highly crystallized components can be achieved [5, 6, 24–26]. However, recent works that investigated phase separation of components in the vertical direction (i.e., normal to the film and electrode surface) revealed that the BHJ structures fabricated by these methods do not have optimized morphology: a nonuniform vertical distribution exists with P3HT phase dominant near the cathode and PCBM phase dominant near the anode [27–30]. Such nonuniform distribution is opposite to the ideal solar cell structure that requires a donor-rich phase near the anode and an acceptor-rich phase near the cathode, and therefore unfavorable to charge transport to the electrodes. In addition, both annealing processes require relatively long processing time (e.g., tens of minutes for TA [6] or a few hours for SAA [5]) and spin casting deposition can only be applied to small and rigid substrate, making these approaches not suitable to practical large-area and mass production of polymer solar cells. Even though high-speed fabrication processes on flexible substrates have been reported, their device performances still cannot compare to those of spin casting-based polymer solar cells, followed by further post-treatments [31–33]. This is because the traditional high-speed roll-to-roll coating method may not provide sufficient annealing time for crystallization and hence result in lower device efficiency [31–33].

Here, we introduce a novel route that allows *e*vaporation of *s*olvent through *s*urface *en*capsulation and with *i*nduced *al*ignment (ESSENCIAL) of polymer chains by applied pressure [34]. The essence of this approach is to utilize a gas-permeable cover layer for solvent evaporation that simultaneously protects the otherwise free surface and induces shear flow of the blend solution by an applied pressure. The process leads to optimized morphology, more uniform distribution, and improved crystallinity of the components favorable for charge generation and transportation that cannot be achieved by conventional TA and SAA methods. Comparisons among structures fabricated by different methods are made by measurement of quantum efficiency, absorbance, X-ray photoelectron spectroscopy (XPS), and hole and electron mobilities. Furthermore, the effect of domain features of the components on efficient exciton dissociation is studied using atomic force microscopy (AFM) and photoluminescence (PL). The power conversion efficiency is obtained by J–V measurement based on isolated cathode geometry in order not to overestimate the efficiency commonly occurring in devices previously reported by using crossbar electrode geometry [35, 36]. Our results reveal that this new ESSENCIAL method not only induces much uniformly distributed interpenetrating continuous pathways having finer nanodomains with high crystallinity, but also is applicable to high-speed dynamic process that is ultimately demonstrated in a roll-to-roll process while preserving high device performances.

The ESSENCIAL process is depicted in Figure 6.2a. P3HT:PCBM blend is used as a model system, and the blend solution is placed on a transparent anode substrate and capped with a gas-permeable silicone film where pressure is applied. The induced shear flow can help align the polymer chains while the solvent evaporates through the silicone film. The solidified blend layer remains on the substrate after removing the silicone film. Interestingly, poly(3,4-ethylenedioxythiophene):poly(styrene sulfonate) (PEDOT:PSS) layer, widely used on top of transparent anode in organic solar cells, is not indispensable to this processing, and will be further explained later with roll-to-roll application. The thickness of the active layer can be controlled by adjusting the solution concentration and the applied pressure, and

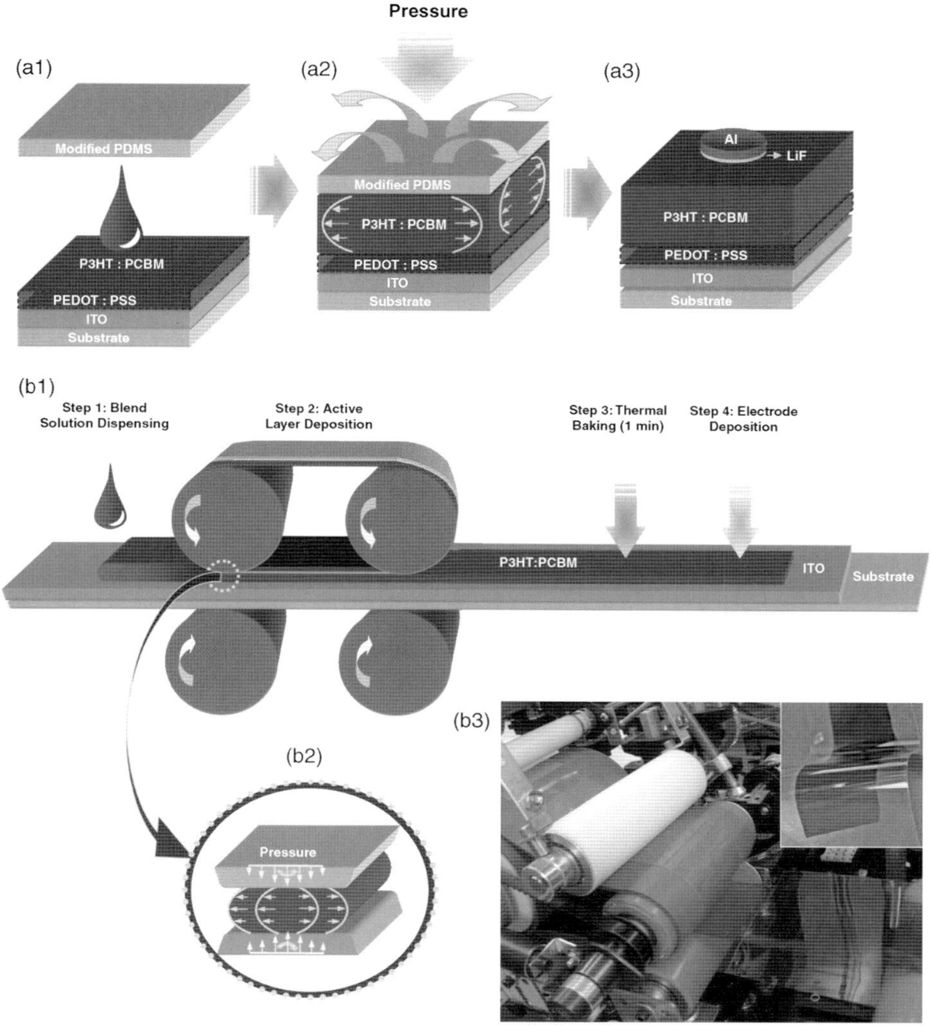

6.2 Optimization of Bulk Heterojunction Composite Nanostructures | 175

Figure 6.2 (a) Schematic of the ESSENCIAL process for fabricating polymer solar cells: (a1) applying blend solution; (a2) active layer formation during solvent evaporation under pressure; (a3) isolated island-type electrode deposition on top of polymer blend film after removing the PDMS stamp. Note that PEDOT: PSS layer is not indispensable to this processing as described in the text. (b) Roll-to-roll processing for polymer solar cells: (b1) schematic of roll-to-roll process for polymer solar cell fabrication; (b2) a schematic to depict the squeezed flow behavior of solution during dynamic roll-to-roll process. The thickness of liquid coating is affected by concentration of solution, roller pressure, and rolling speed according to the dynamic elastic contact model. (b3) A photograph of the roll-to-roll apparatus and process. The inset image is the resultant flexible polymer solar cell before electrode deposition. Reproduced with permission from Ref. [34]. Copyright 2010, Wiley-VCH Verlag GmbH.

the evaporation and solidification time can be reduced to a few seconds by controlling the applied pressure.

First, the effects of different processing methods on the crystallinity of the conjugate polymer are investigated by absorbance spectroscopy. The chain ordering of the conjugate polymer in a BHJ structure is one of the essentials to achieve improved crystallinity for high-efficiency solar cell, because the improved organization of polymer chains facilitates hole transport and the long conjugation length enhances the absorption of light resulting in efficient exciton generation [3]. Thus, the conjugate polymer with high crystallinity such as P3HT is advantageous. The absorption spectrum of the blend film fabricated by the ESSENCIAL method is compared with that made by the spin casting method in Figure 6.3a. To further evaluate the efficiency of the ESSENCIAL method, the samples treated by TA and SAA after spin casting, which are generally used to improve the crystallinity of the P3HT, are also examined. A shear stress applied to polymer solution, which causes the organization of polymer, across entire depth between two plates is much more effective than that between a plate and air surface (e.g., spin casting), which is continuously decreased from the plate and is subsequently zero at air interface [37]. Therefore, as expected, enhanced vibronic peaks in the absorbance spectra as well as significant redshift, which indicate a higher degree of ordering of P3HT chains [38], are observed in the ESSENCIAL sample. In addition, this processing can be completed in just a few seconds, but the enhancement found in this sample is much higher than that in samples thermally annealed for 20 min and is even comparable to that of the solvent-assisted annealed sample in which P3HT crystals are slowly grown for 2 h. Such property permits this method to be applied to high-speed roll-to-roll process and can produce well-ordered P3HT domains.

In addition to improved polymer crystallinity, the nanodomains of each blend component should be well connected in order for holes from charge-separated excitons to be effectively transported to anode, and the electrons to cathode through continuous pathways. However, the nonuniform distribution of the donor and acceptor components found in the spin-casted sample, even after TA or SAA, is not helpful to the effective charge transport to the electrodes [27–30]. This nonuniform distribution of components in the vertical direction is a consequence of the surface energy difference between P3HT ($26.9 \, \text{mN m}^{-2}$) and PCBM ($37.8 \, \text{mN m}^{-2}$), which pushes P3HT to the low surface energy air surface to minimize the overall free energy [28–30]. In contrast, much uniform vertical distribution is expected for films prepared by the ESSENCIAL process, because the gas-permeable silicone film effectively provides a higher surface energy than that of air surface. XPS results shown in Figure 6.3b clearly illustrate these trends. Though the weight ratio of PCBM to P3HT in the blend solution is 1 : 1, the weight ratios of PCBM to P3HT of thermal and solvent-assisted annealed samples measured at the top surface are 0.411 and 0.488, respectively, which indicates a large accumulation of P3HT at the top. But the ESSENCIAL sample produced much more balanced value (0.855), which implies more uniform distribution of components in the vertical direction.

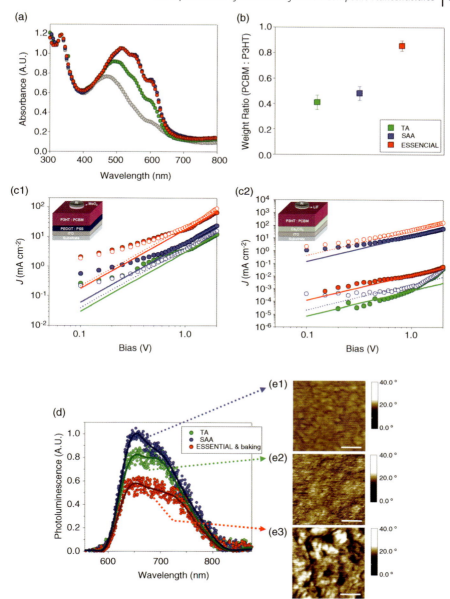

Figure 6.3 Characteristics of bulk heterojunction structures obtained by different processing methods. Gray, green, blue, and red represent the blend films made by spin casting, thermal annealing, solvent-assisted annealing, and ESSENCIAL process, respectively. (a) Absorption spectra of the blend films. The spectra have been normalized to the PCBM peak around 325 nm. (b) Weight ratio of PCBM to P3HT calculated by X-ray photoelectron spectroscopy results for different processing methods. The error bars represent standard deviation. (c) Measured log J–log V plots under dark conditions for hole- and electron-only

In order to confirm that the uniformly distributed components in the ESSENCIAL-treated sample are truly beneficial to effective charge transport by providing more continuous pathways through the film, we construct hole- and electron-only devices in order to evaluate the charge transport properties in the phase-separated blend film [39, 40]. The hole-only device is fabricated by replacing LiF with high-work-function molybdenum oxide (MoO_3) to block the injection of electrons from the Al cathode, and electron-only device by using low-work-function cesium carbonate (Cs_2CO_3) to replace PEDOT:PSS to block the hole injection from the ITO anode [40]. Both hole and electron mobilities are calculated using space–charge limited current (SCLC) model at low voltage (Eq. (6.1)) [41]:

$$J = \frac{9}{8}\varepsilon_0\varepsilon_r\mu\frac{V^2}{L}, \tag{6.1}$$

where $\varepsilon_0\varepsilon_r$ is the permittivity of the component, μ is the carrier mobility, and L is the thickness. As shown in Figure 6.3c1, the hole mobility of the thermally annealed sample shows the lowest value ($\mu_h \sim 1.57 \times 10^{-4}\,cm^2\,V^{-1}\,s^{-1}$) due to relatively poor crystallinity of polymers in such samples. In comparison, the hole mobility of the ESSECIAL-treated sample ($\mu_h \sim 1.15 \times 10^{-3}\,cm^2\,V^{-1}\,s^{-1}$) is much higher, and also higher than the solvent-assisted annealed sample ($\mu_h \sim 3.29 \times 10^{-4}\,cm^2\,V^{-1}\,s^{-1}$), even though the two types of samples show similar polymer crystallinity in the absorbance measurement (Figure 6.3a). This means that the uniformly distributed P3HT polymer domains facilitate the hole transport and the most optimized hole transporting pathways are obtained in the ESSENCIAL sample across the entire depth of the device. Meanwhile, it is known that insufficient crystallinity of P3HT in the thermally annealed sample is due to the rapidly grown large PCBM crystal domains [42], which hampers further crystallization of P3HT. But in the solvent-assisted annealed sample, the mild growth of PCBM provides sufficient time for P3HT to be fully crystallized, leading to much balanced crystalline morphologies of the two domains [43]. Therefore, contrary to the rapidly grown large aggregation of PCBM that produces poor electron pathways and significantly lower electron mobility in the thermally annealed sample, the well-balanced PCBM pathway developed in the solvent-assisted annealed sample show higher electron mobility ($\mu_e \sim 4.95 \times 10^{-4}\,cm^2\,V^{-1}\,s^{-1}$) as shown in Figure 6.3c2. On the other hand, an interesting effect is observed for various samples after

devices. Symbols are experimental data and lines are the fit to the experimental data by the SCLC model showing linear line with slope = 2 at the log scale. Open symbols and dotted lines represent further thermally annealed results. The applied bias voltage is corrected for the built-in potential due to the difference in work function of the two electrodes. Insets are schematics of the device structures: (c1) hole-only devices; (c2) electron-only devices. (d) Photoluminescence spectra of blend films. The ESSENCIAL sample was further treated by heat: filled symbols and lines represent experimental data and their polynomial regressions, respectively. (e) Atomic force microscope phase images. The images in (e1)–(e3) correspond to solvent-assisted annealed sample, thermally annealed sample, and heat-treated ESSENCIAL sample, respectively. The white scale bars represent 50 nm. Reproduced with permission from Ref. [34]. Copyright 2010, Wiley-VCH Verlag GmbH.

further TA: for the ESSENCIAL sample where the electron mobility is drastically increased to the highest value (from 3.61×10^{-7} to 1.46×10^{-3} cm^2 V^{-1} s^{-1}), but the solvent-assisted annealed sample treated by further TA shows reduced electron mobility similar to that of the thermally annealed sample. We believe that PCBM molecules in the ESSENCIAL sample, which are not well organized to form efficient electron pathways before TA, are effectively crystallized, and due to the suppression by the uniformly distributed P3HT polymers they are not overgrown to large aggregates even after thermal treatment. Consequently, well-organized PCBM pathways are formed among P3HT polymers to give optimized interpenetrating structures after the thermal treatment. However, in the case of the solvent-assisted annealed sample, nonuniformly accumulated PCBM molecules can easily assemble to large PCBM aggregates due to weaker suppression effect by the P3HT during the TA, which inevitably produce poor electron pathways and result in depletion of PCBM in regions within the network structure. In addition, the effects of further heat treatment are not significant to the hole mobilities of the solvent-assisted annealed sample (from 3.29×10^{-4} to 2.20×10^{-4} cm^2 V^{-1} s^{-1}) and the ESSENCIAL-treated sample (from 1.15×10^{-3} to 1.26×10^{-3} cm^2 V^{-1} s^{-1}) due to sufficiently crystallized P3HTs obtained under both processing conditions (Figure 6.3c1). Consequently, the most optimized transport pathways for both charge carriers are achieved in the heat-treated ESSENCIAL sample, and well-balanced mobilities ($\mu_e/\mu_h \sim 1.16$) are obtained. Even though the electron mobilities of the thermally annealed device and the solvent-assisted annealed device with further heat treatment are not well matched with the SCLC model, the significantly lower expected values (10^{-8} to 10^{-7} cm^2 V^{-1} s^{-1}) do not affect any conclusions here. The carrier mobilities depending on different processing methods are summarized in Table 6.1.

As a last aspect, the domain size in the BHJ structures and their effects on the exciton quenching are investigated using AFM and PL shown in Figure 6.3d and e. Compared with the nonuniform mixture where one phase is dominant at one surface, more uniformly mixed donor and acceptor phases throughout the film are expected to have finer interpenetrating nanodomains that are advantageous to efficient dissociation of photogenerated excitons, and hence result in suppressed PL from the donor polymer. It has been reported that the PL of annealed sample is enhanced as compared with just spin-casted film, because the higher crystallinity induced by annealing gives relatively poor exciton dissociation due to the reduction

Table 6.1 The calculated carrier mobilities depending on the different processing methods.

Method	Carrier mobility (10^{-4} cm^2 V^{-1} s^{-1})		Ratio (μ_e/μ_h)
	Electron (μ_e)	Hole (μ_h)	
TA	—	1.57	—
SAA	4.95	3.29	1.50
ESSENCIAL[a]	14.60	12.60	1.16

a) ESSENCIAL sample was further treated by heat.

of interfacial area between the donor and acceptor domains [44]. However, the improved charge transport of the annealed samples due to increased crystallinity can offset the poor exciton dissociation effect, still producing high-efficiency solar cells [45]. Therefore, the solvent-assisted annealed sample having higher crystallinity than the thermally annealed sample shows well-defined domains in AFM phase images, and this improved crystallinity induces the enhancement of PL in Figure 6.3d. As for the heat-treated ESSENCIAL sample, finer interpenetrating networks than the solvent-assisted annealed sample are expected due to the more uniform distribution of the blend, and well-defined nanodomains are much more discernable in AFM phase images. These uniformly distributed and fine interpenetrating nanodomains not only permit good charge pathways, but also facilitate the efficient exciton dissociation, therefore suppressing the PL from the donor, and consequently give the lowest PL in Figure 6.3d.

The device performances including power conversion efficiency and J–V curves measured under AM 1.5 G simulated sunlight (at 100 mW cm^{-2} intensity) are summarized in Figure 6.4. In device fabrication, isolated island-type metallic cathode is used to exclude the overestimation of the photocurrent commonly occurring in devices using crossbar-type electrodes [35, 36]. In our former report, PCE can be overestimated as much as by 30%, and can be even higher depending on electrode design and measurement conditions [36]. The devices fabricated by the ESSENCIAL method followed by baking show the highest PCE due to the optimized domain morphologies and charge pathways that result in both increased short-circuit current and fill factor (Figure 6.4c and d). The favorable morphology also leads to the lowest series resistance ($\sim$1.2 Ω cm^2) as expected. The improvement of PCE is confirmed by improved external quantum efficiency in the range from 350 to 650 nm. The detailed measured values are shown in Figure 6.4b.

Before advancing toward the roll-to-roll fabrication, we would like to point out another interesting property of the ESSENCIAL-based device, which is advantageous for high-speed fabrication. Traditionally, PEDOT:PSS has been widely used as a buffer layer between ITO anode and the active organic semiconductor to improve the performances of polymer solar cells. Apart from functioning as a hole transporting layer, another important role of this PEDOT:PSS layer is to provide efficient electron blocking [46] to prevent electron leakage from the BHJ acceptors [47]. If PEDOT:PSS is not used, our experiments on the annealed devices after spin casting show the significant drop in fill factor (e.g., from 65.1 to 54.8%) that results in much reduction in PCE. However, the ESSENCIAL-based device shows only small drop in fill factor (e.g., from 69.1 to 67.3%) that results in negligible effect on PCE. This effect is consistent with the improved morphology in ESSENCIAL devices discussed above. In the devices with spin-casted film, large amount of PCBM is assembled near the ITO anode, causing electron leakage. However, the uniform distribution of components in ESSENCIAL-based devices drastically reduces these electron leakage pathways. Therefore, this observation alone can be an important evidence for the uniform distribution of components in the blend film fabricated by ESSENCIAL. Moreover, avoiding the use of PEDOT:PSS can significantly reduce the processing time as it eliminates the PEDOT:PSS coating and

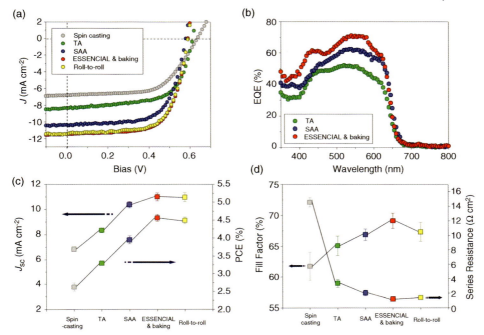

Figure 6.4 Device performances depending on different processing methods. Colors of symbols are as in Figure 6.3 and yellow color is added for devices made by the roll-to-roll process. All data were measured at AM 1.5 G/100 mW cm^{-2}. (a) J–V plots. (b) External quantum efficiency. (c) Power conversion efficiency and short circuit current density (J_{sc}). The error bars represent standard deviation. (d) Fill factor (FF) and series resistance (R_s). Average solar cell characteristics are summarized as follows: spin casting ($J_{sc} = 6.67$ mA cm^{-2}, $V_{oc} = 0.65$ V, FF = 61.7%, $R_s = 14.5$ Ω cm^2, PCE = 2.50%); thermal annealing ($J_{sc} = 8.14$ mA cm^{-2}, $V_{oc} = 0.61$ V, FF = 65.1%, $R_s = 3.3$ Ω cm^2, PCE = 3.23%); solvent-assisted annealing ($J_{sc} = 10.11$ mA cm^{-2}, $V_{oc} = 0.58$ V, FF = 66.9%, $R_s = 2.0$ Ω cm^2, PCE = 3.84%); heat-treated ESSENCIAL ($J_{sc} = 10.68$ mA cm^{-2}, $V_{oc} = 0.60$ V, FF = 69.1%, $R_s = 1.2$ Ω cm^2, PCE = 4.46%); roll-to-roll process ($J_{sc} = 10.59$ mA cm^{-2}, $V_{oc} = 0.60$ V, FF = 67.3%, $R_s = 1.4$ Ω cm^2, PCE = 4.40%). The PV cells fabricated by roll-to-roll processing were prepared without PEDOT:PSS layer. Reproduced with permission from Ref. [47]. Copyright 2008, Wiley-VCH Verlag GmbH.

the baking step to remove residual H$_2$O molecules from PEDOT:PSS used in conventional fabrications, which is especially attractive for high-speed roll-to-roll processing. The advantages of the ESSENCIAL method to induce superior BHJ morphology in short processing time pave the way for the fabrication of scalable high-efficiency polymer solar cells. Here, we demonstrate fabrication of polymer solar cells using a roll-to-roll apparatus composed of dual rollers and tensioned belt covered with gas-permeable silicone film, which enables coating of polymer blend with uniform thickness and fast solvent evaporation in a continuous fashion (Figure 6.2b). According to the dynamic elastic contact model developed for the roll-to-roll nanoimprinting [48] (Figure 6.2b2), the thickness of coated active layer can be controlled by the solution concentration and roller pressure, which is the same condition as the small-scale ESSENCIAL experiment discussed earlier, and the film uniformity can be preserved by the belt tension during the solvent evaporation process. Figure 6.2b3 and inset image show 3 in. wide uniform BHJ active layer film made of P3HT:PCBM blend on ITO-coated PET substrate for flexible solar cells using the continuous roll-to-roll process. After 1 min baking, LiF and Al cathode are deposited, and a PCE (~4.5%) comparable with the small-scale ESSENCIAL experiment is achieved using the roll-to-roll process as shown in Figure 6.4.

6.3
Fabrication of Sub-20 nm Scale Semiconducting Polymer Nanostructure

Vertically oriented sub-20 nm nanohole and nanopillar structures made in conjugated polymers are highly desirable structures. To achieve high-efficiency OSCs, it is most essential to produce interpenetrating networks of an electron-donor and an electron-acceptor material. It is desirable to limit the size of each domain within the exciton diffusion length, which has been measured to be 4–20 nm [49, 50]. By making such controlled nanoscale morphology between the donor and the acceptor, and having the donor–acceptor interface vertically oriented to the cathode and the anode [51], the excitons can be fully dissociated to electrons and holes, and can be efficiently transported to the electrodes before recombination. Here, as an initial effort to produce the ideal interdigitated interface structure, we attempt to imprint directly a regioregular poly(3-hexylthiophene) (rr-P3HT) having high crystallinity. This material is one of the most widely used donor materials in making the OSCs. Figure 6.5 shows that the fabrication of sub-20 nm hole- and pillar-type

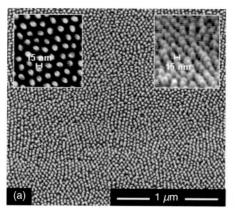
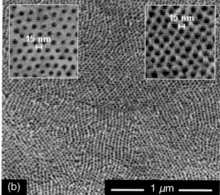
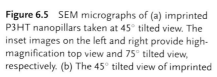

Figure 6.5 SEM micrographs of (a) imprinted P3HT nanopillars taken at 45° tilted view. The inset images on the left and right provide high-magnification top view and 75° tilted view, respectively. (b) The 45° tilted view of imprinted P3HT nanoholes. The left and right inset images are high-magnification top view and 45° tilted view of the structure, respectively. Reproduced with permission from Ref. [52]. Copyright 2009, ACS Publications.

nanostructures in conjugated polymer is possible using the SiO$_2$ molds of both polarity patterns. The nanoimprint process, carried out at 185 °C and 750 psi, is successful despite that the high crystallinity of the rr-P3HT may prevent the polymer chains from flowing easily. The aspect ratio of the imprinted P3HT nanostructures can be controlled by adjusting the imprint temperature, time, and pressure.

6.3.1
Nanoimprint Mold Fabrication

To produce these dense nanoscale patterns, we describe our effort to fabricate an appropriate nanoimprint mold that can imprint over relatively large areas. Though electron beam lithography and focused ion beam (FIB) lithography can be used to produce structures of such dimensions, the time-consuming process and the high cost associated with such techniques seriously restrict the large-scale production of high-density nanopatterns. As alternatives, nanotemplate approaches using track-etched membrane [53] and anodized aluminum oxide (AAO) [54] have been extensively used to produce tens of nanometer-scale patterns. However, the poor adhesion of template to a substrate, the harsh processing conditions to remove the template, or the resolution of the patterns often limits these pattern transferring techniques.

Recently, the self-assembly behavior of diblock copolymers has drawn significant attentions for nanotemplate fabrication [55]. It is well known that the self-organization of block copolymer can access complex nanostructures. The density and the dimension of such self-organized nanostructures are usually beyond the reach of typical conventional top-down nanofabrication techniques [54]. Furthermore, various techniques that have been developed to control the orientation of the nanoscale morphology in the thin film make it possible for these self-organized nanostructures to be used as templates for various applications [55–57]. Despite the numerous advantages of these nanotemplates fabricated by block copolymer self-assembly, the long processing time needed to develop the self-assembled structures has drastically reduced the potential impact of this versatile nanopatterning technique. To overcome this limitation, we introduce a reliable and practical method of fabricating SiO$_2$ molds for nanoimprint lithography using the block copolymer template, which are then used in NIL for high-speed nanopatterning [52, 58, 59]. Such an approach represents a big step forward to mass production of nanostructures with dimension, density, and areal coverage only accessible by the block copolymer self-assembly process.

NIL is a promising lithographic technique capable of replicating large-area nanostructures with resolution down to a few nm. Though there have been previous reports to fabricate SiO$_2$ or Si nanostructures using self-assembled block copolymer template, the processing methods were not straightforward for both pillar and hole polarity patterns and none has succeeded in applying those nanostructures to NIL [60, 63]. In this work, polymer nanotemplate originating from PS–PMMA diblock copolymer is successfully transferred to high-density, high aspect ratio sub-20 nm

SiO_2 nanopillar and nanohole structures over a large area using the novel processing techniques. Not only can NIL create resist patterns, as in a lithographic process, but it can also imprint functional device structures in various polymers, which can lead to a wide range of applications [64]. Imprinting results show that densely packed sub-20 nm nanopillar and nanohole polymer patterns can be easily fabricated on arbitrary substrate. Ultimately, it is possible to develop sub-20 nm conjugated polymer patterns on ITO transparent electrode, which are promising structures for interdigitated OSC structure.

The overall processes for fabricating the NIL molds are described in Figure 6.6. A SiO_2 layer, which is thermally grown on Si wafer surface, is first treated by polystyrene-r-poly(methyl methacrylate) (PS-r-PMMA) random copolymer to produce the neutral surface to polystyrene and poly(methyl methacrylate). Subsequently, PS-b-PMMA diblock copolymer that has 0.7 volume fraction of PS is spin-coated and annealed above the glass transition temperature of the copolymer for the self-assembly process to proceed. This process results in the equilibrated morphology of hexagonally packed cylinders. The identical interfacial tension provided by the neutral PS-r-PMMA layer to each block of the PS-b-PMMA copolymer on the surface eventually leads to vertically oriented cylindrical PMMA nanodomains surrounded by PS. This morphological structure can be used to fabricate PS nanotemplate with hole arrays by removing the PMMA cylindrical nanodomain utilizing the etching selectivity between PS and PMMA to O_2 plasma etching. The resultant PS template with nanohole array patterns is shown in Figure 6.7a.

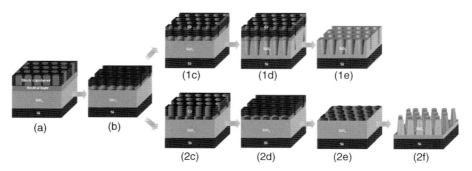

Figure 6.6 Schematic of the process for fabricating both nanohole and nanopillar array patterns in SiO_2. (a) Cylindrical PS-b-PMMA morphology is developed on the substrate surface modified by the neutral PS-r-PMMA layer. (b) PMMA nanodomains are selectively removed by O_2 plasma etching. (1c–e) Nanohole array fabrication process: (1c) Cr is selectively deposited on polymer template using shadow evaporation; (1d) SiO_2 layer is etched using Cr mask by RIE; (1e) Cr mask and polymer template are removed for SiO_2 and form nanohole arrays. (2c–f) Nanopillar arrays fabrication process: (2c) Cr is deposited over the polymer template; (2d) top Cr layer is removed by Cr etching RIE; (2e) Cr nanodots remained after liftoff; (2f) SiO_2 layer is etched using Cr mask by RIE and Cr masks are subsequently removed. Reproduced with permission from Ref. [52]. Copyright 2009, ACS Publications.

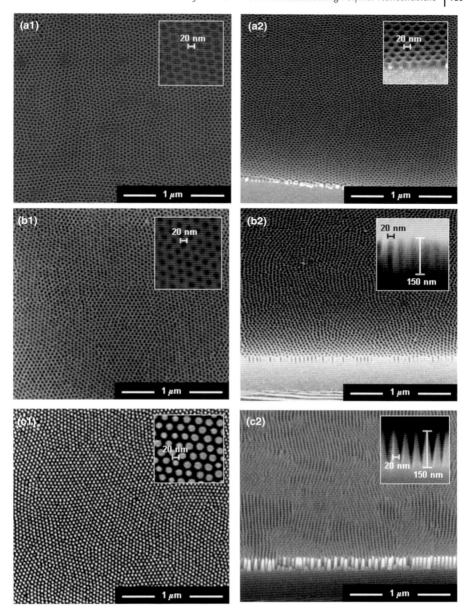

Figure 6.7 SEM micrographs of various nanostructures (left column: top view; right column: perspective view at 45° tilt angle). (a) PS template fabricated by etching away the PMMA nanodomains of the self-assembled PS-b-PMMA hexagonal patterns. (b) SiO$_2$ nanohole structure. (c) SiO$_2$ nanopillar structure. All the inset images are high-magnification SEM images, and the inset images in (b2) and (c2) show the cross-sectional view of the nanostructure in the fabricated SiO$_2$ mold. Reproduced with permission from Ref. [52]. Copyright 2009, ACS Publications.

There have been previous attempts to directly use the PS template as an etching mask to transfer the block copolymer patterns to other materials [60, 61]; however, the soft PS mask is not strong enough as an etch mask to produce structures with high aspect ratios. Here by using a few nm Cr deposited at an angle to the substrate (i.e., shadow evaporation), we can successfully transfer the block copolymer patterns into SiO_2 layer having almost 1 : 10 aspect ratio. Cr is chosen because it is extremely resistant to the RIE gases used to etch oxide materials. Cr is selectively deposited on the top surface of PS template by angled deposition and this Cr layer acts as the actual mask to etch SiO_2 layer rather than PS. The resultant SiO_2 hole arrays with 20 nm diameter and 150 nm depth are shown in Figure 6.7b.

The PS nanotemplate from PS-*b*-PMMA block copolymer patterns is also used to fabricate SiO_2 nanopillars. Again Cr is selected as an etch mask. Here we adds a Cr RIE etching step before liftoff in organic solvent, as shown in processing flows in Figure 6.6. This step is to ensure that the liftoff can be completed easily and the Cr dots remain at the bottom of the nanoholes. Well-defined and uniform SiO_2 nanopillars, shown in Figure 6.6c, fabricated by further etching of SiO_2 layer after the liftoff are clear evidence of this successful process, and we are able to fabricate densely packed SiO_2 nanopillar structures with 1 : 10 aspect ratio. Because of the freedom to insert a hard etch mask for the fabrication of both polarity patterns, the aspect ratio and the shape of the SiO_2 nanostructure are easily controllable, and the pattern transfers to the substrate materials other than SiO_2 are also possible, simply by choosing the appropriate metal masks that are resistant to the RIE chemistry for etching the target layer.

The nanoimprint molds of both pattern polarities made in SiO_2 described above can be used to fabricate various functional nanostructures. For example, using the nanopillar mold fabricated here, it is straightforward to reproduce the nanohole template similar to that produced by the block copolymer self-assembly process in essentially any polymer in a short time by using the NIL technique. As shown in Figure 6.8a, a PMMA template with densely packed nanoholes is generated by a simple thermal nanoimprinting step using nanopillar mold at a temperature of 180 °C and a pressure of 600 psi within a few minutes. To use them as masks for further pattern transferring into a substrate or to produce a porous PMMA membrane, the residual layer of the imprinted PMMA pattern can be easily removed by a subsequent O_2 plasma after an angled deposition of a metal mask such as Cr on top of the imprinted PMMA to reinforce the etching resistance, as shown in Figure 6.8b.

6.4
Conclusions

The novel ESSENCIAL-based technique shows the most optimum BHJ morphology compared with conventional methods and permits us to realize high-speed roll-to-roll process for mass production of low-cost high-efficiency polymer PV cells. Furthermore, for the ideal polymer PV cell structures, nanopillar- and nanohole-type

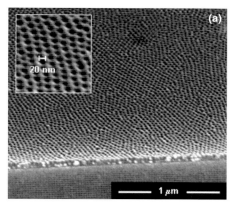 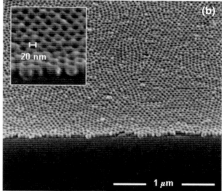

Figure 6.8 SEM micrographs of (a) PMMA nanotemplate imprinted by SiO$_2$ nanopillar mold and (b) PMMA nanotemplate of which the residual layer was removed by O$_2$ plasma using Cr mask deposited by angled evaporation. Both the inset images are the high-magnification SEM images. All the images are 45° tilted views. Reproduced with permission from Ref. [52]. Copyright 2009, ACS Publications.

sub-20 nm polymer semiconducting nanostructures are introduced based on a self-assembled block copolymer template and nanoimprint lithography. This is the first step toward creating an ideal OPV structure with interdigitated donor–acceptor nanodomain morphology.

References

1 Peumans, P., Yakimov, A., and Forrest, S. R. (2003) Small molecular weight organic thin-film photodetectors and solar cells. *J. Appl. Phys.*, **93**, 3693–3723.
2 Hoppe, H. and Sariciftci, N.S. (2004) Organic solar cells: an overview. *J. Mater. Res.*, **19**, 1924–1945.
3 Coakley, K.M. and McGehee, M.D. (2004) Conjugated polymer photovoltaic cells. *Chem. Mater.*, **16**, 4533–4542.
4 Tang, C.W. (1986) Two-layer organic photovoltaic cell. *Appl. Phys. Lett.*, **48**, 183–185.
5 Li, G., Shrotriya, V., Huang, J., Yao, Y., Moriarty, T., Emery, K., and Yang, Y. (2005) High-efficiency solution processable polymer photovoltaic cells by self-organization of polymer blends. *Nat. Mater.*, **4**, 864–868.
6 Ma, W., Yang, C., Gong, X., Lee, K., and Heeger, A.J. (2005) Thermally stable, efficient polymer solar cells with nanoscale control of the interpenetrating network morphology. *Adv. Funct. Mater.*, **15**, 1617–1622.
7 Kim, J.Y., Lee, K., Coates, N.E., Moses, D., Nguyen, T.-Q., Dante, M., and Heeger, A.J. (2007) Efficient tandem polymer solar cells fabricated by all-solution processing. *Science*, **317**, 222–225.
8 Park, S.H., Roy, A., Beaupre, S., Cho, S., Coates, N., Moon, J.S., Moses, D., Leclerc, M., Lee, K., and Heeger, A.J. (2009) Bulk heterojunction solar cells with internal quantum efficiency approaching 100%. *Nat. Photon.*, **3**, 297–302.
9 Gilles, D., Markus, C.S., and Christoph, J.B. (2009) Polymer–fullerene bulk-heterojunction solar cells. *Adv. Mater.*, **21**, 1323–1338.
10 Karg, S., Riess, W., Dyakonov, V., and Schwoerer, M. (1993) Electrical and optical characterization of poly(phenylene-

vinylene) light emitting diodes. *Synth. Met.*, **54**, 427–433.
11 Yu, G., Zhang, C., and Heeger, A.J. (1994) Dual-function semiconducting polymer devices: light-emitting and photodetecting diodes. *Appl. Phys. Lett.*, **64**, 1540–1542.
12 Yu, G., Pakbaz, K., and Heeger, A.J. (1994) Semiconducting polymer diodes: large size, low cost photodetectors with excellent visible–ultraviolet sensitivity. *Appl. Phys. Lett.*, **64**, 3422–3424.
13 Marks, R.N., Halls, J.J.M., Bradley, D.D.C., Friend, R.H., and Holmes, A.B. (1994) The photovoltaic response in poly (*p*-phenylene vinylene) thin-film devices. *J. Phys.: Condens. Matter*, **6**, 1379–1394.
14 Pope, M. and Swenberg, C.E. (1999) *Electronic Processes in Organic Crystals and Polymers*, 2nd edn, Oxford University Press, New York.
15 Sariciftci, N.S. (1997) *Primary Photoexcitations in Conjugated Polymers: Molecular Exciton Versus Semiconductor Band Model*, World Scientific, Singapore.
16 Chandross, M., Mazumdar, S., Jeglinski, S., Wei, X., Vardeny, Z.V., Kwock, E.W., and Miller, T.M. (1994) Excitons in poly (*para*-phenylenevinylene). *Phys. Rev. B*, **50**, 14702–14705.
17 Campbell, I.H., Hagler, T.W., Smith, D.L., and Ferraris, J.P. (1996) Direct measurement of conjugated polymer electronic excitation energies using metal/polymer/metal structures. *Phys. Rev. Lett.*, **76**, 1900–1903.
18 Knupfer, M. (2003) Exciton binding energies in organic semiconductors. *Appl. Phys. A*, **77**, 623–626.
19 Halls, J.M., Pichler, K., Friend, R.H., Moratti, S.C., and Holmes, A.B. (1996) Exciton diffusion and dissociation in a poly(*p*-phenylenevinylene)/C_{60} heterojunction photovoltaic cell. *Appl. Phys. Lett.*, **68**, 3120–3122.
20 Pettersson, L.A.A., Roman, L.S., and Inganäs, O. (1999) Modeling photocurrent action spectra of photovoltaic devices based on organic thin films. *J. Appl. Phys.*, **86**, 487–496.
21 Theander, M., Yartsev, A., Zigmantas, D., Sundström, V., Mammo, W., Anderson, M.R., and Inganäs, O. (2000) Photoluminescence quenching at a polythiophene/C_{60} heterojunction. *Phys. Rev. B*, **61**, 12957–12963.
22 Savenije, T.J., Warman, J.M., and Goossens, A. (1998) Visible light sensitisation of titanium dioxide using a phenylene vinylene polymer. *Chem. Phys. Lett.*, **287**, 148–153.
23 Haugeneder, A., Neges, M., Kallinger, C., Spirkl, W., Lemmer, U., Feldman, J., Scherf, U., Harth, E., Gugel, A., and Mullen, K. (1999) Exciton diffusion and dissociation in conjugated polymer/fullerene blends and heterostructures. *Phys. Rev. B*, **59**, 15346–15351.
24 Yu, G. and Heeger, A.J. (1995) Charge separation and photovoltaic conversion in polymer composites with internal donor/acceptor heterojunctions. *J. Appl. Phys.*, **78**, 4510–4515.
25 Yu, G., Gao, J., Hummelen, J.C., Wudl, F., and Heeger, A.J. (1995) Polymer photovoltaic cells: enhanced efficiencies via a network of internal donor–acceptor heterojunctions. *Science*, **270**, 1789–1791.
26 Halls, J.J.M., Walsh, C.A., Greenham, N.C., Marseglia, E.A., Friend, R.H., Moratti, S.C., and Holmes, A.B. (1995) Efficient photodiodes from interpenetrating polymer networks. *Nature*, **376**, 498–500.
27 Campoy-Quiles, M., Ferenczi, T., Agostinelli, T., Etchegoin, P.G., Kim, Y., Anthopoulos, T.D., Stavrinou, P.N., Bradley, D.D.C., and Nelson, J. (2008) Morphology evolution via self-organization and lateral and vertical diffusion in polymer:fullerene solar cell blends. *Nat. Mater.*, **7**, 158–164.
28 Yao, Y., Hou, J., Xu, Z., Li, G., and Yang, Y. (2008) Effects of solvent mixtures on the nanoscale phase separation in polymer solar cells. *Adv. Funct. Mater.*, **18**, 1783–1789.
29 Xu, Z., Chen, L.-M., Yang, G., Huang, C.-H., Hou, J., Wu, Y., Li, G., Hsu, C.-S., and Yang, Y. (2009) Vertical phase separation in poly(3-hexylthiophene):fullerene derivative blends and its advantage for inverted structure solar cells. *Adv. Funct. Mater.*, **19**, 1227–1234.
30 Germack, D.S., Chan, C.K., Hamadani, B.H., Richter, L.J., Fischer, D.A., Gundlach, D.J., and DeLongchamp, D.M.

(2009) Substrate-dependent interface composition and charge transport in films for organic photovoltaics. *Appl. Phys. Lett.*, **94**, 233303.

31 Krebs, F.C., Gevorgyan, S.A., and Alstrup, J. (2009) A roll-to-roll process to flexible polymer solar cells: model studies, manufacture and operational stability studies. *J. Mater. Chem.*, **19**, 5442–5451.

32 Krebs, F.C. (2009) All solution roll-to-roll processed polymer solar cells free from indium-tin-oxide and vacuum coating steps. *Org. Electron.*, **10**, 761–768.

33 Blankenburg, L., Schultheis, K., Schache, H., Sensfuss, S., and Schrödner, M. (2009) Reel-to-reel wet coating as an efficient up-scaling technique for the production of bulk-heterojunction polymer solar cells. *Sol. Energy Mater. Sol. Cells*, **93**, 476–483.

34 Park, H.J., Kang, M.-G., Ahn, S.H., and Guo, L.J. (2010) Facile route to polymer solar cells with optimum morphology readily applicable to roll-to-roll process without sacrificing high device performances. *Adv. Mater.*, **22**, E247–E253.

35 Cravino, A., Schilinsky, P., and Brabec, C.J. (2007) Characterization of organic solar cells: the importance of device layout. *Adv. Funct. Mater.*, **17**, 3906–3910.

36 Kim, M.-S., Kang, M.-G., Guo, L.J., and Kim, J. (2008) Choice of electrode geometry for accurate measurement of organic photovoltaic cell performance. *Appl. Phys. Lett.*, **92**, 133301.

37 Kim, S.-S., Na, S.-I., Jo, J., Tae, G., and Kim, D.-Y. (2007) Efficient polymer solar cells fabricated by simple brush painting. *Adv. Mater.*, **19**, 4410–4415.

38 Sunderberg, M., Inganas, O., Stafstrom, S., Gustafsson, G., and Sjogren, B. (1989) Optical absorption of poly(3-alkylthiophenes) at low temperatures. *Solid State Commun.*, **71**, 435–439.

39 Mihailetchi, V.D., Koster, L.J.A., Blom, P.W.M., Melzer, C., Boer, B.D., Duren, J.K.J., and Janssen, R.A.J. (2005) Compositional dependence of the performance of poly(p-phenylene vinylene):methanofullerene bulk-heterojunction solar cells. *Adv. Funct. Mater.*, **15**, 795–801.

40 Shrotriya, V., Yao, Y., Li, G., and Yang, Y. (2006) Effect of self-organization in polymer/fullerene bulk heterojunctions on solar cell performance. *Appl. Phys. Lett.*, **89**, 063505.

41 Lampert, M.A. and Mark, P. (1970) *Current Injection in Solids*, Academic Press, New York.

42 Swinnen, A., Haeldermans, I., Ven, M.V., D'Haen, J., Vanhoyland, G., Aresu, S., D'Oliesslaeger, M., and Manca, J. (2006) Tuning the dimensions of C_{60}-based needlelike crystals in blended thin films. *Adv. Funct. Mater.*, **16**, 760–765.

43 Jo, J., Kim, S.-S., Na, S.-I., Yu, B.-K., and Kim, D.-Y. (2009) Time-dependent morphology evolution by annealing processes on polymer:fullerene blend solar cells. *Adv. Funct. Mater.*, **19**, 866–874.

44 Kim, Y., Cook, S., Tuladhar, S.M., Choulis, S.A., Nelson, J., Durrant, J.R., Bradley, D.D.C., Giles, M., Mcculloch, I., Ha, C.-S., and Ree, M. (2006) A strong regioregularity effect in self-organizing conjugated polymer films and high-efficiency polythiophene:fullerene solar cells. *Nat. Mater.*, **5**, 197–203.

45 Li, G., Shrotriya, V., Yao, Y., Huang, J., and Yang, Y. (2007) Manipulating regioregular poly(3-hexylthiophene):[6,6]-phenyl-C_{61}-butyric acid methyl ester blends – route towards high efficiency polymer solar cells. *J. Mater. Chem.*, **17**, 3126–3140.

46 Koch, N., Elschner, A., and Johnson, R.L. (2006) Green polyfluorene–conducting polymer interfaces: energy level alignment and device performance. *J. Appl. Phys.*, **100**, 024512.

47 Irwin, M.D., Buchholz, D.B., Hains, A.W., Chang, R.P.H., and Marks, T.J. (2008) p-Type semiconducting nickel oxide as an efficiency-enhancing anode interfacial layer in polymer bulk-heterojunction solar cells. *Proc. Natl. Acad. Sci. USA*, **105**, 2783–2787.

48 Ahn, S.H. and Guo, L.J. (2009) Large-area roll-to-roll and roll-to-plate nanoimprint lithography: a step toward high-throughput application of continuous nanoimprinting. *ACS Nano*, **3**, 2304–2310.

49 Savenije, T.J., Warman, J.M., and Goossens, A. (1998) Visible light sensitisation of titanium dioxide using a phenylene vinylene polymer. *Chem. Phys. Lett.*, **287**, 148–153.

50 Haugeneder, A., Neges, M., Kallinger, C., Spirkl, W., Lemmer, U., Feldman, J., Scherf, U., Harth, E., Gugel, A., and Mullen, K. (1999) Exciton diffusion and dissociation in conjugated polymer/fullerene blends and heterostructures. *Phys. Rev. B*, **59**, 15346–15351.

51 Kim, M.-S., Kim, J.-S., Cho, J., Stein, M., Guo &, L.J., and Kim, J. (2007) Flexible conjugated polymer photovoltaic cells with controlled heterojunctions fabricated using nanoimprint lithography. *Appl. Phys. Lett.*, **90**, 123113.

52 Park, H.J., Kang, M.-G., and Guo, L.J. (2009) Large area high density sub-20nm SiO_2 nanostructures fabricated by block copolymer template for nanoimprint lithography. *ACS Nano*, **3**, 2601–2608.

53 Martin, C.R. (1994) Nanomaterials: a membrane-based synthetic approach. *Science*, **266**, 1961–1966.

54 Shimizu, T., Xie, T., Nishikawa, J., Shingubara, S., Senz, S., and Gösele, U. (2007) Synthesis of vertical high-density epitaxial Si(100) nanowire arrays on a Si(100) substrate using an anodic aluminum oxide template. *Adv. Mater.*, **19**, 917–920.

55 Hawker, C.J. and Russell, T.P. (2005) Block copolymer lithography: merging "bottom-up" with "top-down" processes. *MRS Bull.*, **30**, 952–966.

56 Huang, E., Rockford, L., Russell, T.P., and Hawker, C.J. (1998) Nanodomain control in copolymer thin films. *Nature*, **395**, 757–758.

57 Ryu, D.Y., Shin, K., Drockenmuller, E., Hawker, C.J., and Russell, T.P. (2005) A generalized approach to the modification of solid surfaces. *Science*, **308**, 236–239.

58 Chou, S.Y., Krauss, P.R., and Renstrom, P.J. (1996) Imprint lithography with 25-nanometer resolution. *Science*, **272**, 85–87.

59 Guo, L.J. (2007) Nanoimprint lithography: methods and material requirements. *Adv. Mater.*, **19**, 495–513.

60 Jeong, S.-J., Xia, G., Kim, B.H., Shin, D.O., Kwon, S.-H., Kang, S.-W., and Kim, S.O. (2008) Universal block copolymer lithography for metals, semiconductors, ceramics, and polymers. *Adv. Mater.*, **20**, 1898–1904.

61 Black, C.T., Guarini, K.W., Breyta, G., Colburn, M.C., Ruiz, R., Sandstrom, R.L., Sikorski, E.M., and Zhang, Y. (2006) Highly porous silicon membrane fabrication using polymer self-assembly. *J. Vac. Sci. Technol. B*, **24**, 3188–3191.

62 Gowrishankar, V., Miller, N., McGehee, M.D., Misner, M.J., Ryu, D.Y., Russell, T.P., Drockenmuller, E., and Hawker, C.J. (2006) Fabrication of densely packed, well-ordered, high-aspect-ratio silicon nanopillars over large areas using block copolymer lithography. *Thin Solid Films*, **513**, 289–294.

63 Black, C.T., Ruiz, R., Breyta, G., Cheng, J.Y., Colburn, M.C., Guarini, K.W., Kim, H.-C., and Zhang, Y. (2007) Polymer self assembly in semiconductor microelectronics. *IBM J. Res. Dev.*, **51**, 605–633.

64 Guo, L.J. (2004) Recent progress in nanoimprint technology and its applications (Topical review). *J. Phys. D*, **37**, R123–R141.

7
Thin-Film Transistors Based on Polythiophene/Insulating Polymer Composites with Enhanced Charge Transport

Longzhen Qiu, Xiaohong Wang, and Kilwon Cho

7.1
Introduction

The discovery of high conductivity in doped polyacetylene (PA) by Shirakawa, MacDiarmid, and Heeger in 1977 generated enormous interest in π-conjugated polymers [1]. These polymers are very different from the traditional polymers that are generally regarded as insulators with surface resistivity of 10^{12}–10^{16} ohms per square, and belong to a new class of materials known today as "conducting polymers" or "synthetic metals." They combine flexibility and ease of processing of common plastic with metal-type electrical conductivity, and can be used as antistatic materials, electromagnetic shielding materials, and possibly replacements for metal wires. Therefore, Heeger, MacDiarmid, and Shirakawa were the Nobel laureates in 2000 "for the discovery and development of conductive polymers."

Though early work has been focused on increasing the conductivity of conjugated polymers by highly doping the polymers, the more recent dramatic development in the field of conjugated polymers took off after the discovery of semiconducting properties in undoped conjugated polymers. One of the key milestones was the first observation of field-effect characteristics in thin-film transistors (TFTs) based on polythiophene reported by Tsumura *et al.* in 1986 [2]. It demonstrated, for the first time, semiconducting device operation in undoped conjugated polymers and led to other useful semiconductor device properties such as electroluminescence in conjugated polymer light-emitting diodes [3] and photoinduced charge transfer in molecular composites of conjugated polymers and fullerenes [4]. Since then, an explosion of published work on polymeric semiconductors has been driven by many new opportunities offered by these unique materials.

Compared to their inorganic equivalents, such as silicon or germanium, semiconducting polymers with a wide range of functionalities as building blocks are chemically versatile, and can be synthesized in large quantities. Furthermore, their solution processability enables fast and cost-effective manufacturing of electronic devices over large areas using established printing technologies (e.g., spin coating, screen printing, inkjet printing) [5–9]. The use of semiconducting polymers as the active layer of TFTs opens up the possibility of fabricating lightweight, flexible, low-cost, and large-area electronic devices including logic circuits, flat panel displays (FPDs), sensors, and radio-frequency identification (RFID) tags, and thus has stimulated much progress in research and technically relevant applications [10–12]. The first TFT based on semiconducting polymer had a field mobility of about $1 \times 10^{-5}\,\mathrm{cm^2\,V^{-1}\,s^{-1}}$ and was still vastly inferior to the inorganic TFT based on amorphous silicon ($0.1–1\,\mathrm{cm^2\,V^{-1}\,s^{-1}}$). After decades of intensive research in synthesis chemistry, a variety of conjugated polymers, such as regioregular poly(3-hexylthiophene) (P3HT) [13], poly(9,9-dioctylfluorene-*co*-bithiophene) (F8T2) [9], poly-(3,3′-didodecylquaterthiophene) (PQT-12) [14], poly(2,5-bis(3-alkylthiophen-2-yl)thieno[3,2-*b*]thiophene) (PBTTT) [15], and cyclopentadithiophene benzothiadiazole copolymer (CDT-BTZ) [16], have been developed to build OTFT devices with high performances. The maximum field mobility of polymeric TFTs has reached $3.3\,\mathrm{cm^2\,V^{-1}\,s^{-1}}$, which is comparable with the mobilities of amorphous silicon [16].

Alongside with the synthesis of new conjugated polymers, a series of studies have been focused on improving the properties of known materials. A widely used method for this purpose is to blend several materials with desired properties. Polymer blends can be designed to integrate the advantageous properties of each component and sometimes to generate new properties not present in the individual components. The blending technique has been used for a long time in the case of nonconjugated polymers to improve their physical properties, especially mechanical properties. For conjugated polymers, blending also presents an alternative and facile way to tune and optimize the electronic properties of devices. For example, donor and acceptor semiconductors have been blended together to dissociate the excitons produced by photoexcitation in bulk heterojunction solar cells (BHSCs) [4, 17–21]. Two different conjugated polymers have also been chosen to give efficient injection of electrons and holes at opposite electrodes in polymeric light-emitting diodes (PLEDs) [22–28]. Blends of conjugated polymers can of course be used in OTFTs. The ambipolar OTF based on binary blends of p-type and n-type conjugated polymers or oligomers is a good representative example of such devices [29–33]. Recently, there has been significant progress in the development of OTFTs based on semiconducting blends.

In this chapter, we focus on OTFT devices based on semiconducting/insulating polymer composites. Because polythiophenes, for example, poly(3-alkylthiophene)s (P3ATs), are among one of the most promising and interesting polymer semiconductors, we pay close attention to the literature work in this chapter. In Section 7.2, the construction of OTFTs and their important characteristics will be introduced. In Section 7.3, the challenges of polymer blend-based OTFTs and the strategies to solve them will be briefly discussed. Next, we review the works on the applications of polymer blends with vertical stratified structure (Section 7.4) and with embedded semiconducting nanowires (Section 7.5) in OTFTs. In addition, concluding remarks and future research perspectives in these fields are suggested in Section 7.6.

7.2
Fundamental Principle and Operating Mode of OTFTs

OTFT inherits the design architecture from its inorganic counterpart metal-oxide-semiconductor field-effect transistor (MOSFET) [34]. Figure 7.1a schematically shows an OTFT device. This device consists of three electrodes (gate, source, and drain electrodes), an insulator (gate dielectric), and an organic semiconductor. The source and drain are in contact with the semiconductor film and the gate electrode is separated from the semiconductor by an insulating layer, i.e., the gate dielectric.

An OTFT device is essentially a voltage-controlled device. It can switch the transistor between "on" and "off" states by biasing the gate electrode. The current–voltage (I–V) characteristics of a p-type polymeric TFT are shown in Figure 7.1b. When there is no voltage bias applied on the gate electrode ($V_{GS} = 0$), almost no current flows between source and drain because the thermally induced charge carriers distribute homogeneously in the whole semiconductor layer at very low density. The OTFT is in the "off" state. However, when a negative gate voltage is applied ($V_{GS} < 0$), a large number of positive charge carriers, i.e., holes, are accumulated in semiconductor layer at its interface with the dielectric layer. The application of a sufficiently high drain–source bias ($V_{DS} < 0$) switches the transistor to its "on" state. The concentration of the charge carries in the semiconductor can be precisely controlled by small variation of the gate bias. Correspondingly, for an n-type polymer transistor, a positive gate bias ($V_{GS} > 0$) and a positive drain–source bias ($V_{DS} > 0$) must be applied to switch the transistor to its "on" state.

There are several important parameters used to characterize organic thin-film transistors, including mobility (also known as charge carrier mobility or field-effect mobility, and has units of cm^2 V^{-1} s^{-1}), the threshold voltage (i.e., the voltage required to turn "on" the transistor), and the on/off current ratio. They can be determined from the transfer and output characteristics of the transistor. The mobility in the saturation regime can be calculated by the equation $I_{DS} = (WC_i/2L)\mu(V_{GS} - V_T)^2$, where W is the channel width, L is the channel length, C_i is

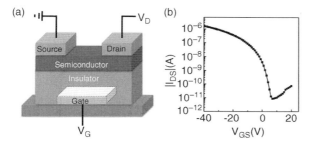

Figure 7.1 (a) Schematic representation of an organic thin-film transistor. (b) Typical transfer characteristics of a p-type organic thin-film transistor.

the capacitance per unit area of the insulating layer, V_T is the threshold voltage, V_G is the gate voltage, and μ is the field-effect mobility.

7.3
Strategies for Preparing High-Performance OTFTs Based on Semiconducting/Insulating Blends

Semiconducting and insulating polymer blends have attracted increasing interest because they can combine electronic properties of semiconducting polymers with the low-cost and excellent mechanical characteristics of the insulating polymers. The solution processability of semiconducting and insulating polymers enables easy preparation of blends by dissolving them in a common solvent. However, the use of these blends in TFTs would be expected to diminish electronic performance relative to that of pure polymers, because the inherent nonconductivity of insulating polymer tends to dilute the conductive network of semiconducting polymer if the semiconducting and insulating polymers are perfectly mixed on a molecular level.

Due to their low entropy of mixing, polymer blends always phase separate into relatively pure phases of the two components. The morphology and phase behavior of the blends significantly affect electric and optoelectric properties. Presently, two strategies have been employed to eliminate the passive effect of insulating polymers and to achieve good electronic properties at low loading of semiconducting polymers in TFT devices based on semiconducting/insulating polymer blends. One is forming self-stratified structure via vertical phase separation in polymer blend thin film, and the other is inducing the semiconductors to form nanofibrillar networks embedded in insulating matrix.

7.4
Blend Films with Vertical Stratified Structure

As discussed in Section 7.2, charge transport in TFTs occurs between the source and drain electrodes, which are usually in the same plane. Therefore, a continuous semiconducting layer at the channel region is needed for TFT operation. Controlling blends to vertically phase separate to form stratified structure presents an effective way to keep the connectivity of semiconducting layer in blended films. In the next section, phase behavior in polymer blend films will be reviewed followed by examples of semiconducting/insulating polymer blends with vertical stratified structure.

7.4.1
Phase Behavior of Polymer Blends

Preparation of vertical stratified blend films relies upon an understating and control of the phase separation process. The Flory–Huggins theory expresses a fundamental insight into the nature of polymer mixing [35–37]:

$$\Delta G = RT\left(\frac{\varrho_A}{M_A}\phi_A \ln \phi_A + \frac{\varrho_B}{M_B}\phi_B \ln \phi_B + \chi \frac{\varrho_A}{M_A}\phi_A\phi_B\right),$$

where ΔG is the Gibbs free energy, R is the universal gas constant, T is the absolute temperature, ϱ, ϕ, and M refer to the density, volume fraction, and molecular weight, respectively, of each polymer component, and χ is the interaction parameter. If there is no specific interaction between the polymer components such as hydrogen bonding interaction or charge transfer interaction, chemical similarity between the two polymers can only result in a positive interaction parameter. So enthalpy, the third item in the parentheses, is generally unfavorable for polymer mixing. In a blend system of small molecules, the mixing entropy, the first two items in the parentheses, plays an important role for achieving a homogenous mixture. However, due to the high molecular weight of polymer, the contribution of mixing entropy to the total free energy is much smaller for the polymer system than the small-molecule system. Therefore, polymer blends have a strong tendency to demix into separate phases.

A phase separation diagram of a binary polymer blend in the bulk is depicted in Figure 7.2a. It represents an "upper critical solution temperature" (UCST) system. At high temperatures, the blend is homogeneously mixed. As the temperature decreases, the system begins to decompose into macroscopically large domains being rich in A or B polymers. Phase separation proceeds either by spinodal decomposition or by nucleation and growth mechanism. When the blend composition is between binodal and spinodal curves, the system is in metastable region and can be stable within small fluctuations in composition. Phase separation then undergoes a nucleation and growth mechanism, in which the formation of a droplet with a certain minimum radius is required before further phase separation (Figure 7.3a) [38]. The domain size and the composition fluctuation are determined by kinetics and thermodynamics, respectively. Spinodal decomposition, on the other hand, occurs when the system is in the region under spinodal curve. If the temperature is reduced setting the initially homogeneous system into the two-

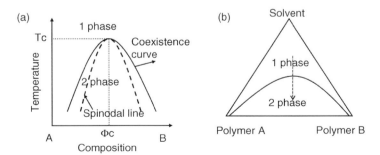

Figure 7.2 (a) A phase diagram of a binary polymer blend, showing the coexistence curve (solid line) and the spinodal line (dash line). (b) A ternary phase diagram of a polymer blend solution, showing the process of solvent evaporation and phase separation.

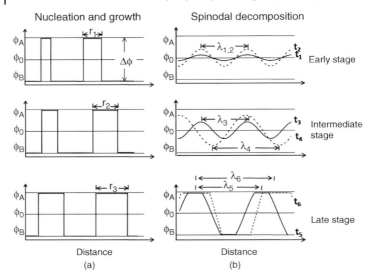

Figure 7.3 Composition fluctuations at early, intermediate, and late stages of phase separation undergoing nucleation and growth (a) and spinodal decomposition (b) mechanism.

phase region, the composition fluctuations become unstable and phase separation starts. Because the system is unstable, there is no energy barrier for the growing composition fluctuations. According to Cahn's theory [39], this process belongs to the linear regime; that is, phase separation is caused by a thermodynamic driving force that initiates a diffusion flux against the concentration gradient. In other words, the unmixing process is treated entirely as diffusion controlled. Therefore, in the early stage of spinodal decomposition, the wavelength almost remains constant, while the amplitude of the composition fluctuations increases with time. In the last stages, nonlinear effects become operative; thus, both amplitude and size of concentration fluctuations grow as shown schematically in Figure 7.3b [38].

In a solution process of blends, the film formation should be described by a ternary phase diagram including two polymers and a solvent (Figure 7.2b). At low concentrations, both polymers are dissolved in the solvent and the system is in one-phase region of the phase diagram. When the polymer concentration increases by solvent evaporation, the composition point shifts vertically toward the solid–solid axis, and at some stage the system will cross the binodal line and come to a point where the interactions between the polymer pairs become strong enough to make the concentrated solution start to phase separate. Therefore, the factors directing morphology evolution in blend solutions include both intrinsic properties of the system such as the polymer–polymer interaction and the blend ratio of polymers and the externally imposed conditions such as the solvent evaporation rate.

The phase separation in blend thin films also proceeds by the well-described bulk mode of nucleation and growth and spinodal decomposition. However, the conditions of phase separation during the formation of a blend thin film are significantly different from bulk conditions. In the bulk, the composition fluctuations grow with randomly directed wave vectors leading to an isotropic factor. In thin films, the presence of film/air and film/substrate interfaces breaks the system symmetry and makes the composition fluctuations parallel, $q_{//}$, and perpendicular, $q_\perp$, to the substrate surface grow at different rates. In general, due to its short fluctuation wavelength, composition fluctuations parallel to substrate surface, $q_{//}$, are favored by mass transport limitations, which results in lateral phase separation. However, if one of the components in blends is preferentially attracted to the film/air and/or film/substrate interface, the composition fluctuations normal to the substrate surface, $q_\perp$, are increased, which leads to a preference for vertical phase separation.

Spin coating is a very widely used method for making highly uniform, submicrometer polymer films. Because spin coating is a nonequilibrium process, phase separation during the spin coating process is very complex. The outcome morphology is very sensitive to the solvent used and the precise spinning conditions. Using *in situ* light scattering measurements, Heriot and Jones [40] found that the film first splits into two layers in the initial step of spin coating, and then the polymer/polymer interface develops an instability that grows to such an extent that the highest liquid/liquid interface protrusions touch the top of the film, at which point that film laterally phase separates (Figure 7.4). The instability at the polymer/polymer interface may arise from a solvent concentration gradient through the film that is caused by a faster evaporation rate at the film surface than in the bulk. Therefore, a relatively low solvent evaporation rate favors the formation of a vertical segregated structure.

In practice, it is not easy to control the phase separation process in polymer blend films. The phase separation process in the polymer blends is very complicated, and the final morphology in the blend films is highly sensitive to many factors, such as solvent evaporation rate, solubility parameter, film–substrate interaction, surface tension of each component, and film thickness. Thus, the vertical phase separation can take place only under very extreme conditions [40, 41], and, alternatively, lateral phase separation is more typical than vertical phase separation because the forces that contribute to the formation of lateral structures minimize the interface area.

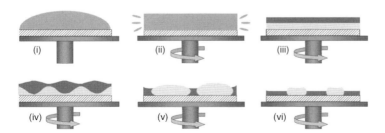

Figure 7.4 A schematic model describing the film formation during the spin coating process. After the initial spin-off stage where both polymer and solvent are removed (i, ii), the film separates into two layers (iii) and the film thins owing to solvent evaporation only. The interface between the polymers destabilizes (iv) and the film phase separates laterally (v, vi). (Reproduced with permission from Ref. [40]. Copyright 2005, Nature Publishing Group.)

Therefore, there are not many reports of vertically separated semiconducting/insulating polymer blends. However, the achievements in the limited reports have displayed the versatility of possible applications of semiconducting/insulating polymer blends in OTFT devices.

7.4.2
One-Step Formation of Semiconducting and Insulating Layers in OTFTs

A major application for OTFTs is in low-end high-volume microelectronics such as smart cards, displays, sensors, and electronic barcodes. To achieve these goals, the OTFT fabrication should be optimized for economical and high-throughput techniques. All solution-processable TFTs, in which all components including semiconductor, dielectric, and electrodes are prepared via solution processing methods, are well desirable. However, when both semiconductor and dielectric layers are formed by the solution process, the underlying polymer layer may be swollen or dissolved by the solvent during the upper layer formation [42, 43]. Because the charge transport is located on the first few molecular layers near the gate dielectric layer, a well-defined interface between the organic semiconductor and the gate insulator is crucial to the device performance [44]. Semiconducting/insulating polymer blends with vertical stratified structure can be used to address this problem.

One of the first examples was described by Chua *et al.* [45]. The used blends consist of poly(9,9-dialkyfluorene-*alt*-triarylamine) (TFB) and cross-linkable dielectric bisbenzocyclobutene (BCB) derivative. The large Flory–Huggins interaction parameter between BCB and TFB (e.g., $\chi_{BCB/TFB} = 3$) provides the strong driving force for complete phase separation. The low cohesive energy density of siloxane group, together with the small molecular weight, ensures that the BCB molecules overcome the sluggish chain entanglements and congregate rapidly on the top surface during film formation. Despite the deliberate design of blend system to favor vertical phase separation in thermodynamics, the morphology of final film depends strongly on solvent evaporation rate during the spin coating process (Figure 7.5). Fast solvent evaporation leads to rapid quenching into phase separation and gives rise to lateral microscale phase-separated domains. When the solvent evaporation is too slow, the binary fluid states have sufficient time to destabilize and break up to form colossal lateral domains. Spinodal-decomposition-produced bilayered structure with a defect-free and ultrathin BCB underlayer was obtained by controlling the rate of solvent evaporation via regulating the solvent vapor pressure during spin coating. After rapid thermal curing at 290 °C, over 90% BCB monomers were cross-linked and gave a BCB network polymer. The obtained self-organized semiconductor–dielectric bilayered blends were then used for low-voltage OTFTs. There are several unique advantages of this self-organization method. First, the semiconductor–dielectric interface is formed in one step without being exposed to ambient conditions, thereby significantly reducing the possibility of atmospheric contamination (a serious consideration in chemically amplified resists). Second, the process is conformal and thus is compatible with other solution processing methods, including inkjet printing

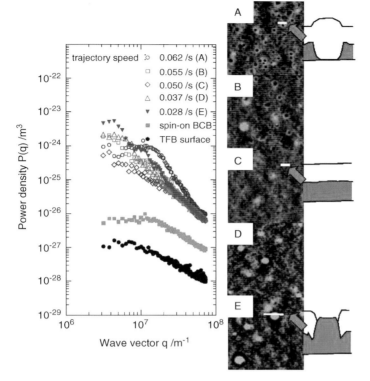

Figure 7.5 AFM images of the interface structures and their one-dimensional isotropic power density $P(q)$ versus wave vector q plots, fabricated at phase separation speed (v_{tr}) between 0.062 and 0.028 s^{-1}. The interface was exposed by a selective solvent (9:1 (v/v) hexane/methyl isobutyl ketone) to remove the pre-cross-linked BCB. The schematic cross sections of the bilayers for various v_{tr} are also shown. (Reproduced with permission from Ref. [45]. Copyright 2003, John Wiley & Sons, Inc.)

and screen printing. However, the high temperature required for BCB cross-linking would prevent their practical applications. Hence, a more facile method using simple polymer dielectrics that do not need thermal cross-linking to provide vertically stratified structures is highly appreciated.

In a recent study, Qiu et al. [46] observed that surface-directed vertical phase separation occurred in a blend of P3HT and poly(methyl methacrylate) (PMMA) when spin coated on substrate with hydrophilic surface. In contrast to the TFB/BCB system, the films have a semiconductor-top and dielectric-bottom bilayer structure because the relatively hydrophobic P3HT component is preferentially segregated at the air–film interface while the relatively hydrophilic PMMA component is preferentially adsorbed onto the hydrophilic silicon substrate during spin coating (Figure 7.6a). The PMMA layer operated either as a dielectric treatment or as the dielectric itself. Therefore, field-effect characteristics can be observed on TFT devices based on self-stratified P3HT/PMMA (5:95) without using any other

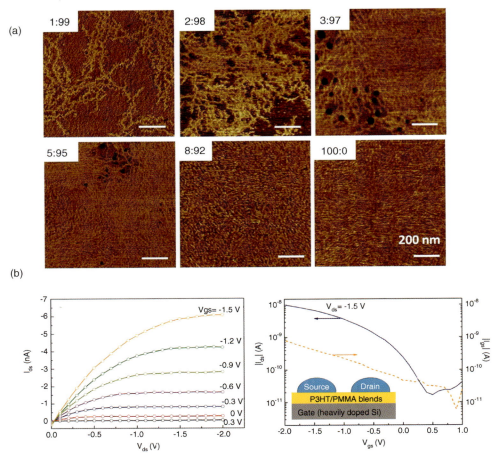

Figure 7.6 (a) AFM phase images of P3HT/PMMA films with different P3HT/PMMA ratios obtained on bare silicon substrates. (b) Output (left) and transfer (right) characteristics of low-voltage-driven devices based on P3HT/PMMA (5:95) blends without any other dielectric: channel length $L = 30$ mm, channel width $W = 1$ mm, determined PMMA layer thickness is about 60 nm. *Inset*: Schematic view of the fabricated low-voltage-driven devices. (Reproduced with permission from Ref. [46]. Copyright 2008, John Wiley & Sons, Inc.)

dielectic, as shown in Figure 7.6b. With V_d below -2 V and V_g in the range of 1 to -2 V, the drain current can be modulated by $\sim 10^3$ with one order lower gate leakage current, and the saturation field-effect mobility was 3×10^{-3} cm^2 V^{-1} s^{-1}. Similar phase behavior was observed in triethylsilylethynyl anthradithiophene (TESADT)/PMMA blend as well [47]. Solvent annealing in this case was used to enhance the vertical phase separation and allow crystallization of the TESADT on the top. OTFTs fabricated in this way exhibited a high field-effect mobility (up to 0.47 cm^2 V^{-1} s^{-1}) with no hysteresis. Furthermore, the application of the one-step approach for the fabrication of all-organic/all-solution-processed OTFTs on flexible substrates has also been confirmed.

7.4.3
Improved Environmental Stability

A major problem in the use of OTFTs is the instability of organic semiconductors to light, atmospheric oxygen, and humidity, or a combination of these stress factors, which limits the shelf life of these components. The exposure of OTFT devices to the environment typically results in devices with high off currents, large and unstable onset voltages, and large subthreshold slopes. Encapsulation of the devices can prevent the penetration of air and humidity and thus effectively enhances the environmental stability of OTFTs.

Arias demonstrated a self-encapsulating OTFT by depositing a blend of semiconductor and encapsulant from solution [48]. When a blend of P3HT or regioregular poly(3,3′′′-didodecylquaterthiophene) (PQT-12) with PMMA in dichlorobenzene was spun onto a substrate treated with octyltrichlorosilane (OTS), instead of the semiconductor-top and dielectric-bottom bilayer structure observed on hydrophilic substrate [46], a bilayer structure with overall encapsulation of PQT-12 or P3HT layers with PMMA layer is obtained. A surface-directed spinodal decomposition mechanism is proposed. By measuring contact angles of PQT-12 and PMMA solutions on surfaces of SiO_2, OTS-treated SiO_2, PQT-12 film, and PMMA film, the largest contrast in wetting properties of PQT-12 and PMMA solutions is observed on OTS-treated SiO_2 surfaces. Therefore, when the polymer blend solution is deposited, preferential absorption of PQT-12 on the OTS-modified substrate is energetically favored over PMMA. As phase separation occurs with the solvent evaporation, the large energy required to form an OTS/PMMA interface laterally prevents domain formation, and the film segregates vertically. Incorporating the self-encapsulated PQT-12/PMMA blends as active layers in OTFTs, the exposure of the semiconductor to the environment is minimized and an improved environmental stability in TFT devices is obtained. As shown in Figure 7.7, the degradation of homo-PQT-12 device is very pronounced after exposure to air for 20 h in the dark. The whole curve is shifted giving a higher onset voltage and higher off current. In contrast, the transfer characteristics do not show any change during the period of 2 days, illustrating the improved environmental stability of PQT-based devices.

7.4.4
Patterned Domains of Polymer Blends

To produce electronic circuitries based on organic devices, it is also crucial to pattern the organic semiconductors on micrometer scale. Spin coating, as a deposition technique used to prepare uniform thin films, has been successfully implemented for the production of organic electronics. However, the films prepared by spin coating are always continuous and need to be patterned using other

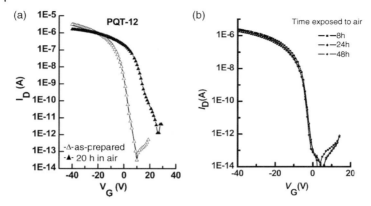

Figure 7.7 (a) Transfer characteristic of homo-PQT TFTs measured under nitrogen flow (open triangles) and after exposure to air for 20 h (closed triangles). (b) Transfer characteristics for TFTs prepared using blends of PQT/PMMA (2/1) measured in air every 8 h for 2 days. (Reproduced with permission from Ref. [48]. Copyright 2006, John Wiley & Sons, Inc.)

techniques. Böltau et al. [49] have demonstrated that the phase-separated domains in a polymer blend thin film can be guided by preferential surface segregation. When a polymer blend is spin cast onto a surface with a prepatterned variation of surface energies, patterns on the substrate surface can be readily transferred to the polymer film. Such surface-directed patterning phase separation of polymer blends combines film deposition and patterning in a one-step process, and thus might enable an effective method for the mass production of organic electronics.

Ginger's group [50, 51] used dip-pen nanolithography (DPN) to generate 16-mercaptohexadecanoic acid (MHA) monolayer templates with backfills of benzenethiol (BZT) for guiding pattern formation in spin-cast polymer blend films (Figure 7.8a). The blend system employed was a blend of polystyrene (PS) with conjugated polymer P3TH. During spin coating of the blend solution, both lateral and vertical phase separation occurred leading to the formation of cylindrical P3HT-rich domains on top of the MHA dots surrounded by vertically separated bilayers with PS in contact with the BZT-coated surface (Figure 7.8c–f). A mechanism of pattern formation has been ascribed to nucleated dewetting of the bottom PS layer at the MHA sites. The final size of the polymer patterns depends both on the template diameter and on the P3HT concentration. By comparing the results obtained from five solutions with relative P3HT/PS concentrations ranging from 60/40 to 20/80 (w/w) and seven substrates with varying MHA template diameters, a sigmoidal relationship between the probability of the formation of P3HT-rich domains and template diameters and P3HT concentration was obviously observed. The minimum polymer patterns with feature size down to 150 nm were achieved using surface templates about 50 nm in size (Figure 7.8f).

In another work, Jaczewska [52] demonstrated that the strategy of pattern-induced self-organization is presented for the spin-cast polythiophene/polystyrene

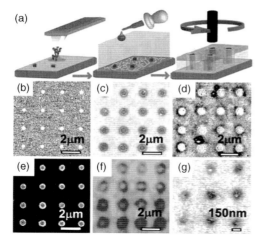

Figure 7.8 (a) Monolayers written by DPN are used for template phase separation in polymer blend films. (b) AFM lateral force image of MHA dots on a BZT background. (c–f) Images of P3HT/PS blend spin coated onto this template. (c) AFM height image (5 nm z-scale). (d) AFM lateral force image. (e) Conducting AFM image (20 nA z-scale). (f) Fluorescence image. (g) AFM height image of 150 nm diameter P3HT domains (2 nm z-scale) on smaller templates at lower P3HT concentration. (Reproduced with permission from Ref. [50]. Copyright 2005, American Chemical Society.)

blends. For example, the P3DDT/PS blends can be ordered by gold with microprinted hexadecanethiol monolayer patterns, the P3HT/PS blends can be aligned by oxidized silicon with silane patterns, and the poly(3-butylthiophene) (P3BT)/PS can be ordered by oxidized silicon with gold patterns. The forces driving pattern-directed self-organization of the polymer blends are discussed based on complementary studies of preferential surface segregation, observed for blend films spin cast on homogeneous surfaces that correspond to the different regions of the surface templates.

A real application of pattern-induced self-organization of polymer blends in OTFT devices was first demonstrated by Salleo and Arias [53]. The material employed was PQT-12/PMMA blend. As mentioned above, during spin coating of this blend, a vertically separated bilayer with PMMA encapsulated on the top surface was obtained with an OTS-treated substrate, while lateral phase separation was observed on untreated substrate. When the substrate was patterned with OTS monolayer, both lateral and vertical phase separation occurred leading to the PQT-12-rich phase preferentially forming on the OTS-treated region and overall encapsulation of the devices with PMMA (Figure 7.9). This system, therefore, not only produces an array of OTFTs, but also improves air stability due to the self-encapsulation. Processing and drying of the film must occur relatively rapidly since such film morphology is metastable. Solvent annealing leads to a loss of vertical phase separation and results in a homogenous, bulk demixed morphology.

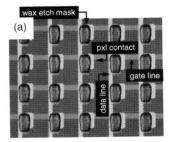 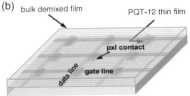

Figure 7.9 Fabrication of a self-assembled OTFT array. In (a), the bright-field wax mask is used as an etch mask to pattern the OTS layer. After spin coating the PQT-12:PMMA blend and drying, phase separation leads to the microstructure sketched. In (b), a PQT-12 thin layer phase separates on the dielectric surface in the channel regions. The wax mask is designed to ensure that the entire PQT-12 film is gated. The rest of the film, which encapsulates and isolates the single transistors, is made of PQT-12 droplets in a PMMA matrix. (Reproduced with permission from Ref. [53]. Copyright 2007, John Wiley & Sons, Inc.)

7.4.5
Improved Charge Carrier Mobility

In order for thin-film transistors to be successfully used as driving circuits in display applications, they need to exhibit high current output, low operating voltage, and high current ratio of "on" and "off" states. Several approaches such as more regioregular molecules, substrate surface modification, controlled solvent evaporation, and post-deposition annealing have been used to improve the performance of TFTs based on the homo-polythiophene system [44, 54–59]. TFT devices based on polythiophene/insulating polymer composites would be expected to show diminished electronic performance related to pure polythiophene. However, a few recent results demonstrate the opposite that blending of polythiophene with an insulating polymer can improve the device performance.

Cho and coworkers have examined the electric characteristics of OTFTs based on P3HT/PMMA blends [46]. The dependence of the field-effect mobility on the P3HT content is shown in detail in Figure 7.10. It is interesting that the blend films with P3HT content ranging from 20 to 5 wt%, in which a good bilayer structure has formed, show much higher charge carrier mobility than homo-P3HT. The reason for this improvement can be ascribed to the modification of the hydrophilic surface of the SiO_2 dielectric by the relatively hydrophobic PMMA. This modification may increase the concentration of the highly oriented P3HT crystal at the critical buried interface between P3HT and the dielectric, and results in an improvement of the TFT performance in the same way as in dielectric surfaces treated with self-assembled monolayers (SAMs) [44, 60–62].

Similar results were also reported on PQT-12/PS blends. Katz's group [63] observed that devices based on PQT-12/PS (2/5) blend films showed field-effect mobility of $8 \times 10^{-2}\,cm^2\,V^{-1}\,s^{-1}$ and an on/off ratio of 10^5 compared to field-effect mobility of $4 \times 10^{-3}\,cm^2\,V^{-1}\,s^{-1}$ and on/off ratio of 4×10^4 for devices based on

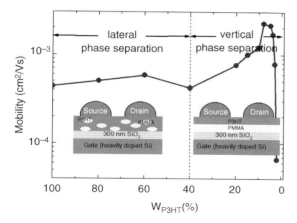

Figure 7.10 Average field-effect mobility measured from saturation region as a function of the P3HT content in the blends. *Inset*: Schematic cross section of the blend TFT. (Reproduced with permission from Ref. [46]. Copyright 2008, John Wiley & Sons, Inc.)

homo-PQT-12. Structure analysis reveals that phase separation leads to a high PQT-12 concentration and more ordering of the top layer of the PQT-12/PS composite, with a mostly PS composite on the bottom layer. Furthermore, when the blends are annealed at the glass temperature of PS, molecular motions near the PQT-12/PS interface could result in a locally annealed, more crystalline PQT-12 domain, which is very beneficial for high mobility.

The possibility of formation of sandwich-like structure has been demonstrated [64]. By spin coating a dimethylsiloxane (DMS) and F8T2 blend with a curing agent onto a silicon oxide substrate, the hydrophilic DMS layer is preferentially adsorbed on the SiO_2 substrate. After thermal curing, the bottom DMS layer becomes a hydrophobic polydimethylsiloxane (PDMS) layer. Meanwhile, the hydrophobic PDMS molecules may move upward to the air/film interface. Finally, a PDMS/F8T2/PDMS sandwich structure forms. These films show similar FET mobility to polymeric semiconductors on octadecyltrichlorosilane-treated SiO_2 dielectric layers indicating that the PDMS bottom layers can modify the film/substrate interface. However, the improvement of environmental stability is rare, which means the passivation effect of PDMS at air/film interface is poor. The reason may be that the coverage of PDMS layers on top surface is complete.

In a recent work, a blend consisting of P3HT and PMMA has been used as the semiconducting layer in water-gated OTFTs [65]. The devices can be operated at very low voltage (below 1 V) with reasonably high mobility (up to $0.15 \text{ cm}^2 \text{V}^{-1} \text{s}^{-1}$). The mobility is not affected when increasing the proportion of PMMA up to 70%, while the off current is significantly lowered, thus resulting in an enhancement in the on/off ratio. The reason could be that the formation of insulating domains decreases the leakage current, or/and the lowering of the P3HT concentration in the film decreases the charge density at the semiconductor/water interface.

7.4.6
Crystallization-Induced Vertical Phase Segregation

Crystallization-induced phase segregation indeed provides a new method to construct vertically stratified structures. If the component is crystalline, the crystallization enthalpy can provide an extra driving force for phase separation. Goffri et al. [66] examined OTFTs consisting of crystalline–crystalline blend of regioregular P3HT and polyethylene (PE). Figure 7.11 quantitatively shows the phase diagram of a 10–90 P3HT–PE polymer blend/xylene system. At low concentrations, all P3HT–PE blends formed homogeneous solutions at elevated temperatures. However, the significantly different solubility of the two components in xylene leads to separated phases in concentrated systems. In mixtures comprising up to approximately 40 wt% of P3HT–PE blend, the PE component crystallized before P3HT, while at higher polymer contents, the consequence was reversed. This special phase behavior naturally allows for two different processing schemes in which either the P3HT or the PE crystallizes first, as shown in Figure 7.11. In scheme I, when the solvent was removed at a temperature (125 °C) above the crystallization temperature of PE, the P3HT component crystallized before PE was further expelled to the film/substrate interface during the following crystallization of PE, and thus blend films with vertically stratified structures were obtained. The concentration of semiconductor in the obtained blends can be reduced to a value as low as 3 wt% without any degradation in device performance. Furthermore, the active

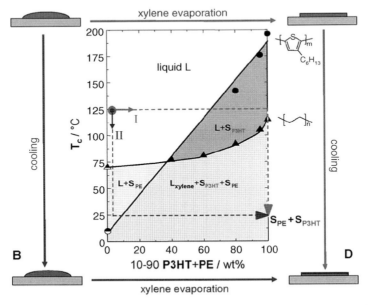

Figure 7.11 Phase diagram of P3HT/PE (1/9) blend and xylene. (Reproduced with permission from Ref. [66]. Copyright 2006, Nature Publishing Group.)

P3HT layer at the interface was effectively encapsulated by the crystalline PE phase on the top, resulting in significantly improved environmental stability. In contrast, when a film was cast at room temperature (scheme II), PE crystallized before P3HT, which resulted in poor device performance ($\mu_{FET} < 8 \times 10^{-6}\,cm^2\,V^{-1}\,s^{-1}$; on/off ratio <100, low P3HT crystallinity, and thickness-independent P3HT distribution in the blend films.

The same group [67] also reported that semiconducting diblock copolymer consisting of PE and P3HT segments (P3HT-*b*-PE) demonstrated a crystallization-induced phase segregation behavior, high charge carrier mobility, and on–off ratios even at 90 wt% insulating polyethylene moiety. In addition, the diblock copolymers displayed outstanding flexibility and toughness with elongations at break exceeding 600% and true tensile strengths around 70 MPa, opening the path toward constructing robust and truly flexible electronic components (Figure 7.12).

The blends or copolymers consisting of P3HT and PE present high-performance and cost-effective materials with improved environmental and mechanical properties. However, the high temperature required for crystallization-induced phase segregation would prevent their practical applications. To solve this problem, Zhao *et al.* [68] demonstrated a facile strategy for constructing crystallization-induced vertical stratified structures in double crystalline P3HT/poly(ethylene glycol) (PEG) systems at room temperature. The key was to ensure the priority of P3HT crystallization sequence using marginal solvents. The following crystallization of PEG drove P3HT crystallites to the interface, resulting in the formation of vertical stratified microstructures with highly crystalline P3HT network.

7.5
Blened Films with Embedded P3HT Nanowires

Controlling blended films to form vertical phase-separated structures allows the connectivity of the semiconducting layer at the channel region between the source and drain electrodes to be maintained, thus retaining the electronic properties of the semiconducting/insulating polymer blends, even at very low semiconductor content levels. However, this morphology depends strongly on the film deposition conditions, and is practically hard to achieve. Therefore, the development of a more facile and general method for the realization of high-performance, low-semiconductor-cost devices would be significant, from both technological and academic standpoint.

It is well known that the percolation threshold is strongly dependent on the aspect ratio. One-dimensional (1D) nanostructures, such as nanowires, nanotubes, and nanoribbons, have geometries that are favorable for the maintenance of connectivity at low content of active materials. Carbon nanotubes (CNTs) are a good representative example. Electrical percolation can be achieved in polymer composites with well-dispersed single-walled carbon nanotubes (SWCNTs) at levels as low as 0.03 wt% [69]. Recently, it has been reported that the formation of semiconducting nanofibers facilitates percolation in semiconducting/insulating polymer blends.

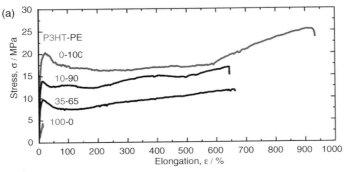

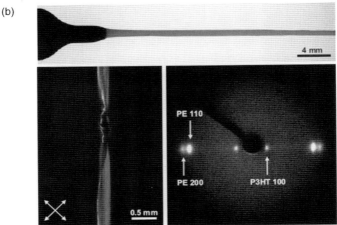

Figure 7.12 (a) Stress–elongation curves recorded at room temperature of elevated temperature cast films of 10–90 and 35–65 P3HT–PE, as well as reference P3HT and PE homopolymers. (b) *Top panel*: Optical micrograph of part of a dog-bone-shaped tensile test sample of 35–65 P3HT–PE stretched 600%, featuring classical neck formation (black–gray transition) due to plastic deformation and resulting oriented, highly birefringent polymer structure. *Bottom left*: A knotted oriented tape, illustrating the excellent toughness of the material (crossed polarizers, directions indicated by arrows and grazing incident light). *Bottom right*: Wide-angle X-ray diffraction patterns of a film of 35–65 P3HT–PE elongated 600% revealing a high degree of uniaxial order and crystallinity of the stretched diblock copolymer film. Tensile deformation direction is vertical. (Reproduced with permission from Ref. [67]. Copyright 2007, John Wiley & Sons, Inc.)

7.5.1
P3AT Nanowires

Polymer semiconductors often have a semicrystalline structure due to their relatively rigid π-conjugated backbone. Compared with two-dimensional (2D) lamellar crystals observed in the crystallization of typical insulating polymers such as polyethylene, π-conjugated polymer semiconductors generally tend to crystallize into 1D nanostructures, such as nanowires; this is induced by both an attractive π–π interaction between polymer backbones, and the crystallization of alkyl side chains. Interst in the growth of crystalline nanowires from π-conjugated polymers has focused on the regioregular P3ATs, due to their great promise as p-type semiconductors for solution-processable, low-cost organic field-effect transistors (OFETs) [5, 70, 71] and solar cells [72–74]. P3AT nanowires are obtained from dilute solutions with limited solubility, in which P3AT molecules tend to aggregate in a face-to-face stacking pattern to minimize unfavorable interactions between the solvent and the aromatic main chains. In 1993, Ihn *et al.* were the first to demostrate the preparation of P3HT nanowires from a dilute *p*-xylene and cyclohexanone solution [75]. These nanowires had a width (*d*) of approximately 15 nm, a length (*L*) exceeding 10 μm, and an aspect ratio (*L/d*) larger than 670. As determined using XRD and TEM, the P3HT chains adopted a face-to-face packing style along the nanowire axis. By taking into account the contour length of P3HT (∼65 nm, calculated from the number-average molecular weight (M_n)) and the width of the nanowires (∼15 nm, corresponding to ∼40 thiophene units), it was confirmed that the polymer chains folded to accommodate the molecular packing within the nanorwires. It has been reported that the morphology of P3AT nanowires depends on factors such as the length of the alkyl side chains, the regioregularity, the molecular weight, the polymer concentration in solution, the crystallization temperature, the solvent quality, and the cooling rate [76–78].

The charge transport in P3AT nanowires synthesized via self-assembly has been studied by fabricating and testing OTFT devices constructed from a single nanowire, or a network of nanowires [79, 80]. A field-effect mobility of $0.02\,\text{cm}^2\,\text{V}^{-1}\,\text{s}^{-1}$ and on/off ratios of ∼10^6 were determined for the P3HT nanowires. The conductance and turn-on voltage were also measured for the single nanowires and nanowire networks as a function of the charge density, temperature, and substrate surface energy. The charge transport mechanism could be interpreted in terms of both the multiple trap and release (MTR) model, and the variable range hopping (VRH) model [80].

7.5.2
Polymer Blends with Embedded P3HT Nanowires

Because polythiophene can self-assemble to form nanowires in dilute solutions, polythiophene nanowire/insulating compounds can be easily achieved by dissolving the insulating components in the same solution. Lu *et al.* [81] were the first to report the preparation of semiconducting/insulating polymer blends with embedded nanowires. Typically, a dilute P3BT solution ($10\,\text{mg}\,\text{ml}^{-1}$) was prepared in the poor solvent, *o*-dichlorobenzene (ODCB), at elevated temperatures. Because of the limited solubility of P3BT in ODCB at room temperature, the cooling of the as-prepared hot P3BT/ODCB solution gradually induced the self-assembly of the P3BT chains into nanowires. Upon addition of PS into the P3BT nanowire suspension solution, and the deposition of the mixed solution via spin coating, a P3BT/PS composite film with an interconnected P3BT nanowire network within the PS matrix was obtained. The conductivity of the P3BT nanowire/PS composite film was enhanced by nearly an order of magnitude, compared with that of pure P3BT. Conductive atomic force microscopy (C-AFM) revealed that the activation energy for charge transportation within the P3BT nanowires was reduced upon the formation of an extremely large interfacial area between the conjugated polymer and the insulating polymer matrix, which contributed to the higher carrier mobility [82].

Cho and coworkers [83] extended the use of semiconducting/insulating polymer blends with embedded nanowires to apply them in OTFT devices. The material system employed was a blend of P3HT with PS. The blends obtained from chloroform (CF) showed typical spinodal decomposition morphologies, and poor P3HT crystallinity, which resulted in the monotonic degradation of electronic performance with increasing PS content. In stark contrast, the blended films obtained using the marginal solvent methylene chloride (CH_2Cl_2) showed a network of superlong and highly crystalline P3HT nanofibers embedded in the insulating PS matrix (Figure 7.13). The use of this material as the active layer in TFT devices allows electronic performances comparable with those of pristine P3HT to be achieved in blends of P3HT and amorphous PS at P3HT content as low as 5 wt%. This structure provides good encapsulation for the active P3HT nanofibers, and, therefore, significantly improves the environmental stability of the devices. In addition, this simple, mild, and reproducible method is quite suitable for the fabrication of large-area, flexible electronic devices.

Recent results demonstrated that the morphology and electronic properties of P3HT nanowires depened on both the solubility of P3HT in the solvent and the aging time of the precursor solution [78, 84]. By comparing the results obtained from a mixed solvent containing chloroform, which is a common solvent for both P3HT and PS, and dioxane (DI), which is a good solvent for PS but a poor solvent for P3HT, sigmoidal relationships between the probability of the formation of

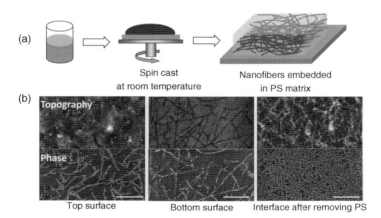

Figure 7.13 (a) Schematic representation of the formation of a nanofibrillar network in a PS matrix. (b) AFM topography (top) and phase (bottom) images of a P3HT/PS (5/95) blend spin cast from CH_2Cl_2: top surface (left), bottom surface (middle), and interface after selectively dissolving PS (right). The scale bar is 500 nm. (Reproduced with permission from Ref. [83]. Copyright 2009, John Wiley & Sons, Inc.)

P3HT nanowires and the DI/CF ratio and aging time were clearly observed. Under optimized conditions, devices based on P3HT/PS blend films containing only 1 wt% P3HT showed field-effect mobilities as high as $1 \times 10^{-2}\,\mathrm{cm^2\,V^{-1}\,s^{-1}}$, values is comparable with those obtained for the pristine P3HT film.

Inkjet printing is an attractive direct patterning technique for the cost-effective fabrication of organic electronic devices, because the designed patterns of ink material can be directly deposited on a substrate, without the need for a prepatterned mask or additional etching processes [9, 85–87]. However, the performance of electronic devices based on inkjet-printed organic semiconductors has been found to be lower than that of devices fabricated using spin casting or vapor deposition processes, due to the uneven morphology and randomly oriented crystalline structures of printed organic semiconducting layers [88]. Cho and coworkers [88] demonstrated the inkjet printing of P3HT/PS blends containing semiconducting nanowires from a mixed solvent composed of chlorobenzene (CB) and cyclohexanone (CHN) (4/1 v/v) (Figure 7.14). The inkjet-printed

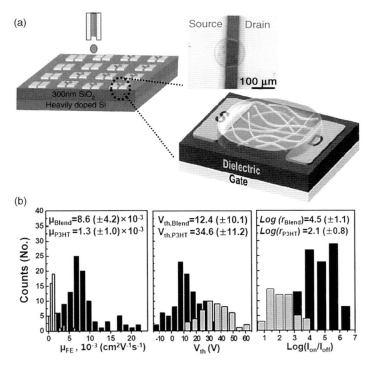

Figure 7.14 (a) Schematic diagram of the fabrication process and optical microscopy image of an inkjet-printed single-droplet transistor based on the P3HT/PS blend. P3HT/PS blend films printed from a CB/CHN mixed solvent had unique structures, with P3HT nanowires dispersed in a PS matrix. (b) Summary of the device characteristics for 100 transistors based on P3HT nanowire/PS (20 : 80) blends (black) and 50 transistors based on P3HT printed from CB (gray) in different batches. (Reproduced with permission from Ref. [88]. Copyright 2010, John Wiley & Sons, Inc.)

blended films with this unique structure provided effective pathways for charge carrier transport through semiconductor nanowires, and significantly improved the environmental stability. In addition, the single-droplet printed transistors showed a high ratio of mobility to conductivity, dramatically enhancing the on–off characteristics.

7.5.3
Nanowires from Conjugated Block Copolymers

Conjugated block copolymers are promising molecular architectures since they can self-assemble into a number of nanoscale morphologies. Copolymers containing polythiophene and insulating flexible segments can often self-assemble into nanowires from solution. McCullough and coworkers [89] synthesized diblock and triblock copolymers composed of P3HT and PS segments with different block compositions. These conjugated copolymers self-assembled to form well-defined nanowires with widths of 30–40 nm upon solvent evaporation. Recently, Yu *et al.* [90] reported that P3HT-*b*-PS with 85 wt% P3HT displayed significantly enhanced charge transport properties and environmental stability compared with the P3HT homopolymer, with the mobility also increasing by a factor of up to 2. Diblock copolymers of P3HT and poly(methyl acrylate) (PMA) were also synthesized and formed nanofibrillar structures. OTFTs based on P3HT-*b*-PMMA exhibited good charge carrier mobilities approaching that of pure P3HT [91].

Jenekhe and coworkers [92] recently demonstrated that block copoly(3-alkylthiophene)s composed of variously sized side chains, for example, poly(3-hexylthiophene)-*b*-poly(3-cyclohexylthiophene) (P3HT-*b*-P3cHT), self-assembled into nanowires. Interestingly, the morphology could be tuned by changing the block composition. Field-effect transistors fabricated from P3HT-*b*-P3cHT thin films showed a hole mobility of $0.0019\,\text{cm}^2\,\text{V}^{-1}\,\text{s}^{-1}$ that was independent of thermal annealing.

7.5.4
Electrospun Nanowires from Conjugated Polymer Blends

Electrospinning is a simple, inexpensive, and scalable method for the production of long, continuous nanofibers of diverse materials using a high electric field. Electrospun nanofibers are of great interest, because they allow the combination of various materials on the nanometer scale. They can exhibit novel properties and phenomena as a result of intermolecular interactions, confinement effects, and extended chain conformations. Babel *et al.* found that electrospun nanofibers consisting of binary blends of poly[2-methoxy-5-(2-ethylhexoxy)-1,4-phenylenevinylene] (MEH-PPV) and P3HT phase separated to give smaller domain sizes than those found in spin-coated thin films, resulting in significant energy transfer [93]. In contrast, such energy transfer was absent in electrospun nanofibers consisting of blends of MEHPPV and poly(9,9-dioctylfluorene) (PFO), likely because of the larger scale of the phase separation in the nanofibers. The MEH-PPV/PHT blend nanofibers exhibited p-channel transistor characteristics with hole mobilities in the range of $(0.05–1) \times 10^{-4}\,\text{cm}^2\,\text{V}^{-1}\,\text{s}^{-1}$ as nonwoven mats and one order of magnitude higher after correction of the channel area.

Due to their low molecular weight, rigid chains, and limited solubility, the electrospinning of conjugated polymers is not as easy as with conventional insulating polymers. Therefore, insulating polymers are often added to assist in the electrospinning of conjugated polymers. A blend of doped polyaniline (PANI) and poly(ethylene oxide) (PEO) was electrospun into nanofibers with diameters in the range of 120–300 nm [94]. Bottom-gated transistors constructed from the PANI/PEO nanofibers exhibited field-effect characteristics at low source–drain voltages, with a hole mobility of $1.4 \times 10^{-4}\,\text{cm}^2\,\text{V}^{-1}\,\text{s}^{-1}$. Field-effect transistors based on electrospun nanofibers of pure P3HT were reported by the groups of Craighead [95] and Pinto [96]. However, the morphology of the P3HT nanofibers was poor, and many beads were present along the fibers. Insulating polymers such as PEO [97], poly(ε-caprolactone) (PCL) [98, 99], PS, and PMMA [100] have been widely used as carrier polymers to improve electrospinning. The field-effect hole mobilities of electrospun nanofibers of pure P3HT were in the range of 0.017–0.192 $\text{cm}^2\,\text{V}^{-1}\,\text{s}^{-1}$. Blended nanowires also showed field-effect behavior with mobilities of 1.6×10^{-5} to $2 \times 10^{-3}\,\text{cm}^2\,\text{V}^{-1}\,\text{s}^{-1}$. A high mobilities of $2\,\text{cm}^2\,\text{V}^{-1}\,\text{s}^{-1}$ and an on/off current ratio of 10^5 were also reported for OTFTs based on P3HT/PCL-blend nanofibers in which traditional silicon oxide was replaced with a polyelectrolyte gate dielectric (Figure 7.15) [99].

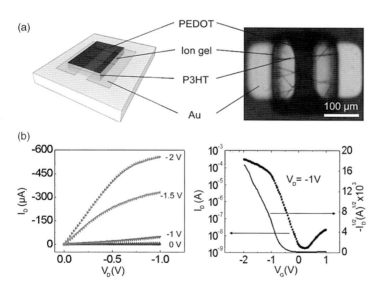

Figure 7.15 (a) Scheme and optical microscopy image showing an ion gel-gated transistor constructed from P3HT nanowires. (b) Output and transfer characteristics of the arrayed transistor. (Reproduced with permission from Ref. [99]. Copyright 2010, American Chemical Society.)

7.6
Conclusions and Outlook

In this chapter, we reviewed recent progress on the use of polythiophene/insulating polymer composites as semiconducting layer of OTFT devices. These composites combining electronic properties of semiconducting polymers with the low-cost and excellent mechanical characteristics of the insulating polymers are promising materials for organic electronics. The solubility of polythiophene and insulating polymer enables facile preparation of blend films via deposition from a common solution. Challenges for OTFTs based on polythiophene/insulating blends deal with the actual trade-off between the electrical performances and other properties because the insulating polymers tend to hinder charge transport in the blended films. Except for the common factors, such as molecular ordering and crystallinity, that affect the electrical performance of polythiophene OTFT devices, phase morphology is another crucial factor for polymer blend-based OTFTs. Blends with vertical stratified structure and with embedded semiconducting nanowires have been proved as effective ways to eliminate the passive effect of insulating polymers and to achieve good electronic properties at low loading of semiconducting polymers. The use of such blends in OTFTs has brought about significantly improved features including material cost, device performance, mechanical properties, environmental stability, device patterning, and one-step formation of gate dielectric and semiconductor. However, the phase separation during solution processes, such as spin coating or printing, is complicated. These techniques are often far from equilibrium. Many factors, such as solvent evaporation rate, solubility parameter, film–substrate interaction, surface tension of each component, and film thickness, affect the final morphology of the blend films. A better understanding and control of phase separation in thin films would be highly desirable. Furthermore, it would be critical to understand the charge transport mechanism in the blends and how the polymer matrix affects it. With continuous emergence of new materials, the development of new blend materials and establishment of criteria for the property optimization are needed for the further improvement in organic electronic devices.

References

1 Shirakawa, H., Lewis, E.J., MacDiarmid, A.G., Chiang, C.K., and Heeger, A. (1977) *J. Chem. Soc., Chem. Commun.*, 578.
2 Tsumura, A., Koezuka, H., and Ando, T. (1986) *Appl. Phys. Lett.*, **49**, 1210.
3 Burroughes, J.H., Bradley, D.D.C., Brown, A.R., Marks, R.N., Mackay, K., Friend, R.H., Burn, P.L., and Holmes, A.B. (1990) *Nature*, **347**, 539.
4 Yu, G., Gao, J., Hummelen, J.C., Wuld, F., and Heeger, A.J. (1995) *Science*, **270**, 1789.
5 Bao, Z., Dodabalapur, A., and Lovinger, A.J. (1996) *Appl. Phys. Lett.*, **69**, 4108.
6 Garnier, F., Hajlaoui, R., Yassar, A., and Srivastava, P. (1994) *Science*, **265**, 1684.
7 Bao, Z.N., Feng, Y., Dodabalapur, A., Raju, V.R., and Lovinger, A.J. (1997) *Chem. Mater.*, **9**, 1299.

8 Rogers, J.A., Bao, Z.N., Makhija, A., and Braun, P. (1999) *Adv. Mater.*, **11**, 741.
9 Sirringhaus, H., Kawase, T., Friend, R.H., Shimoda, T., Inbasekaran, M., Wu, W., and Woo, E.P. (2000) *Science*, **290**, 2123.
10 Yan, H., Chen, Z.H., Zheng, Y., Newman, C., Quinn, J.R., Dotz, F., Kastler, M., and Facchetti, A. (2009) *Nature*, **457**, 679.
11 Allard, S., Forster, M., Souharce, B., Thiem, H., and Scherf, U. (2008) *Angew. Chem., Int. Ed.*, **47**, 4070.
12 Facchetti, A. (2007) *Mater. Today*, **10**, 28.
13 Sirringhaus, H., Brown, P.J., Friend, R.H., Nielsen, M.M., Bechgaard, K., Langeveld-Voss, B.M.W., Spiering, A.J.H., Janssen, R.A.J., Meijer, E.W., Herwig, P., and de Leeuw, D.M. (1999) *Nature*, **401**, 685.
14 Ong, B.S., Wu, Y.L., Liu, P., and Gardner, S. (2004) *J. Am. Chem. Soc.*, **126**, 3378.
15 McCulloch, I., Heeney, M., Bailey, C., Genevicius, K., Macdonald, I., Shkunov, M., Sparrowe, D., Tierney, S., Wagner, R., Zhang, W.M., Chabinyc, M.L., Kline, R.J., McGehee, M.D., and Toney, M.F. (2006) *Nat. Mater.*, **5**, 328.
16 Tsao, H.N., Cho, D.M., Park, I., Hansen, M.R., Mavrinskiy, A., Yoon, D.Y., Graf, R., Pisula, W., Spiess, H.W., and Mullen, K. (2011) *J. Am. Chem. Soc.*, **133**, 2605.
17 Yang, F., Shtein, M., and Forrest, S.R. (2005) *Nat. Mater.*, **4**, 37.
18 Yang, X.N., Loos, J., Veenstra, S.C., Verhees, W.J.H., Wienk, M.M., Kroon, J.M., Michels, M.A.J., and Janssen, R.A.J. (2005) *Nano Lett.*, **5**, 579.
19 Currie, M.J., Mapel, J.K., Heidel, T.D., Goffri, S., and Baldo, M.A. (2008) *Science*, **321**, 226.
20 Bundgaard, E. and Krebs, F.C. (2007) *Sol. Energy Mater. Sol. Cells*, **91**, 954.
21 Arias, A.C., MacKenzie, J.D., Stevenson, R., Halls, J.J.M., Inbasekaran, M., Woo, E.P., Richards, D., and Friend, R.H. (2001) *Macromolecules*, **34**, 6005.
22 Berggren, M., Inganas, O., Gustafsson, G., Rasmusson, J., Andersson, M.R., Hjertberg, T., and Wennerstrom, O. (1994) *Nature*, **372**, 444.
23 Chappell, J., Lidzey, D.G., Jukes, P.C., Higgins, A.M., Thompson, R.L., O'Connor, S., Grizzi, I., Fletcher, R., O'Brien, J., Geoghegan, M., and Jones, R.A.L. (2003) *Nat. Mater.*, **2**, 616.
24 Granstrom, M. and Inganas, O. (1996) *Appl. Phys. Lett.*, **68**, 147.
25 Halls, J.J.M., Walsh, C.A., Greenham, N.C., Marseglia, E.A., Friend, R.H., Moratti, S.C., and Holmes, A.B. (1995) *Nature*, **376**, 498.
26 Yim, K.H., Zheng, Z.J., Friend, R.H., Huck, W.T.S., and Kim, J.S. (2008) *Adv. Funct. Mater.*, **18**, 2897.
27 Shih, P.I., Tseng, Y.H., Wu, F.I., Dixit, A.K., and Shu, C.F. (2006) *Adv. Funct. Mater.*, **16**, 1582.
28 Liedtke, A., O'Neill, M., Wertmoller, A., Kitney, S.P., and Kelly, S.M. (2008) *Chem. Mater.*, **20**, 3579.
29 Babel, A., Zhu, Y., Cheng, K.F., Chen, W.C., and Jenekhe, S.A. (2007) *Adv. Funct. Mater.*, **17**, 2542.
30 Babel, A., Wind, J.D., and Jenekhe, S.A. (2004) *Adv. Funct. Mater.*, **14**, 891.
31 Shi, J.W., Wang, H.B., Song, D., Tian, H., Geng, Y.H., and Yan, D.H. (2007) *Adv. Funct. Mater.*, **17**, 397.
32 Pal, B.N., Trottman, P., Sun, J., and Katz, H.E. (2008) *Adv. Funct. Mater.*, **18**, 1832.
33 Cho, S., Yuen, J., Kim, J.Y., Lee, K., Heeger, A.J., and Lee, S. (2008) *Appl. Phys. Lett.*, **92**, 063505.
34 Horowitz, G. (1998) *Adv. Mater.*, **10**, 365.
35 Flory, P.J. (1941) *J. Chem. Phys.*, **9**, 660.
36 Huggins, M.L. (1941) *J. Chem. Phys.*, **9**, 440.
37 Flory, P.J. (1942) *J. Chem. Phys.*, **1942**, 51.
38 Arias, A.C. (2006) *Polym. Rev.*, **46**, 103.
39 Cahn, J.W. and Hilliard, J.E. (1958) *J. Chem. Phys.*, **28**, 258.
40 Heriot, S.Y. and Jones, R.A.L. (2005) *Nat. Mater.*, **4**, 782.
41 Walheim, S., Böltau, M., Mlynek, J., Krausch, G., and Steiner, U. (1997) *Macromolecules*, **30**, 4995.
42 Park, J., Park, S.Y., Shim, S.O., Kang, H., and Lee, H.H. (2004) *Appl. Phys. Lett.*, **85**, 3283.
43 Park, J., Shim, S.O., and Lee, H.H. (2005) *Appl. Phys. Lett.*, **86**.
44 Kline, R.J., McGehee, M.D., and Toney, M.F. (2006) *Nat. Mater.*, **5**, 222.
45 Chua, L.L., Ho, P.K.H., Sirringhaus, H., and Friend, R.H. (2004) *Adv. Mater.*, **16**, 1609.

46 Qiu, L., Lim, J.A., Wang, X., Lee, W.H., Hwang, M., and Cho, K. (2008) *Adv. Mater.*, **20**, 1141.

47 Lee, W.H., Lim, J.A., Kwak, D., Cho, J.H., Lee, H.S., Choi, H.H., and Cho, K. (2009) *Adv. Mater.*, **21**, 4243.

48 Arias, A.C., Endicott, F., and Street, R.A. (2006) *Adv. Mater.*, **18**, 2900.

49 Böltau, M., Walheim, S., Mlynek, J., Krausch, G., and Steiner, U. (1998) *Nature*, **391**, 877.

50 Coffey, D.C. and Ginger, D.S. (2005) *J. Am. Chem. Soc.*, **127**, 4564.

51 Wei, J.H., Coffey, D.C., and Ginger, D.S. (2006) *J. Phys. Chem. B*, **110**, 24324.

52 Jaczewska, J., Budkowski, A., Bernasik, A., Raptis, I., Moons, E., Goustouridis, D., Haberko, J., and Rysz, J. (2009) *Soft Matter*, **5**, 234.

53 Salleo, A. and Arias, A.C. (2007) *Adv. Mater.*, **19**, 3540.

54 Salleo, A., Chabinyc, M.L., Yang, M.S., and Street, R.A. (2002) *Appl. Phys. Lett.*, **81**, 4383.

55 Jung, Y., Kline, R.J., Fischer, D.A., Lin, E.K., Heeney, M., McCulloch, L., and DeLongchamp, D.M. (2008) *Adv. Funct. Mater.*, **18**, 742.

56 Sirringhaus, H., Wilson, R.J., Friend, R.H., Inbasekaran, M., Wu, W., Woo, E.P., Grell, M., and Bradley, D.D.C. (2000) *Appl. Phys. Lett.*, **77**, 406.

57 Kim, D.H., Jang, Y., Park, Y.D., and Cho, K. (2005) *Langmuir*, **21**, 3203.

58 Kim, D.H., Jang, Y., Park, Y.D., and Cho, K. (2006) *Macromolecules*, **39**, 5843.

59 Kim, D.H., Park, Y.D., Jang, Y.S., Yang, H.C., Kim, Y.H., Han, J.I., Moon, D.G., Park, S.J., Chang, T.Y., Chang, C.W., Joo, M.K., Ryu, C.Y., and Cho, K.W. (2005) *Adv. Funct. Mater.*, **15**, 77.

60 Veres, J., Ogier, S., Lloyd, G., and de Leeuw, D. (2004) *Chem. Mater.*, **16**, 4543.

61 Grecu, S., Roggenbuck, A., Opitz, A., and Brutting, W. (2006) *Org. Electron.*, **7**, 276.

62 Kim, D.H., Lee, H.S., Yang, H.C., Yang, L., and Cho, K. (2008) *Adv. Funct. Mater.*, **18**, 1363.

63 Sun, J., Jung, B.J., Lee, T., Berger, L., Huang, J., Liu, Y., Reich, D.H., and Katz, H.E. (2009) *ACS Appl. Mater. Interfaces*, **1**, 412.

64 Chung, D.S., Lee, D.H., Park, J.W., Jang, J., Nam, S., Kim, Y.H., Kwon, S.K., and Park, C.E. (2009) *Org. Electron.*, **10**, 1041.

65 Kergoat, L., Battaglini, N., Miozzo, L., Piro, B., Pham, M.C., Yassar, A., and Horowitz, G. (2011) *Org. Electron.*, **12**, 1253.

66 Goffri, S., Muller, C., Stingelin-Stutzmann, N., Breiby, D.W., Radano, C.P., Andreasen, J.W., Thompson, R., Janssen, R.A.J., Nielsen, M.M., Smith, P., and Sirringhaus, H. (2006) *Nat. Mater.*, **5**, 950.

67 Muller, C., Goffri, S., Breiby, D.W., Andreasen, J.W., Chanzy, H.D., Janssen, R.A.J., Nielsen, M.M., Radano, C.P., Sirringhaus, H., Smith, P., and Stingelin-Stutzmann, N. (2007) *Adv. Funct. Mater.*, **17**, 2674.

68 Zhao, K., Ding, Z.C., Xue, L.J., and Han, Y.C. (2010) *Macromol. Rapid Commun.*, **31**, 532.

69 Chatterjee, T., Yurekli, K., Hadjiev, V.G., and Krishnamoorti, R. (2005) *Adv. Funct. Mater.*, **15**, 1832.

70 Sirringhaus, H., Tessler, N., and Friend, R.H. (1998) *Science*, **280**, 1741.

71 Dimitrakopoulos, C.D. and Malenfant, P.R.L. (2002) *Adv. Mater.*, **14**, 99.

72 Coakley, K.M. and McGehee, M.D. (2004) *Chem. Mater.*, **16**, 4533.

73 Gunes, S., Neugebauer, H., and Sariciftci, N.S. (2007) *Chem. Rev.*, **107**, 1324.

74 Thompson, B.C. and Frechet, J.M.J. (2008) *Angew. Chem., Int. Ed.*, **47**, 58.

75 Ihn, K.J., Moulton, J., and Smith, P. (1993) *J. Polym. Sci. B*, **31**, 735.

76 Lu, G.H., Li, L.G., and Yang, X.N. (2008) *Macromolecules*, **41**, 2062.

77 Liu, J.H., Arif, M., Zou, J.H., Khondaker, S.I., and Zhai, L. (2009) *Macromolecules*, **42**, 9390.

78 Samitsu, S., Shimomura, T., and Ito, K. (2008) *Thin Solid Films*, **516**, 2478.

79 Merlo, J.A. and Frisbie, C.D. (2003) *J. Polym. Sci. B*, **41**, 2674.

80 Merlo, J.A. and Frisbie, C.D. (2004) *J. Phys. Chem. B*, **108**, 19169.

81 Lu, G.H., Tang, H.W., Qu, Y.P., Li, L.G., and Yang, X.N. (2007) *Macromolecules*, **40**, 6579.

82 Lu, G.H., Tang, H.W., Huan, Y.A., Li, S.J., Li, L.G., Wang, Y.Z., and Yang, X.N. (2010) *Adv. Funct. Mater.*, **20**, 1714.

83 Qiu, L.Z., Lee, W.H., Wang, X.H., Kim, J.S., Lim, J.A., Kwak, D., Lee, S., and Cho, K. (2009) *Adv. Mater.*, **21**, 1349.

84 Qiu, L.Z., Wang, X., Lee, W.H., Lim, J.A., Kim, J.S., Kwak, D., and Cho, K. (2009) *Chem. Mater.*, **21**, 4380.

85 Lim, J.A., Lee, H.S., Lee, W.H., and Cho, K. (2009) *Adv. Funct. Mater.*, **19**, 1515.

86 Lim, J.A., Lee, W.H., Lee, H.S., Lee, J.H., Park, Y.D., and Cho, K. (2008) *Adv. Funct. Mater.*, **18**, 229.

87 Singh, M., Haverinen, H.M., Dhagat, P., and Jabbour, G.E. (2010) *Adv. Mater.*, **22**, 673.

88 Lim, J.A., Kim, J.H., Qiu, L., Lee, W.H., Lee, H.S., Kwak, D., and Cho, K. (2010) *Adv. Funct. Mater.*, **20**, 3292.

89 Liu, J.S., Sheina, E., Kowalewski, T., and McCullough, R.D. (2002) *Angew. Chem. Int. Ed.*, **41**, 329.

90 Yu, X., Xiao, K., Chen J.H., Lavrik, N.V., Hong, K.L., Sumpter, B.G., and Geohegan, D.B. (2011) *Acs Nano*, **5**, 3559.

91 Sauve, G. and McCullough, R.D. (2007) *Adv. Mater.*, **19**, 1822.

92 Wu, P.T., Ren, G.Q., Kim, F.S., Li, C.X., Mezzenga, R., and Jenekhe, S.A. (2010) *J. Polym. Sci. A*, **48**, 614.

93 Babel, A., Li, D., Xia, Y.N., and Jenekhe, S.A. (2005) *Macromolecules*, **38**, 4705.

94 Pinto, N.J., Johnson, A.T., MacDiarmid, A.G., Mueller, C.H., Theofylaktos, N., Robinson, D.C., and Miranda, F.A. (2003) *Appl. Phys. Lett.*, **83**, 4244.

95 Liu, H.Q., Reccius, C.H., and Craighead, H.G. (2005) *Appl. Phys. Lett.*, **87**, 253106.

96 Gonzalez, R. and Pinto, N.J. (2005) *Synth. Met.*, **151**, 275.

97 Pinto, N.J., Carrasquillo, K.V., Rodd, C.M., and Agarwal, R. (2009) *Appl. Phys. Lett.*, **94**, 073505.

98 Lee, S., Moon, G.D., and Jeong, U. (2009) *J. Mater. Chem.*, **19**, 743.

99 Lee, S.W., Lee, H.J., Choi, J.H., Koh, W.G., Myoung, J.M., Hur, J.H., Park, J.J., Cho, J.H., and Jeong, U. (2010) *Nano Lett.*, **10**, 347.

100 Chen, J.Y., Kuo, C.C., Lai, C.S., Chen, W.C., and Chen, H.L. (2011) *Macromolecules*, **44**, 2883.

8
Semiconducting Organic Molecule/Polymer Composites for Thin-Film Transistors
Jeremy N. Smith, John G. Labram, and Thomas D. Anthopoulos

8.1
Introduction

Small-molecule organic semiconductors have been studied extensively as solution-processed thin films [1], thermally evaporated films [2], and in single crystalline form [3]. Often they have excellent electrical properties as a result of the high levels of crystallinity and close π-orbital overlap between molecules. These intermolecular interactions are relatively weak and lead to narrow electronic bands; however, in contrast to polymeric semiconductors, it is possible to significantly reduce the density of states arising from disorder in the material. Unfortunately, suitable processing to achieve this is not always simple or scalable to large-area electronics. For example, the highest charge carrier mobilities have been reported in small-molecule single crystals, which cannot straightforwardly be used to make large arrays of electronic devices. Thin films of such materials tend to suffer from grain boundary limited charge transport [4], which can significantly reduce the maximum measured carrier mobility. There may also be issues with film uniformity and/or anisotropic material properties. Conversely, it is usually very easy to form a uniform polymer film that is largely isotropic and results in very low device-to-device variation over large areas. However, in general charge carrier mobilities in polymers are lower than those in many small molecules [5]. This is as a result of the inherent disorder in polymeric materials and the fact that it is almost impossible to form a completely crystalline film.

This chapter will therefore explore how composites of small molecules and polymers have been employed to combine some of the advantageous properties of each material type. The combination of inherently high mobility of small molecules and the processability of polymers is one promising route to control. In addition, interesting electronic properties can be achieved, for example, by blending n- and p-type materials, an extremely useful feature for fabricating integrated logic circuits based on complementary or complementary-like logic.

Organic field-effect transistors (OFETs) are important both in organic electronics applications and as a probe for the electronic properties of the materials system. In

Semiconducting Polymer Composites: Principles, Morphologies, Properties and Applications, First Edition.
Edited by Xiaoniu Yang.
© 2012 Wiley-VCH Verlag GmbH & Co. KGaA. Published 2012 by Wiley-VCH Verlag GmbH & Co. KGaA.

particular, the field-effect charge carrier mobility is a useful measure of performance. For display backplanes [6] that control current-driven devices, a larger OFET mobility results in higher currents for a given transistor size. In many other circuits such as organic radio-frequency identification (RFID) [7] devices, higher mobilities equate to faster dynamic circuit operation. The OFET characteristics can also be used to understand charge transport in organics and in the case of composite systems can be used to study effects such as film morphology [8], charge percolation [9], and phase separation [10]. Such systems are common in organic photovoltaic [11] and organic light-emitting diode [12] (OLED) applications but are less frequently applied in OFETs due to their inherently higher complexity compared to single-component films. Despite this, there are many interesting features of the composite molecule/polymer OFET that have been studied.

8.1.1
OFET Device Operation

The basic operation of a thin film OFET can be modeled using a similar method to its inorganic equivalent [13]. Various device architectures are possible employing a variety of materials; however, all OFETs consist of metallic source and drain electrodes for injection/extraction of charge to/from the semiconductor thin film, and a third gate electrode separated from the semiconducting channel by a dielectric. The latter induces accumulation or depletion of charges at the semiconductor–dielectric interface by the application of an electric field. A schematic of this device is shown in Figure 8.1. If the geometric capacitance of the dielectric, C_i, is known, then the surface charge density can be estimated and from this the current that flows from source to drain for a given drain voltage, V_d. This current, I_d, is given by

$$I_d = \frac{W}{L} \mu C_i \left[(V_g - V_T) V_d - \frac{V_d^2}{2} \right], \tag{8.1}$$

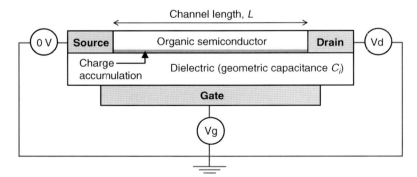

Figure 8.1 Schematic representation of an organic field-effect transistor with typical measurement voltages. The accumulation layer forms at the semiconductor–dielectric interface due to the gate voltage, V_g. The drain voltage, V_d, is then used to extract charges from the semiconductor.

where W and L are the transistor channel width and length, respectively, V_g is the gate voltage, V_T is the threshold voltage, and μ is the charge carrier mobility. By measuring I_d as a function of V_g (transfer characteristics) and V_d (output characteristics), it is possible to extract mobility, threshold voltage, and current on/off ratio as well as analyze charge injection and charge trapping issues. The latter two effects along with electric field- and carrier density-dependent mobility will lead to a deviation from the ideal expressed in Eq. (8.1). However, the mobility can usually be estimated using the following equations in the linear operating regime, where $V_d \ll V_g$, and in the saturation regime, where $V_d \geq V_g$:

$$\mu_{\text{lin}} = \frac{L}{W C_i V_d} \left(\frac{\partial I_{d,\text{lin}}}{\partial V_g} \right),$$

$$\mu_{\text{sat}} = \frac{L}{W C_i} \left(\frac{\partial^2 I_{d,\text{sat}}}{\partial V_g^2} \right) = \frac{2L}{W C_i} \left(\frac{\partial \sqrt{I_{d,\text{sat}}}}{\partial V_g} \right)^2. \qquad (8.2)$$

The actual device structure and materials can play a significant role in the OFET performance due to the position of the source and drain contacts, energy level offsets between metal and semiconductor molecular orbitals, and the nature of the dielectric–semiconductor interface. This can be especially true in the case of blended semiconductors where phase separation can lead to different concentrations of the components at the different interfaces. By changing the contact positions, four device architectures can be achieved (Figure 8.2), which are either staggered or coplanar depending on whether the dielectric is separated from the

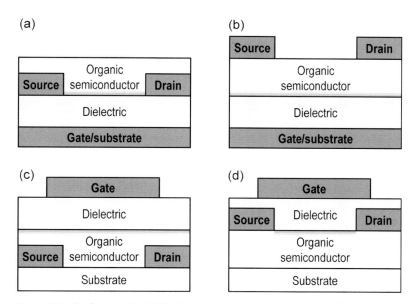

Figure 8.2 The four possible OFET device architectures: (a) bottom-gate, bottom-contact, (b) bottom-gate, top-contact, (c) top-gate, bottom-contact, and (d) top-gate, top-contact. (a) and (d) are considered coplanar while (b) and (c) are staggered.

source and drain by the semiconducting layer or not. Staggered architectures can be advantageous from an injection perspective due to gate field-enhanced injection [14] while the choice of position for the source, drain, and dielectric can be vital in controlling interface properties. Ease of manufacture also plays a role with bottom-gate, bottom-contact designs being simple to process especially in more complex circuits [15].

8.1.2
Small-Molecule/Polymer Film Morphology

When two or more components are present in a thin film, there is the potential for phase separation and therefore a wide range of microstructural variation. In addition, for solution-processed materials, the solvent acts as a further component that is controllably removed during fabrication. Therefore, in the case of small-molecule/polymer systems, there can be several factors that determine the overall phase behavior: first, the thermodynamics of mixing between components and the balance of entropic and enthalpic contributions to the free energy of mixing, ΔG_{mix}; second, the interaction between solution and substrate or atmosphere interfaces; and finally, the kinetics of solvent evaporation and changes to the solution that this induces, such as viscosity variation or phase separation within the solution. It should be noted that these processes are often far from thermodynamic equilibrium leading to a film microstructure that can sometimes be difficult to predict.

Liquid–liquid phase separation is often the first process to occur during solution processing. In general, this can be through a nucleation and growth mechanism [16] or a spinodal decomposition mechanism [17]. Given the fast rates involved in methods such as spin coating, the latter is more common and has been observed in several polymer systems. For a two-component system with a composition c, a typical phase diagram is shown in Figure 8.3. The free energy, ΔG_{mix}, varies with composition leading to three distinct regions: complete solution, a metastable region, and an unstable region. The latter means that the solution is unstable to any small fluctuation in composition and so spontaneous and rapid phase separation occurs – this is termed spinodal decomposition. If the solution is metastable, nucleation and growth of a second phase is possible. The composition fluctuations in the case of spinodal decomposition can be described by characteristic waves with randomly oriented wave vectors, q. Wavelengths longer than a critical value will be unstable and diffusion of material will bring the local concentration back to a stable regime. As phase separation progresses, the average q reduces and the microstructure coarsens. The mechanism is complicated by the addition of a solvent; however, in this case rather than quenching into a two-phase region by reducing the temperature, a change in composition caused by solvent evaporation drives the phase separation. In OFET devices, we need to consider thin films rather than bulk behavior. The two material interfaces break the symmetry of the system and thus q values are no longer randomly oriented. Instead, wave vectors normal to the surface may dictate the final microstructure leading to a vertically varying composition profile.

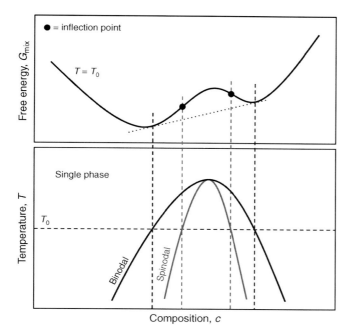

Figure 8.3 Free energy of mixing, ΔG_{mix}, at a particular temperature T_0 and how this translates into a phase diagram for a two-component system as a function of composition. The phase diagram shows the single-phase region, a metastable region between the spinodal and binodal lines, and the unstable region inside the spinodal.

In polymer/polymer systems, the entropy change on mixing is usually negligible especially for high molecular weights and noncrystalline systems [18]. Therefore, mixing is only favorable when there is a strong enthalpic interaction between components and so phase separation is common. This is true not only for polymer melts but also for polymers in solution even at relatively low concentrations [19]. If one of the components is replaced by a small molecule, then there can be significant entropy changes that may favor mixing of the components. On the other hand, the introduction of a highly crystalline small molecular component, or even semi-crystalline polymer, can lead to an enthalpic driving force for solid–liquid phase separation due to the strong intermolecular interactions of the crystalline component. The point at which crystallization of a component occurs during processing will also significantly affect the film microstructure. Generally, this will be controlled by the composition of the blend, the temperature, and the solubilities of the various components. For example, by allowing one component to solidify first, it may be expelled to the film interface as a separate phase on solidification of the remaining material. An alternative phase separation mechanism is through film annealing after solution processing. In this case, solid–solid separation occurs again often as a result of the crystallization of one or more of the components. This can be useful as a way of controlling film microstructure without having to consider the detailed kinetics of the solution process.

8.2
Unipolar Films for OFETs

Small-molecule/polymer blends where the molecular component is either n- or p-type are commonly used to enhance processability in organic electronics. The small molecule will inherently have a high charge carrier mobility, such as those based on acenes or oligothiophenes, and can then be combined with polymers that are easy to process and/or cheap to manufacture. Examples include simple engineering polymers such as polystyrene (PS) and poly(methyl methacrylate) (PMMA) as well as semiconductors that tend to have mobilities much lower than the small molecule but are enhanced by blending, such as polytriarylamines and polythiophenes. One of the main issues when employing these blends for OFETs is the percolation of charge carriers between source and drain. Whether charges are transported predominantly through high-mobility pathways and whether injection occurs into the polymer or the small molecule are important factors in the overall device performance.

This section will therefore examine several unipolar blend systems that have been employed to improve OFET mobility, to enhance processability especially with a view to printing devices, to study the links between morphology and charge transport, and finally to enable novel fabrication techniques such as gate dielectric self-assembly. We will broadly divide the section into blends based on oligothiophenes and those based on acenes since this covers a large fraction of the materials and concepts investigated to date.

8.2.1
Oligothiophene/Polymer Blends

Oligothiophene molecules have been studied extensively as organic semiconductors, originally in thermally evaporated films and more recently processed from solution, a common example being α-sexithiophene [20]. Such materials can be solution processed by the addition of solubilizing side chains, usually in the form of long alkyl chains. The molecule α,ω-dihexylquaterthiophene (DH4T) has a mobility of $0.23\,\text{cm}^2\,\text{V}^{-1}\,\text{s}^{-1}$ in single grains within an evaporated film [21] as well as is readily soluble. However, lower mobilities are always obtained in solution-processed films and films where grain boundaries are present; therefore, an early approach by Russell *et al.* [22] was to utilize a blend of DH4T with poly(3-hexylthiophene) (P3HT). Maximum mobilities were still only around $0.01\,\text{cm}^2\,\text{V}^{-1}\,\text{s}^{-1}$ with 60 wt% D4HT; however, this was limited by the low P3HT mobility and in fact the work demonstrates a useful percolation model that can be used to understand how two-phase films may behave as the concentration of small molecule is increased. It was observed that there is a critical concentration of DH4T at 29 wt%. When changing from 10 to 50 wt% small molecule, there was an order of magnitude change in measured OFET mobility. This coincided with a change in film microstructure – at low DH4T concentration there is minimal agglomeration of material whereas as the concentration increases, well-defined DH4T-rich crystallites appear to be embedded in a bulk, polymer-rich phase. When percolation of carriers in this high-mobility phase becomes significant, the device properties change dramatically. The logarithm of the mobility was fitted using a simple sigmoidal function in order to extract the percolation threshold, c_0, where

$$\mu = \mu_{\text{polymer}} + \frac{\mu_{\text{max}}}{1 + \exp[(c_0 - c)/d]}. \tag{8.3}$$

In this case, d is a fitting parameter relating to the width of the sigmoidal function, μ_{polymer} is the mobility of the polymer alone, and μ_{max} is the maximum achievable mobility in the blend. The latter will ideally be the intrinsic mobility of the high-mobility small molecule; however, this is rarely the case. The aim is to achieve a μ_{max} that is greater than simply processing the small molecule with no polymer matrix. This will also be lower than the intrinsic mobility due to the problems discussed earlier with thin-film formation of highly crystalline materials.

The model of percolation itself is relatively straightforward and is based on the fraction of a particular phase within a particular conduction pathway. The mobility within the pathway is given by

$$\mu_{\text{path}}(f) = \frac{\mu_H}{1 + f(\mu_H/\mu_L - 1)} \approx \frac{\mu_H}{1 + f(\mu_H/\mu_L)}, \quad \text{when} \quad \mu_H \gg \mu_L, \tag{8.4}$$

where H and L refer to the high- and low-mobility phases, respectively, and f is the fraction of low-mobility phase within the pathway. The two phases in a composite will generally be a polymer-rich phase and a small molecule-rich phase where the former will have a considerably lower mobility than the latter. In this case, $\mu_{\text{path}}(f)$ becomes almost independent of μ_H, and additionally μ_L is approximately equal to the polymer mobility, μ_{polymer}. Percolation theory predicts that the high-mobility pathways with low values of f dominate transport. In the DH4T:P3HT blends, it was found that f values of around 4 wt% were consistent with the observed device mobilities. Therefore, by calculating $\mu_{\text{path}}(f)$ as a function of μ_L, it was possible to show that modest improvements to the polymer mobility can potentially lead to very high-performance blend devices (see Figure 8.4). However, even using insulating polymers where μ_L will be very low, by decreasing f at the conduction interface high-mobility OFETs are feasible. Examples of this include many of the more recent acene-based composite devices that will be discussed in the next section, where f can be reduced to <1 wt%.

The DH4T:P3HT system is not the only one employing oligothiophenes as the small-molecule component. It has been shown that even without the solubilizing side chains, the quaterthiophene (4T) molecule can be processed from solution by blending with linear high-density polyethylene (HDPE) [23]. Normally such materials are challenging to process due to melting temperatures above 200 °C, very limited solubility at room temperature, and low-viscosity solutions. In contrast, the mechanical properties of engineering polymers such as HDPE are excellent for large-area and flexible films. Careful control of crystallization allowed blend OFETs to function even with 4T concentrations as low as 10 wt%. This makes processing not only simpler but also cheaper since very small amounts of the semiconductor are needed and the remainder of the film is a ubiquitous, mass-produced polymer.

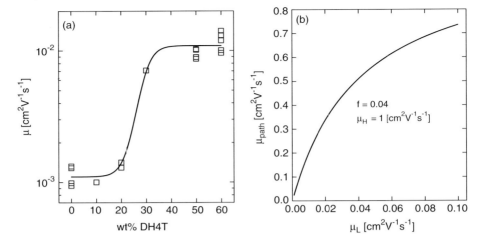

Figure 8.4 (a) Mobility as a function of composition for DH4T:P3HT blend OFETs. The data are fitted using Eq. (8.3). (b) The predicted increase in percolation pathway mobility as the mobility of the polymer, μ_L, is increased. Values of f and μ_H are based on the DH4T:P3HT system. (Adapted from Ref. [22] with permission of the American Institute of Physics.)

Both examples so far have employed normally p-type materials; however, it is also possible to synthesize oligothiophenes that are good electron conductors by the addition of perfluorophenacyl end groups [24]. By blending this molecule with a suitable fluorinated polythiophene, solution-cast OFETs were realized with the potential for use in printing and other solution-based processes. Despite a lower performance in the blend (0.01 cm^2 V^{-1} s^{-1}) than in the drop-cast small molecule (0.21 cm^2 V^{-1} s^{-1}), it is likely that optimizing the choice of polymer and crystallization conditions would significantly improve performance. As the following will describe, this has been done in the case of 4T:HDPE.

A fundamental feature of the 4T:HDPE blend is the crystallinity of the polymer, which can be as high as 90%, essentially leading to a crystalline–crystalline system. If one can control the order of crystallization within the film during spin coating, then crystallization-induced phase separation can lead to a high degree of vertical phase separation with very distinct high- and low-mobility phases. Just as in the DH4T:P3HT OFETs, if there is very little polymer at the conduction interface, then transport is governed by the oligothiophene. For this to occur, it is crucial that the semiconductor crystallizes first and is thus expelled to the film interfaces when the insulator crystallizes. The advantage of the small-molecule/polymer system is that usually the former will crystallize more readily than the latter, which leads to the "correct" vertical phase separation for OFET applications. In the study by Wolfer et al., the processing conditions were used to create morphological differences in the thin films and observe their effects on electronic performance. Rapid solvent removal at 130 °C followed by solidification was compared to quenching and room-temperature solvent evaporation. The former led to a finer microstructure with uniform 4T crystallites, while the latter yielded a coarser microstructure. In addition, two different 4T crystal structures were measured using wide-angle X-ray diffraction; however, this is unlikely to have influenced charge transport as much as the microstructural variation. By constructing the binary phase diagram from differential scanning calorimetry results, the point of liquid miscibility can be estimated and correlated to the critical concentration of small molecule needed for phase separation and therefore electronic percolation. Thus, in the films crystallized at 130 °C, OFET mobilities were $\sim 10^{-4}$ $cm^2 V^{-1} s^{-1}$ as long as the 4T concentration was large enough to induce liquid–liquid phase separation. At lower concentrations and for the room-temperature cast samples, mobilities were considerably lower. The lack of phase separation in the former case and the large-scale but nonuniform phase separation in the latter are the main reasons for poor device performance. The maximum blend mobility is comparable to OFETs prepared from thermally evaporated 4T and demonstrates that, with a detailed understanding of crystallization, even hard to process oligomers can function in a composite.

8.2.2
Acene/Polymer Blends

This section will focus on another highly successful group of molecular semiconductors, namely, acenes. Pentacene has become a very common and well-understood material in organic electronics with thermally evaporated thin films giving OFET mobilities of 1.5 $cm^2 V^{-1} s^{-1}$ on SiO_2 [25] and up to 6 $cm^2 V^{-1} s^{-1}$ on polymeric dielectrics [26]. In addition, some of the highest reported single-crystal OFET mobilities are obtained with another acene, rubrene [27]. The problem with pentacene is its low solubility. Several approaches have been made to overcome this: first, the use of high-temperature, low-concentration solution processing [28]; second, the production of pentacene precursors that can be thermally converted to pentacene after film deposition [29]; and finally, the synthesis of novel substituted pentacene-based molecules with increased solubility such as 6,13-bis(triisopropylsilylethynyl)pentacene (TIPS-pentacene) [1]. Rubrene does not have an issue with low solubility; however, along with the other substituted acenes, it is still not always straightforward to form uniform thin films from solution or to control crystallization during printing. Therefore, the principal aim of blending soluble acenes with polymers is to maintain a high-performance system while improving processability.

The first example we will look at here is the rubrene hypereutectic mixture where a glass-forming small molecular component is added to allow processing in an amorphous state [30]. In itself, this is a molecular–molecular blend; however, optimal behavior is obtained by the addition of high molecular weight atactic polystyrene. Thin films are readily processed from toluene solutions with rubrene concentrations in the range of 30–60 wt% of the solid material. The polymer allows improved wetting and improved mechanical robustness of the solid film. Amorphous films are then annealed at around 240 °C, which is lower than the melting temperature but just above the eutectic temperature. This allows crystallization of the rubrene into a planar polycrystalline film with highly birefringent grains and (a, b)-plane normals parallel to the substrate surface. This is ideal for charge transport and OFETs with mobilities of up to $0.7\,\text{cm}^2\,\text{V}^{-1}\,\text{s}^{-1}$ and excellent switch-on characteristics were measured.

Taking this approach further, it is possible to use a polymer combined with a soluble acene alone. Two common materials that were developed by Anthony et al. and have been highly successful in their own right as organic semiconductors are TIPS-pentacene [1] and 2,8-difluoro-5,11-bis(triethylsilylethynyl)anthradithiophene [31] (diF-TESADT). Both have high intrinsic mobilities ($>1\,\text{cm}^2\,\text{V}^{-1}\,\text{s}^{-1}$), are relatively air stable, and are readily soluble in many organic solvents. They can also be blended with polymers and retain a very high charge carrier mobility while improving film-forming properties both during spin coating and inkjet printing. These concepts were first reported in the patent literature [32] and then by Hamilton et al. with hole mobilities over $2\,\text{cm}^2\,\text{V}^{-1}\,\text{s}^{-1}$ for OFETs based upon diF-TESADT [33].

TIPS-pentacene has been studied extensively with insulating polymers to better understand the acene/polymer system. One of the main features of these blends, and the reason that such high mobilities can be obtained even with significant concentration of insulator, is vertical phase separation. Generally, it is possible to enhance the acene concentration at the dielectric–semiconductor interface and thus improve performance. Several methods have been used to probe vertical phase separation: first, dual-gate OFETs with both top and bottom gates within a single device; second, depth profiling with secondary ion mass spectrometry (SIMS); third, surface-sensitive techniques such as X-ray photoemission spectroscopy (XPS); and finally, neutron reflectivity (NR) measurements. Dual-gate devices with TIPS-pentacene and poly(α-methylstyrene) (PαMS) have shown improved top gate performance in terms of mobility, threshold voltage, and hysteresis [33]. Although some of these effects may be explained by different dielectric interfaces, it is likely that phase separation of the acene to the top of the film is occurring. Additionally by studying blends of diF-TESADT and a polytriarylamine (PTAA) with varying concentration of small molecule, a percolation model similar to that used for oligothiophene blends (Section 8.2.1) could be fitted to the data [9]. In this case, a percolation threshold of 39 wt% diF-TESADT was observed and mobility increased by over two orders of magnitude. It is clear that changes to the crystallinity of the film, as with the oligothiophene blends, are very important. This effect was monitored by differential scanning calorimetry (DSC) and polarized optical microscopy. The sharp percolation threshold could be correlated to a change from amorphous to crystalline film (Figure 8.5). Fitting Eq. (8.3) and extracting values of f suggested that above the threshold only $\sim$0.3 wt% of the polymer-rich phase was present in the conduction pathways. Maximum values of mobility for the blends are very high, especially when compared to unblended small molecules, suggesting that the acene-rich phase is almost unaffected electronically by the polymer and is able to dominate charge transport in the OFET. Again this would be consistent with vertical phase separation but does not demonstrate it directly. It has also been suggested that phase separation not only allows the formation of highly acene-rich conduction pathways but also helps to expel impurities from these pathways in a mechanism

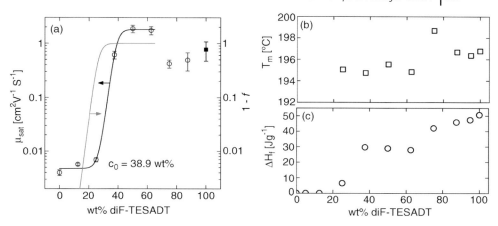

Figure 8.5 (a) Percolation behavior in a diF-TESADT:PTAA blend showing a threshold at 39 wt% small molecule. Also plotted is the fraction of high-mobility phase present in the conduction pathways, $1-f$. (b) The melting temperature and (c) the enthalpy of fusion for the same blends, estimated from DSC measurements, demonstrating the increase in crystallinity beyond the percolation threshold. (Adapted from Ref. [9] with permission of Elsevier B.V.)

similar to zone refinement. In particular, photo-oxidative products of aged TESADT had very little effect on blend OFETs compared to those without a polymer matrix [34].

SIMS measurements by Hamilton *et al.* directly indicate accumulation of TIPS-pentacene within the top 15 nm of a blend film and results by Ohe *et al.* [10] show accumulation at both top and bottom interfaces (Figure 8.6). Similarly for TESADT:

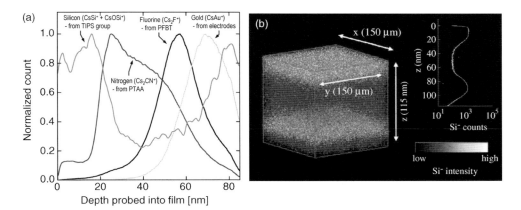

Figure 8.6 SIMS data for TIPS-pentacene blends demonstrating vertical phase separation. (a) A TIPS-pentacene:PTAA blend from Ref. [33], and (b) a TIPS-pentacene:PαMS blend from Ref. [10]. Silicon ions were used as an indicator of the presence of the TIPS group in both cases. Panel (b) shows a 3D map of Si⁻ intensity. (Reproduced with permission of Wiley-VCH Verlag GmbH and American Institute of Physics.)

PMMA blends, XPS combined with argon milling indicated vertical phase separation that can be enhanced through vapor annealing [35]. In addition, it is possible to fabricate high-performance circuits such as ring oscillators despite employing a bottom-gate, bottom-contact architecture [36]. It is therefore suggested that the polymer dielectric in this case changes the interaction between substrate and solution during spin casting leading to changes to the vertical phase separation [37]. The exact nature of the phase separation clearly depends on several factors, namely, the substrate surface energy, the rate of solvent evaporation, the diffusion of small molecule within the polymer, and the molecular weight of the polymer. High boiling point solvents are often used to reduce solvent evaporation rate and allow time for liquid–liquid phase separation to occur. Depending on the nature of the substrate, there may also be the formation of a wetting layer at the dielectric interface. This explains the presence of high acene concentrations at the bottom interface in some cases. The generally higher solubility and diffusivity of the acene may explain its accumulation at the top interface along with the improvement upon vapor annealing. As solvent evaporates from the top of the film, the concentration increase favors the presence of the more soluble component.

Several groups have investigated the role of the polymer component on phase separation and OFET performance. Kang et al. [38] have studied TIPS-pentacene: PαMS blends using neutron reflectivity measurement and grazing incident X-ray diffraction (GIXD) with two different molecular weights of PαMS. Fitting to the NR data allowed the determination of a composition depth profile and it was found that significant vertical phase separation occurred only in the low molecular weight sample after it was annealed above its glass transition temperature (Figure 8.7a). The high molecular weight sample always showed vertical phase separation and

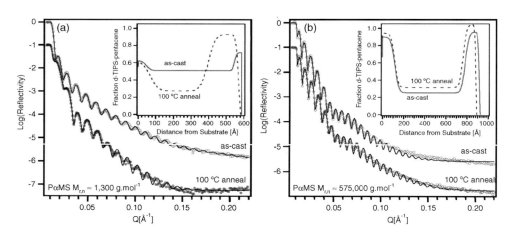

Figure 8.7 Neutron reflectivity measurements on deuterated TIPS-pentacene:PαMS blends. PαMS had molecular weights of (a) 1.3 kg mol^{-1} and (b) 575 kg mol^{-1} and results are shown for before and after annealing. The vertical composition profile is calculated from the NR data and shows the fraction of TIPS-pentacene. (Reproduced from Ref. [38] with permission of the American Chemical Society.)

annealing had very little effect. TIPS-pentacene accumulated at both interfaces as observed before; however, it is clear that if the molecular weight of the polymer is not high enough, phase separation is not thermodynamically favored. These results were confirmed by Ohe et al. [39] using SIMS measurements. A range of molecular weights from 2 to 800 kDa was studied and the OFET mobilities correlated to the calculated ΔG_{mix} for each blend. At a molecular weight of around 10 kDa, $\Delta G_{mix} = 0$; therefore, only above this weight is phase separation possible and thence high-mobility OFETs (Figure 8.7b). Diffraction studies in both cases confirm that the highly crystalline phase with (00ℓ)-planes parallel to the substrate surface forms in the phase-separated blends, again ideal for OFET performance.

The choice of polymer matrix in many of these acene blends is often based on its low-cost or ease of processability, but also its amorphous nature. The advantage of using an amorphous polymer is that the control of crystallization is less challenging than when compared to, for example, the oligothiophene/crystalline polymer blends in Section 8.2.1. Some typical polymers that have been studied are therefore PαMS, polytriarylamines, poly(4-vinylbiphenyl) [40], amorphous fluorene–triarylamine copolymers, PTAA, and even amorphous polythiophenes [41]. By moving from insulating polymers to semiconducting ones, mobilities can be enhanced; for example, TIPS-pentacene blended with PαMS gives a mobility of 0.69 cm^2 V^{-1} s^{-1}, whereas with PTAA this increases to 1.1 cm^2 V^{-1} s^{-1} [33]. Even higher mobilities can be achieved with diF-TESADT blends. A diF-TESADT:PTAA blend has an OFET mobility of 2.4 cm^2 V^{-1} s^{-1} and by replacing the PTAA with poly(dioctylfluorene-co-dimethyltriarylamine), which has an order of magnitude higher intrinsic mobility, the blend performance can be enhanced to produce average saturation mobilities of 4.2 cm^2 V^{-1} s^{-1} [42]. These data can be explained by the improved conduction pathways created by the polymer as predicted by Eq. (8.4), as well as improved injection from the source–drain contacts into the polymer. The effect of replacing the amorphous polymer with a semicrystalline one was investigated by Madec et al. [43] in TIPS-pentacene systems. As with the diF-TESADT:PTAA blend, it was observed that acene concentrations of >40–50 wt% were needed to obtain vertical phase separation and good device performance for the amorphous PαMS. Amorphous polystyrene, however, led to no phase separation at all. Low acene concentration resulted in a lack of liquid–liquid phase separation and only very small crystallites embedded in a polymer matrix at best. However, by switching to a semicrystalline, isotactic polystyrene, phase separation to both top and bottom interfaces could be achieved with compositions as low as 10 wt% TIPS-pentacene. This supports the theories of crystalline–crystalline blend behavior discussed in Section 8.2.1. However, it should be noted that if the polymer is "too crystallizable," it is possible for it to crystallize first leading to film microstructures dominated by the polymer and OFET mobilities closer to that of the neat polymer than that of the acene molecule.

Most of the discussion thus far has focused on thin films produced by spin casting. However, one goal of these composite systems, and the acene blends in particular, is to improve processability during printing, especially inkjet printing. TIPS-pentacene-based OFETs fabricated by inkjet-printed semiconductor have been reported with the best mobilities in the range of 0.2–0.3 cm^2 V^{-1} s^{-1} [44], but also mobilities as low as 10^{-4} cm^2 V^{-1} s^{-1}. In order to improve upon this, the solvent evaporation and wetting of the ink must be carefully tailored to control crystal growth. Some problems with using simple acene-based inks are the nonuniformity between devices, the effects of Marangoni flow during droplet drying, and the brittle nature of the solid semiconductor. The first two can be partly solved by using heated substrates and/or dual-component solvents. This allows the solvent evaporation rate and flow within the droplet to be controlled. The other approach, which also improves the mechanical properties of the semiconductor, is to blend the acene with a polymer. Madec et al. [45] reported TIPS-pentacene/amorphous polystyrene blend OFETs printed from anisole or anisole:acetophenone solutions. The effect of multilayer printing as well as dual-solvent printing on OFET mobility, mobility variance, and film crystallinity was observed. Using two or more layers reduces device-to-device variation while still maintaining relatively high mobilities; alternatively, the dual-solvent system improves the characteristics of a single-layer device giving a maximum mobility of 0.11 cm^2 V^{-1} s^{-1} and an average mobility of 0.046 cm^2 V^{-1} s^{-1}. Remarkably, this also seems to reduce the crystallinity of the blend (as estimated from XPS) but as previously discussed the key to high blend performance is local crystallinity at the dielectric interface that may well be improved by the introduction of the higher boiling point solvent, acetophenone.

A final novel application of composite systems is to use phase separation as a self-assembly mechanism for OFET fabrication. This potentially allows the formation of dielectric and semiconductor layer with a single step, thus reducing the processing time and costs. Several polymer–polymer blends have been successful in this regard due to the large driving force for phase separation. However, it is also possible to use a TESADT:PMMA blend to create a layered structure suitable for combining dielectric and semiconductor [35]. The PMMA-rich phase contains a low enough concentration of acene to act as the dielectric partly due to enhanced phase separation by vapor annealing. A TESADT-rich phase forms on the top surface and acts as the conduction layer while in between an intermixed region is created. Fully organic OFETs were fabricated on flexible substrates with conducting polymer electrodes and an average OFET mobility of 0.38 cm^2 V^{-1} s^{-1} was obtained.

8.3
Polymer/Fullerene Ambipolar OFETs

While the majority of OFET-based research is targeted toward improvements in device performance [46–49], OFETs have also proven to be excellent test beds for studying the characteristics of semiconducting material systems and evaluating semiconductor performance. OFETs have been used to probe many relevant phenomena observed in organic semiconductors, such as charge transport mechanisms [50–52], charge trapping behavior [53, 54], photophysical processes [55, 56], and morphological properties [57–59].

Here the discussion of semiconductor composite-based OFETs is extended to polymer:fullerene systems. As will be discussed in Chapter 16, such blends have been observed to give rise to so-called ambipolar OFETs, which possess many attractive properties with regard to integrated circuits. In addition to their use in integrated circuits, polymer:fullerene OFETs have also been employed to study the morphological properties of polymer:fullerene blends. While knowledge about the morphology of such blends is important for microelectronic purposes, it is with regard to organic photovoltaic solar cells (OPVs) that such information is most relevant. Polymer:fullerene composite solar cells are the subject of Chapter 12. The morphology of polymer:fullerene blends is absolutely crucial to the power conversion efficiency of OPVs, and is a significant field of research [60–63].

This section focuses on the relationship between the microstructure of polymer: fullerene blends and the measured properties of OFETs formed from these systems. Despite the experimental data presented within this chapter being restricted to polymer:fullerene blends, it will become apparent that the phenomena and techniques discussed here are relevant for other semiconductor composites also.

8.3.1
Polymer:Fullerene Blend Morphology

Postproduction annealing is known to lead to improved power conversion efficiencies in certain polymer:fullerene bulk heterojunction (BHJ) solar cells [64–71]. This is attributed to a redshift in the absorption spectrum of certain polymer:fullerene blends [67] and improved charge carrier transport properties of the two phases [66, 69, 72, 73]. These improvements are believed to arise from a thermally induced phase separation of the polymer and fullerene components [65–67, 72–76]. Upon annealing, the polymer is believed to become more crystalline, while the fullerene molecules are reported to form clusters [67, 73–75]. This phase segregation is believed to improve the charge transport of holes in the polymer and electrons in fullerenes [69], aiding charge dissociation. This process is illustrated diagrammatically in Figure 8.8c.

In the system of blended P3HT and [6,6]-phenyl-C_{61}-butyric acid methyl ester ([60]PCBM), for example (see Figure 8.9 for structures), this phase segregation has been observed by Campoy-Quiles *et al.* using optical microscopy [65]. Figure 8.8a and b show a film of a 1 : 1 (wt %) blend of P3HT:[60]PCBM deposited onto fused silica substrates before and after thermal annealing. Crystallites with dimensions on the order of microns are clearly visible and were identified as [60]PCBM-rich entities. Other techniques such as transmission electron microscopy (TEM) have resolved fullerene crystallites on much smaller scales [76], indicating that the phase segregation process can induce a wide range of micromorphologies. Independent of their size, however, at higher temperatures, the melting of these [60] PCBM crystallites has been observed using optical microscopy, which has been attributed to a eutectic phase behavior [71].

The three-terminal structure of OFETs allows the selective injection and transport of carriers into the different components of the polymer:fullerene blend [78].

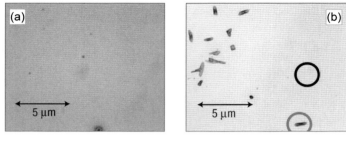

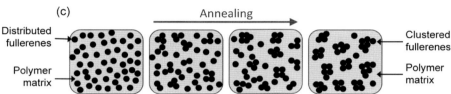

Figure 8.8 Optical microscope images of P3HT:[60]PCBM blends (1:1 wt%) on fused silica (a) before and (b) after annealing under atmospheric-pressure N_2. The annealing time and temperature are not specified in this case, although similar phase segregation has been observed after annealing at temperatures of 140 °C for 30 min [77]. (c) Schematic representation of two-dimensional cross section of a polymer:fullerene blend undergoing phase segregation. (Adapted from Ref. [65] with permission of Nature Publishing Group).

This unique property allows the long-range microstructure of each component to be probed independently, in the absence of carrier recombination. The work function of gold (4.8 eV) is well aligned with the highest occupied molecular orbital (HOMO) of many polymers (see Figure 8.9). There is an offset between the work function of gold and the lowest unoccupied molecular orbital (LUMO) of [60]PCBM. However, electron injection and transport is known to take place from gold

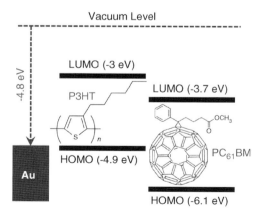

Figure 8.9 Energy levels and molecular structures of P3HT and [60]PCBM, relative to the vacuum level and the work function of gold.

into [60]PCBM because of a reduction in the potential barrier, due to the formation of interface dipoles [79]. Therefore, holes and electrons can easily be injected and transported in the polymer and [60]PCBM networks, respectively, but not vice versa. As a consequence, many polymer:[60]PCBM OFETs are ambipolar and through appropriate biasing conditions [80] selective injection and transport of holes and/or electrons can take place into the appropriate materials [78]. Hence, by measuring the mobility at different biasing conditions, the average long-range crystallinity and percolation pathways of each material network can then be probed independently.

If a variable range hopping (VRH) charge transport mechanism is assumed, the mobility of charge carriers depends exponentially on the distance between adjacent molecules in the film [51]. Therefore, changes in the packing/structure of the film should be reflected in the measured field-effect mobility. It has been shown [81] that charge transport in OFETs occurs within the first few nanometers from the semiconductor–dielectric interface, suggesting that any changes in observed mobility will be representative of changes at this interface.

It should be noted that previous studies have shown that when using gold source and drain electrodes, hole injection and transport can take place in [60]PCBM [79]. However, in this case the mobility is significantly lower than that of holes in P3HT [82], for example, and hence one can assume that hole field-effect mobility measurements made on a 1:1 (wt%) P3HT:[60]PCBM blend should be representative of the P3HT network only, with a negligible modification due to the [60]PCBM network. The same can also be said of most other p-type polymers.

8.3.1.1 Solvent and Polymer Molecular Weight

The effects of solvent and polymer molecular weight on the measured hole and electron mobilities of polymer:fullerene blends were studied by Morana *et al.* in 2007 [57]. A bottom-contact, bottom-gate (BCBG) OFET structure was used in their study. It was found that by changing the molecular weight of 95% regioregular P3HT in P3HT:[60]PCBM blends from $\sim 25\,000$ to $\sim 160\,000\,\mathrm{g\,mol^{-1}}$ the electron mobility dropped from 5×10^{-4} to $6 \times 10^{-6}\,\mathrm{cm^2\,V^{-1}\,s^{-1}}$, while the hole mobility rose from 5×10^{-6} to $2 \times 10^{-4}\,\mathrm{cm^2\,V^{-1}\,s^{-1}}$. This was when the solutions were formed in the solvent *o*-xylene. However, when the solvent used was chlorobenzene, the relationship is inversed. These data are summarized in Table 8.1 and illustrate how sensitive the mobilities of the charge carriers are to such changes.

By thermally annealing these blends at $130\,°\mathrm{C}$ for 5 min, significant changes in mobility were again observed (see Table 8.1). In general, the mobilities of both charge carriers are observed to increase after annealing, with the exception of the electrons in the lower molecular weight polymer blends deposited from chlorobenzene. The mobility of electrons was in this case observed to fall from 1.6×10^{-4} to $7.2 \times 10^{-5}\,\mathrm{cm^2\,V^{-1}\,s^{-1}}$. The increase in mobility was attributed to the crystallization of the two phases, consistent with studies carried out using other devices [69]. The decrease observed in the case mentioned above was attributed to a change in the vertical concentration profile of the system.

Table 8.1 Mobility of holes and electrons in P3HT, [60]PCBM, and P3HT:[60]PCBM blend OFETs using different solvents and polymer molecular weights.

Material	Electron mobility ($cm^2 V^{-1} s^{-1}$)		Hole mobility ($cm^2 V^{-1} s^{-1}$)	
	o-Xylene	Chlorobenzene	o-Xylene	Chlorobenzene
Pristine [60]PCBM	2×10^{-2} (nA)	8×10^{-3} (nA)	—	—
High-MW P3HT	—	—	2.4×10^{-3} (nA)	5×10^{-4} (nA)
Low-MW P3HT	—	—	8×10^{-4} (nA)	6×10^{-4} (nA)
High-MW P3HT:[60]PCBM (1:1)	6×10^{-6} (nA)	1.5×10^{-4} (nA)	1.8×10^{-4} (nA)	4.8×10^{-5} (nA)
	3.8×10^{-5} (AC)	1.1×10^{-3} (AC)	1.3×10^{-3} (AC)	4×10^{-4} (AC)
Low-MW P3HT:[60]PCBM (1:1)	5×10^{-4} (nA)	1.6×10^{-4} (nA)	5×10^{-6} (nA)	4×10^{-5} (nA)
	1.1×10^{-3} (AC)	7.2×10^{-5} (AC)	5.5×10^{-5} (AC)	3.1×10^{-4} (AC)

Non-annealed devices are labeled "nA" and annealed devices are labeled "AC".
Adapted from Ref. [57] with permission of John Wiley & Sons, Inc.

8.3.1.2 Blend Composition

The effect of blend composition on the hole and electron mobilities measured using OFETs was first published in 2004 [83]. It was found that the mobilities of holes and electrons were strongly dependent upon the weight ratio of blends of poly-[2-methoxy-5-(2′-ethylhexyloxy)-1,4-(1-cyanovinylene)phenylene (MEH-PPV) and [60]PCBM. This is shown in Figure 8.10c. The mobility of electrons is not measurable until a ratio of 4 : 6 (MEH-PPV:[60]PCBM) is reached, after which the mobility increases strongly with increasing fullerene content. This was attributed to improvements in the percolating pathways for electrons in such systems.

The hole mobility in such blends is observed to increase with increasing polymer content (as one would intuitively expect) and then peaks at a ratio of 6 : 4 (MEH-PPV:[60]PCBM) before falling. A similar set of results has also been obtained using poly[2-methoxy-5-(3′,7′-dimethyloctyloxy-p-phenylene vinylene)] (MDMO-PPV) (see Figure 8.10d) [59]. Although not mentioned at the time, this observation has since been explained by the intercalation of fullerene molecules between the polymer side chains and a subsequent extension of the polymer backbone [84–86]. This can explain why the hole mobility of PPV:PCBM blends increases with increasing fullerene content.

A study carried out using P3HT:[60]PCBM blends produced results that are again same as one would expect intuitively when considering the percolation and crystallization properties of the two components (see Figure 8.11a) [87]. A more in-depth study reported 2 years later shows how this composition dependence is affected by annealing at 150 °C for various periods of time (see Figure 8.11b) [59]. These data show again that in general the hole mobility of P3HT:[60]PCBM OFETs decreases with increasing [60]PCBM content, yet after annealing for long periods of time the hole mobility drops by roughly one order of magnitude.

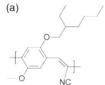
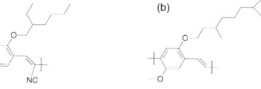
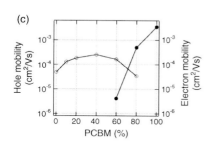
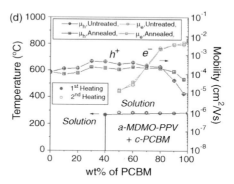

Figure 8.10 Molecular structures of (a) MEH-PPV and (b) MDMO-PPV. (c) Field-effect mobility of holes and electrons in MEH-PPV:[60]PCBM blend OFETs in various blend ratios. (d) Field-effect mobility of holes and electrons in MDMO-PPV:[60]PCBM blend OFETs in various blend ratios before and after annealing at 100 °C for 30 min. The data at the bottom of the figure, corresponding to the left axis, are corresponding phase data obtained from X-ray scattering data and thermal analysis. (Reproduced from Ref. [83] with permission of the Material Research Society.)

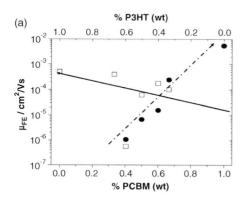
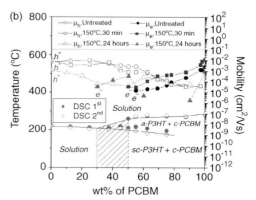

Figure 8.11 (a) Field-effect mobility of holes and electrons in P3HT:[60]PCBM blend OFETs of various blend ratios. Panel (b) additionally shows the effect of annealing at 150 °C for various periods of time. (Reproduced from Ref. [59] with permission of American Chemical Society.) (Reproduced from Ref. [87] with permission of Elsevier B.V.)

238 | *8 Semiconducting Organic Molecule/Polymer Composites for Thin-Film Transistors*

The electron mobility was observed to increase strongly with fullerene content. After annealing at 150 °C for 24 h, the electron mobility was observed to increase for some blend ratios and fall for others. The fall, and observation of so-called "mobility gap" where there was no measurable n-channel, was attributed to large phase-separated [60]PCBM domains that were not connected to the rest of the system.

8.3.1.3 Temperature- and Time-Dependent Annealing

While the studies cited above have considered the effect of annealing, the ranges of annealing temperatures and times have been relatively restricted. In 2011, a study by Labram *et al.* investigated how the mobility of holes and electrons evolved with various annealing temperatures and times [77]. Figure 8.12a shows the average mobility of four P3HT:[60]PCBM (1 : 1 wt%) blend OFETs plotted as a function of annealing temperature and time. Figure 8.12b shows the corresponding differential scanning calorimetry thermogram. The measurements were made under N_2 at room temperature after cumulatively annealing the OFETs at temperatures between 60 and 300 °C in 30 min steps. The OFET structures were BCBG, using gold source and drain electrodes and an HMDS-treated SiO_2 gate dielectric. Figure 8.12c–e shows the corresponding plots for similar pristine P3HT and [60]PCBM OFETs.

Upon annealing the blend OFETs at temperatures of 140 °C, the hole mobility is observed to increase and the electron mobility to fall drastically. The increase in hole mobility was attributed to an increase in P3HT crystallinity and hence

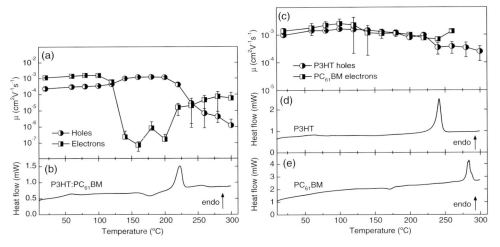

Figure 8.12 (a) Average field-effect mobility of holes and electrons in four 1:1 (wt%) P3HT:[60]PCBM blend ambipolar BCBG OFETS measured at room temperature as a function of annealing temperature. (b) DSC heating thermogram of 1:1 (wt%) P3HT:[60]PCBM blend films. (c) Average field-effect mobilities measured from several neat P3HT and [60]PCBM BCBG OFETs as a function of annealing temperature. Corresponding DSC thermograms of (d) P3HT and (e) [60]PCBM. (Adapted from Ref. [77] with permission of John Wiley & Sons, Inc.)

improvement in interchain transport, while the fall in electron mobility was attributed to a reduction in percolation pathways, caused by the large-scale clustering of [60]PCBM molecules. For devices annealed at 220 °C, the hole mobility drops and the electron mobility partially recovers. This was attributed to a melting of the polymeric component at temperatures above 205 °C (the eutectic temperature) [71] and a resulting redistribution of fullerene molecules.

The observed reduction in electron field-effect mobility was then used as an indicator of when the phase segregation process illustrated in Figure 8.8c was taking place. Single OFETs were annealed under high vacuum at temperatures of 130 and 160 °C, and the field-effect mobilities of holes and electrons were measured. The resulting data are plotted in Figure 8.13. Since in this case the measurements are made while at elevated temperatures, the effects of temperature-dependent transport [51, 88] had also to be considered.

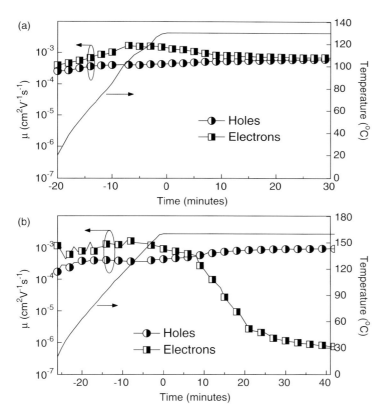

Figure 8.13 Field-effect mobility of holes and electrons in a 1:1 (wt%) P3HT:[60]PCBM blend ambipolar bottom-gate, bottom-contact OFET as a function of time, while being annealed at (a) 130 °C and (b) 160 °C. $L = 5\,\mu m$, $W = 10\,mm$ for both OFETs. (Reproduced from Ref. [77] with permission of John Wiley & Sons, Inc.)

When annealed at 130 °C (Figure 8.13a), a small increase and subsequent fall in electron mobility takes place. The increase was attributed to the evaporation of oxygen and water. The fall was assigned to the possible onset of formation of fullerene clusters. During annealing at 160 °C (Figure 8.13b), the hole and electron mobilities remain constant for approximately 10 min, after which the electron mobility begins to drop rapidly, indicating the loss of percolation due to the aggregation of fullerene-rich crystallites.

8.3.1.4 Effect of Fullerene Molecular Weight

Despite P3HT:[60]PCBM being the standard "benchmark" system for OPV studies, many high-performance OPVs are reported to be increasingly employing [6,6]-phenyl-C_{71}-butyric acid methyl ester ([70]PCBM) as opposed to [60]PCBM as the acceptor molecule in OPV blends [89–91]. Due to its lower symmetry, the absorption spectrum of [70]PCBM is redshifted significantly with respect to the more commonly used [60]PCBM [92], giving it an improved overlap with the AM 1.5 solar spectrum [93]. With certain high-performance polymer:fullerene systems requiring a relatively high concentration of fullerene acceptor [90, 91], this extra absorption is expected to result in significant improvements in solar cell efficiencies under AM 1.5 conditions. It has previously been shown that despite these differences in optical properties, the charge transport characteristics of [70]PCBM are very similar to those of [60]PCBM [94].

In another study by Labram et al., the mobility evolution of P3HT:[70]PCBM and P3HT:[60]PCBM blends was compared upon annealing [95]. It was found that the temperature dependences of both hole and electron mobilities are similar in both systems, but with a higher prerequisite annealing temperature required for a fall in electron mobility to be observed in the case of P3HT:[70]PCBM (see Figure 8.14). It

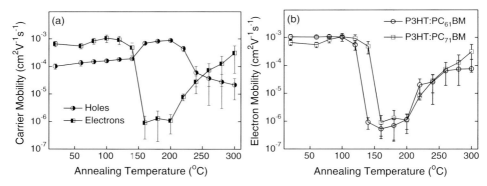

Figure 8.14 (a) Average saturation regime field-effect mobility of holes and electrons in four 1:1 (wt %) P3HT:[60]PCBM blend ambipolar bottom-contact, bottom-gate OFETs measured at room temperature after being annealed at various temperatures for 30 min. (b) Average electron saturation regime field-effect mobility from (a) plotted alongside similar electron field-effect mobility measurements made on four 1:1 (wt%) P3HT:[60]PCBM blend OFETs. (Adapted from Refs [77, 95] with permission of John Wiley & Sons, Inc.)

was suggested that the higher prerequisite annealing temperature for phase segregation in the P3HT:[70]PCBM system is due to an increase in required thermal energy. Due to their increased volume, the frictional drag of [70]PCBM molecules was anticipated to be higher than that of [60]PCBM molecules when dispersed in P3HT.

The reports cited above illustrate the various differences between how the electron mobility is believed to change in such polymer:fullerene blends upon annealing. It appears that parameters such as polymer molecular weight [57], transistor dielectric [96], solvent [57], annealing temperature [77], annealing time [59, 77], fullerene molecular weight [95], and blend composition [59, 83, 87] are all likely to affect the system in some way. These differences are likely to be manifest as changes in the performance of the corresponding OPVs. With such a large parameter space to optimize, it is hence important to use as many tools as possible to understand the system. OFETs offer a unique and innovative way to study such systems.

Being able to monitor the two semiconductor systems independently of each other, for example, is quite difficult to do via any other technique. Similarly, being able to probe the bottom interface of a film in isolation has previously been quite challenging. It can be argued that such techniques could prove highly useful in the future processing and optimization procedure for OPVs.

8.3.2
Polymer:Fullerene Bilayer Diffusion

Due to the structural complexity of the BHJ system, there have recently been several reports devoted to the study of equivalent bilayer structures [97–104]. Using deposition techniques such as stamp transfer [97, 100, 104] and orthogonal solvents [98, 102, 103], several groups have now reported bilayer structures based upon the benchmark system of P3HT:[60]PCBM. It has been shown that [60]PCBM will diffuse into P3HT upon annealing. This is a phenomenon distinct from the widely reported polymer crystallization and fullerene clustering process generally associated with improvements in OPV power conversion efficiency [66, 68, 69]. Until recently, this process was largely unknown and hence absent from any OPV processing and optimization procedure. Understanding and controlling this process is of great importance to the performance and stability of OPV devices.

Figure 8.15a shows the concentration of deuterated [60]PCBM at a distance from the top surface of a bilayer P3HT:[60]PCBM heterostructure as a function of annealing temperature [104]. The substrate was a silicon wafer and the annealing time was 5 min in each case. Figure 8.15b shows a scanning electron microscope image of an unannealed P3HT:[60]PCBM heterostructure, illustrating the well-defined interface between the two semiconductors.

Bilayer P3HT:[60]PCBM OFET mobility measurements have similarly been shown to be capable of studying this diffusion process [101]. By employing a BCBG architecture (see Figure 8.16a) with a vacuum-deposited [60]PCBM layer, it was found that the OFETs were initially p-type only (see Figure 8.16b). This is believed to be because only P3HT is in contact with the source and drain electrodes and with

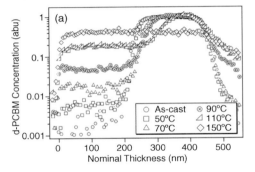
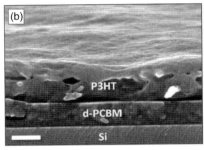

Figure 8.15 (a) Concentration profiles of ^{2}H in bilayer samples of P3HT and d-[60]PCBM annealed for 5 min at different temperatures as measured using dynamic secondary ion mass spectrometry (DSIMS). The thicknesses of the layers were normalized such that 0 nm is the free surface of the film and 450 nm is the substrate. (b) Cross-sectional scanning electron microscope (SEM) image of P3HT:d-[60]PCBM bilayer film fabricated on an SiO$_2$:Si substrate. (Adapted from Ref. [104] with permission of John Wiley & Sons, Inc.)

the semiconductor–dielectric interface, and because carrier transport takes place only within the first few nanometers of the semiconductor–dielectric interface [81]. Bilayer ambipolar systems have, however, previously been demonstrated using BCBG architectures [105] where the electrodes have been in direct contact with one semiconductor only. This difference in behavior was attributed to differences in the energy levels of the semiconductors and electrodes.

The mobility evolution of this system was studied as a function of annealing temperature and time. Figure 8.17a shows the average field-effect mobility of five bilayer P3HT:[60]PCBM OFETs measured at room temperature under atmospheric

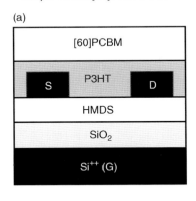
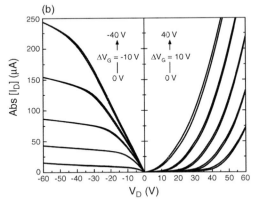

Figure 8.16 (a) Schematic diagram of BCBG bilayer OFET illustrating the position of the source (S), drain (G), and gate (G) electrodes relative to the semiconductor:semiconductor and semiconductor:dielectric interfaces. (b) Output characteristics of bilayer P3HT:[60]PCBM BCBG OFET measured before annealing. This device had a channel length and width of 7.5 and 10 000 μm, respectively, and P3HT and [60]PCBM thicknesses of 8 and 31 nm, respectively. (Adapted from Ref. [101] with permission of American Physical Society.)

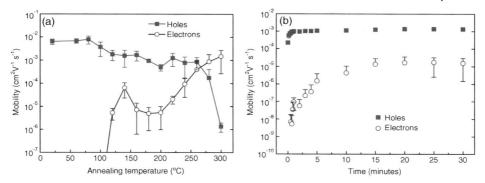

Figure 8.17 (a) Average saturation regime field-effect mobility of holes and electrons in five bilayer P3HT:[60]PCBM OFETs measured at room temperature after being annealed at various temperatures for 30 min. The thicknesses of the P3HT and [60]PCBM layers were 8 and 40 nm, respectively. (b) Average saturation regime field-effect mobility of holes and electrons in five bilayer P3HT:[60]PCBM OFETs measured under ambient pressure N_2 after being annealed at 130 °C, plotted as a function of time. The thicknesses of the P3HT and [60]PCBM layers were 25 and 31 nm, respectively. (Adapted from Ref. [101] with permission of American Physical Society.)

pressure N_2 after being annealed at increasing temperatures. Each annealing step was for 30 min and the annealing was cumulative, so, for example, the devices annealed at 80 °C for 30 min had also been annealed at 60 °C for 30 min, and so on. Up to an annealing temperature of 120 °C, no electron channel is measurable, after which it rises sharply before falling back to $\sim 10^{-5}\,\mathrm{cm^2\,V^{-1}\,s^{-1}}$ upon annealing at 160 °C. The rise in electron mobility is consistent with the description of fullerenes diffusing from the top layer through the P3HT and forming an electron percolation pathway at the semiconductor–dielectric interface. The fall was attributed to the fullerene clustering process illustrated in Figure 8.8c, where the percolation pathways for electrons are disrupted. At temperatures of 220 °C and above, the electron mobility is observed to rise. This was interpreted as being due to a melting of the P3HT:[60]PCBM system and subsequent redistribution of [60]PCBM molecules.

The time dependence of this process was also studied using OFETs. Five similar bilayer BCBG P3HT:[60]PCBM OFETs were annealed at a constant temperature of 130 °C for various periods of time. These devices were then cooled to room temperature and their mobilities were measured. These data are shown in Figure 8.17b. The electron mobility is not measurable for the first 20 s of annealing, after which it rises rapidly before slowly saturating to a constant value. This was again attributed to the diffusion of [60]PCBM from the top layer, through the P3HT, and finally forming a stable distribution.

8.3.2.1 Modeling Fullerene Diffusion

Using a simple model based upon the one-dimensional diffusion equation, it was shown [101] that the areal concentration of [60]PCBM molecules C at a distance x from the semiconductor–dielectric interface could be described as a function of

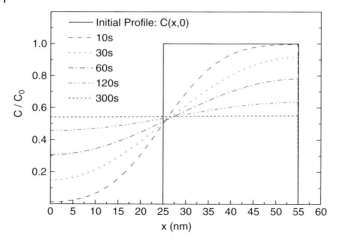

Figure 8.18 Initial 2D concentration profile (C/C_0) of [60]PCBM in bilayer P3HT:[60]PCBM OFET with P3HT and [60]PCBM thicknesses of 25 and 31 nm, respectively, and [60]PCBM concentration profile calculated using Eq. (8.1), evaluated at various times. The parameters used in this calculation were as follows: $x_p = 25$ nm, $x_c = 55$ nm, and $D = 5$ nm^2 s^{-1}. (Adapted from Ref. [101] with permission of American Physical Society.)

time t, by the following equation:

$$C(t, x) = C_0\left(1 - \frac{x_p}{x_c}\right) - C_0 \sum_{n=1}^{\infty} \frac{2}{n\pi} \sin\left(\frac{n\pi x_p}{x_c}\right) \exp\left(-\frac{n^2\pi^2 Dt}{x_c^2}\right) \cos\left(\frac{n\pi x}{x_c}\right). \quad (8.4)$$

This equation has been plotted as a function of x in Figure 8.18 for various values of t. Here C_0 is the two-dimensional concentration of [60]PCBM in the top (pristine) [60]PCBM layer before annealing, x_p is the distance of the P3HT–[60]PCBM interface from the substrate, x_c is the distance of the top [60]PCBM interface from the substrate (i.e., the thickness of the P3HT and [60]PCBM layers combined), and D is the diffusion coefficient of [60]PCBM in P3HT at the given annealing temperature.

Since the OFET channel is located at the semiconductor–dielectric interface, the concentration calculated at $x = 0$ can be assumed to be representative of that measured using bilayer BGBC OFETs. Setting $x = 0$ gives

$$C(t) = C_0\left(1 - \frac{x_p}{x_c}\right) - C_0 \sum_{n=1}^{\infty} \frac{2}{n\pi} \sin\left(\frac{n\pi x_p}{x_c}\right) \exp\left(-\frac{n^2\pi^2 Dt}{x_c^2}\right). \quad (8.5)$$

To relate the measured electron field-effect mobility to the concentration of fullerene molecules at the semiconductor–dielectric interface, an implementation of percolation theory was employed [106]. Details of the derivation are given elsewhere [101], but it was found that the following equation gave an adequate description of the relationship (when fullerene aggregation is neglected):

$$\mu_e(c) = \mu_{e0} 2^{1.43} \left[\max(p - p_c)^{1.43} \exp\left(-\frac{a(p, c) - 1}{\alpha}\right)\right]. \quad (8.6)$$

Here μ_e is the electron field-effect mobility, c is the fullerene site occupation probability ($c = C/C_0$), p is the bond occupation probability, p_c is the threshold bond occupation probability for percolation, α is the wavefunction overlap parameter of the fullerene molecules, and a is given by the following function:

$$a(p,c) = \sqrt{\frac{\ln(1-\sqrt{p})}{\ln(1-c)}}. \tag{8.7}$$

By converting the measured electron field-effect mobilities in Figure 8.17a into concentrations using Eq. (8.6), the solution to the diffusion equation (Eq. (8.5)) could be fitted to the data and an approximate diffusion coefficient of $D = 5 \text{ nm}^2 \text{ s}^{-1}$ could be extracted for [60]PCBM in P3HT at a temperature of 130 °C.

8.4 Conclusions

This chapter has highlighted some features of small-molecule/polymer composites for OFET devices, both in terms of direct applications and as a probe for blend microstructure and charge transport. The potential to combine the processability and mechanical properties of polymers with the electronic properties of molecular semiconductors is extremely useful since it is not always easy to find a single material that can meet all of the requirements for a particular OFET application.

Several unipolar systems have been described mainly based upon oligothiophenes or acenes as the molecular component; however, the concepts are generally applicable to many such composites. We have seen that the control of phase separation during spin coating or printing is a key aspect of these films. An improved understanding of the link between the OFET properties and the film morphology is developing especially with regard to electronic percolation pathways, the role of crystallinity, and the vertical composition profile in a thin film. Most solution processes are far from equilibrium, so many factors have to be considered in order to obtain the required microstructure, for example, blend composition, temperature, substrate surface energies, solvent evaporation rates, and solvent–solute interactions. However, there has been some excellent progresses in achieving high-mobility OFETs that have low device-to-device variation and can even be used in more industrial processes such as inkjet printing.

It has additionally been shown that the microstructure and diffusion properties of polymer:fullerene blends can be probed using OFETs. Given the data presented within this chapter, it is clear that the measured mobilities, and hence the crystallinity and percolation properties of the components of these blends, are highly dependent upon many processing factors. These include polymer molecular weight, transistor dielectric, solvent, annealing temperature, annealing time, fullerene molecular weight, and blend composition.

Nonetheless, such measurements could prove valuable with regard to future processing and optimization procedures for polymer:fullerene organic photovoltaic cells. The unique nature of a technique to probe the time-dependent development of each component of the film individually, for example, has many potential applications.

Overall, field-effect transistors have been shown to have a wide variety of impressive and novel applications with regard to semiconductor–composite systems, not just for device applications but also for fundamental studies of widely used semiconductor systems.

References

1 Anthony, J.E., Brooks, J.S., Eaton, D.L., and Parkin, S.R. (2001) *J. Am. Chem. Soc.*, **123**, 9482.
2 Katz, H.E. (1997) *J. Mater. Chem.*, **7**, 369.
3 Podzorov, V., Sysoev, S.E., Loginova, E., Pudalov, V.M., and Gershenson, M.E. (2003) *Appl. Phys. Lett.*, **83**, 3504.
4 Horowitz, G. and Hajlaoui, M.E. (2000) *Adv. Mater.*, **12**, 1046.
5 Sirringhaus, H. (2005) *Adv. Mater.*, **17**, 2411.
6 Gelinck, G.H., Huitema, H.E.A., Veenendaal, E.v., Cantatore, E., Schrijnemakers, L., Putten, J.B.P.H.v.d., Geuns, T.C.T., Beenhakkers, M., Giesbers, J.B., Huisman, B.-H., Meijer, E.J., Benito, E.M., Touwslager, F.J., Marsman, A.W., van Rens, B.J.E., and de Leeuw, D.M. (2004). *Nat. Mater.*, **3**, 106.
7 Subramanian, V., Frechet, J.M.J., Chang, P.C., Huang, D.C., Lee, J.B., Molesa, S. E., Murphy, A.R., Redinger, D.R., and Volkman, S.K. (2005) *Proc. IEEE*, **93**, 1330.
8 Tsao, H.N., Cho, D., Andreasen, J.W., Rouhanipour, A., Breiby, D.W., Pisula, W., and Müllen, K. (2009) *Adv. Mater.*, **21**, 209.
9 Smith, J., Heeney, M., McCulloch, I., Malik, J.N., Stingelin, N., Bradley, D.D. C., and Anthopoulos, T.D. (2011) *Org. Electron.*, **12**, 143.
10 Ohe, T., Kuribayashi, M., Yasuda, R., Tsuboi, A., Nomoto, K., Satori, K., Itabashi, M., and Kasahara, J. (2008) *Appl. Phys. Lett.*, **93**, 053303.
11 Hoppe, H. and Sariciftci, N.S. (2006) *J. Mater. Chem.*, **16**, 45.
12 Moons, E. (2002) *J. Phys.: Condens. Matter*, **14**, 12235.

13 Horowitz, G. and Delannoy, P. (1991) *J. Appl. Phys.*, **70**, 469.
14 Richards, T.J. and Sirringhaus, H. (2007) *J. Appl. Phys.*, **102**, 094510.
15 Gelinck, G.H., Geuns, T.C.T., and de Leeuw, D.M. (2000) *Appl. Phys. Lett.*, **77**, 1487.
16 Balsara, N.P., Lin, C., and Hammouda, B. (1996) *Phys. Rev. Lett.*, **77**, 3847LP.
17 Jones, R.A.L., Norton, L.J., Kramer, E.J., Bates, F.S., and Wiltzius, P. (1991) *Phys. Rev. Lett.*, **66**, 1326.
18 Flory, P.J. (1953) *Principles of Polymer Chemistry*, Cornell University Press.
19 Dobry, A. and Boyer-Kawenoki, F. (1947) *J. Polym. Sci.*, **2**, 90.
20 Torsi, L., Dodabalapur, A., Rothberg, L.J., Fung, A.W.P., and Katz, H.E. (1998) *Phys. Rev. B*, **57**, 2271.
21 Katz, H.E., Lovinger, A.J., and Laquindanum, J. (1998) *Chem. Mater.*, **10**, 457.
22 Russell, D.M., Newsome, C.J., Li, S.P., Kugler, T., Ishida, M., and Shimoda, T. (2005) *Appl. Phys. Lett.*, **87**, 222109.
23 Wolfer, P., Müller, C., Smith, P., Baklar, M.A., and Stingelin-Stutzmann, N. (2007) *Synth. Met.*, **157**, 827.
24 Letizia, J.A., Facchetti, A., Stern, C.L., Ratner, M.A., and Marks, T.J. (2005) *J. Am. Chem. Soc.*, **127**, 13476.
25 Lin, Y.-Y., Gundlach, D.J., Nelson, S.F., and Jackson, T.N. (1997) *IEEE Electron. Device Lett.*, **18**, 606.
26 Kelley, T.W., Baude, P.F., Gerlach, C., Ender, D.E., Muyres, D. Haase, M.A., Vogel, D.E., and Theiss, S.D. (2004) *Chem. Mater.*, **16**, 4413.
27 Sundar, V.C., Zaumseil, J., Podzorov, V., Menard, E., Willett, R.L., Someya, T.,

Gershenson, M.E., and Rogers, J.A. (2004) *Science*, **303**, 1644.

28 Minakata, T. and Natsume, Y. (2005) *Synth. Met.*, **153**, 1.

29 Weidkamp, K.P., Afzali, A., Tromp, R.M., and Hamers, R.J. (2004) *J. Am. Chem. Soc.*, **126**, 12740.

30 Stingelin-Stutzmann, N., Smits, E., Wondergem, H., Tanase, C., Blom, P., Smith, P., and de Leeuw, D. (2005) *Nat. Mater.*, **4**, 601.

31 Subramanian, S., Park, S.K., Parkin, S.R., Podzorov, V., Jackson, T.N., and Anthony, J.E. (2008) *J. Am. Chem. Soc.*, **130**, 2706.

32 Ogier, S.D., Veres, J., and Zeidan, M. (2007) Patent No. WO/2007/082584. Electronic component or device useful in ink jet printing comprises gate electrode, source electrode, drain electrode and organic semiconducting material between source and drain electrode where device has specific channel length.

33 Hamilton, R., Smith, J., Ogier, S., Heeney, M., Anthony, J.E., McCulloch, I., Bradley, D.D.C., Veres, J., and Anthopoulos, T.D. (2009) *Adv. Mater.*, **21**, 1166.

34 Chung, Y.S., Shin, N., Kang, J., Jo, Y., Prabhu, V.M., Satija, S.K., Kline, R.J., DeLongchamp, D.M., Toney, M.F., Loth, M.A., Purushothaman, B., Anthony, J.E., and Yoon, D.Y. (2010) *J. Am. Chem. Soc.*, **133**, 412.

35 Lee, W.H., Lim, J.A., Kwak, D., Cho, J.H., Lee, H.S., Choi, H.H., and Cho, K. (2009) *Adv. Mater.*, **21**, 4243.

36 Smith, J., Hamilton, R., Heeney, M., de Leeuw, D.M., Cantatore, E., Anthony, J.E., McCulloch, I., Bradley, D.D.C., and Anthopoulos, T.D. (2008) *Appl. Phys. Lett.*, **93**, 253301.

37 Smith, J., Hamilton, R., McCulloch, I., Heeney, M., Anthony, J.E., Bradley, D.D.C., and Anthopoulos, T.D. (2009) *Synth. Met.*, **159**, 2365.

38 Kang, J., Shin, N., Jang, D.Y., Prabhu, V.M., and Yoon, D.Y. (2008) *J. Am. Chem. Soc.*, **130**, 12273.

39 Ohe, T., Kuribayashi, M., Tsuboi, A., Satori, K., Itabashi, M., and Nomoto, K. (2009) *Appl. Phys. Express*, **2**, 121502.

40 Kwon, J.-H., Shin, S.-I., Choi, J., Chung, M.-H., Ryu, H., Zschieschang, U., Klauk, H., Anthony, J.E., and Ju, B.-K. (2009) *Electrochem. Solid-State Lett.*, **12**, H285.

41 Lee, J., Kwon, J., Lim, J., and Lee, C. (2010) *Mol. Cryst. Liq. Cryst.*, **519**, 179.

42 Smith, J. (2011) *High-performance organic field effect transistors for large-area microelectronics.* Ph.D. thesis, Imperial College London.

43 Madec, M.-B., Crouch, D., Llorente, G.R., Whittle, T.J., Geogheganc, M., and Yeates, S.G. (2008) *J. Mater. Chem.*, **18**, 3230.

44 Lee, S.H., Choi, M.H., Han, S.H., Choo, D.J., Jang, J., and Kwon, S.K. (2008) *Org. Electron.*, **9**, 721.

45 Madec, M.-B., Smith, P.J., Malandraki, A., Wang, N., Korvink, J.G., and Yeates, S.G. (2010) *J. Mater. Chem.*, **20**, 9155.

46 Anthopoulos, T.D., Singh, B., Marjanovic, N., Sariciftci, N.S., Ramil, A.M., Sitter, H., Colle, M., and de Leeuw, D.M. (2006) *Appl. Phys. Lett.*, **89**, 213504.

47 Hamilton, R., Smith, J., Ogier, S., Heeney, M., Anthony, J.E., McCulloch, I., Veres, J., Bradley, D.D.C., and Anthopoulos, T.D. (2009) *Adv. Mater.*, **21**, 1166.

48 Yan, H., Chen, Z.H., Zheng, Y., Newman, C., Quinn, J.R., Dotz, F., Kastler, M., and Facchetti, A. (2009) *Nature*, **457**, 679.

49 Halik, M., Klauk, H., Zschieschang, U., Schmid, G., Dehm, C., Schutz, M., Maisch, S., Effenberger, F., Brunnbauer, M., and Stellacci, F. (2004) *Nature*, **431**, 963.

50 Horowitz, G., Hajlaoui, R., Fichou, D., and El Kassmi, A. (1999) *J. Appl. Phys.*, **85**, 3202.

51 Vissenberg, M.C.J.M. and Matters, M. (1998) *Phys. Rev. B*, **57**, 12964.

52 Nelson, S.F., Lin, Y.Y., Gundlach, D.J., and Jackson, T.N. (1998) *Appl. Phys. Lett.*, **72**, 1854.

53 Salleo, A. and Street, R.A. (2003) *J. Appl. Phys.*, **94**, 471.

54 Hallam, T., Lee, M., Zhao, N., Nandhakumar, I., Kemerink, M., Heeney, M., McCulloch, I., and Sirringhaus, H. (2009) *Phys. Rev. Lett.*, **103**, 256803.

55 Lloyd-Hughes, J., Richards, T., Sirringhaus, H., Johnston, M.B., and Herz, L.M. (2008) *Phys. Rev. B*, **77**, 125203.

56 Cho, S., Yuen, J., Kim, J.Y., Lee, K., and Heeger, A.J. (2007) *Appl. Phys. Lett.*, **90**, 063511.

57 Morana, M., Koers, P., Waldauf, C., Koppe, M., Muehlbacher, D., Denk, P., Scharber, M., Waller, D., and Brabec, C. (2007) *Adv. Funct. Mater.*, **17**, 3274.

58 Smith, J., Hamilton, R., McCulloch, I., Stingelin-Stutzmann, N., Heeney, M., Bradley, D.D.C., and Anthopoulos, T.D. (2010) *J. Mater. Chem.*, **20**, 2562.

59 Kim, J.Y. and Frisbie, D. (2008) *J. Phys. Chem. C*, **112**, 17726.

60 Hoppe, H. and Sariciftci, N.S. (2006) *J. Mater. Chem.*, **16**, 45.

61 Chen, L.M., Hong, Z.R., Li, G., and Yang, Y. (2009) *Adv. Mater.*, **21**, 1434.

62 Giridharagopal, R. and Ginger, D.S. (2010) *J. Phys. Chem. Lett.*, **1**, 1160.

63 Brabec, C.J., Heeney, M., McCulloch, I., and Nelson, J. (2011) *Chem. Soc. Rev.*, **40**, 1185.

64 Padinger, F., Rittberger, R.S., and Sariciftci, N.S. (2003) *Adv. Funct. Mater.*, **13**, 85.

65 Campoy-Quiles, M., Ferenczi, T., Agostinelli, T., Etchegoin, P.G., Kim, Y., Anthopoulos, T.D., Stavrinou, P.N., Bradley, D.D.C., and Nelson, J. (2008) *Nat. Mater.*, **7**, 158.

66 Yan, H., Yoon, M.-H., Facchetti, A., and Marks, T.J. (2005) *Appl. Phys. Lett.*, **87**, 183501.

67 Chirvase, D., Parisi, J., Hummelen, J.C., and Dyakonov, V. (2004) *Nanotechnology*, **15**, 1317.

68 Kim, Y., Choulis, S.A., Nelson, J., Bradley, D.D.C., Cook, S., and Durrant, J.R. (2005) *Appl. Phys. Lett.*, **86**, 063502.

69 Mihailetchi, V.D., Xie, H.X., de Boer, B., Koster, L.J.A., and Blom, P.W.M. (2006) *Adv. Funct. Mater.*, **16**, 699.

70 Vanlaeke, P., Swinnen, A., Haeldermans, I., Vanhoyland, G., Aernouts, T., Cheyns, D., Deibel, C., D'Haen, J., Heremans, P., Poortmans, J., and Manca, J.V. (2006) *Sol. Energy Mater. Sol. Cells*, **90**, 2150.

71 Muller, C., Ferenczi, T.A.M., Campoy-Quiles, M., Frost, J.M., Bradley, D.D.C., Smith, P., Stingelin-Stutzmann, N., and Nelson, J. (2008) *Adv. Mater.*, **20**, 3510.

72 Camaioni, N., Ridolfi, G., Casalbore-Miceli, G., Possamai, G., and Maggini, M. (2002) *Adv. Mater.*, **14**, 1735.

73 Yang, X.N., van Duren, J.K.J., Janssen, R.A.J., Michels, M.A.J., and Loos, J. (2004) *Macromolecules*, **37**, 2151.

74 Hoppe, H., Niggemann, M., Winder, C., Kraut, J., Hiesgen, R., Hinsch, A., Meissner, D., and Sariciftci, N.S. (2004) *Adv. Funct. Mater.*, **14**, 1005.

75 Erb, T., Zhokhavets, U., Gobsch, G., Raleva, S., Stuhn, B., Schilinsky, P., Waldauf, C., and Brabec, C.J. (2005) *Adv. Funct. Mater.*, **15**, 1193.

76 Yang, X.N., Loos, J., Veenstra, S.C., Verhees, W.J.H., Wienk, M.M., Kroon, J.M., Michels, M.A.J., and Janssen, R.A.J. (2005) *Nano Lett.*, **5**, 579.

77 Labram, J.G., Domingo, E.B., Stingelin, N., Bradley, D.D.C., and Anthopoulos, T.D. (2011) *Adv. Funct. Mater.*, **21**, 356.

78 Meijer, E.J., de Leeuw, D.M., Setayesh, S., Van Veenendaal, E., Huisman, B.H., Blom, P.W.M., Hummelen, J.C., Scherf, U., and Klapwijk, T.M. (2003) *Nat. Mater.*, **2**, 678.

79 Anthopoulos, T.D., Tanase, C., Setayesh, S., Meijer, E.J., Hummelen, J.C., Blom, P.W.M., and de Leeuw, D.M. (2004) *Adv. Mater.*, **16**, 2174.

80 Smits, E.C.P., Anthopoulos, T.D., Setayesh, S., van Veenendaal, E., Coehoorn, R., Blom, P.W.M., de Boer, B., and de Leeuw, D.M. (2006) *Phys. Rev. B*, **73**, 205316.

81 Horowitz, G. (2004) *J. Mater. Res.*, **19**, 1946.

82 Sirringhaus, H., Brown, P.J., Friend, R.H., Nielsen, M.M., Bechgaard, K., Langeveld-Voss, B.M.W., Spiering, A.J.H., Janssen, R.A.J., Meijer, E.W., Herwig, P., and de Leeuw, D.M. (1999) *Nature*, **401**, 685.

83 Hoppe, H. and Sariciftci, N.S. (2004) *J. Mater. Res.*, **19**, 1924.

84 Mayer, A.C., Toney, M.F., Scully, S.R., Rivnay, J., Brabec, C.J., Scharber, M., Koppe, M., Heeney, M., McCulloch, I., and McGehee, M.D. (2009) *Adv. Funct. Mater.*, **19**, 1173.

85 Cates, N.C., Gysel, R., Beiley, Z., Miller, C.E., Toney, M.F., Heeney, M., McCulloch, I., and McGehee, M.D. (2009) *Nano Lett.*, **9**, 4153.

86 Cates, N.C., Gysel, R., Dahl, J.E.P., Sellinger, A., and McGehee, M.D. (2010) *Chem. Mater.*, **22**, 3543.

87 von Hauff, E., Parisi, J., and Dyakonov, V. (2006) *Thin Solid Films*, **511**, 506.
88 Horowitz, G., Hajlaoui, R., and Delannoy, P. (1995) *J. Phys. III*, **5**, 355.
89 Chen, H.Y., Hou, J.H., Zhang, S.Q., Liang, Y.Y., Yang, G.W., Yang, Y., Yu, L.P., Wu, Y., and Li, G. (2009) *Nat. Photon.*, **3**, 649.
90 Park, S.H., Roy, A., Beaupre, S., Cho, S., Coates, N., Moon, J.S., Moses, D., Leclerc, M., Lee, K., and Heeger, A.J. (2009) *Nat. Photon.*, **3**, 297.
91 Liang, Y., Xu, Z., Xia, J., Tsai, S.-T., Wu, Y., Li, G., Ray, C., and Yu, L. (2010) *Adv. Mater.*, **22**, E135.
92 Wienk, M.M., Kroon, J.M., Verhees, W.J.H., Knol, J., Hummelen, J.C., van Hal, P.A., and Janssen, R.A.J. (2003) *Angew. Chem., Int. Ed.*, **42**, 3371.
93 Rostalski, J. and Meissner, D. (2000) *Sol. Energy Mater. Sol. Cells*, **61**, 87.
94 Wobkenberg, P.H., Bradley, D.D.C., Kronholm, D., Hummelen, J.C., de Leeuw, D.M., Colle, M., and Anthopoulos, T.D. (2008) *Synth. Met.*, **158**, 468.
95 Labram, J.G., Kirkpatrick, J., Bradley, D.D.C., and Anthopoulos, T.D. (2011) *Adv. Energy Mater.*, 1, 1176.
96 Germack, D.S., Chan, C.K., Hamadani, B.H., Richter, L.J., Fischer, D.A., Gundlach, D.J., and DeLongchamp, D.M. (2009) *Appl. Phys. Lett.*, **94**, 233303.
97 Ferenczi, T.A.M., Nelson, J., Belton, C., Ballantyne, A.M., Campoy-Quiles, M., Braun, F.M., and Bradley, D.D.C. (2008) *J. Phys.: Condens Matter*, **20**, 475203.
98 Ayzner, A.L., Tassone, C.J., Tolbert, S.H., and Schwartz, B.J. (2009) *J. Phys. Chem. C*, **113**, 20050.
99 Collins, B.A., Gann, E., Guignard, L., He, X., McNeill, C.R., and Ade, H. (2010) *J. Phys. Chem. Lett.*, **1**, 3160.
100 Wang, D.H., Choi, D.G., Lee, K.J., Im, S.H., Park, O.O., and Park, J.H. (2010) *Org. Electron.*, **11**, 1376.
101 Labram, J.G., Kirkpatrick, J., Bradley, D.D.C., and Anthopoulos, T.D. (2011) *Phys. Rev. B*, **84**, 075344.
102 Lee, K.H., Schwenn, P.E., Smith, A.R.G., Cavaye, H., Shaw, P.E., James, M., Krueger, K.B., Gentle, I.R., Meredith, P., and Burn, P.L. (2011) *Adv. Mater.*, **23**, 766.
103 Moon, J.S., Takacs, C.J., Sun, Y.M., and Heeger, A.J. (2011) *Nano Lett.*, **11**, 1036.
104 Treat, N.D., Brady, M.A., Smith, G., Toney, M.F., Kramer, E.J., Hawker, C.J., and Chabinyc, M.L. (2011) *Adv. Energy Mater.*, **1**, 82.
105 Dodabalapur, A., Katz, H.E., Torsi, L., and Haddon, R.C. (1995) *Science*, **269**, 1560.
106 Kirkpatrick, S. (1973) *Rev. Mod. Phys.*, **45**, 574.

9
Enhanced Electrical Conductivity of Polythiophene/Insulating Polymer Composite and Its Morphological Requirement

Guanghao Lu and Xiaoniu Yang

9.1
Introduction

Integrating insulating polymers (IPs) into intrinsic semiconducting polymers (SPs) can potentially reduce cost and improve flexibility as well as environmental stability of polymer-based electronics [1]. Even though, one may point out the contradiction between above superior properties and good electrical performance because an insulating polymer matrix usually impairs charge transport. However, in 2006 some research groups discovered an unexpected high field-effect charge carrier mobility in SP/IP composites made of poly(3-hexylthiophene) (P3HT) and polyethylene (PE) or polystyrene (PS), even with very low contents of P3HT (3–5%) [2]. They found that under their experimental conditions composites employing semicrystalline rather than amorphous insulating polymers lead to good electrical performance. They identified improved crystallinity of P3HT in the crystalline insulating matrix and vertically stratified structures, both of which are beneficial for high carrier mobility. On the other hand, Babel and Jenekhe found that the carrier mobility of P3HT/PS composites is larger than that of P3HT/MEH-PPV mixtures (MEH-PPV, poly[2-methoxy-5-(2-ethylhexoxy)-1,4-phenylenevinylene], is another semiconducting polymer) with P3HT content larger than 40%, although their morphologies are almost identical [3]; however, both composites exhibited mobilities lower than that of neat P3HT. It seems that enhanced crystallinity and vertical stratification cannot be the only two crucial factors for enhanced electrical properties, although they are beneficial for it without question. Afterwards, some other independent groups discovered and confirmed substantially enhanced (rather than worsened) field-effect transistor (FET) properties of semiconducting polymer/insulating polymer (SP/IP) composites as compared to the corresponding neat semiconducting polymers [4–6]. Although the mechanism responsible for this exciting phenomenon is not resolved at present, the final composite performance should be determined by the balance between competing processes that lead to an improvement and degradation of the charge transport properties, respectively.

Based on these studies on FETs, one might have a question: can the conductivity of conjugated polymer be enhanced by blending it with insulating polymer? The

Semiconducting Polymer Composites: Principles, Morphologies, Properties and Applications, First Edition.
Edited by Xiaoniu Yang.
© 2012 Wiley-VCH Verlag GmbH & Co. KGaA. Published 2012 by Wiley-VCH Verlag GmbH & Co. KGaA.

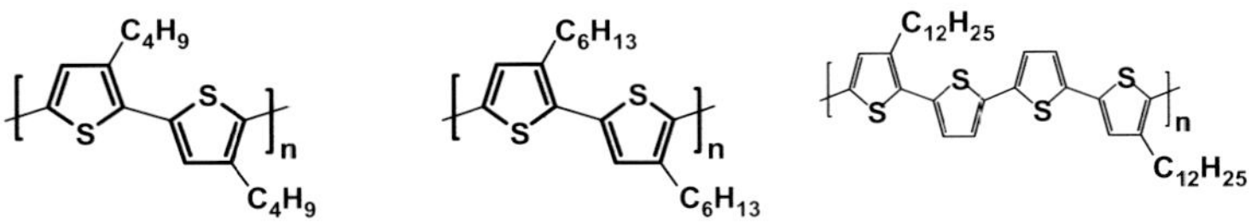

Figure 9.1 Chemical structure of P3BT, P3HT, and PQT-12.

answer is positive, although so far the mechanism and optimization conditions are still not very unclear. Before we have a general overview of this research topic, let us have a look at the conjugated polymers and think about the origin of their conductivity.

The conjugated polymers we will focus on in this chapter are soluble polythiophene as shown in Figure 9.1, which include poly(3-butylthiophene) (P3BT) [7–9], P3HT [7–9], and poly(3, 3‴-didodecylquaterthiophene) (PQT-12) [10]. This class of conjugated polymers includes the benchmark semiconductor materials for flexible electronics.

The intrinsic carrier (for instance, hole) concentration N_{int} in valance band can be given using the Fermi–Dirac distribution function

$$N_{int} = N_v \frac{1}{\exp((E_f - E_v)/kT) + 1},$$

where N_v is the density of states, E_f and E_v are the Fermi level and the valance band level, respectively, k is the Boltzmann constant, and T is the temperature. The intrinsic carrier concentration depends on temperature, implying that without external excitation (for instance, light) the carrier can only be induced by phonon excitation. The intrinsic bandgap for the above conjugated polymers is around 2.0 eV. Taking the bandgap and temperature into account, one might find the very low intrinsic carrier concentration for ideally pure conjugated polymer at room temperature. Therefore, the intrinsic conductivity σ_{int} of conjugated polymer and its SP/IP composite is rather low because

$$\sigma_{int} = eN_{int}\mu_{int},$$

where μ_{int} is the intrinsic mobility of hole. Note here, particularly for polythiophene in this chapter, we do not consider the contribution of electron carriers to the conductivity.

Consequently, in order to study the conductivity of these semiconducting polymers and their SP/IP composite, doping is somehow necessary. p-type doping can be achieved by donating hole carriers by dopants into valence band of conjugated polymer and thus increasing the hole concentration, which contributes to the conductivity. That is to say, doping shifts the Fermi level toward valence band and thus decreases $E_f - E_v$, so one can expect more carriers in the valence band. For the conjugated polymers P3BT, P3HT, and PQT-12 discussed in this chapter, the ionization potentials are all as low as around 5.0 eV, which means the unintentional doping of these materials happens even in conventional preparation procedure. The dopants are probably from catalyst during the synthesis, residual from solvent that was concentrated during solvent evaporation, and oxygen/water absorbed from surrounding environment. Generally speaking, although the doping levels from above sources depend on the material preparation method, they are usually in low level, implying that the conductivity is typically below 10^{-3} S cm^{-1}. For a higher conductivity and so a higher doping level, strong oxidants or molecular dopants are required.

Subsequently, taking the above considerations into account, Lu *et al.* studied the conductivity of P3BT/PS composites prepared from well precrystallized P3BT in solution [11, 12]. It is shown that higher conductivity compared with highly crystalline neat P3BT can be achieved via blending with amorphous insulating polymers. They identified a morphological interpenetrating P3BT network in the optimized composites, and ascribed the enhanced conductivity of P3BT/PS composites to the enhanced charge mobility rather than increased charge concentration, as compared to pure P3BT. From this point of view, it is reasonable to assume that the high FET mobility and enhanced conductivity of SP/IP composites should somehow share the same key mechanism for the charge transport. Therefore, in this chapter, although we mainly focus on the conductivity, the mechanisms employed to interpret the high FET performance of semiconductor/insulator (including small-molecule semiconductor/insulating polymer) composites are also highlighted.

Actually, for conventional FETs with the enhancement operation mode, which is so far the case for the materials discussed in this chapter, conductivity of active film is usually undesired because high bulk conductivity deteriorates the on/off ratio of a device. Therefore, in order to prevent doping by ambient oxygen or radicals on the surface of dielectrics, inert environment and passivated dielectric surfaces are commonly used.

However, until now, most studies from the literature on this unusual electrical property of SP/IP composite focused on the film preparation or device optimization, while only a few are available that dealt with the underlying fundamental mechanism. However, only knowledge of these mechanism(s) will allow us to effectively optimize the performance of such composites and to promote their application in flexible electronics.

9.2
Phase Evolution and Morphology

Generally speaking, one of the most important roles that determine the conductivity of conductor/insulator composite is morphology, which was directly described in percolation models. No doubt that the conductivity of SP/IP composite strongly depends on the spatial distribution of SP component within the SP matrix, which is sensitive to the phase evolution during sample preparation. Therefore, here we would like to exclusively discuss the influence of phase evolution and thus formed morphology on material conductivity.

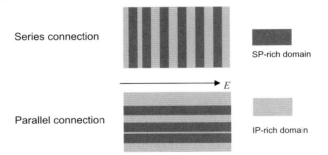

Figure 9.2 Ideal series connection and parallel connection between SP-rich and IP-rich domains.

As a result of phase evolution of SP/IP from either solution or melt, the material shows typical two-phase morphology [1, 13], which corresponds to SP-rich and IP-rich domains. From the microscopic point of view, here we consider two extreme spatial distributions of SP component, relative to the direction of electrical field, which are series connection and parallel connection between SP-rich and IP-rich components (Figure 9.2). For the series connection, the conductivity of composite σ_{total} should obey

$$\frac{f_{SP\text{-rich}}}{\sigma_{SP\text{-rich}}} + \frac{f_{IP\text{-rich}}}{\sigma_{IP\text{-rich}}} = \frac{f_{total}}{\sigma_{total}} \quad (f_{SP\text{-rich}} + f_{IP\text{-rich}} = 1),$$

where f is the volume fraction of respective component. On the other hand, the conductivity of parallel connection is

$$\sigma_{total} = f_{SP\text{-rich}}\sigma_{SP\text{-rich}} + f_{IP\text{-rich}}\sigma_{IP\text{-rich}} \quad (f_{SP\text{-rich}} + f_{IP\text{-rich}} = 1).$$

Actually, for a realistic SP/IP composite, these two connections between SP-rich and IP-rich domains should both contribute to the macroscopic conductivity. The above discussion implies that in cases of $\sigma_{SP\text{-rich}} \gg \sigma_{IP\text{-rich}}$ or $\sigma_{SP\text{-rich}} \ll \sigma_{IP\text{-rich}}$, the spatial distribution of SP-rich domains within IP-rich domains is very crucial for the conductivity.

Let us take P3BT/a-PS (a-PS is amorphous polystyrene) as an example to illuminate the relationship between phase evolution and morphology. To study the phase evolution and corresponding two-phase morphology of P3BT/a-PS composites from solvent o-dichlorobenzene (ODCB), the phase diagram of P3BT/a-PS/ODCB tricomponent system has to be made first (Figure 9.3). From the phase diagram, a liquid–liquid ($L_{P3BT/ODCB} + L_{a\text{-PS}/ODCB}$) phase region can be well identified. For solution-processed P3BT/a-PS composite, P3BT and a-PS should be first dissolved in ODCB at an elevated temperature because ODCB is a poor solvent for P3BT at room temperature. If film preparation conditions are taken into account, two typical phase evolution routes can be figured out, depending on whether phase evolution goes across this liquid–liquid phase region or not. The P3BT/a-PS composite prepared via crystallization of P3BT and subsequent vetrification of a-PS (cooling to room temperature and then solvent evaporation) is homogenous as a result of the absence of liquid–liquid phase region. Three-dimensional reconstruction techniques of electron microscopy

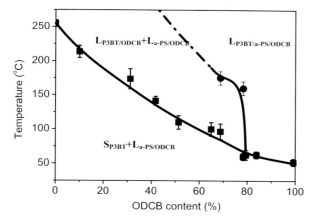

Figure 9.3 Phase diagram of P3BT/a-PS/ODCB (P3BT:a-PS = 1:9, w/w) tricomponent systems (unpublished data).

were used to characterize the detailed three-dimensional tomography of P3BT/a-PS composite prepared along the route that involved crystallization of P3BT and subsequent vetrification of a-PS. In Figure 9.4, three-dimensional electron tomography images were obtained upon continuous tilting experiments within

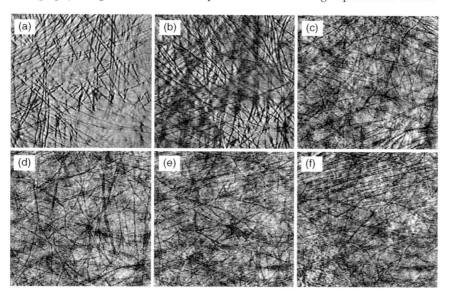

Figure 9.4 Three-dimensional reconstructed electron tomography of semiconducting polymer/insulating polymer composite film. (In order to obtain a well-resolved 3D result with enough and accountable P3BT domain in each 2D cross section to figure out its distribution in the Z direction, here we used P3BT with 50 wt% in PS matrix and film thickness 80–100 nm.) All slices are lying in the horizontal (X, Y) plane of the film at a different depth (Z location): slice (a) is close to the bottom of the film, slices (b–d) are in the middle of the film, and slice (f) is close to the top of the film. The dimensions of the slices are 1700 nm × 1700 nm, and the back fiber-like features are ascribed to semiconducting component P3BT (unpublished data).

TEM, and some slices lying in the horizontal (X, Y) plane of the film at a different depth (Z location) within the film with thickness 80–100 nm are shown. In each 2D slice, the P3BT nanowires are randomly dispersed within a-PS matrix and compose the topological interpenetrating network in 3D space. It should be noted that the P3BT nanowires are somehow homogeneously distributed in the Z direction of the P3BT/a-PS composite film. This morphological feature of P3BT/a-PS composite should be determined by the corresponding phase evolution route that is free of liquid–liquid phase separation. After the crystallization of P3BT within a-PS/ODCB environment, the chain-like nanowires in solution can be treated as the so-called supramolecular chain with super-high "molecular" weight; therefore, its "molecular" motion is pronouncedly limited. As a result, during solvent evaporation and subsequent vetrification of a-PS, P3BT nanowires do not further aggregate despite the thermodynamic immiscibility between these two polymer components. As a consequence, P3BT interpenetrating network within a-PS matrix is achieved after solvent evaporation. These morphological two-phase features supply not only interconnected conducting network, but also huge two-phase interfaces between P3BT and a-PS matrix, both of which contribute to the enhanced conductivity of P3BT/PS composite as compared with pure P3BT, as further discussed later.

Similar phenomenon that the different phase evolution routes (described in respective phase diagram) lead to completely different morphology even for the same materials has been observed in the phase evolution of P3HT/polyethylene composite and P3HT–polyethylene copolymer. Therefore, the phase evolution routes strongly determine the eventual device performance. Particularly, the phase evolution of SP/IP composite with crystalline IP is different from that using amorphous IP matrix, because the SP component can only be confined among crystalline IP domains. For simplicity and clarity, here we take P3BT/PEO (PEO is polyethylene oxide) as an example to illustrate the influence of matrix crystallization on composite morphology [12]. Figure 9.5a7–c gives optical micrographs of P3BT/PEO composite film cast from o-dichlorobenzene solution. The bright-field Optical Microscopy (OM) (Figure 9.5a) and corresponding Polarized Optical Microscopy (POM) (Figure 9.5b) images show clearly that PEO has crystallized into typical spherulites with size above 100 7μm. A zoom-in image of Figure 5a (Figure 5c) gives details into this composite: lots of wires with submicron width are well dispersed within PEO spherulites, which makes these conductive wires continuously connected through out the whole composite. The red color of these submicro-wires in OM confirms they are composed of P3BT. TEM image (Figure 9.5c, inset) implies that the submicro-wires are connected with each other, forming connected network. However, the length scale of this network is much larger than that of P3BT nanowires. C-AFM measurements (Figure 9.5c, inset) confirm that the conducting pathway is composed of submicro-wire, which are composed of bundled P3BT nanowires.

Although some groups have achieved enhanced FET mobility of the SP/IP composite, they somehow proposed different morphological requirements for this enhancement, according to their respective sample preparation conditions. In Table 9.1, the phase evolution routes (tentatively drawn by the author of this chapter,

9.2 Phase Evolution and Morphology | 257

Table 9.1 Phase evolution routes of other SP/IP composites as tentatively concluded from their preparation methods in respective references.

P3HT/HDPE	P3HT/a-PS	P3HT/a-PS	PQT-12/a-PS
PE, P3HT, Melt → Cooling → Crystallization of matrix polymer → Cooling	Polymer solution → Cooling → Solvent evaporation → Vetrification of matrix polymer	Polymer solution → Adding poor solvent → Solvent evaporation → Vetrification of matrix polymer	Polymer solution → Cooling → Solvent evaporation → Vetrification of matrix polymer
Ref. [2]	Refs [5, 11, 12]	Ref. [6]	Ref. [14]

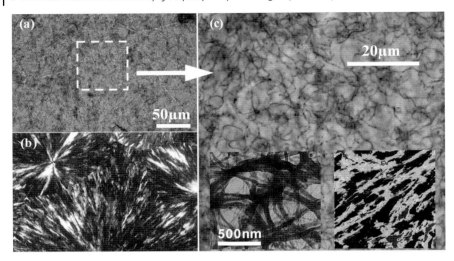

Figure 9.5 Morphology of P3BT/PEO (10 wt% P3BT) composite prepared from solution showing the enriching of conducting phase between insulating crystalline domains. (a) Bright-field OM image; (b) Polarized OM image (the same area as (a)); (c) zoom-in bright-field OM image for the area marked in (a). Insets are corresponding TEM image (left) and C-AFM current image (right, 5 μm∗5 μm; current from low (dark) to high (bright)) in P3BT/PEO composites. Reprinted with permission from ref [12]. Copyright 2010 Wiley-VCH.

based on the preparation methods from the respective references) from recent literatures are put together and it is found that the composites (including the one shown in Figure 9.4) with higher mobility/conductivity share a common morphological feature: semiconductor phase is morphological continuous with fine two-phase structure that supplies huge SP–IP interfacial area, even though they were prepared via different methods. This is highly consistent with the assumption in this chapter.

9.3
Enhanced Conductivity of Conjugated Polymer/Insulating Polymer Composites at Low Doping Level: Interpenetrated Three-Dimensional Interfaces

Although the enhanced FET mobility and the enhanced conductivity of SP/IP composite should be somehow strongly correlated with each other, the former has been discussed in other chapter(s) of this book. Therefore, here we mainly pay attention to the conductivity.

Recently, Lu *et al.* reported substantially enhanced conductivity of P3BT/PS and P3BT/PMMA composites as compared with the corresponding neat semiconducting polymers [11, 12]. The enhanced conductivity can be achieved for P3BT/insulating polymer composites that were unintentionally doped by oxygen/water and/or other impurities in the materials, so the doping level is low. Figure 9.6a is the dependence of conductivities of P3BT/a-PS composites on blending ratios, which shows the

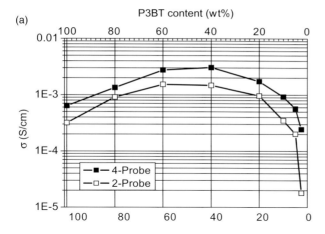

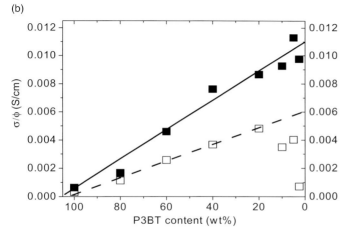

Figure 9.6 (a) Dependence of P3BT/a-PS conductivity on P3BT content in the composites as measured with 4-probe method (solid square) and 2-probe method (open square), respectively; (b) Dependence of σ/φ on P3BT content (φ) in the composites as measured with 4-probe method (solid square) and 2-probe method (open square), respectively. Reprinted with permission from ref [11]. Copyright 2007 American Chemical Society.

substantially increased conductivity upon mixing with insulating PS as compared to the pure P3BT. Moreover, the conductivity of P3BT/PS with 10 wt% P3BT content is still comparable with that of pure P3BT. Their study also shows that the inclusion of insulating PS into P3BT has not introduced additional contact resistance that much as long as the content of P3BT is above 10 wt%. For the composite with very low P3BT content, the conductivity substantially drops down, particularly for those results determined from two-probe method, as a result of less connection between P3BT nanowires. In this situation, contact resistance turns to be highly significant for conductivity of the P3BT/PS composite obtained [11].

Actually, the values of σ/φ (where σ is the conductivity of the composite and φ is the polythiophene content (weight fraction) in the composite) should reflect the true conductivity contribution of P3BT nanowires in insulating polymer matrix. Figure 9.6b shows the dependence of σ/φ of P3BT/PS on P3BT content (φ). The data obtained from four-probe method could be linearly fitted throughout almost the whole content range, while the data from two-probe method can only be linearly fitted as the P3BT content is higher than 20%, as shown in Figure 9.6. The loss of linear relationship at lower P3BT content should be ascribed to the much higher interface resistance at the probes especially for two-probe method. The extrapolation of σ/φ to $\varphi = 0$, namely, the intercept with Y-axis, could be used to describe the conductivity of P3BT individually dispersed in PS matrix provided that the P3BT nanowires are still interconnected with each other. It can be carefully concluded that for an individual conducting P3BT nanowire, the surrounding matrix indeed plays an important role in the charge carrier transport. The conductivity of a single P3BT nanowire in PS matrix is thus at least one order of magnitude larger than that of P3BT nanowire in its own P3BT matrix. However, the increased conductivity of P3BT can only be attributed to enhanced charge carrier mobility since the charge carrier concentration in P3BT nanowires within PS matrix is at least not higher than that in pure P3BT material. In the conventional FET architecture, the interface between semiconductor and insulator (dielectric) can significantly enhance charge carrier mobility of semiconductor [15–17]; the homogeneously dispersed P3BT nanowires in PS matrix provide extremely large interface area between P3BT and PS, which is crucial for the increased conductivity of P3BT/PS composite in macroscopic scale.

9.4
Conductivity of Semiconducting Polymer/Insulating Polymer Composites Doped by Molecular Dopant

Above, we discussed the conductivity of SP/IP composite from unintentional doping or doping by ambient, where the doping level should be in a relatively low level and the conductivity of SP/IP composite is typically $\leq 10^{-3}$ S cm^{-1}. On the other hand, in order to increase the doping level and so increase the conductivity, doping of the SP/IP composite using molecular dopant or oxidant is possible [14].

For example, Sun et al. used solution method to incorporate a dopant tetrafluorotetracyanoquinodimethane (F4TCNQ) [18, 19] into PQT-12/PS composite [14]. They observed that the conductivity of F4TCNQ-doped PQT-12/PS (blending ratio is 2/5) composite approaches that of F4TCNQ-doped pure PQT-12 film (Figure 9.6). However, the conductivity of the doped PQT-12/PS composite prepared using this method is relatively low, which is around 10^{-3} S cm^{-1}. The low conductivity of both composite and pure conjugated polymer is probably because conjugated polymer and dopant form the so-called charge transfer complex in solution [18, 19], and during the solvent evaporation the charge transfer complex preserves and so limits the crystallization of semiconducting polymer. It is noted here that using P3HT and via optimization of the material preparation condition, the conductivity of pure P3HT doped by F4TCNQ can maximally reach 1 S cm^{-1} [18].

Except doping SP/IP composite from solution process, other alternative doping methods are also available that can be used to avoid the negative role of charge transfer complex on the packing of conjugated polymer chain during solvent evaporation. For example, Lu *et al.* used saturated iodine vapor to dope the pre-prepared P3BT/IP composite, which shows that the conductivity of P3BT/PS composite is over 1 S cm^{-1} even for P3BT content as low as 10 wt%. However, this conductivity is somehow still lower than that of pure P3BT, which is probably because that iodine molecule is small enough to diffuse into the composite, and then the dopant can even incorporate between P3BT chains, which thus distorts the crystalline structure and so reduces the overlapping of wavefunction. The iodine within P3HT domains enlarges the scattering of carrier during charge transport.

From the above-mentioned experimental observations, it is rather confusing to compare the doping results from different sample preparation methods and from different dopants. It seems that different dopants act in different ways for the doping of SP/IP composite. This confusion calls for more research on the detail of charge transfer from conjugated polymer to dopant, especially at the presence of insulating polymer.

9.5 Mechanisms for the Enhanced Conductivity/Mobility

So far, it seems that a reliable model should be "universally" successful in interpreting both the enhanced FET mobility of SP/IP composite and the enhanced conductivity, the latter of which is related to doping. It is also reasonable to ascribe the enhanced conductivity to enhanced carrier mobility (the mobility is different from FET mobility, the latter of which is a characteristic of material in the presence of a gate bias). From the recent research, one can carefully expect that the charge transport in SP/IP composite is different from that in pure semiconducting polymer. However, so far the mechanism for this enhanced charge transport is still in debate. Here we mainly review some of these possible mechanisms proposed in recent years. From this point of view, we not only review the conductivity of some conjugated polymer/insulating polymer composite, but also take the enhanced FET performance into account for semiconductor/insulator composites including small-molecule/insulating polymer composites.

9.5.1 Improved Crystallinity and Molecular Ordering

For conjugated polymer, the carrier transport is achieved in the form of hoping among localized states. Therefore, in order to reduce the localization of carrier, high order of molecule packing is preferred. Generally speaking, the chain of commonly used insulating polymer (such as PS, PE, and PMMA) is more flexible than conjugated polymer chain. So in the composite preparation process from solution or melt, the presence of insulating chain supplies a flexible matrix for the crystallization of semiconducting polymer. As a result, the order of semiconducting domain can be higher than that of corresponding pure semiconducting polymer prepared via the same way. This feature was experimentally confirmed in P3HT/polyethylene composite and P3HT–insulator copolymer [2, 20].

So far, several methods have been proposed to enhance the crystallinity of semiconducting phase by (i) employing crystalline insulating polymer as matrix and so the crystallization of matrix endows the composite with a low percolation threshold; (ii) using poor solvent of semiconducting polymer; and (iii) adding poor solvent to polymer solution. However, the relationship between the morphologies and electrical properties has not yet clearly been understood. Actually, the morphological requirements of SP/IP composite for enhanced charge transportation are still not clear. All of these imply that the charge transport in SP/IP composite is somehow different from that in pure semiconducting polymer.

9.5.2
"Self-Encapsulation" Effect

In some composite film prepared from either solution or melt, vertically phase separation between semiconducting component and insulating matrix was also observed [2, 4, 21]. This vertical stratified structure enables the semiconducting domains to connect with each other, forming a continuous two-dimensional film. The vertical phase separation between SP/IP composite relative to substrate can be employed to improve the performance of the FET device. Depending on the surface energy of material–substrate interface, the semiconducting component can enrich either at the material–substrate interface or at the material–air interface, both of which can be employed to improve the device performance [2, 4, 21]. Especially, in the case of semiconducting component encapsulated by insulating layer, higher environment stability of FET device was reported, as compared to corresponding pure conjugated polymer, because the insulating layer acts as effective barriers and so decreases the diffusion of oxygen/water from ambient into semiconductor–dielectric interface, which is the key for the accumulation layer in FETs [22]. Anyway, the life of such devices should somehow depend on the diffusion coefficient of oxygen/water in respective layers, considering the semiconducting film for FET devices is typically around 100 nm thick, which means the choice of insulating layer for self-encapsulation is crucial for the device optimization.

The self-encapsulation effect is successful in interpreting the stability of FET. However, this effect is not suitable to reveal the mechanism of enhanced conductivity of SP/IP composite as compared with corresponding pure semiconducting polymer, because for the conductivity, doping is necessary.

9.5.3
Bulk 3-D Interface and Reduced Polarization of Matrix at Interface

Lu *et al.* [11, 12] realized that the individual P3BT nanowire in P3BT/a-PS composite is identical to the one in neat P3BT in terms of morphology, which implies that in P3BT/a-PS composite the enhanced charge transport should be determined by the electronic/electrical properties of two-phase P3BT/a-PS interfaces rather than the crystallinity of P3BT (Figure 9.7). It is primarily interpreted that in interpenetrating composite, the charge transport occurs via polarons [23] that are likely to be predominantly present near the two-phase interface because of the very fine two-phase structure. From this viewpoint, the polaron should be largely influenced by interface energetic disorder that can be weakened by insulating polymer typically with low polarizability, so the presence of insulating polymer nearby should facilitate the transport of polarons.

Semiconductor–insulator (dielectric) interfaces are crucial for the performance of conventional FETs that are composed of a semiconductor layer for charge transport and an insulator layer as gate dielectric. It is well known that the mobility strongly depends on the polarizability or permittivity of dielectric layer [15, 17]. The interpenetrating bulk heterojunction in P3BT/PS composite is composed of interpenetrating 3D two-phase interfaces rather than the typical 2D interface in typical FET geometry. Therefore, the properties of the 3D interface are expected to play a crucial role in the charge distribution and charge transport [12].

From morphological point of view, P3BT usually adopts nanowire-like appearance with a width of ∼15 nm and a thickness of ∼5 nm. The variation of these two-dimensional parameters is usually very limited and insensitive to crystallization conditions [9, 11, 24]. So during charge transport along nanowire

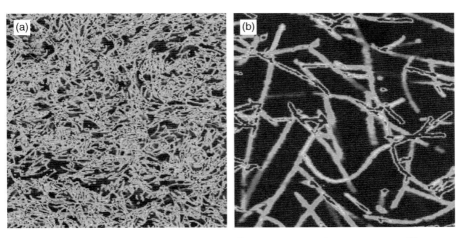

Figure 9.7 Hole-only transportation measurement of P3BT/a-PS (10 wt% P3BT) composite film using Conductive-AFM. (a) Overview (15 μm × 15 μm); (b) zoom-in image (2 μm × 2 μm).

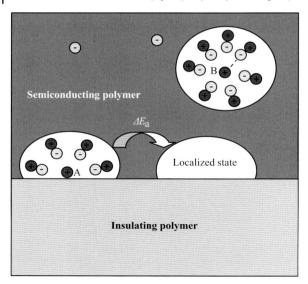

Figure 9.8 Scheme of charge transport at the SP/IP interface.

surrounded by insulating polymer matrix, it can be inevitably influenced by the electronic/electrical properties of SP/IP interface. As a result, the achievement of interconnected semiconducting polymer network in insulating polymer matrix, which maximizes the SP/IP interface, should be the crucial factor for the enhanced mobility/conductivity (Figure 9.8).

Based on the above analysis and interpretation, we can tentatively propose the morphological criteria for enhanced electrical properties: (i) semiconducting phase in composite should be morphologically continuous; (ii) in order to supply enough two-phase interfacial area, the phase separation of SP/IP composite should be on the nanometer scale. These criteria can be used to conduct the further optimization of SP/IP composite for electronic application.

9.5.4
"Zone Refinement" Effect

Small-molecule organic semiconductor blended with insulating polymer also shows high FET performance [25–27], although the mechanism is still not very clear and should strongly correlate with that for conjugated polymer/insulating polymer composite. Interestingly, Chung et al. observed the "zone refinement" effect in triethylsilylethynyl anthradithiophene (TESADT)/poly(α-methylstyrene) (PαMS) composite [26]. Actually, the zone refinement shares the same mechanism as sedimentation and recrystallization, both of which are commonly used in

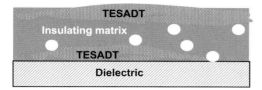

Figure 9.9 Tentative scheme for the degraded TESADT/insulating polymer composite as concluded from Ref. [26]. The white dots represent degraded TESADT molecules.

chemical purification. Because the impurity species is typically a minority component, it is easily dissolved within any other majority components. For the preparation of TESADT/PαMS composites from solution, the authors partly degraded the semiconductor, and it was observed that during solvent evaporation, degraded species enriched within insulating polymer, while "good" molecules crystallized at the film top and bottom interfaces due to vertical phase separation (Figure 9.9).

This effect observed by the authors strongly evidenced the recovery of FET properties by blending with insulating polymer, by using materials with low purity. This is a very interesting phenomenon. On the other hand, people also usually observed the relatively long environment stability of FET for SP/IP composite where degrading/doping of semiconducting polymer happens after the sample preparation, which is different from the zone refinement effect occurring in solution process. Because during the exposure of SP/IP composite to ambient environment, degraded polymer chains could not diffuse into insulating polymer matrix at room temperature, especially below the glass transition temperature of polymer matrix and semiconducting polymer, the long-term stability of SP/IP composite cannot ascribed to the "zone refinement" effect. Moreover, the conductivity of SP/IP composite actually originates from the doped or even degraded species, which means the "zone refinement" effect is probably not the key factor responsible for interpreting the enhanced conductivity.

9.5.5
Reduced Polaron–Dopant Interaction

As we discussed above, the doping of polythiophene/IP composite is somehow inevitable, because (i) the ionization potential is only around 5 eV; (ii) the material can only be prepared from solution, and during solvent evaporation the residual impurity in solvent is concentrated; and (iii) the intrinsic defect in polymer chain or undefined end groups might act as dopant. Certainly, the unintentional doping can be controlled in a low level. Therefore, let us consider the enhanced mobility/conductivity from the role of carrier–dopant interactions.

Here we introduce a Monte Carlo model to consider the influence of carrier–dopant interaction on the charge transport [28].

Microscopically, the carrier hops from one localized state to the neighbor one, and the energy E of localized site can be written as

$$E = E_{int} + E_{coul} + U_{ext},$$

where E_{int}, E_{coul}, and U_{ext} are contributed from intrinsic (without doping) energy, Coulomb interaction, and external field, respectively.

Gaussian distribution function is usually used to represent the density of states (DOS), which is given by

$$g_{int}(E) = \frac{N_t}{\sigma\sqrt{2\pi}} \exp\left[\frac{-(E_{int} - E_0)^2}{2\sigma^2}\right],$$

where N_t is the total state density, σ denotes the width of the distribution, and E_0 is the center of the distribution.

The Coulomb potential E_{coul} of a hole carrier as a result of interaction with dopant anions in the material is

$$E_{coul} = -\sum_i e^2/(4\pi\varepsilon_0\varepsilon a_i),$$

where a_i is the distance between hole and anion i.

The hoping rate v_{ij} (also treated as attempt frequency to hoping) from localized site i to j with energy E_i and E_j can be written by the Miller–Abraham equation

$$v_{ij} = v_0 \exp(-2\gamma R_{ij})\exp(-(E_j - E_i)/kT), \qquad E_i < E_j$$

and

$$v_{ij} = v_0 \exp(-2\gamma R_{ij}), \qquad E_i \geq E_j,$$

where v_0 is a prefactor, R_{ij} is the distance between site i and site j, r^{-1} is the decay length of the wavefunction, and k and T are the Boltzmann constant and temperature, respectively.

Here, we only consider the hoping from localized site to the spatial neighbor sites. It is noted that in pure conjugated polymer, the charge transport is quasi-two or quasi-three dimensional, while in composite it is locally one dimensional.

In pure conjugated polymer, the hoping probability from site m to neighbor site n should be proportional to the attempt frequency to hoping v_{mn}, so the hoping can be mimicked using a series of random events with corresponding probability. As the hoping distance is set to be a constant, locally μ_{mn} is determined by v_{mn}. Because the mobility of these materials is low (as compared to μ_0 which is given in Arrhenius's equation $\mu = \mu_0 \exp(-E_a/kT)$), we can carefully make one order approximation in this model, that is, the mobility μ_{mn} is also approximately proportional to v_{mn}.

Subsequently, the average mobility μ after p times hoping is

$$\frac{p}{\mu} = \sum_1^p \frac{1}{\mu_i}.$$

The carriers including both doping-induced and gate field-induced carriers should obey Fermi–Dirac distribution across DOS.

Monte Carlo simulation shows that, although the local one-dimensional feature of charge transport limits the intrinsic hoping, the doping dramatically increases the mobility as a result of the spatial occupation of insulating matrix that dilutes the dopant anions and weakens the Coulomb interaction.

This model can be used to interpret the high FET performance of SP/IP composite in very low doping level, and the enhanced conductivity of the same composite in medium and high doping state. However, in this model the dopants are assumed to be randomly distributed within insulating matrix, that is, they do not diffuse within the crystallite of conjugated polymer.

Actually, upon blending with insulating polymer, the total density of states of semiconducting polymer is largely reduced, especially for the case that the semiconducting polymer is the minority component. Consequently, the density of traps is reduced, provided the insulating matrix does not change the distribution function of DOS. This point of view is helpful to understand the FET performance of SP/IP composite.

9.6
Perspective

In fact, although the two-phase interface enhances the intraphase charge transportation within semiconducting component, in the meantime it inevitably hinders the interphase hopping/tunneling due to the intrinsic feature of insulating polymer. So the final composite performance should be determined by the balance between competing processes that lead to an improvement and degradation of the transportation properties, respectively. That is, more work may be devoted to the morphological optimization, which is both complicated and interesting. On the other hand, the SP/IP composite is composed of 3D two-phase interface rather than the typical 2D semiconductor–insulator (dielectric) interface in conventional transistor geometry. Therefore, the charge distribution, charge transport, and interpenetrating charge pathway should be different from those in a conventional transistor, and therefore need more devotion.

The composites resulting from integrating insulating polymers into semiconducting polymers were endowed with potentially reduced cost and flexibility as well as environmental stability, all of which are advantages for industrial application of flexible electronics. In addition, as conjugated polymers typically absorb light in the visible range, they are normally colorful. Unfortunately, although this feature is useful when they are used in OLED and polymer solar cells, it reduces potential applicability in devices where optical transparency is needed (e.g., head-up displays). Since insulating polymers are fully transparent, one can expect transparent flexible electronics employing insulating polymer as matrix.

Acknowledgments

This work was financially supported by Alexander von Humboldt Stiftung; we are grateful to the helpful discussion with Prof. Xiaoniu Yang, Prof. Norbert Koch, and Prof. Dieter Neher. We thank Jiayue Chen for her valuable help in editing the manuscript.

References

1 Folkes, M.J. and Hope, P.S. (1993) *Polymer Blends and Alloys*, Blackie Academic & Professional, London.
2 Goffri, S., Müller, C., Stingelin-Stutzmann, N., Breiby, D.W., Radano, C.P., Andreasen, J.W., Thompson, R., Janssen, R.A.J., Nielsen, M.M., Smith, P., and Sirringhaus, H. (2006) *Nat. Mater.*, **5**, 950.
3 Babel, A. and Jenekhe, S.A. (2004) *Macromolecules*, **37**, 9835.
4 Qiu, L., Lim, J.A., Wang, X., Lee, W.H., Hwang, M., and Cho, K. (2008) *Adv. Mater.*, **20**, 1141.
5 Qiu, L., Lee, W.H., Wang, X., Kim, J.S., Lim, J.A., Kwak, D., Lee, S., and Cho, K. (2009) *Adv. Mater.*, **21**, 1349.
6 Qiu, L., Wang, X., Lee, W.H., Lim, J.A., Kim, J.S., Kwak, D., and Cho, K. (2009) *Chem. Mater.*, **21**, 4380.
7 Chen, S.A. and Ni, J.M. (1992) *Macromolecules*, **25**, 6081.
8 Chen, T.-A., Wu, X., and Rieke, R.D. (1995) *J. Am. Chem. Soc.*, **117**, 233.
9 Ihn, K.J., Moulton, J., and Smith, P. (1993) *J. Polym. Sci. B*, **31**, 735.
10 Ong, B.S., Wu, Y., Liu, P., and Gardner, S. (2004) *J. Am. Chem. Soc.*, **126**, 3378.
11 Lu, G.H., Tang, H.W., Qu, Y.P., Li, L.G., and Yang, X.N. (2007) *Macromolecules*, **40**, 6579.
12 Lu, G.H., Tang, H.W., Huan, Y., Li, S.J., Li, L.G., Wang, Y.Z., and Yang, X.N. (2010) *Adv. Funct. Mater.*, **20**, 1714.
13 Utracki, L.A. (1989) *Polymer Alloys and Blends: Thermodynamics and Rheology*, Hanser, New York.
14 Sun, J., Jung, B.-J., Lee, T., Berger, L., Huang, J., Liu, Y., Reich, D.H., and Katz, H.E. (2009) *ACS Appl. Mater. Interfaces*, **1**, 412.
15 Veres, J., Ogier, S.D., Leeming, S.W., Cupertino, D.C., and Khaffaf, S.M. (2003) *Adv. Funct. Mater.*, **13**, 199.
16 Stassen, A.F., Boer, R.W.I.d., Iosad, N.N., and Morpurgo, A.F. (2004) *Appl. Phys. Lett.*, **85**, 3899.
17 Hulea, I.N., Fratini, S., Xie, H., Mulder, C.L., Iossad, N.N., Rastelli, G., Ciuchi, S., and Morpurgo, A.F. (2006) *Nat. Mater.*, **5**, 982.
18 Aziz, E.F., Vollmer, A., Eisebitt, S., Eberhardt, W., Pingel, P., Neher, D., and Koch, N. (2007) *Adv. Mater.*, **19**, 3257.
19 Pingel, P., Zhu, L., Park, K.S., Vogel, J.-O., Silvia Janietz, z., Kim, E.-G., Rabe, J.P., Brédas, J.-L., and Koch, N. (2010) *J. Phys. Chem. Lett.*, **1**, 2037.
20 Sauvé, G. and McCullough, R.D. (2007) *Adv. Mater.*, **19**, 1822.
21 Arias, A.C., Endicott, F., and Street, R.A. (2006) *Adv. Mater.*, **18**, 2900.
22 Sirringhaus, H. (2005) *Adv. Mater.*, **17**, 2411.
23 Silinsh, E.A. and Čápek, V. (1994) *Organic Molecular Crystals: Interaction, Localization, and Transport Phenomena*, AIP Press, New York.
24 Yang, H., Shin, T.J., Yang, L., Cho, K., Ryu, C.Y., and Bao, Z. (2005) *Adv. Funct. Mater.*, **15**, 671.
25 Hamilton, R., Smith, J., Ogier, S., Heeney, M., Anthony, J.E., McCulloch, I., Veres, J., Bradley, D.D.C., and Anthopoulos, T.D. (2009) *Adv. Mater.*, **21**, 1166.
26 Chung, Y.S., Shin, N., Kang, J., Jo, Y., Prabhu, V.M., Satija, S.K., Kline, R.J., DeLongchamp, D.M., Toney, M.F., Loth, M.A., Purushothaman, B., Anthony, J.E., and Yoon, D.Y. (2011) *J. Am. Chem. Soc.*, **133**, 412.
27 Kang, J., Shin, N., Jang, D.Y., Prabhu, V.M., and Yoon, D.Y. (2008) *J. Am. Chem. Soc.*, **130**, 12273. 28 Unpublished results.

10
Intrinsically Conducting Polymers and Their Composites for Anticorrosion and Antistatic Applications
Yingping Li and Xianhong Wang

10.1
ICPs and Their Composites for Anticorrosion Application

10.1.1
Introduction

The corrosion of metal has been described as extractive metallurgy in reverse [1], where the large free energy difference drives the metal back to its original oxidized state. Thus, the corrosion of metal is a thermodynamically favorable process. It is estimated that the annual cost of corrosion and its control for a developed country accounts for approximately 3–4% of the country's gross domestic product. As far as the highway bridges in the United States are concerned, a recent study indicates that the annual direct cost of corrosion is $5.9–9.7 billion [2]; if indirect factors are included, this cost can be as much as 10 times higher [3].

Various strategies have been developed to control the dynamics of corrosion by slowing the rate of corrosion. A common approach is applying one or more layers of coating on the metal surface, where the coating serves as a barrier to separate the metal from the environment. Active ingredients are often added to ensure the reduction of the corrosion rate once the barrier has been breached. The effective ingredients generally contain toxic heavy metal such as chromate (Cr(VI)) [4].

Intrinsically conducting polymers (ICPs) are electroactive long-range conjugated polymers. They generally possess reversible redox performance, while metal corrosion is also a redox process; therefore, it is possible that ICPs may find application for metal anticorrosion. It is true since the early report for corrosion inhibition performance of ICPs such as polyaniline (PANI) by DeBerry [5]. After more than 20 years of development, now ICPs have received much attention, since they may be a kind of alternative anticorrosion agents instead of the toxic heavy metal in anticorrosion coating, no matter they are used alone or as composite with substrate resin.

Several approaches have been used to create coatings made of ICPs, including casting soluble ICPs to form coatings, electrochemically depositing ICPs on the metal substrate, or blending with insulated polymers to form electrically conductive composite coatings [6, 7]. However, the application of ICPs is limited, not only for

Semiconducting Polymer Composites: Principles, Morphologies, Properties and Applications, First Edition.
Edited by Xiaoniu Yang.
© 2012 Wiley-VCH Verlag GmbH & Co. KGaA. Published 2012 by Wiley-VCH Verlag GmbH & Co. KGaA.

their high cost, but also for their strong intermolecular interaction leading to insolubility or difficulty to be dispersed in common solvents. Moreover, pure ICP coatings usually have poor adhesion to substrate [8], and electrochemically deposited ICP coatings are cumbersome and virtually impossible on huge structures such as ships, bridges, and pipelines [9]. In contrast with pure ICP coatings and electrochemically deposited ICP coatings, the electrically conductive composite coatings can take advantage of the good mechanical properties and processability of conventional polymers [10], which allows electrically conductive composite coatings to have more practical sense for general corrosion control. In this chapter, we will focus on the most concerned issues in ICP-based metal anticorrosion coating, especially anticorrosion mechanism and anticorrosion efficiency, trying to point out that ICP-based coating, as represented by conductive polyaniline coating, is emerging new generation of eco-friendly metal anticorrosion coating.

10.1.2
Protection Mechanism

Many protection mechanisms related to ICPs have been proposed in the literature, and in some cases opposing evidence exists. This confusion may be attributed to the wide variations in experimental procedures used (coating type, substrate preparation, corrosive environment, test method). In the following, we just list the main mechanisms that have appeared in the literature. Further details can be found in the review by Spinks *et al.* [7]. However, the anodic protection mechanism will be discussed at length because it is the most commonly accepted mechanism in previous studies.

10.1.2.1 Anodic Protection Mechanism
The anodic protection mechanism can be dated back to the work of DeBerry [5]. He found that electrochemically deposited polyaniline-coated 410 or 430 stainless steel (SS) samples remained passive from several hours to at least 1200 h after immersion in dilute sulfuric acid solutions, while bare anodically passivated 410 or 430 specimens became active within minutes. More interesting feature is that the coated PANI can even repair the breakdown passive film by aggressive chloride ion. Hermas *et al.* [11] electrochemically deposited PANI film on 304 stainless steel in sulfuric acid solution with aniline monomer. The resulting PANI-coated SS was found to maintain passive state in a deaerated 1 M H_2SO_4 at 45 °C for several weeks. It was further confirmed that the passive layers underlying PANI film had higher chromium content compared with the air-formed and anodically formed oxide films on SS by the X-ray photoelectron spectroscopy (XPS) technique. Other conductive polymers, such as polypyrrole (PPy) [12–16], poly(*o*-anisidine) [17], poly(*o*-ethoxyaniline) [18], and poly(*o*-phenylenediamine) [19], when electrochemically deposited on stainless steel [20] and iron [21, 22], were also found to maintain the potential of the metal substrate in passive region. However, it should be noted that the metal substrate has been passivated during electrodeposition, and the subsequent protection effect depends on the nature of this passive layer. The conducting

polymer may act as a stabilizer of this layer. Wessling [23] suggested that the pure metal surface without electrochemical passivation can also be passivated by PANI film, where the corrosion potential (E_{corr}) of PANI-coated steel in 3% NaCl increased by as much as +800 mV compared with that of the bare steel in the same solution, and a passive metal oxide layer could be observed by scanning electron microscopy (SEM) after the PANI coating was peeled off. The ICP-containing coatings made from blending of conductive polymers and substrate polymers directly deposited on metals can also shift the E_{corr} to positive direction [24–38]. Talo et al. [25] reported a large (+500 mV) increase in E_{corr} over bare steel, and in this case PANI was blended with epoxy. de Souza [28] deposited a blend formed by camphor sulfonate-doped polyaniline and poly(methyl methacrylate) (PMMA) on iron, copper, and silver. The stable open-circuit potential (OCP) was found to shift about +160, +130, and 70 mV compared with its naked counterparts in the same solution, respectively. The OCP shift was also observed when PANI–TiO_2 composite-containing were applied to magnesium ZM21 alloy [39]. It is worth noting that the shift of OCP is not just a sign of the oxide film formation, but the oxide film growth rate can also be obtained by the analysis of the OCP evolution. Recently, the author's group [40] has successfully measured the growth rates of the oxide films formed at the polyaniline/steel interface under different conditions (pH, NaCl concentration, and temperature) through OCP measurements. Figure 10.1 shows the dependence of oxide film growth rates on temperatures (T). An excellent linear relationship exists between the natural logarithm and $1/T$ (the inset). According to the Arrhenius equation, an active energy value of $-39.8\,kJ\,mol^{-1}$ was obtained, indicating that the oxide film growth was under diffusion control.

Wessling [41, 42] postulated that ICPs could stabilize the potential of a metal in a passive region, maintaining a protective oxide layer on the metal. In this model, oxygen reduction on the polymer film replenished the polymer charge consumed by metal dissolution, thereby stabilizing the potential of the metal in the passive

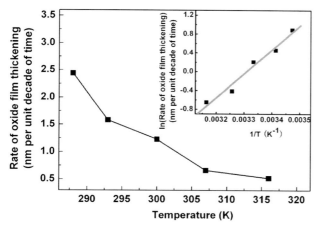

Figure 10.1 Dependence of oxide film growth rates on temperatures in 0.01 M NaCl solution with pH 6.14. The inset is the plot of the natural logarithm of the growth rate against $1/T$ [40].

region and minimizing the rate of metal dissolution. Equations (10.1) and (10.2) depict the reactions of ICP reduction/metal oxidation and ICP oxidation by atmospheric or dissolved oxygen, respectively.

$$\frac{1}{n}M + \frac{1}{m}ICPs^{m+} + \frac{y}{n}H_2O \rightarrow \frac{1}{n}M(OH)_y^{(n-y)+} + \frac{1}{m}ICPs^0 + \frac{y}{n}H^+ \quad (10.1)$$

$$\frac{m}{4}O_2 + \frac{m}{2}H_2O + ICPs^0 \rightarrow ICPs^{m+} + mOH^- \quad (10.2)$$

The reaction state of ICPs has been further studied by many other researchers [43, 44]. We [45] disclosed a "quasi-reversible" change of oxidation degree of polyaniline base (EB) via UV–Vis spectra during 2000 h of observation, where the reduced iron powder was stirred with EB (mole ratio of EB/Fe = 1/10) in 3.5 wt% NaCl (Figure 10.2). Gustavsson et al. [46] reported that both dedoping and reduction of the polyaniline salt (ES) occurred under ambient conditions and the reduced PANI could be reoxidized once the oxygen was introduced.

However, there are observations failing to support the anodic passivation protection mechanism. Cook et al. [47, 48] believed that the passivation of metal by ICPs was "unusual," since carbon steel did not usually passivate in acid or near-neutral chloride environments. Rammelt et al. [49] reported that polymethylthiophene (PMT) cannot passivate mild steel in 0.1 M LiClO$_4$ (pH 5.3) by galvanic coupling the separated PMT/Pt electrode and mild steel electrode. Tallman et al. [50] also reported that PPy cannot passivate aluminum alloy AA 2024-T3 in 0.35 wt% (NH$_4$)$_2$SO$_4$ + 0.05 wt% NaCl by galvanic coupling techniques. Similar phenomena exist in PANI galvanic coupling with aluminum alloy AA 2024-T3 in 3.5% NaCl [51]. Zhu et al. [52] have investigated the galvanic interaction between the PANI film electrode and ferrous metals in 0.5 M Na$_2$SO$_4$ (pH 1). PANI film slightly accelerated the corrosion of 20 A carbon steel when the area ratio of PANI to steel

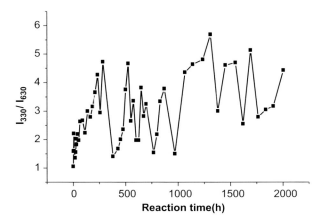

Figure 10.2 Time dependence of absorption intensity ratio of 330 nm to 630 nm (I_{330}/I_{630}) in UV–Vis spectra of EB/NMP solution (mole ratio of EB/Fe = 1/10, measured in 3.5 wt% NaCl solution at 25 °C) [45].

Table 10.1 Passivation region and passivation state current for iron and steel in 1 M H_2SO_4 [54].

Type of iron or steel	Passivation region (V, versus SCE)	Passivation state current (mA cm^{-2})
Pure iron wire (99.99%)	+0.55 to +1.15	0.5
Pure iron strip (99.87%)	+0.50 to +0.95	2.03
Mild steel (1018)	+0.50 to +0.95	30
Stainless steel (304)	−1.0 to +0.85	1.40×10^{-3}
Stainless steel (410)	−0.1 to +1.00	4.70×10^{-3}
Stainless steel (430)	−0.2 to +1.00	9.5×10^{-2}

was less than 25 : 1; however, an equal-area PANI film could maintain 2Cr13 stainless steel in a passivated state.

It is well known that two requirements must be satisfied to achieve the anodic passivation for metal in corrosive environment: first, the metal should have an active–passive property in the studied environment, and, second, the free corrosion potential E_{corr} or OCP should exceed the passivation potential (E_p). Table 10.1 shows the passivation regions of different steels in 1 M H_2SO_4, while Table 10.2 gives the reduction potentials of some ICPs at pH 7. Though the given potentials in these two tables are measured under different conditions, we can qualitatively judge that the stainless steel can be passivated by ICPs in common environments. If the E_p of stainless steel is low, the ICPs can provide enough anodic potential to passivate stainless steel. In this case, the protection of stainless steel by ICPs is same as anodic protection provided by the electrochemical technique. It was confirmed by Gasparac and Martin [53] that a chemically deposited EB film could passivate stainless steel in 1 M H_2SO_4. However, for the metals with high passivation

Table 10.2 Reduction potentials for various redox couples [6].

Redox couple[a]	Potential (V, versus SCE at pH 7)
Mg^{2+}/Mg	−2.60
Al_2O_3/Al	−2.20
Zn^{2+}/Zn	−1.00
Fe^{2+}/Fe	−0.86
H_2/H_2O	−0.65
O_2/H_2O	0.58
$Cr_2O_7^{2-}/Cr_2O_3$	0.18
Polypyrrole[b]	−0.34 to 0.06
Polythiophene[b]	0.56 – 0.96
Polyaniline[b]	0.04–0.76

a) The values were recalculated in that the V versus SHE was replaced by V versus SCE.
b) For ICPs, an approximate range of potential was provided in comprehensive consideration of the data from the publications, since the actual potentials depend on the dopant (counterion) and doping level, the electrolyte, and other experimental variables.

potentials (such as mild steel or pure iron) or the non-passive metals (such as aluminum and magnesium alloy) in acid or near-neutral chloride corrosion environment, the ICPs may fail to passivate them like in the case of stainless steel. The protection of these metals will be discussed later.

10.1.2.2 Inhibitory Protection Mechanism

Dopant ions are the anions incorporated into ICPs during the oxidation process, which can be released upon reduction of ICPs [55]. Many studies showed the influence of the dopant ions on corrosion [56–62]. The anticorrosion performance of two formulations fabricated by sulfonic and phosphonic acid-doped PANI was studied by Kinlen et al. [57]. Salt fog exposure tests indicated that the phosphonic acid-doped PANI was more effective for corrosion protection than sulfonic acid-doped analogue, and the scanning reference electrode technology (SRET) data supported the salt fog results in that the sulfonic acid dopants exhibited an increasing galvanic activity with time, while the phosphonic acid dopants showed a decrease in activity with time, indicating that the passivation arised from the anodization of the metal by PANI and an insoluble iron–dopant salt formed at the metal surface. Souza et al. [30] presented the electrochemical behavior of an acrylic resin blend formed by camphorsulfonic acid (CSA)-doped PANI and poly(methyl methacrylate). When the blend was used for corrosion protection of iron in chloride solutions, a passivating complex of iron–dopant was formed as further confirmed by Raman and FTIR spectroscopies.

The essence of inhibitory protection mechanism is that the dopant ions can be released through reduction of ICPs [55] or ion-exchange process [63]. The released dopant ions can then form a second physical barrier to prevent the penetration of aggressive ions [30], or as an oxygen reduction inhibitor [63]. Once the dopant ion was selected as a known corrosion inhibitor, the ICP-containing coating became an inhibitor releasing coating. Such coatings are sometimes referred to as "smart" coatings [59, 60, 62, 63] in that the anion is released on demand, that is, when damage on the coating results in metal oxidation and polymer reduction. Based on the inhibitory conception, ICPs doped with inhibitory anion may be used as pigments in coatings, as summarized in Table 10.3.

However, reports exist that the conducting composite coatings containing both undoped ICPs and doped ICPs bearing aggressive dopant ion [37] could provide good corrosion protection to metal substrate. We [70] observed that EB/epoxy resin (ER) composite coatings (EB/ER) provided efficient corrosion protection to mild steel no matter in neutral, basic, or acidic 3.5% NaCl media, and these coatings could also protect mild steel coupled with copper in 3.5% NaCl solution [71]; the structure of the mild steel/copper couple is shown in Figure 10.3, and the results of visual observation are shown in Figure 10.4.

10.1.2.3 Cathodic Protection Mechanism

ICPs and their derivatives can exist in several redox states: oxidized or p-doped state, where electrons are removed from the polymer backbone; neutral or undoped states are typically insulating or semiconducting; and reduced or n-doped state

Table 10.3 Summary of ICPs doped with different dopant ions.

Conducting composite coatings	Observations	Reference
Poly(methyl methacrylate)/PANI salts (dopant: camphorsulfonic acid or phenylphosphonic acid)	Raman microscopy technique showed the presence of a second passive layer formed by dopant ions released from PANI salts and different metal cations (Fe, Zn, Cu, Ni, or Pt)	[64]
Insulating matrix polymer/PPy salts (dopant anions: $S_2O_8^{2-}$, MoO_4^{2-}, $[PMo_{12}O_{40}]^{3-}$)	A potential-driven anion release from the PPy coating that results in an inhibition of the corrosion process taking place in the defect	[65]
Acrylic resin/PANI salts (dopant: $C_6H_5PH(O)OH$, $C_6H_5CH=CHPH(O)OH$, 2-Cl-$C_2H_5PH(O)OH$ and H_2SO_4)	PANI doped with different anions containing phosphorus presents a better protection compared with PANI doped with H_2SO_4	[24]
Polyurethane resin/PANI salt (dopant: methyl orange and camphorsulfonic acid)	Camphorsulfonic acid was found to be a better dopant than methyl orange	[66]
Various resins/PANI–dioctyl phosphonate	5 phr PANI salt-containing epoxy coating offers best protective performance	[67]
Epoxy resin/PANI–amino trimethylene phosphonic acid	Charge transfer resistance values increased with time	[68]
Polyvinyl butyral-*co*-vinyl alcohol-*co*-vinyl acetate/PANI salts (dopant: H_3PO_4, camphorsulfonic acid, *para*-toluenesulfonic acid, and phenylphosphonic acid)	Inhibit delamination with efficiency order: *para*-toluenesulfonic acid < camphorsulfonic acid < H_3PO_4 ≪ phenylphosphonic acid	[69]

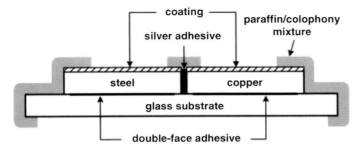

Figure 10.3 Scheme of an EB/ER blend-coated steel–copper couple [71].

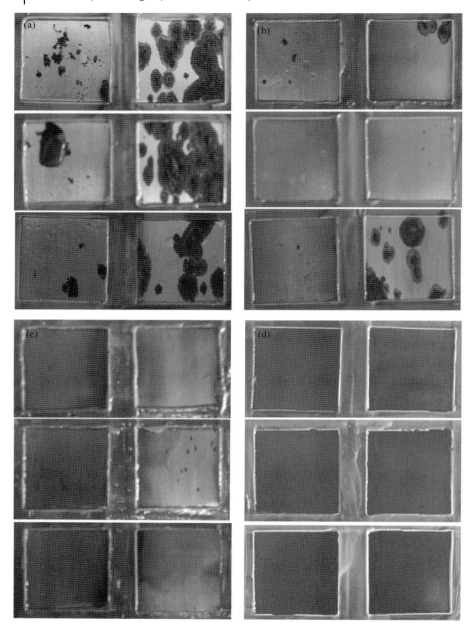

Figure 10.4 Visual observations of coated steel–copper couples after 30 days of immersion in 3.5% NaCl solution at 40°C: ER coating (a), EB/ER coatings with EB contents of 1 wt% (b), 5 wt% (c), and 10 wt% (d). Thicknesses of the coatings were 20 μm. Three replicates were shown for each coating. The left part of each sample is coated steel, and the right part is coated copper [71].

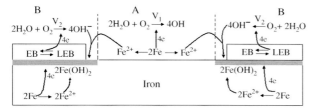

Figure 10.5 Oxygen scavenging protection model of EB on the exposed steel at coating defect [73].

where electrons are added to the polymer backbone. Yan *et al.* [72] reported that the n-doped conducting polymer coating poly(2,3-dihexylthieno[3,4-*b*]pyrazine (PC$_6$TP) could cathodically protect the exposed alloy (AA 2024-T3) by electrochemical way at the coating defect. By means of scanning vibrating electrode technique (SVET), the current density map for AA 2024-T3 coated with n-doped PC$_6$TP showed a rather intense reduction current in the defect after 5 min of immersion in dilute Harrison's solution (DHS), while the oxidation current was distributed more or less evenly over the polymer surface. The neutral PC$_6$TP also cathodically protected the exposed alloy in the defect at the expense of alloy dissolution at the reactive S-phase intermetallics.

A recent work [73] by our group shows that neutral polyaniline (EB) can also protect the exposed mild steel at coating defect in 3.5 wt% NaCl solution, but this protection is achieved by a chemical way, and not by an electrochemical way, since EB is an insulator, where EB acts as an oxygen scavenging agent to protect exposed steel at coating defect through catalytic reduction of oxygen on EB film. A scheme of this protection mechanism is shown in Figure 10.5. Based on the cathodic protection mechanism, a new strategy was developed [74], where polypyrrole aluminum flake composite (PAFC) was synthesized by chemical oxidation of pyrrole in the presence of aluminum flake, which was then used to formulate a corrosion inhibiting primer. An enhanced anticorrosion performance was observed for AA 2024-T3 compared to the Al flake primer. The cathodic protection for AA 2024-T3 was confirmed by galvanic coupling measurement.

10.1.2.4 Comprehensive Understanding on Protection Mechanism of ICPs

There are wide variations in experimental procedures in the literature, such as coating type, substrate, corrosion environment, and test methods. Therefore, various corrosion protection mechanisms for ICPs have been proposed in the past 30 years, and sometimes contradictory viewpoints exist. Till now, thanks to the extensive research on protection mechanisms, we may have a chance to gain a comprehensive understanding of protection mechanism for ICPs or ICP-containing coatings.

The core issue in corrosion protection mechanism lies in the unique redox property of ICPs as shown in Eqs. (10.1) and (10.2), and the following three important points could be summarized:

1) **Reduction potential of ICPs:** the ICPs can be reduced, and a reduction potential (E_{ICP}) exists. The value of this reduction potential varies depending on the

dopant and doping level, the electrolyte, and other experimental variables. A passive potential (E_p) exists for a metal (especially for ferrous metal) that has active-/passive property, and the actual E_p depends on the nature of the corrosive environment, such as pH, oxygen concentration, aggressive ions, and so on. The electrochemical anodic passivation is achieved once E_{ICP} exceeds E_p in certain corrosive environment, which occurs frequently for stainless steel, since stainless steel usually possesses a low E_p in common corrosive environment (Table 10.1), but only a few cases for iron [75] or carbon steel [52], since they possess high E_p or usually cannot be passivated in neutral or acid solution containing aggressive ion.

2) **Oxidation of metals (Eq. (10.1))**: for ferrous metals that cannot be anodically passivated or for more active metals such as aluminum alloy or magnesium alloy, a chemical interaction occurs between ICPs and metal once the corrosive medium penetrates coating, that is, the ICPs can oxide metal. It is meaningful, for the metal cations can be formed locally, and then react with water on the interface without transportation, which is beneficial for the formation of a dense oxide layer. Sometimes this protective layer is formed by the help of released dopant ion (depositing with the metal cations to form dopant–metal salt layer or adsorbing on the surface to inhibit corrosion, since the dopant ion can be released upon reduction of ICPs). The shift of corrosion potential (E_{corr}) of ICP-coated metal is considered as the evidence of the existence of this protective oxide layer. It should be noted that the formation of protective oxide layer is different from the anodic protection in the case of stainless steel, but these two cases are often mixed up in the literature. A good adhesion of ICP coating is important for oxide film formation, since the coating plays a role of template during the oxide film formation process.

3) **Reduction of oxygen (Eq. (10.2))**: the role of accelerating reduction of oxygen by ICPs in corrosion protection has been disclosed recently by our group. ICPs can act as an oxygen scavenging agent, which is similar to zinc-rich primer.

In summary, the different protection mechanisms, such as anodic protection, inhibitory protection, or cathodic protection, are just the different expressions of the unique redox property of ICPs under different conditions, especially for different corrosion environments and metal kinds.

10.1.3
Matrix Resin of Conducting Composite Coating

Common matrix resins for blending with conducting polymers are widely used in traditional anticorrosion coatings, such as epoxy resin [25, 27, 32, 34, 68, 70, 71, 76], polyacrylic-based resin [24, 39, 48, 77, 78], and poly(methyl methacrylate) [30, 60, 62, 64]. The feature of matrix resin as well as the amount of ICPs is important to the anticorrosion performance of conductive composite coating. Samui and Phadnis [67] blended various amounts of dioctyl phosphate (DOPH)-doped PANI with different polymeric matrices (epoxy resin, polyurethane resin, styrene–butyl acrylate copolymer, soy alkyd resin, and silicone alkyd). The impedance measurement in combination with humidity cabinet exposure study and field exposure study in surf zone indicated that 5 phr PANI–DOPH-containing epoxy resin coating offered protective performance for longer period compared with other resins and other loadings.

Due to increasing concerns on environment, the eco-friendly coatings, such as coatings from renewable resources and waterborne coatings, have been developed to replace petroleum-based or solvent-based coatings. A new coating strategy combining the eco-friendly coatings and ICPs has been proposed, and the anticorrosion performance of the resulting coatings has been evaluated by several groups. Ahmad and coworkers [66, 79, 80] studied the corrosion protection performance of a series of vegetable oil-based alkyd resin/PANI composite coatings on mild steel. The results were positive and promising. Bagherzadeh *et al*. [81] found that the anticorrosion performance of a two-component water-based epoxy resin coating system was improved by adding only 0.02 wt% nano-PANI. Ahmad *et al*. [82] investigated the corrosion protection behavior of waterborne resorcinol formaldehyde (RF)-cured composite coatings of poly(1-naphthylamine) (PNA)/poly(vinyl alcohol) (PVA) on mild steel, and even superior corrosion protective performance was observed in different corrosive media when compared with the reported solvent-based conductive polymer coatings.

10.1.4 Processing Methods

Mechanical blending method employing ICP powder as pigment additive to formulate conductive composite coatings has been widely adopted [83]. However, the dispersion of ICPs in conductive composite coatings is generally poor simply by mechanical blending, since ICPs are insoluble in common solvents widely used for anticorrosion coating. Improving dispersion of ICPs in polymer matrix is meaningful at several aspects: improving anticorrosion performance of conductive composite coatings and reducing the negative effect of ICPs on polymer matrix (such as brittleness, adhesion to the substrate, and penetration of corrosive medium), as well as the cost reduction of the expensive ICPs. Methods to improve dispersion of ICPs in polymer matrix can be summarized as follows:

1) **Introduction of side groups:** the side groups can improve solubility of the ICPs and the miscibility with substrate resins. Iribarren *et al*. [84] added soluble poly (3-alkylthiophene) into alkyd, epoxy, and polyurethane paints, and the conductive composite coatings gave improved corrosion resistance under ICP loading as low as 0.3% (w/w). The Young's modulus and the elongation at break were significantly better than those films generated with the original paints. In addition, PANI derivatives were also studied and showed similar results [85].
2) **Doping with functional dopants:** the properties of the ICPs, such as solubility in organic solvent [86] or water [87–91], can be modified by doping with appropriate dopant. Camphorsulfonic acid is one of such dopants. CSA-doped PANI can be dissolved in *m*-cresol. A number of researchers [30, 60, 62, 64] blended

PANI–CSA/m-cresol solution with PMMA/m-cresol solution to prepare PANI–CSA/PMMA coatings to investigate the corrosion protection property. In some cases, the dopant can also be polymer matrix. Oliveira et al. [92] prepared poly(methyl methacrylate-co-acrylic acid) (PMMA-co-AA)-doped PANI by using the template-guided method. The doped PANI (PANI–PMMA-co-AA) was soluble in ethyl acetate, and the solution was used for dip coating plates of aluminum alloy.

3) **In situ polymerization:** *in situ* chemical polymerization method is commonly used to fabricate conductive composite coatings. The typical procedure can be described as follows: monomer for conducting polymer is added into host polymer (acting as polymer matrix in conductive composite coatings) dispersion [93] or latex [94]. Then initiator solution is added and the monomer polymerizes. ICP/polymer composites can be obtained after purification treatment. Electrochemical polymerization method has also been adopted to formulate conductive composite coatings. Ding et al. [95, 96] electrodeposited PANI–thiokol rubber (TR) composite coating on mild steel electrode from a solution containing aniline, acetonitrile, trifluoroacetic acid, trichloroacetic acid, and thiokol rubber. The adhesion to the mild steel, as well as the corrosion protection of the PANI/TR coating, was more satisfactory than that of PANI coating.

10.1.5
Conclusions and Perspectives

The study of corrosion protection of metals by ICPs has been continuing for nearly 30 years. Though a consensus about the protection mechanism has not been reached until now, the products of ICP paints have been developing. For example, many anticorrosion applications have been realized by Dr. Wessling in Ormecon Chemie GmbH & Co at Ammersbek, Germany, who is the pioneer in this area, in the name of ORMECON®, a trademark now owned by Enthone (see Cookson Electronics web site http://www.enthone.com). Also a series of products have been developed and applied in practical projects in our lab by over 15 years of continuing fundamental study, such as container ship (Figure 10.6), guardrail of expressway bridges (Figure 10.7), and pipelines in coal-powered plants (Figure 10.8). Moreover, it is widely accepted, since the pioneering work of Dr. Wessling who claimed organic nanometal anticorrosion coatings based on polyaniline, that better dispersion of ICPs in the matrix resin is most important to further improve the anticorrosion performance. A recent breakthrough from Ancatt Inc. in the United States indicated that nanopolyaniline-based coating gave nearly 13,000 h of salt fog (ASTM B 117) tests for iron/steel, and over 10,000 h of salt fog tests for aluminum (see Ancatt Inc. web site http://www.ancatt.com). Many methods have been developed to improve the dispersion, but most of them are usually cumbersome. A more convenient method should be developed to meet the industrial need. In another consideration, introduction of inorganic pigment to the conductive composite coating is a promising way to greatly improve the anticorrosion ability through synergic effect between ICPs and inorganic ingredients [97–101].

Figure 10.6 Test coating of anticorrosion polyaniline coating on a 2000 ton container ship. The coating is still effective till now since December 2005.

Figure 10.7 Test coating of anticorrosion polyaniline coating in guardrail of expressway bridges in Hunan Province, China. The coating is still effective since November 2006.

Figure 10.8 Test coating of anticorrosion polyaniline coating in pipelines of a coal-powered plant in Hu Nan Hua Run Li Yu Jiang Power Ltd Company. The coating is still effective since December 2005.

10.2
Antistatic Coating

10.2.1
Introduction

Polymers such as polyesters and polyolefins are widely used in various fields such as packaging materials, textiles, electrical/electronic parts, and automotive parts. However, they are insulating materials with surface resistivity over $10^{15}\,\Omega/\square$. Static charge that easily builds up on such parts by contact and/or rubbing may cause sparking, leading to electrostatic damage in sensitive semiconductor devices and interference on circuit operation [102]. To avoid static charge build-up, a common approach in electronic industry is that the surface resistivity should be controlled below $10^{10}\,\Omega/\square$. It is even below $10^{4}\,\Omega/\square$ for electrostatic dissipation application. Antistatic materials are added during the molding process or simply applied to the surface of the products in a finishing process to control the surface resistivity [103]. Generally, antistatic materials can be divided into nonconductive and conductive additives; the former mainly include surfactants and other hydrophilic substances such as anionic, cationic, amphoteric (amine oxides and betaine-type antistatics), nonionic [104], and so on, while the latter include carbon [105–107], metal powder [108, 109], conductive fibers [110–112], conductive mica [113], metallic oxide [114, 115], ICPs [116–120], and so on. Though surfactants or hydrophilic substances are cheap and convenient to use, their antistatic effect is only realized by the equilibrium moisture adsorbed. Thus, satisfactory antistatic effect is difficult to achieve under low humidity. Moreover, the surfactant can be removed by rubbing or washing back and forth, leading to worsening of the antistatic effect easily [103]. In contrast, the intrinsically conductive additives can grant the matrix long conducting property, and the conductivity can be controlled by changing the concentration of the fillers. Among conductive additives, high concentration of metal powder is needed to achieve good electrical conductivity, which makes the system heavy and inflexible. Carbon black can reinforce the polymer especially elastomer matrix preserving its flexibility and light weight, without adversely affecting the environmental and thermal stability of the polymer matrix [118], but it will darken the surface of polymer products, which makes coloring of polymers difficult [103]. It can also be removed easily by rubbing.

The ICPs can change the electrical conductivity of polymer matrix from semiconductor to conductor by controlling the loading of ICPs, accompanied by light color change of the matrix, or even no color change. Although ICPs are difficult to process due to its insolubility and infusibility, the antistatic investigation of ICPs has been extensively conducted for their easy synthesis, low cost, high conductivity, and environmental stability. Currently, the ICPs have evolved as one of the main long-life antistatic materials. This chapter will focus on the antistatic application of ICPs, based on the description of synthesis and processing of various ICPs.

10.2.2
Synthesis of Processable ICPs

Most ICPs are neither infusible nor insoluble in common organic solvents, which makes processing of ICPs into electrically conductive composite on an industrial scale a tough task [121]. ICPs in their neutral form, such as polyaniline emeraldine base, are soluble in polar solvents such as N-methylpyrrolidone, N,N-dimethylpropylene urea, and dimethyl sulfoxide [122], and the solution can then be doped by special dopants [123]. The freestanding film of neutral or doped polymer can be casted from these solutions. However, currently only a small number of neutral or doped polymers can be dissolved in polar solvents, even though the polar solvents are difficult to evaporate. In order to process in common organic solvents or water, substituted conducting polymer was synthesized. Alkyl [124–127] or alkoxy [128, 129] ring-substituted aniline and thiophene [130, 131] are usually used to synthesize corresponding polymers, but most of the polymers, especially for polyaniline, possess lower conductivity when compared with their unsubstituted counterparts. Cao et al, [86] disclosed counterion-induced processability of doped polyaniline using a functionalized protonic acid to protonate polyaniline. The resulting conducting polymer materials can be blended with conventional polymer by melt processing or solution processing, and transparent films, sheets, fibers, bulk parts, and so on can be obtained, exhibiting relatively high level of electrical conductivity at low volume fraction of the polyaniline loading. In the meantime, the good mechanical properties are maintained, nearly equivalent to those of the host bulk polymer.

Although the counterion-induced processability has solved the solubility problem of ICPs, especially for polyaniline, the maximal solubility is low; for example, the maximum solubility of the polyaniline doped by dodecylbenzenesulfonic acid in xylene was found to be less than 0.5% (w/w) [132]. According to the threshold theory, the critical concentration at which the conductivity of composite experiences breakthrough increase decreases with the reduction of conductive filler size. Thus, reducing particle size of synthesized conducting polymers is another strategy in antistatic study. Emulsion polymerization [133–136] is found to be a very effective method to reduce conducting polymer particle size. In a typical emulsion polymerization of aniline, aniline with protonic acid and oxidant is dissolved in a mixture of water and a nonpolar or weakly polar solvent, for example, xylene, chloroform, or toluene, to form an emulsion, and protonic acids having substantial emulsifying properties in weakly polar liquid are employed, for example, dodecylbenzenesulfonic acid. Another advantage of emulsion polymerization is that polymer matrix can be dissolved in water or nonpolar solvent during polymerization, and electrically conductive composite can be directly prepared therefrom. Ruckenstein and Yang [137] described a method for the preparation of electrically conductive polyaniline–polystyrene composite from an emulsion, where a solution of sodium dodecylsulfate (SDS) in water constituted the continuous phase and a solution of aniline and polystyrene in benzene acted as the dispersed phase. Polymerization of aniline took place by introducing an oxidant dissolved in an aqueous solution of HCl under vigorous stirring in the emulsion. The conductivity of the obtained composite can reach as high as $3–5\, \text{S}\, \text{cm}^{-1}$. Yang and Ruckenstein [138] also prepared polyaniline/poly(alkyl methacrylate) composites using an emulsion pathway including two steps. First, an emulsion with the appearance of a gel was prepared by dispersing a chloroform solution of poly(alkyl methacrylate) and aniline in an aqueous solution of SDS. Second, an oxidant (sodium persulfate) dissolved in an aqueous solution of HCl was added dropwise into the emulsion with vigorous stirring in order to polymerize the aniline as well as to dope the polyaniline formed. The composite can be processed by either cold or hot pressing. Hot pressing composite can have electrical conductivity as high as $2\, \text{S}\, \text{cm}^{-1}$ with good mechanical properties, while the cold pressing composite possessed somewhat higher conductivity.

10.2.3
Processing of ICPs for Antistatic Application

As far as processing of ICPs is concerned, various methods have been developed to prepare electrically conductive composites [139], among which melt processing is quite popular for thermoplastics [140, 141]. Martins and De Paoli [142] prepared polystyrene and polyaniline blends in a double-screw extruder, giving a conductive thermoplastic with electrical conductivity of 10^{-6} to $10^{-2}\, \text{S}\, \text{cm}^{-1}$. In most cases, conducting polymers are incompatible with thermoplastic, so a functional dopant is needed to improve compatibility. For example, dodecylbenzenosulfonic acid

(DBSA)-doped polyaniline is usually used to blend with thermoplastic to prepare conductive composite [143, 144], where DBSA plays a role of dopant as well as compatibilizer.

The disadvantage of melt mixing is the low mechanical strength and high viscosity of the prepared compounds that contain a high level of additives. Moreover, antistatic application only concerns surface resistivity, bulky conductivity is not necessary, which may lead to an increase in cost. Therefore, although melt processing is widely used in thermoplastics processing, it is seldom used for antistatic purpose. One successful example for the melt processing product is Panipol® of Panipol Oy Ltd.

Solution/dispersion method is widely accepted for conductive coating, where conductive ICPs are dissolved or dispersed with polymer matrix in organic solvents or aqueous media. The conductive coating can be obtained by casting or printing the solution or dispersion in the substrate and removing the solvent by drying [145–147]. Most ICPs in antistatic application have been realized by the solution/dispersion method. A successful example is poly(3,4-ethylenedioxythiophene) (PEDOT) in aqueous solution, where the bicyclic ethylenedioxythiophene monomer was invented by the Bayer Corporate Research laboratories, and the PEDOT in polystyrene sulfonic acid aqueous solution (PEDOT–PSS) was sold under the trade name Baytron®, now playing an important role in antistatic, electric, and electronic applications due to its colorless, high electrical conductivity (10^3–10^6 $\Omega/\square$) and long-term stability under application conditions.

Another successful example is a commercialized product of dodecylbenzene sulfonic acid-doped polyaniline (PANI–DBSA) in xylene solution/dispersion in combination with matrix such as polyurethane, polyester, and so on. Thanks to the counterion-induced processability of polyaniline, pale green color antistatic coating with surface resistivity of 10^3–10^6 $\Omega/\square$ can be produced (Figure 10.9), which has shown great success in antistatic application in electronic industry, mainly due to its good transparency, low cost, and long-term stability.

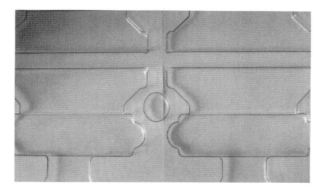

Figure 10.9 Pale green PANI–DBSA-based antistatic coating for electronic industry developed in the author's lab.

In situ polymerization to prepare ICP-based conductive composite is also used, where polymer matrix is first dissolved or dispersed in a suitable solvent, and then the monomers of conducting polymers (such as aniline, thiophene, and pyrrole) are polymerized in the same medium [148–150]. A different procedure is usually adopted to obtain antistatic fabrics: conducting polymers are *in situ* deposited on the surface of the fibers, rather than blending with it. The direct deposition of conducting polymers on fabrics can be roughly classified into two types: the fabrics are immersed in a solution containing an oxidizing agent, a dopant, and monomer [151–154], or the fabrics are immersed in a solution containing an oxidizing agent and a dopant, followed by drying and exposing the fabrics to monomer vapor [155–157]. The resultant conductive fabrics have a high grade of fastness to rubbing under dry conditions, though both types of fabrics have a low grade of fastness to rubbing under wet conditions [117]. Egami *et al.* [117] point out that this may be caused by the uneven distribution of conducting polymer on the fabrics, because the excess solution on fabrics leads to uneven adsorption of the oxidizing agent and dopant on the fabrics during drying, which in turn leads to uneven deposition of the ICPs. Therefore, the excess solution was pressed from the fabrics before drying and then after exposing the fabrics to monomer vapor, a conductive fabric with a high grade of fastness to rubbing as well as low resistivity was obtained.

10.2.4
Water-Based Polyaniline and Its Complex

With increasing environmental concerns, the development of water-based ICPs and their complex becomes a necessity. Extensive research has been conducted in this field, especially for polyaniline. Generally, water-based polyaniline can be obtained by three methods: introduction of water-soluble substituent to backbone to gain water-soluble polyaniline [158, 159], emulsion polymerization to obtain water-dispersed polyaniline [160–163], and the counterion-induced processability of doped polyaniline using a hydrophilic protonic acid [164, 165]. The first two methods have a common problem of purification and efficiency. In contrast, the pure polyaniline is used in counterion-induced processability of doped polyaniline. The purification and efficiency problems do not exist with this method, so it is the most promising way to obtain water-based polyaniline.

Wallace and coworkers [166] successfully prepared water-soluble polyaniline by doping polyaniline base (EB) with calix-[4]-*p*-tetrasulfonic acid hexahydrate and calix-[6]-*p*-hexasulfonic acid. This solution shows certain stability even in alkaline medium (pH 12). In the author's laboratory, sulfate [164] and phosphate [165] containing hydrophilic ethyleneoxide oligomer tail ($-(CH_2CH_2O)_m-$) are used to dope EB to obtain water-soluble conductive polyaniline. The solution-cast films of doped polyaniline exhibit good conducting and electrochemical properties. A series of polyanilines doped by phosphates with different side chain lengths ($m = 16, 12, 7, 4, 2,$ and 1) [167] were studied. It was found that the electrical conductivity increased from 5.9×10^{-6} to $27\,\mathrm{S\,cm^{-1}}$, while the mean particle size of doped polyaniline decreased from 950 to 42 nm with decrease of *m* value (see Figure 10.10). A film

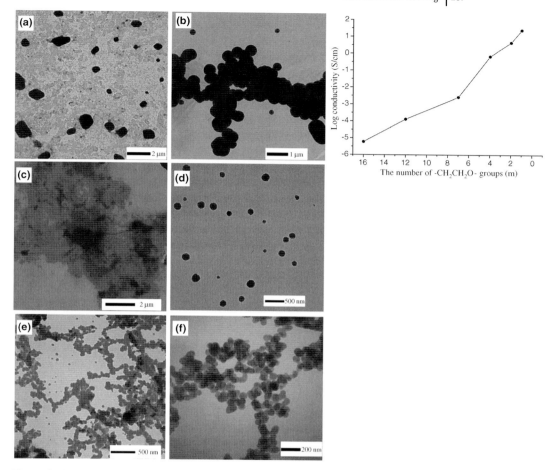

Figure 10.10 TEM image and conductivity of polyaniline aqueous solution/dispersion doped by mixture of monoester $[CH_3(OCH_2CH_2)_mOH]P(O)(OH)_2$ and diester $[CH_3(OCH_2CH_2)_mOH]_2P(O)(OH)$ with $m = 16$ (a), 12 (b), 7 (c), 4 (d), 2 (e), and 1 (f) [167].

with high electrical conductivity can be easily obtained by casting from the water solution/dispersion, and by means of the plasticization effect due to phosphate ester dopant, the obtained film was flexible. However, the water resistance of the film was not satisfactory. When the film was soaked in water, it rapidly broke into pieces, accompanied by sudden decline in the electrical conductivity. In order to solve this problem, a strategy of confining water-dispersed polyaniline in network was proposed in our laboratory. An early work was to prepare semi-interpenetrating networks of polyaniline and melamine–urea resin [168] and polyaniline/silica hybrids [169, 170]. A sol–gel process to prepare polyaniline/silica hybrids is shown in Figure 10.11.

The polyaniline/silica hybrids show distinctly improvable water resistance, but high-quality polyaniline hybrid films can hardly be made because of the inherent

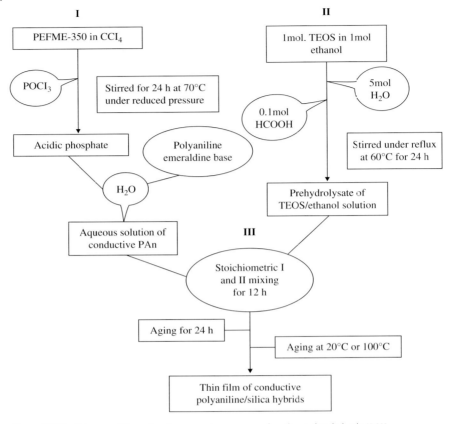

Figure 10.11 Scheme of the sol–gel process to prepare polyaniline/silica hybrids [169].

brittleness and severe shrinkage of the silica network due to the poor compatibility between polyaniline and inorganic matrix. In order to improve multiphase compatibility, interactions such as hydrogen bonding [91, 171], covalent bonding [88], and electrostatic interactions [172, 173] between polyaniline and matrix were introduced to the blend. The introduction of different interactions not only improved the water resistance, but also improved the stability of polyaniline in alkaline environment. For an electrostatic interaction conductive hybrid, the electroactivity of polyaniline was extended even to pH 14 alkaline medium [174] (Figure 10.12).

10.3
Summary

Conducting polymers are usually infusible and insoluble, which results in difficulty in processing of these conductive materials even when blended with other polymer

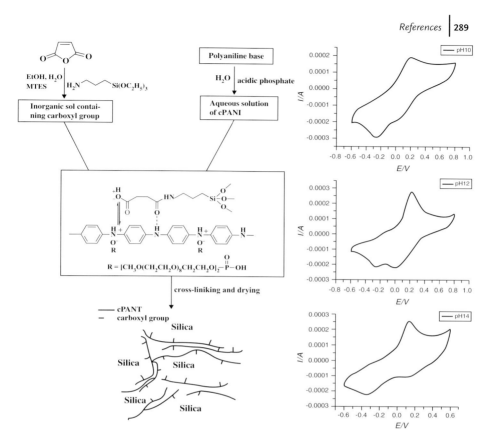

Figure 10.12 Scheme of the electrostatic interaction polyaniline/silica hybrid and the corresponding cyclic voltammograms in alkaline medium [174].

matrices for antistatic application. However, with the development of processing techniques, the above problem is being conquered and the conducting polymers are promising conductive materials in antistatic application due to their easy synthesis, low density, light color, and controllable conductivity, among which the water-based conductive coatings are extremely attractive, deserving further effort for practical antistatic application.

References

1 Jones, D.A. (1996) *Principles and Prevention of Corrosion*, Prentice Hall, New York.
2 Koch, G.H., Brongers, P.H., Thompson, N.G., Virmani, Y.P., and Payer, J.H. (2002) *Corrosion Costs and Prevention Strategies in the United States*. Publication No. FHWA-RD-01-156, U.S. Department of Transportation Federal Highway Administration, Washington, DC.
3 Yunovich, M., Thompson, N.G., and Virmani, Y.P. (2003) *Life cycle cost analysis for reinforced concrete bridge decks*. Paper No. 03309. Presented at

CORROSION/03, March 10–14, San Diego, CA.

4 Cohen, S.M. (1995) Review – replacements for chromium pretreatments on aluminum. *Corrosion*, **51**, 71–78.

5 Deberry, D.W. (1985) Modification of the electrochemical and corrosion behavior of stainless steels with an electroactive coating. *J. Electrochem. Soc.*, **132**, 1022–1026.

6 Tallman, D.E., Spinks, G., Dominis, A., and Wallace, G.G. (2002) Electroactive conducting polymers for corrosion control. Part 1. General introduction and a review of non-ferrous metals. *J. Solid State Electrochem.*, **6**, 73–84.

7 Spinks, G.M., Dominis, A.J., Wallace, G.G., and Tallman, D.E. (2002) Electroactive conducting polymers for corrosion control. Part 2. Ferrous metals. *J. Solid State Electrochem.*, **6**, 85–100.

8 Araujo, W.S., Margarit, I.C.P., Ferreira, M., Mattos, O.R., and Lima Neto, P. (2001) Undoped polyaniline anticorrosive properties. *Electrochim. Acta*, **46**, 1307–1312.

9 Li, P., Tan, T.C., and Lee, J.Y. (1997) Corrosion protection of mild steel by electroactive polyaniline coatings. *Synth. Met.*, **88**, 237–242.

10 Cao, Y., Smith, P., and Heeger, A.J. (1993) Counter-ion induced processibility of conducting polyaniline. *Synth. Met.*, **57**, 3514–3519.

11 Hermas, A.A., Nakayama, M., and Ogura, K. (2005) Enrichment of chromium-content in passive layers on stainless steel coated with polyaniline. *Electrochim. Acta*, **50**, 2001–2007.

12 Hermas, A.A., Nakayama, M., and Ogura, K. (2005) Formation of stable passive film on stainless steel by electrochemical deposition of polypyrrole. *Electrochim. Acta*, **50**, 3640–3647.

13 Kowalski, D., Ueda, M., and Ohtsuka, T. (2008) The effect of ultrasonic irradiation during electropolymerization of polypyrrole on corrosion prevention of the coated steel. *Corros. Sci.*, **50**, 286–291.

14 Kowalski, D., Ueda, M., and Ohtsuka, T. (2007) Corrosion protection of steel by bi-layered polypyrrole doped with molybdophosphate and naphthalenedisulfonate anions. *Corros. Sci.*, **49**, 1635–1644.

15 Kowalski, D., Ueda, M., and Ohtsuka, T. (2007) The effect of counter anions on corrosion resistance of steel covered by bi-layered polypyrrole film. *Corros. Sci.*, **49**, 3442–3452.

16 Kowalski, D., Ueda, M., and Ohtsuka, T. (2010) Self-healing ion-permselective conducting polymer coating. *J. Mater. Chem.*, **20**, 7630–7633.

17 Ozyilmaz, A.T., Ozyilmaz, G., and Yigitoglu, O. (2010) Synthesis and characterization of poly(aniline) and poly(o-anisidine) films in sulphamic acid solution and their anticorrosion properties. *Prog. Org. Coat.*, **67**, 28–37.

18 Kraljic Rokovic, M., Kvastek, K., Horvat-Radosevic, V., and Duic, L. (2007) Poly (*ortho*-ethoxyaniline) in corrosion protection of stainless steel. *Corros. Sci.*, **49**, 2567–2580.

19 Hermas, A.A., Wu, Z.X., Nakayama, M., and Ogura, K. (2006) Passivation of stainless steel by coating with poly(*o*-phenylenediamine) conductive polymer. *J. Electrochem. Soc.*, **153**, B199–B205.

20 Kraljic, M., Mandic, Z., and Duic, L. (2003) Inhibition of steel corrosion by polyaniline coatings. *Corros. Sci.*, **45**, 181–198.

21 Bernard, M.C., Joiret, S., Hugot-Le Goff, A., and Phong, P.V. (2001) Protection of iron against corrosion using a polyaniline layer. I. Polyaniline electrodeposit. *J. Electrochem. Soc.*, **148**, B12–B16.

22 Nguyen Thi Le, H., Garcia, B., Deslouis, C., and Le Xuan, Q. (2001) Corrosion protection and conducting polymers: polypyrrole films on iron. *Electrochim. Acta*, **46**, 4259–4272.

23 Wessling, B. (1994) Passivation of metals by coating with polyaniline – corrosion potential shift and morphological changes. *Adv. Mater.*, **6**, 226–228.

24 Plesu, N., Ilia, G., Pascariu, A., and Vlase, G. (2006) Preparation, degradation of polyaniline doped with organic phosphorus acids and corrosion essays of polyaniline–acrylic blends. *Synth. Met.*, **156**, 230–238.

25 Talo, A., Passiniemi, P., Forsen, O., and Ylasaari, S. (1997) Polyaniline/epoxy coatings with good anti-corrosion properties. *Synth. Met.*, **85**, 1333–1334.

26 Seegmiller, J.C., da Silva, J.E.P., Buttry, D.A., de Torresi, S.I.C., and Torresi, R.M. (2005) Mechanism of action of corrosion protection coating for AA2024-T3 based on poly(aniline)–poly (methylmethacrylate) blend. *J. Electrochem. Soc.*, **152**, B45–B53.

27 Sathiyanarayanan, S., Jeyaram, R., Muthukrishnan, S., and Venkatachari, G. (2009) Corrosion protection mechanism of polyaniline blended organic coating on steel. *J. Electrochem. Soc.*, **156**, C127–C134.

28 de Souza, S. (2007) Smart coating based on polyaniline acrylic blend for corrosion protection of different metals. *Surf. Coat. Technol.*, **201**, 7574–7581.

29 Sathiyanarayanan, S., Azim, S.S., and Venkatachari, G. (2007) A new corrosion protection coating with polyaniline–TiO_2 composite for steel. *Electrochim. Acta*, **52**, 2068–2074.

30 Souza, S.d., Silva, J.E.P.d., Torresi, S.I.C. r.d., Temperini, M.L.A., and Torresi, R.M. (2001) Polyaniline based acrylic blends for iron corrosion protection. *Electrochem. Solid-State Lett.*, **4**, B27–B30.

31 Sathiyanarayanan, S., Azim, S.S., and Venkatachari, G. (2006) Corrosion resistant properties of polyaniline–acrylic coating on magnesium alloy. *Appl. Surf. Sci.*, **253**, 2113–2117.

32 Sathiyanarayanan, S., Azim, S., and Venkatachari, G. (2008) Performance studies of phosphate-doped polyaniline containing paint coating for corrosion protection of aluminium alloy. *J. Appl. Polym. Sci.*, **107**, 2224–2230.

33 Sathiyanarayanan, S., Azim, S.S., and Venkatachari, G. (2008) Corrosion protection of magnesium alloy ZM21 by polyaniline-blended coatings. *J. Coat. Technol. Res.*, **5**, 471–477.

34 Tiitu, M., Talo, A., Forsén, O., and Ikkala, O. (2005) Aminic epoxy resin hardeners as reactive solvents for conjugated polymers: polyaniline base/epoxy composites for anticorrosion coatings. *Polymer*, **46**, 6855–6861.

35 Sathiyanarayanan, S., Syed Azim, S., and Venkatachari, G. (2009) Corrosion protection of galvanized iron by polyaniline containing wash primer coating. *Prog. Org. Coat.*, **65**, 152–157.

36 Syed Azim, S., Sathiyanarayanan, S., and Venkatachari, G. (2006) Anticorrosive properties of PANI–ATMP polymer containing organic coating. *Prog. Org. Coat.*, **56**, 154–158.

37 Samui, A.B., Patankar, A.S., Rangarajan, J., and Deb, P.C. (2003) Study of polyaniline containing paint for corrosion prevention. *Prog. Org. Coat.*, **47**, 1–7.

38 Adhikari, A., Claesson, P., Pani, J., Leygraf, C., Deidinaitei, A., and Blomberg, E. (2008) Electrochemical behavior and anticorrosion properties of modified polyaniline dispersed in polyvinylacetate coating on carbon steel. *Electrochim. Acta*, **53**, 4239–4247.

39 Sathiyanarayanan, S., Azim, S.S., and Venkatachari, G. (2007) Corrosion protection of magnesium ZM 21 alloy with polyaniline–TiO_2 composite containing coatings. *Prog. Org. Coat.*, **59**, 291–296.

40 Li, Y., Zhang, H., Wang, X., Li, J., and Wang, F. (2011) Growth kinetics of oxide films at the polyaniline/mild steel interface. *Corros. Sci.*, **53**, 4044–4049.

41 Wessling, B. (1998) Dispersion as the link between basic research and commercial applications of conductive polymers (polyaniline). *Synth. Met.*, **93**, 143–154.

42 Wessling, B. (1996) Corrosion prevention with an organic metal (polyaniline): surface ennobling, passivation, corrosion test results. *Werkst. Korros. (Mater. Corros.)*, **47**, 439–445.

43 Nguyen Thi Le, H., Bernard, M.C., Garcia-Renaud, B., and Deslouis, C. (2004) Raman spectroscopy analysis of polypyrrole films as protective coatings on iron. *Synth. Met.*, **140**, 287–293.

44 Uehara, K., Ichikawa, T., Serikawa, T., Yoshikawa, S., Ehara, S., and Tsunooka, M. (1998) Redox reaction at the two-layer interface between aluminum and electropolymerized poly(3-methylthiophene) thin solid films. *Thin Solid Films*, **322**, 198–205.

45 Lu, J.L., Liu, N.J., Wang, X.H., Li, J., Jing, X.B., and Wang, F.S. (2003) Mechanism and life study on polyaniline anti-corrosion coating. *Synth. Met.*, **135–136**, 237–238.

46 Gustavsson, J.M., Innis, P.C., He, J., Wallace, G.G., and Tallman, D.E. (2009) Processable polyaniline–HCSA/poly(vinyl acetate-*co*-butyl acrylate) corrosion protection coatings for aluminium alloy 2024-T3: a SVET and Raman study. *Electrochim. Acta*, **54**, 1483–1490.

47 Cook, A., Gabriel, A., and Laycock, N. (2004) On the mechanism of corrosion protection of mild steel with polyaniline. *J. Electrochem. Soc.*, **151**, B529–B535.

48 Cook, A., Gabriel, A., Siew, D., and Laycock, N. (2004) Corrosion protection of low carbon steel with polyaniline: passivation or inhibition? *Curr. Appl. Phys.*, **4**, 133–136.

49 Rammelt, U., Nguyen, P.T., and Plieth, W. (2003) Corrosion protection by ultrathin films of conducting polymers. *Electrochim. Acta*, **48**, 1257–1262.

50 Yan, M., Tallman, D.E., and Bierwagen, G.P. (2008) Role of oxygen in the galvanic interaction between polypyrrole and aluminum alloy. *Electrochim. Acta*, **54**, 220–227.

51 Cogan, S.F., Gilbert, M.D., Holleck, G.L., Ehrlich, J., and Jillson, M.H. (2000) Galvanic coupling of doped poly-aniline and aluminum alloy 2024-T3. *J. Electrochem. Soc.*, **147**, 2143–2147.

52 Zhu, H., Zhong, L., Xiao, S., and Gan, F. (2004) Accelerating effect and mechanism of passivation of polyaniline on ferrous metals. *Electrochim. Acta*, **49**, 5161–5166.

53 Gasparac, R. and Martin, C.R. (2001) Investigations of the mechanism of corrosion inhibition by polyaniline – polyaniline-coated stainless steel in sulfuric acid solution. *J. Electrochem. Soc.*, **148**, B138–B145.

54 Ahmad, N. and MacDiarmid, A.G. (1996) Inhibition of corrosion of steels with the exploitation of conducting polymers. *Synth. Met.*, **78**, 103–110.

55 Syritski, V., Öpik, A., and Forsén, O. (2003) Ion transport investigations of polypyrroles doped with different anions by EQCM and CER techniques. *Electrochim. Acta*, **48**, 1409–1417.

56 Kinlen, P.J., Menon, V., and Ding, Y.W. (1999) A mechanistic investigation of polyaniline corrosion protection using the scanning reference electrode technique. *J. Electrochem. Soc.*, **146**, 3690–3695.

57 Kinlen, P.J., Ding, Y., and Silverman, D.C. (2002) Corrosion protection of mild steel using sulfonic and phosphonic acid-doped polyanilines. *Corrosion*, **58**, 490–497.

58 Dominis, A.J., Spinks, G.M., and Wallace, G.G. (2003) Comparison of polyaniline primers prepared with different dopants for corrosion protection of steel. *Prog. Org. Coat.*, **48**, 43–49.

59 Kendig, M., Hon, M., and Warren, L. (2003) 'Smart' corrosion inhibiting coatings. *Prog. Org. Coat.*, **47**, 183–189.

60 Torresi, R.M., Souza, S.d., Silva, J.E.P.d., and Torresi, S.I.C.d. (2005) Galvanic coupling between metal substrate and polyaniline acrylic blends: corrosion protection mechanism. *Electrochim. Acta*, **50**, 2213–2218.

61 Gabriel, A., Laycock, N.J., McMurray, H.N., Williams, G., and Cook, A. (2006) Oxidation states exhibited by in-coating polyaniline during corrosion-driven coating delamination on carbon steel. *Electrochem. Solid State Lett.*, **9**, B57–B60.

62 Pereira da Silva, J.E., Córdoba de Torresi, S.I., and Torresi, R.M. (2005) Polyaniline acrylic coatings for corrosion inhibition: the role played by counter-ions. *Corros. Sci.*, **47**, 811–822.

63 Kendig, M. and Hon, M. (2004) Environmentally triggered release of oxygen-reduction inhibitors from inherently conducting polymers. *Corrosion*, **60**, 1024–1030.

64 da Silva, J.E.P., de Torresi, S.I.C., and Torresi, R.M. (2007) Polyaniline/poly(methylmethacrylate) blends for corrosion protection: the effect of passivating dopants on different metals. *Prog. Org. Coat.*, **58**, 33–39.

65 Paliwoda-Porebska, G., Stratmann, M., Rohwerder, M., Potje-Kamloth, K., Lu, Y., Pich, A.Z., and Adler, H.J. (2005) On the development of polypyrrole coatings with

self-healing properties for iron corrosion protection. *Corros. Sci.*, **47**, 3216–3233.

66 Riaz, U., Ahmad, S.A., Ashraf, S.M., and Ahmad, S. (2009) Effect of dopant on the corrosion protective performance of environmentally benign nanostructured conducting composite coatings. *Prog. Org. Coat.*, **65**, 405–409.

67 Samui, A.B. and Phadnis, S.M. (2005) Polyaniline–dioctyl phosphate salt for corrosion protection of iron. *Prog. Org. Coat.*, **54**, 263–267.

68 Azim, S.S., Sathiyanarayanan, S., and Verikatachari, G. (2006) Anticorrosive properties of PANI–ATMP polymer containing organic coating. *Prog. Org. Coat.*, **56**, 154–158.

69 Williams, G., Gabriel, A., Cook, A., and McMurray, H.N. (2006) Dopant effects in polyaniline inhibition of corrosion-driven organic coating cathodic delamination on iron. *J. Electrochem. Soc.*, **153**, B425–B433.

70 Chen, Y., Wang, X.H., Li, J., Lu, J.L., and Wang, F.S. (2007) Long-term anticorrosion behaviour of polyaniline on mild steel. *Corros. Sci.*, **49**, 3052–3063.

71 Chen, Y., Wang, X.H., Li, J., Lu, J.L., and Wang, F.S. (2007) Polyaniline for corrosion prevention of mild steel coupled with copper. *Electrochim. Acta*, **52**, 5392–5399.

72 Yan, M.C., Tallman, D.E., Rasmussen, S.C., and Bierwagen, G.P. (2009) Corrosion control coatings for aluminum alloys based on neutral and n-doped conjugated polymers. *J. Electrochem. Soc.*, **156**, C360–C366.

73 Li, Y., Zhang, H., Wang, X., Li, J., and Wang, F. (2011) Role of dissolved oxygen diffusion in coating defect protection by emeraldine base. *Synth. Met.*, **161**, 2312–2317.

74 Yan, M., Vetter, C.A., and Gelling, V.J. (2010) Electrochemical investigations of polypyrrole aluminum flake coupling. *Electrochim. Acta*, **55**, 5576–5583.

75 Nguyen, T.D., Nguyen, T.A., Pham, M.C., Piro, B., Normand, B., and Takenouti, H. (2004) Mechanism for protection of iron corrosion by an intrinsically electronic conducting polymer. *J. Electroanal. Chem.*, **572**, 225–234.

76 Sathiyanarayanan, S., Muthkrishnan, S., and Venkatachari, G. (2006) Corrosion protection of steel by polyaniline blended coating. *Electrochim. Acta*, **51**, 6313–6319.

77 Truong, V.T., Lai, P.K., Moore, B.T., Muscat, R.F., and Russo, M.S. (2000) Corrosion protection of magnesium by electroactive polypyrrole/paint coatings. *Synth. Met.*, **110**, 7–15.

78 Yfantis, A., Paloumpa, I., Schmeißr, D., and Yfantis, D. (2002) Novel corrosion-resistant films for Mg alloys. *Surf. Coat. Technol.*, **151–152**, 400–404.

79 Ahmad, S., Ashraf, S.M., and Riaz, U. (2005) Corrosion studies of polyaniline/coconut oil poly(esteramide urethane) coatings. *Polym. Adv. Technol.*, **16**, 541–548.

80 Alam, J., Riaz, U., and Ahmad, S. (2009) High performance corrosion resistant polyaniline/alkyd ecofriendly coatings. *Curr. Appl. Phys.*, **9**, 80–86.

81 Bagherzadeh, M.R., Mahdavi, F., Ghasemi, M., Shariatpanahi, H., and Faridi, H.R. (2010) Using nanoemeraldine salt–polyaniline for preparation of a new anticorrosive water-based epoxy coating. *Prog. Org. Coat.*, **68**, 319–322.

82 Ahmad, S., Ashraf, S.M., Riaz, U., and Zafar, S. (2008) Development of novel waterborne poly(1-naphthylamine)/poly(vinylalcohol)–resorcinol formaldehyde-cured corrosion resistant composite coatings. *Prog. Org. Coat.*, **62**, 32–39.

83 Armelin, E., Pla, R., Liesa, F., Ramis, X., Iribarren, J.I., and Alemán, C. (2008) Corrosion protection with polyaniline and polypyrrole as anticorrosive additives for epoxy paint. *Corros. Sci.*, **50**, 721–728.

84 Iribarren, J.I., Ocampo, C., Armelin, E., Liesa, F., and Alemán, C. (2008) Poly(3-alkylthiophene)s as anticorrosive additive for paints: influence of the main chain stereoregularity. *J. Appl. Polym. Sci.*, **108**, 3291–3297.

85 Moraga, G.A., Silva, G.G., Matencio, T., and Paniago, R.M. (2006) Poly(2,5-dimethoxy aniline)/fluoropolymer blend coatings to corrosion inhibition on stainless steel. *Synth. Met.*, **156**, 1036–1042.

86 Cao, Y., Smith, P., and Heeger, A.J. (1992) Counter-ion induced processibility

of conducting polyaniline and of conducting polyblends of polyaniline in bulk polymers. *Synth. Met.*, **48**, 91–97.

87 Laska, J. and Widlarz, J. (2003) Water soluble polyaniline. *Synth. Met.*, **135–136**, 261–262.

88 Wang, Q.G., Wang, X.H., Li, J., Zhao, X.J., and Wang, F.S. (2005) Water-borne conductive polyaniline doped by acidic phosphate ester containing polysilsesquioxane precursor. *Synth. Met.*, **148**, 127–132.

89 Wang, Q., Liu, N., Wang, X., Li, J., Zhao, X., and Wang, F. (2003) Conductive hybrids from water-borne conductive polyaniline and (3-glycidoxypropyl) trimethoxysilane. *Macromolecules*, **36**, 5760–5764.

90 Luo, J., Zhang, H., Wang, X., Li, J., and Wang, F. (2007) Stable aqueous dispersion of conducting polyaniline with high electrical conductivity. *Macromolecules*, **40**, 8132–8135.

91 Luo, J., Wang, X., Li, J., Zhao, X., and Wang, F. (2007) Conductive hybrid film from polyaniline and polyurethane–silica. *Polymer*, **48**, 4368–4374.

92 Oliveira, M.A.S., Moraes, J.J., and Faez, R. (2009) Impedance studies of poly(methylmethacrylate-*co*-acrylic acid) doped polyaniline films on aluminum alloy. *Prog. Org. Coat.*, **65**, 348–356.

93 Wang, Y.Y. and Jing, X.L. (2004) Preparation of an epoxy/polyaniline composite coating and its passivation effect on cold rolled steel. *Polym. J.*, **36**, 374–379.

94 Mirmohseni, A., Valiegbal, K., and Wallace, G.G. (2003) Preparation and characterization of a polyaniline/poly(butyl acrylate–vinyl acetate) composite as a novel conducting polymer composite. *J. Appl. Polym. Sci.*, **90**, 2525–2531.

95 Ding, K.Q., Jia, Z.B., Mab, W.S., Jiang, D.L., Zhao, Q., Cao, N.L., and Tong, R. (2003) Corrosion protection of mild steel with polyaniline–thiokol rubber composite coatings. *Prot. Met.*, **39**, 71–76.

96 Ding, K.Q., Jia, Z.B., Ma, W.S., Tong, R.T., and Wang, X.K. (2002) Polyaniline and polyaniline–thiokol rubber composite coatings for the corrosion protection of mild steel. *Mater. Chem. Phys.*, **76**, 137–142.

97 Akbarinezhad, E., Ebrahimi, M., Sharif, F., Attar, M.M., and Faridi, H.R. (2011) Synthesis and evaluating corrosion protection effects of emeraldine base PAni/clay nanocomposite as a barrier pigment in zinc-rich ethyl silicate primer. *Prog. Org. Coat.*, **70**, 39–44.

98 Radhakrishnan, S., Siju, C.R., Mahanta, D., Patil, S., and Madras, G. (2009) Conducting polyaniline–nano-TiO_2 composites for smart corrosion resistant coatings. *Electrochim. Acta*, **54**, 1249–1254.

99 Meroufel, A., Deslouis, C., and Touzain, S. (2008) Electrochemical and anticorrosion performances of zinc-rich and polyaniline powder coatings. *Electrochim. Acta*, **53**, 2331–2338.

100 Chang, K.-C., Jang, G.-W., Peng, C.-W., Lin, C.-Y., Shieh, J.-C., Yeh, J.-M., Yang, J.-C., and Li, W.-T. (2007) Comparatively electrochemical studies at different operational temperatures for the effect of nanoclay platelets on the anticorrosion efficiency of DBSA-doped polyaniline/Na^+-MMT clay nanocomposite coatings. *Electrochim. Acta*, **52**, 5191–5200.

101 Zaarei, D., Sarabi, A.A., Sharif, F., and Kassiriha, S.M. (2008) Structure, properties and corrosion resistivity of polymeric nanocomposite coatings based on layered silicates. *J. Coat. Technol. Res.*, **5**, 241–249.

102 Kobayashi, T., Wood, B.A., Takemura, A., and Ono, H. (2006) Antistatic performance and morphological observation of ternary blends of poly(ethylene terephthalate), poly(ether esteramide), and Na-neutralized poly(ethylene-*co*-methacrylic acid) copolymers. *J. Electrostat.*, **64**, 377–385.

103 Chen, K., Xiong, C., Li, L., Zhou, L., Lei, Y., and Dong, L. (2009) Conductive mechanism of antistatic poly(ethylene terephthalate)/ZnOw composites. *Polym. Compos.*, **30**, 226–231.

104 Pionteck, J. and Wypych, G. (2007) *Handbook of Antistatics*, ChemTec Publishing.

105 Wang, L. and Zhao, S. (2010) Study on the structure–mechanical properties relationship and antistatic characteristics of SSBR composites filled with SiO_2/CB. *J. Appl. Polym. Sci.*, **118**, 338–345.

106 Pötschke, P., Abdel-Goad, M., Pegel, S., Jehnichen, D., Mark, J.E., Zhou, D., and Heinrich, G. (2009) Comparisons among electrical and rheological properties of melt-mixed composites containing various carbon nanostructures. *J. Macromol. Sci. A*, **47**, 12–19.

107 Schwarz, M.-K., Bauhofer, W., and Schulte, K. (2002) Alternating electric field induced agglomeration of carbon black filled resins. *Polymer*, **43**, 3079–3082.

108 Kelly, F.M. and Johnston, J.H. (2011) Colored and functional silver nanoparticle–wool fiber composites. *ACS Appl. Mater. Interfaces*, **3**, 1083–1092.

109 Ke-bing, Z. and Zhi-jie, L. (2009) Study of nano-copper antistatic preservative material and its performance packaging engineering. *Packaging Eng.*, **30**, 34–36.

110 Choi, M.H., Jeon, B.H., and Chung, I.J. (2000) The effect of coupling agent on electrical and mechanical properties of carbon fiber/phenolic resin composites. *Polymer*, **41**, 3243–3252.

111 Thongruang, W., Spontak, R.J., and Balik, C.M. (2002) Correlated electrical conductivity and mechanical property analysis of high-density polyethylene filled with graphite and carbon fiber. *Polymer*, **43**, 2279–2286.

112 Byrne, M.T. and Gun'ko, Y.K. (2010) Recent advances in research on carbon nanotube–polymer composites. *Adv. Mater.*, **22**, 1672–1688.

113 Zhonghua, C., Ying, T., and Fei, Y. (2009) Properties of an environmentally friendly waterborne antistatic anticorrosion paint. *J. Chin. Soc. Corros. Protect.*, **29**, 113–118.

114 Löbl, H.P., Huppertz, M., and Mergel, D. (1996) ITO films for antireflective and antistatic tube coatings prepared by d.c. magnetron sputtering. *Surf. Coat. Technol.*, **82**, 90–98.

115 Ma, C.-C.M., Chen, Y.-J., and Kuan, H.-C. (2006) Polystyrene nanocomposite materials – preparation, mechanical, electrical and thermal properties, and morphology. *J. Appl. Polym. Sci.*, **100**, 508–515.

116 Belaabed, B., Lamouri, S., Naar, N., Bourson, P., and Hamady, S.O.S. (2010) Polyaniline-doped benzene sulfonic acid/epoxy resin composites: structural, morphological, thermal and dielectric behaviors. *Polym. J.*, **42**, 546–554.

117 Egami, Y., Suzuki, K., Tanaka, T., Yasuhara, T., Higuchi, E., and Inoue, H. (2011) Preparation and characterization of conductive fabrics coated uniformly with polypyrrole nanoparticles. *Synth. Met.*, **161**, 219–224.

118 Bhadra, S., Singha, N.K., and Khastgir, D. (2009) Dielectric properties and EMI shielding efficiency of polyaniline and ethylene 1-octene based semi-conducting composites. *Curr. Appl. Phys.*, **9**, 396–403.

119 Bhandari, H., Bansal, V., Choudhary, V., and Dhawan, S.K. (2009) Influence of reaction conditions on the formation of nanotubes/nanoparticles of polyaniline in the presence of 1-amino-2-naphthol-4-sulfonic acid and applications as electrostatic charge dissipation material. *Polym. Int.*, **58**, 489–502.

120 Sudha, J.D. and Sivakala, S. (2009) Conducting polystyrene/polyaniline blend through template-assisted emulsion polymerization. *Colloid Polym. Sci.*, **287**, 1347–1354.

121 Skotheim, T.A. (1998) *Handbook of Conducting Polymers*, 2nd edn, Marcel Dekker, Inc.

122 Tzou, K.T. and Gregory, R.V. (1995) Improved solution stability and spinnability of concentrated polyaniline solutions using N,N'-dimethyl propylene urea as the spin bath solvent. *Synth. Met.*, **69**, 109–112.

123 Tzou, K. and Gregory, R.V. (1993) A method to prepare soluble polyaniline salt solutions – *in situ* doping of PANI base with organic dopants in polar solvents. *Synth. Met.*, **53**, 365–377.

124 Leclerc, M., Guay, J., and Dao, L.H. (1989) Synthesis and characterization of poly(alkylanilines). *Macromolecules*, **22**, 649–653.

125 Wei, Y., Focke, W.W., Wnek, G.E., Ray, A., and Macdiarmid, A.G. (1989) Synthesis and electrochemistry of alkyl ring-substituted polyanilines. *J. Phys. Chem.*, **93**, 495–499.

126 Swaruparani, H., Basavaraja, S., Basavaraja, C., Huh, D.S., and Venkataraman, A. (2010) A new approach to soluble polyaniline and its copolymers with toluidines. *J. Appl. Polym. Sci.*, **117**, 1350–1360.

127 Sahin, M., Ozcan, L., Ozcan, A., Usta, B., Sahin, Y., and Pekmez, K. (2010) The substituent effects on the structure and surface morphology of polyaniline. *J. Appl. Polym. Sci.*, **115**, 3024–3030.

128 Macinnes, D. and Funt, B.L. (1988) Poly-*ortho*-methoxyaniline – a new soluble conducting polymer. *Synth. Met.*, **25**, 235–242.

129 Han, C.C., Yang, K.F., Hong, S.P., Balasubramanian, A., and Lee, Y.T. (2005) Syntheses and characterizations of aniline/butylthioaniline copolymers: comparisons of copolymers prepared by the new concurrent reduction and substitution route and the conventional oxidative copolymerization method. *J. Polym. Sci. Polym. Chem.*, **43**, 1767–1777.

130 McCullough, R.D., Lowe, R.D., Jayaraman, M., and Anderson, D.L. (1993) Design, synthesis, and control of conducting polymer architectures –structurally homogeneous poly(3-alkylthiophenes). *J. Org. Chem.*, **58**, 904–912.

131 Yoshino, K., Nakajima, S., and Sugimoto, R. (1987) Fusibility of polythiophene derivatives with substituted long alkyl chain and their properties. *Jpn. J. Appl. Phys. Part 2*, **26**, L1038–L1039.

132 Cao, Y., Smith, P., and Heeger, A.J. (1993) *Processible forms of electrically conductive polyaniline*. USPTO, Uniax Corporation, Santa Barbara, CA.

133 Österholm, J.-E., Cao, Y., Klavetter, F., and Smith, P. (1994) Emulsion polymerization of aniline. *Polymer*, **35**, 2902–2906.

134 Wu, C.-H., Don, T.-M., and Chiu, W.-Y. (2011) Characterization and conversion determination of stable PEDOT latex nanoparticles synthesized by emulsion polymerization. *Polymer*, **52**, 1375–1384.

135 Yang, J. and Weng, B. (2009) Inverse emulsion polymerization for high molecular weight and electrically conducting polyanilines. *Synth. Met.*, **159**, 2249–2252.

136 Kinlen, P.J., Liu, J., Ding, Y., Graham, C.R., and Remsen, E.E. (1998) Emulsion polymerization process for organically soluble and electrically conducting polyaniline. *Macromolecules*, **31**, 1735–1744.

137 Ruckenstein, E. and Yang, S. (1993) An emulsion pathway to electrically conductive polyaniline–polystyrene composites. *Synth. Met.*, **53**, 283–292.

138 Yang, S. and Ruckenstein, E. (1993) Processable conductive composites of polyaniline/poly(alkyl methacrylate) prepared via an emulsion method. *Synth. Met.*, **59**, 1–12.

139 Bhadra, S., Khastgir, D., Singha, N.K., and Lee, J.H. (2009) Progress in preparation, processing and applications of polyaniline. *Prog. Polym. Sci.*, **34**, 783–810.

140 Martins, C.R., Rubinger, C.P.L., Costa, L.C., and Rubinger, R.M. (2008) Dielectric properties of ternary melt processed blends. *J. Non-Cryst. Solids*, **354**, 5323–5325.

141 Yong, K.C. and Saad, C.S.M. (2010) High temperature-mechanical mixing to prepare electrically conductive sulfur-vulcanised poly(butadiene-*co*-acrylonitrile)–polyaniline dodecylbenzenesulfonate blends. *J. Rubber Res.*, **13**, 1–17.

142 Martins, C.R. and De Paoli, M.A. (2005) Antistatic thermoplastic blend of polyaniline and polystyrene prepared in a double-screw extruder. *Eur. Polym. J.*, **41**, 2867–2873.

143 Zhang, Q.H., Wang, X.H., Chen, D.J., and Jing, X.B. (2004) Electrically conductive, melt-processed ternary blends of polyaniline/dodecylbenzene sulfonic acid, ethylene/vinyl acetate, and low-density polyethylene. *J. Polym. Sci. B*, **42**, 3750–3758.

144 Martins, C.R., Faez, R., Rezende, M.C., and De Paoli, M.A. (2006) Reactive processing and evaluation of butadiene–styrene copolymer/polyaniline

144 conductive blends. *J. Appl. Polym. Sci.*, **101**, 681–685.
145 Barra, G.M.O., Jacques, L.B., Oréfice, R. L., and Carneiro, J.R.G. (2004) Processing, characterization and properties of conducting polyaniline–sulfonated SEBS block copolymers. *Eur. Polym. J.*, **40**, 2017–2023.
146 Barra, G.M.O., Matins, R.R., Kafer, K.A., Paniago, R., Vasques, C.T., and Pires, A. T.N. (2008) Thermoplastic elastomer/polyaniline blends: evaluation of mechanical and electromechanical properties. *Polym. Test.*, **27**, 886–892.
147 Muller, D., Garcia, M., Salmoria, G.V., Pires, A.T.N., Paniago, R., and Barra, G. M.O. (2011) SEBS/PPy.DBSA blends: preparation and evaluation of electromechanical and dynamic mechanical properties. *J. Appl. Polym. Sci.*, **120**, 351–359.
148 Malmonge, L.F., Lopes, G.d.A., Langiano, S.d.C., Malmonge, J.A., Cordeiro, J.M.M., and Mattoso, L.H.C. (2006) A new route to obtain PVDF/PANI conducting blends. *Eur. Polym. J.*, **42**, 3108–3113.
149 Soares, B.G., Amorim, G.S., Souza, J.F.G., Oliveira, M.G., and Silva, J.E.P.d. (2006) The *in situ* polymerization of aniline in nitrile rubber. *Synth. Met.*, **156**, 91–98.
150 Sudha, J.D., Sivakala, S., Patel, K., and Nair, P.R. (2010) Development of electromagnetic shielding materials from the conductive blends of polystyrene polyaniline–clay nanocomposite. *Compos. Part A*, **41**, 1647–1652.
151 Ferrero, F., Napoli, L., Tonin, C., and Varesano, A. (2006) Pyrrole chemical polymerization on textiles: kinetics and operating conditions. *J. Appl. Polym. Sci.*, **102**, 4121–4126.
152 Lin, T., Wang, L.J., Wang, X.G., and Kaynak, A. (2005) Polymerising pyrrole on polyester textiles and controlling the conductivity through coating thickness. *Thin Solid Films*, **479**, 77–82.
153 Varesano, A., Aluigi, A., Florio, L., and Fabris, R. (2009) Multifunctional cotton fabrics. *Synth. Met.*, **159**, 1082–1089.
154 Dall'Acqua, L., Tonin, C., Peila, R., Ferrero, F., and Catellani, M. (2004) Performances and properties of intrinsic conductive cellulose–polypyrrole textiles. *Synth. Met.*, **146**, 213–221.
155 Kaynak, A., Najar, S.S., and Foitzik, R.C. (2008) Conducting nylon, cotton and wool yams by continuous vapor polymerization of pyrrole. *Synth. Met.*, **158**, 1–5.
156 Najara, S.S., Kaynak, A., and Foitzik, R.C. (2007) Conductive wool yarns by continuous vapour phase polymerization of pyrrole. *Synth. Met.*, **157**, 1–4.
157 Dall'Acqua, L., Tonin, C., Varesano, A., Canetti, M., Porzio, W., and Catellani, M. (2006) Vapour phase polymerisation of pyrrole on cellulose-based textile substrates. *Synth. Met.*, **156**, 379–386.
158 Hua, M.Y., Su, Y.N., and Chen, S.A. (2000) Water-soluble self-acid-doped conducting polyaniline: poly(aniline-*co*-*N*-propylbenzenesulfonic acid–aniline). *Polymer*, **41**, 813–815.
159 Wei, X.L., Wang, Y.Z., Long, S.M., Bobeczko, C., and Epstein, A.J. (1996) Synthesis and physical properties of highly sulfonated polyaniline. *J. Am. Chem. Soc.*, **118**, 2545–2555.
160 Sulimenko, T., Stejskal, J., Křivka, I., and Prokeš, J. (2001) Conductivity of colloidal polyaniline dispersions. *Eur. Polym. J.*, **37**, 219–226.
161 Chattopadhyay, D. and Mandal, B.M. (1996) Methyl cellulose stabilized polyaniline dispersions. *Langmuir*, **12**, 1585–1588.
162 Stejskal, J., Kratochvil, P., and Helmstedt, M. (1996) Polyaniline dispersions. 5. Poly(vinyl alcohol) and poly(*N*-vinylpyrrolidone) as steric stabilizers. *Langmuir*, **12**, 3389–3392.
163 Banerjee, P., Bhattacharyya, S.N., and Mandal, B.M. (1996) Poly(vinyl methyl ether) stabilized colloidal polyaniline dispersions. *Langmuir*, **12**, 1406–11406.
164 Sun, Z.C., Wang, X.H., Li, J., Zhang, J. Y., Yu, L., Jing, X.B., Wang, F.S., Lee, C. J., and Rhee, S.B. (1999) Preparation and properties of water-based conducting polyaniline. *Synth. Met.*, **102**, 1224–1225.
165 Geng, Y.H., Sun, Z.C., Li, J., Jing, X.B., Wang, X.H., and Wang, F.S. (1999) Water soluble polyaniline and its blend films

prepared by aqueous solution casting. *Polymer*, **40**, 5723–5727.

166 Davey, J.M., Too, C.O., Ralph, S.F., Kane-Maguire, L.A.P., Wallace, G.G., and Partridge, A.C. (2000) Conducting polyaniline/calixarene salts: synthesis and properties. *Macromolecules*, **33**, 7044–7050.

167 Zhang, H., Lu, J., Wang, X., Li, J., and Wang, F. (2011) From amorphous to crystalline: practical way to improve electrical conductivity of water-borne conducting polyaniline. *Polymer*, **52**, 3059–3064.

168 Wang, Y.J., Wang, X.H., Zhao, X.J., Li, J., Mo, Z.S., Jing, X.B., and Wang, F.S. (2002) Conducting polyaniline confined in semi-interpenetrating networks. *Macromol. Rapid Commun.*, **23**, 118–121.

169 Wang, Y.J., Wang, X.H., Li, J., Mo, Z.S., Zhao, X.J., Jing, X.B., and Wang, F.S. (2001) Conductive polyaniline/silica hybrids from sol–gel process. *Adv. Mater.*, **13**, 1582–1585.

170 Wang, Y.J., Liu, N.J., Lu, J.L., Liu, H., Li, J., Jing, X.B., Wang, F.S., and Wang, X.H. (2003) Confining conducting polyaniline in a stable inorganic network. *Chin. J. Polym. Sci.*, **21**, 603–608.

171 Wang, Q.G., Liu, N.J., Wang, X.H., Li, J., Zhao, X.J., and Wang, F.S. (2003) Conductive hybrids from water-borne conductive polyaniline and (3-glycidoxypropyl)trimethoxysilane. *Macromolecules*, **36**, 5760–5764.

172 Luo, J., Wang, X.-h., Li, J., Zhao, X.-J., and Wang, F.-S. (2007) Electrostatic interaction hybrids from water-borne conductive polyaniline and inorganic precursor containing carboxyl group. *Chin. J. Polym. Sci.*, **25**, 181–186.

173 Luo, J., Wang, Q., Wang, X., Li, J., Zhao, X., and Wang, F. (2007) Water-resistant conducting hybrids from electrostatic interactions. *J. Polym. Sci. Polym. Chem.*, **45**, 1424–1431.

174 Luo, J., Wang, X., Li, J., Zhao, X., and Wang, F. (2007) Extending electrochemical activity of polyaniline to alkaline media via electrostatic interaction and sol–gel route. *Electrochem. Commun.*, **9**, 1175–1179.

11
Conjugated–Insulating Block Copolymers: Synthesis, Morphology, and Electronic Properties

Dahlia Haynes, Mihaela C. Stefan, and Richard D. McCullough

11.1
Introduction

Conjugated polymers have been vigorously studied as semiconducting materials for a variety of applications such as polymer field-effect transistors (OFETs) [1–10], photo- and electroluminescent diodes [11–15], organic photovoltaics (OPVs) [16–23], and sensing devices in medical and biological fields [24–26]. Nevertheless, there are still many challenges in the field of semiconducting conjugated materials that inhibit the rate of advancement for organic electronics. The extended π–π network in conjugated polymer systems often times results in high rigidity, and thus causes poor flexibility and low solubility in organic solvents. This tends to limit their applicability and processability toward the design of nanostructured devices that depend on highly optimal configuration and reproducible molecular organization at nanometer and submicron scales. Researchers have attempted to resolve some of these issues by manipulation of the molecular structure, for example, introducing flexible side chains or groups to the backbone of the polymers to improve solubility [2, 27, 28]. In addition, efforts have been made to alter the supramolecular nanostructure and morphological properties via covalent attachment of an insulating block. The device configuration and processing can also be modified through copolymerization procedures [29–35].

The need for organic electronic materials has pushed for the study of conjugated–insulating block copolymers that can overcome the limiting factors associated with conjugated polymers, ultimately yielding low-cost, flexible, and lightweight devices. This block copolymer approach also allows for control over micrometer and submicrometer patterns and morphological structures via self-assembly caused by the interplay between the connected blocks and their distinct chemical differences. As such, the self-assembled components can strongly affect the internal processes associated with energy transfer and conversion in semiconducting polymeric devices and scaffolds. Of the many types of conjugated–insulating block copolymers, rod–coil block copolymers have been recognized as a useful synthon in bringing about desired morphologies via self-assembly and photophysical properties [32–40]. In contrast, rod–rod copolymers have been less

Semiconducting Polymer Composites: Principles, Morphologies, Properties and Applications, First Edition.
Edited by Xiaoniu Yang.
© 2012 Wiley-VCH Verlag GmbH & Co. KGaA. Published 2012 by Wiley-VCH Verlag GmbH & Co. KGaA.

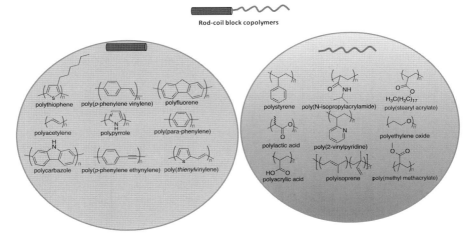

Figure 11.1 Selected examples of conjugated rod and coil structures of semiconducting block copolymers.

studied but recent efforts show that these systems can also offer an attractive route to highly optimal nanostructured devices. Covalent attachment of flexible coil-like polymers such as polymethacrylates, polyesters, and polyethylene derivatives or rigid rod polymers (other conjugated polymers) to semiconducting polymers offers a way to produce a new class of materials capable of self-assembling into a variety of morphologies due to the difference in stiffness between the two domains, aggregation and π–π interactions. Several classes of rod–coil block copolymers have been reported and are typically characterized depending on on the conjugated block. This chapter seeks to provide a broad overview of both pioneering and recent work in conjugated–insulating block copolymers, particularly focusing on rod–coil and rod–rod block copolymers, with emphasis on their syntheses, morphology, and electronic properties (Figure 11.1).

11.2
Oligo- and Polythiophene Rod–Coil Block Copolymers

One of the most versatile conjugated materials to date that have received considerable attention is thiophene-based oligomers and polymers. Due to the relatively facile synthetic modifications of their corresponding monomers, oligothiophenes and polythiophenes have become leaders in the race toward efficient organic semiconducting materials. The comprehensive research on oligothiophenes has been used as a foundation to elucidate relationships between conjugated linear systems and the influence on their electronic properties by structural variation. Furthermore, to understand structure–property relationships, many polymer systems have to rely on low molecular weight counterparts, that is, oligomers, to act as model systems, thus gaining insight into the correlation between chain length and their corresponding properties. In addition, these oligomers are more soluble than their high molecular weight counterparts and characterization is rendered much more feasible. Modeling calculations can then offer hypotheses on the correlation between chain length and electronic properties. For instance, a variety of small-chain (repeat units up to 12) substituted and nonsubstituted thiophene-based oligomers have been explored for semiconducting applications [41–48]. Over the years, many synthetic strategies have been employed in the development of well-defined oligomers of thiophenes and excellent reviews on the evolution of synthetic scopes have been published [49, 50]. Oligothiophenes have typically been synthesized by oxidative homocoupling procedures [51–53], ring closure from acyclic precursors [54, 55], and metal-catalyzed C–C coupling reactions [56–59]. As thiophene chemistry progressed, the syntheses of polythiophenes were embarked upon due to promise of enhanced electrical performance with increasing conjugation length.

Yamamato [60] and Dudek [61] initially reported unsubstituted polythiophenes; however, due to the lack of solubility and the impurities generated in the polymerization, alternative strategies were developed to improve upon solubility, electronic properties, and stability. As the need for soluble thiophenes became more apparent, the attachment of flexible side chains to the polymer backbone became the most viable method for improved processability. Initially, the synthesis of regioirregular poly(3-alkylthiophene)s (P3ATs) demonstrated an increase in the solubility of these materials; nonetheless, due to the orientation and location of the side chains in the polymer chain, twisting of the backbone led to a loss of π-conjugation and significant loss in electronic properties [62–64]. The McCullough method reported in 1992 gave head-to-tail regioregular P3ATs (rr-P3ATs), which led to lower structural defects [65]. Since then, two other methods were also employed to synthesize rr-P3ATs: the Rieke method [66–69] and the Grignard metathesis (GRIM) method [27, 28] (Figure 11.2). All three methods involve Ni-catalyzed cross-coupling

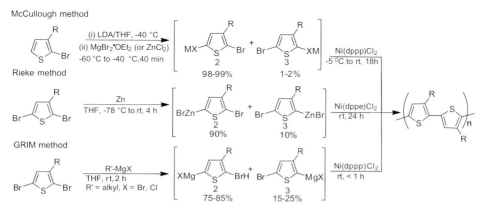

Figure 11.2 Common synthetic procedures for regioregular poly(3-alkylthiophene)s.

reactions and advantages such as the low catalyst loadings and high regioregularity compared to earlier synthetic reports. More details on synthetic strategies for regioregular polythiophenes can be found in the works of McCullough and Yokozawa [70–75].

The syntheses of rod–coil block copolymers usually occur via two different types of "living" polymerization techniques. Controlled radical polymerization (CRP) methods, such as atom transfer radical polymerization (ATRP), reversible addition–fragmentation chain transfer polymerization (RAFT), and nitroxide-mediated polymerization (NMP), have been employed for the synthesis of P3HT diblock copolymers (Figures 11.3 and 11.4) [76–78]. For oligo- and polythiophenes, the synthetic strategies often involve postpolymerization and *in situ* methods to introduce either a macroinitiator or a functional group capable of covalent linkages with another functional group of a coiled counterpart. One of the first rod–coil rr-P3AT-based block copolymers was developed by McCullough *et al.* and showed a facile end group functionalization to form a polythiophene block macroinitiator on which polystyrene or polymethacrylates were attached by using ATRP (Figure 11.3) [79–82]. The end group modification of P3ATs paved the way for a variety of rod–coil block copolymers to be generated using other combinations of polymerization methods such as anionic polymerization [83–86] and RAFT [87, 88] and click chemistry [89, 90].

Alternatively, coupling of two individually formed polymers that have specific functional end groups generated block copolymers. For example, Chao and coworkers recently reported the synthesis of aldehyde-terminated P3HT polymers coupled with insulating polymeric anions to form coil–rod–coil block copolymers [91]. The synthetic procedure allowed for controlled coupling reactions of polymers with low polydispersities and interesting morphologies (Figure 11.5).

Ring-opening cationic polymerization of tetrahydrofuran and 2-ethyl-2-oxazoline using triflate ester-terminated P3HT macroinitiator was also reported (Figure 11.6) [92, 93]. The diblock copolymer containing P3HT and elastomeric polytetrahydrofuran (P3HT–PTHF) displayed nanofibrillar morphology. Nanofibrillar morphology was also observed for the diblock copolymers of P3HT with poly(2-ethyl-2-oxazoline) (P3HT–PEOXA), where the density of the nanofibrils was found to depend on the content of PEOXA insulating coil segment (Figure 11.7). Both P3HT–PTHF and P3HT–PEOXA diblock copolymers displayed solvatochromism and optoelectronic properties comparable to those of the P3HT homopolymer precursor [92, 93].

Meijer reported the synthesis of P3HT–polycyclooctene diblock copolymer by ruthenium-mediated ring-opening metathesis polymerization (ROMP) of cyclooctene using an allyl-terminated P3HT as a chain transfer agent. The polycyclooctene block was hydrogenated with *p*-toluenesulfonyl hydrazide to generate the P3HT–polyethylene conjugated–insulating rod–coil diblock copolymer (Figure 11.8) [94].

The versatile synthetic routes for block copolymer formation allow access to fascinating materials that can form supramolecular nanostructures due to their self-assembling ability. Various research groups have exploited these synthetic

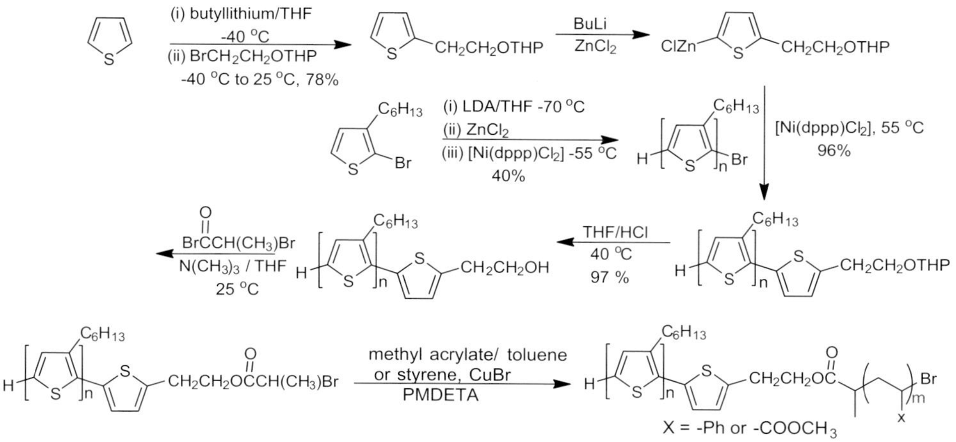

Figure 11.3 Synthesis of rr-P3HT-*b*-PS and rr-P3HT-*b*-PMMA copolymers using postpolymerization functionalization and atom transfer radical polymerization.

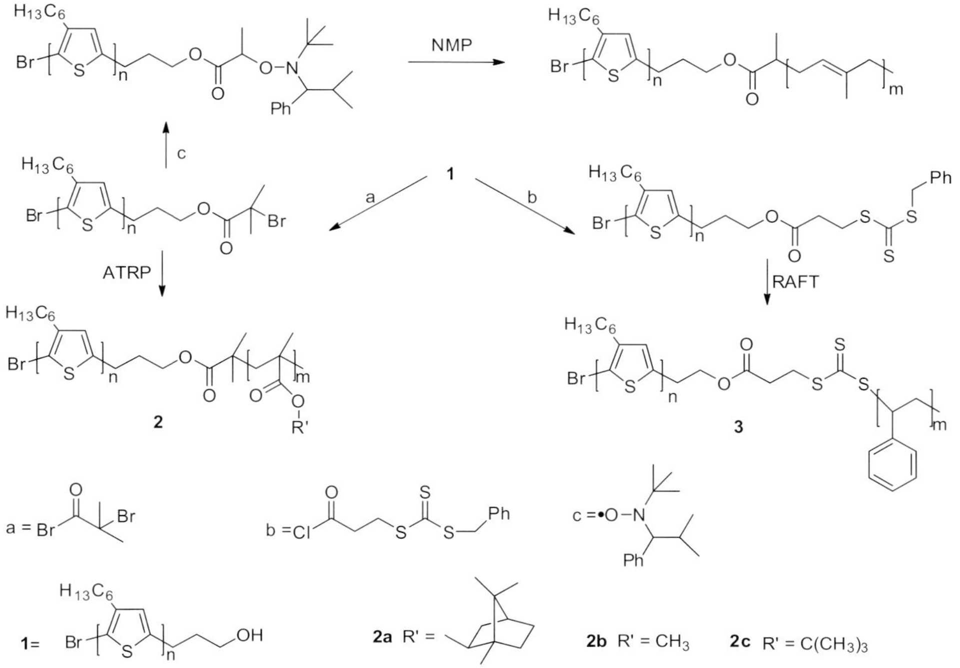

Figure 11.4 Synthesis of rr-P3HT diblock copolymers by using a combination of living GRIM and CRP techniques.

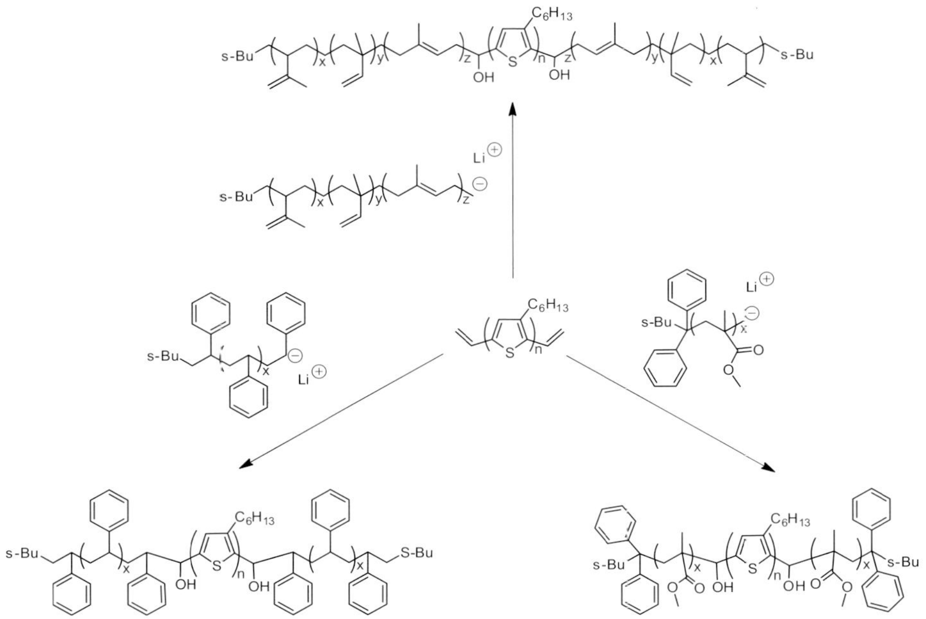

Figure 11.5 Synthesis of P3HT triblock copolymers through anionic coupling.

strategies for rod–coil block copolymers to increase the efficacy of semiconducting conjugated materials, leading to control over an array of different morphologies and modifiable electronic properties. The electronic, optical, and redox properties of oligo- and polythiophenes can be vastly changed upon addition of a flexible coil block. As an example, Chen and coworkers demonstrated interesting electroluminescent properties for poly(3-hexylthiophene)-b-poly(2-(dimethylamino)ethyl methacrylate) rod–coil block copolymers [95]. The P3HT-

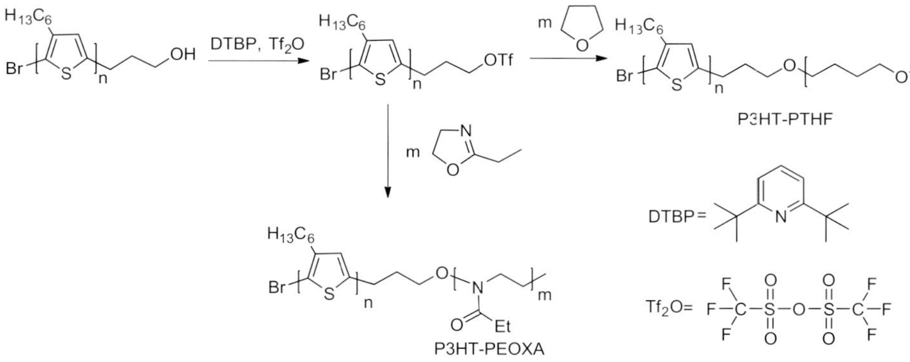

Figure 11.6 Synthesis of diblock copolymers containing P3HT by a combination of GRIM and cationic ring-opening polymerization.

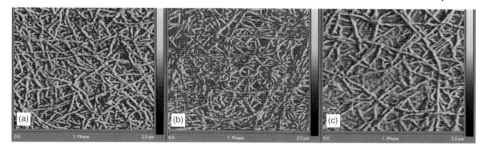

Figure 11.7 TMAFM phase images of poly(3-hexylthiophene)-*b*-poly(2-ethyl-2-oxazoline) (P3HT–PEOXA) diblock copolymers: (a) 5 mol% PEOXA; (b) 15 mol% PEOXA; (c) 30 mol% PEOXA. Reproduced with permission from Ref. [93]. Copyright 2010, Wiley-VCH Verlag GmbH.

b-PDMAEMA copolymers displayed thermal, pH-responsive, and solvatochromic properties. Upon heating and cooling, the micellar size of the block copolymers varied and changed significantly upon water intake in mixed solvent systems. The resultant emission color changes and shifts were evidenced in absorption spectra measurements (Figure 11.9). The intermolecular aggregation of the rod–coil block copolymers strongly depends on external stimuli, such as temperature, pH, and solvent conditions, making these materials promising for sensor applications.

In OPVs, rr-P3HTs are the most common donor (p-type) materials used in conjunction with fullerene derivatives for the bulk heterojunction (BHJ) layer. In this blended system, when light is absorbed by the rr-P3HT, the energy excites electron–hole pairs, which are then transferred to the interface between the rr-P3AT and the acceptor material. Separation of the electron–hole pair then occurs at the interface between the donor and the acceptor. Subsequently, the holes migrate through the donor P3HT material to the negative electrode while the electrons are carried through the acceptor fullerenes to the positive electrode, thus creating photovoltage, that is, power [96]. A major limitation of this methodology is the lack of control and reproducibility in the BHJ active layer that is commonly associated with

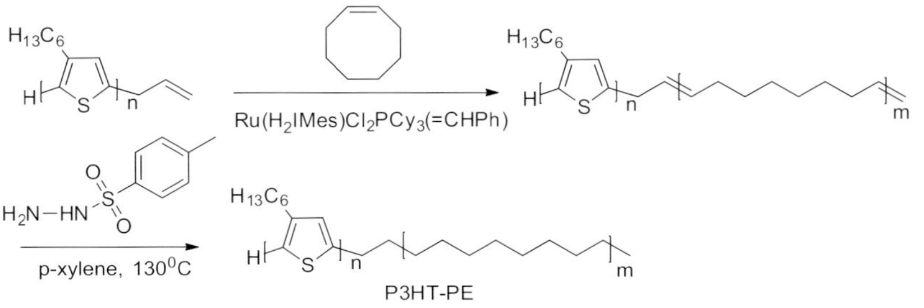

Figure 11.8 Synthesis of P3HT–PE diblock copolymer by a combination of GRIM, ROMP, and hydrogenation.

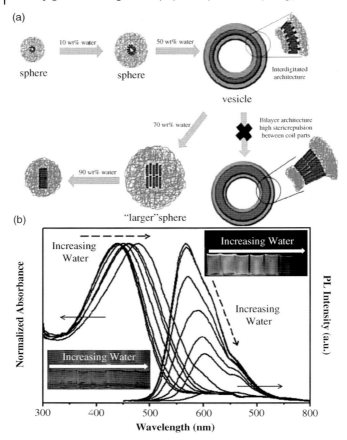

Figure 11.9 (a) Diagram depicting micelle formation upon water uptake in mixed THF/water solvent system. (b) Absorption and luminescent spectral changes with an inset depicting colorimetric variations. Reproduced with permission from Ref. [95]. Copyright 2011, Wiley-VCH Verlag GmbH.

standard blending techniques [97]. In addressing this issue, self-assembly from block copolymer templates can provide a route to well-organized nanostructures having high molecular level precision and minimal structural defects that were proven to be detrimental to organic electronic applications. Controlling the supramolecular self-assembly via molecular composition and functionality offers a unique strategy to provide targeted functional patterns in a reproducibly controlled fashion. More specifically, the transport of holes and electrons can be significantly enhanced if the donor and acceptor components are self-assembled in such a way that gives an optimal morphology.

A unique approach toward the optimization of nanostructured materials, given that the domains of the rod and coil can be engineered specifically to give the most

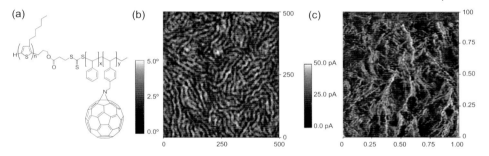

Figure 11.10 (a) Chemical structure of P3HT–PS–PS-*graft*-fullerene rod–coil polymer. (b) 500 nm × 500 nm AFM phase image of thin film of triblock copolymer spin coated from chlorobenzene. (c) C-AFM image depicting current of P3HT regions. Reproduced with permission from Ref. [98]. Copyright 2010, Wiley-VCH Verlag GmbH.

ideal pathway for charge transport and dissociation via the utilization of rod–coil-based polymers, has been realized. The use of semiconducting rod–coil-based block polymers for solar cells was demonstrated by Nyugen and coworkers, who developed a triblock copolymer bearing a polythiophene block, polystyrene spacer, and a shorter polystyrene segment with pendent fullerene groups labeled P3HT-*b*-P(S89BAz11)-C$_{60}$ (Figure 11.10a) [98]. The nonconjugated polystyrene spacer limits the detrimental recombination processes that may occur between the holes and electrons of the P3HT and fullerene derivatives, respectively, known to affect efficiency in electronic devices. Fiber-like structures were observed by atomic force microscopy for the block copolymer, with the interfibrillar domains consisting of the soft polystyrene segments (Figure 11.10a). Conductive AFM (C-AFM) techniques were used to probe local conductivity of the thin film and the high-current regions correspond to P3HT segments, while the darker regions to the softer polystyrene segments (Figure 11.10c). The morphology depicted in both the standard AFM and C-AFM images shows that nanoscale phase separation is achieved, thus making these materials suitable for the BHJ layer in organic solar cells.

The development of rr-P3HT block copolymers for OFETs has also been extensively studied as their self-assembling properties can offer many topologies that can be beneficial to device configurations. Homopolymers of rr-P3HTs typically have relatively high hole mobilities, usually in the range of 0.01–0.1 cm^2 V^{-1} s^{-1}, depending on regioregularity, molecular weight, and side and end group functionality [99–101]. The preparation of rod–coil block copolymers has been used to improve upon the mobility of P3HT thin films in field-effect transistors. Geohegan and coworkers reported a polystyrene-*b*-poly(3-hexylthiophene) (PS-*b*-P3HT) copolymer that exhibited hole mobilities up to two orders of magnitudes higher than those of P3HT homopolymers [102]. The thin-film mobilities for 85 wt% of P3HT in the copolymer were evaluated and determined to be 0.086 cm^2 V^{-1} s^{-1} compared to those of P3HT (0.046 cm^2 V^{-1} s^{-1}). In addition, Suave and McCullough demonstrated high field-effect mobilities of P3HT-*b*-PMMA block copolymers and the formation of nanofibrillar morphology due to the immiscibility of the

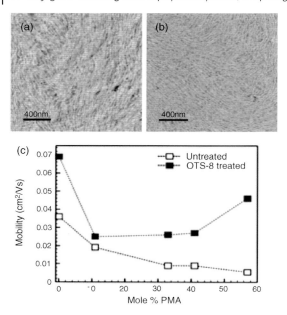

Figure 11.11 AFM phase images of rr-P3HT-b-PMMA containing 57 mol% PMA: (a) on untreated SiO$_2$; (b) on SiO$_2$ treated with OTS-8 devices. (c) Average mobility, measured in saturation mode, as a function of PMMA content of the diblock copolymers. Reprinted (adapted) with permission from Ref. [103]. Copyright 2007, Wiley-VCH Verlag GmbH.

rod and coil segments (Figure 11.11) [103]. The use of semiconducting rod–coil block copolymers to alter the nanostructure of conjugated rod segments has been widely successful and ongoing research in this area continues to be pursued.

11.3
Poly(*p*-phenylene vinylene) Block Copolymers

The development of other π-conjugated rod–coil block copolymer systems has been explored toward the optimization of semiconducting conjugated–insulating materials for optoelectronic applications. Among them, poly(*p*-phenylene vinylene)s (PPVs) are viewed as ideal systems due to their unique luminescent properties leading to exhaustive investigations into applications for OLEDs [104, 105], solar cells [106, 107], and chemical sensors [108]. Significant strides focusing on the synthesis of structurally defined PPVs have been made throughout the years and will continue to play an essential role toward the development of nanostructured materials. A brief overview of the progress of PPV development will be provided, followed by the design and property modifications in rod–coil copolymer systems.

The most common synthetic routes for traditional PPVs are the Gilch, Witting condensation, and Wessling methods (Figure 11.12) [2]. For multifunctional and

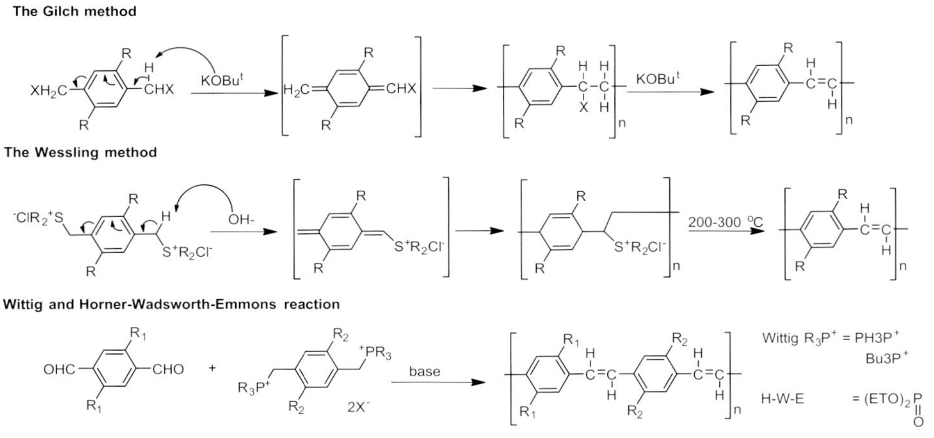

Figure 11.12 Common synthetic routes to traditional poly(p-phenylene vinylene)s [2].

more exotic PPV systems, the use of Heck, Stille, and Suzuki coupling reactions has been explored to couple olefin derivatives to aromatic halides. The Gilch route explores the polymerization of substituted benzenes in the presence of potassium *t*-butoxide base. The mechanism is believed to undergo a radical transformation that can sometimes lead to structural defects along the backbone due to possible head-to-head configurations. Solubility issues also occur during the polymerization and can affect the final polymeric product properties, prompting researchers to explore alternatives to PPV syntheses. Cao and coworkers demonstrated a novel way to limit the structural defects along the PPV backbone due to the incorporation of tolane-bis-benzyl groups by designing spatially adept monomers [109].

The Wessling method involves a water-soluble and processable polyelectrolyte precursor. The reaction proceeds via conversion of the sulfonium salt benzene dihalide derivative to the activated intermediate monomer, followed by chain growth polymerization via thermal treatment. This method is also believed to occur via radical chain growth and has played a special role in the development of PPVs containing nanoparticles. However, the aforementioned methods tend to be limited by the type of phenylene rings that can be incorporated into the backbone and the stereochemical nature of the vinylene bonds along the polymer chain. The Wittig and the Horner–Wadsworth–Emmons condensation reactions circumvent these issues by offering increased synthetic flexibility to include both *cis*- and *trans*-vinylene compounds and regioselective incorporation of the phenylene units. The reaction occurs via a step growth condensation between aromatic aldehydes and bisphosphonium ylide monomers. This technique offers the attractiveness of mild conditions and cheap accessible reagents but still only yields low-to-moderate molecular weights of the final product. The modified reaction, known as the Horner–Wadsworth–Emmons reaction, employs phosphonate esters, which gives better yields and has easily removable by-products.

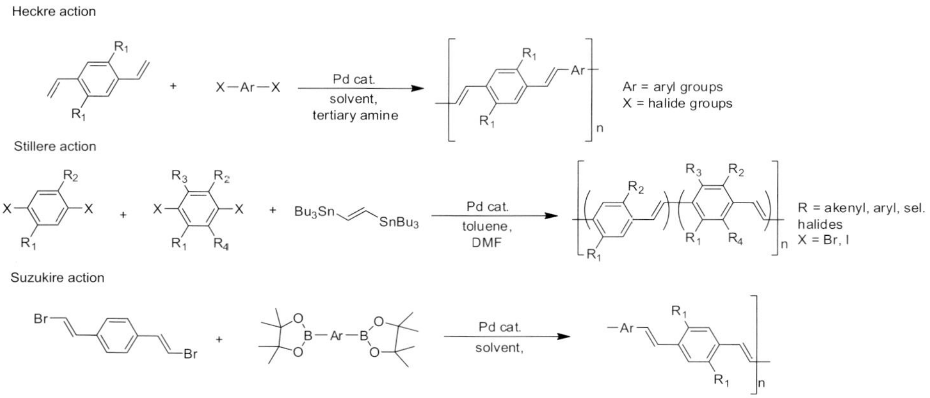

Figure 11.13 Selected examples of Heck, Stille, and Suzuki reactions [2, 109–112].

Further modifications toward the synthesis of substituted PPVs have been recently investigated and involve C–C coupling reactions in the presence of palladium (Pd) catalysts, that is, the Heck [110, 111], Stille [112], and Suzuki [113] reactions (Figure 11.13). These alternative methods rely on the formation of aryl–aryl bonds as opposed to the generation of olefinic double bonds discussed earlier in this section. In the Heck reaction, aryl halides are coupled with alkene derivatives in the presence of Pd catalysts forming the traditional aryl–vinylene bond and allow for modifiable functional PPVs, increasing their applicability in many optoelectronic devices. If greater regiospecific control is desired, the use of Stille or Suzuki coupling reactions for the generation of substituted PPVs has been investigated. In the Stille reaction, the aryl halide is coupled with alkenyl stannanes; in the Suzuki method, the reaction involves coupling of halide derivatives and boron-containing compounds. Despite the progress in the syntheses of substituted functional PPVs, optoelectronic applications require the design of highly nanostructured architectures that can optimize device performance through the use of copolymerization techniques.

The PPV building blocks for rod–coil polymers undergo similar approaches to that of polythiophene segments. A recent example of end group functionalization of OPVs with defined molecular weights was achieved by Yu and coworkers, using an orthogonal approach incorporating two types of reactions: the Horner–Wadsworth–Emmons and the Heck reaction to grow the copolymer chain [114]. Their approach eliminates the need for protective group chemistry, in turn allowing a much more facile route to controlled PPV growth. Diblock copolymers containing PPVs and coil segments are becoming increasingly complex and exotic in nature. For instance, the synthesis of a hybrid rod–coil copolymer consisting of an oligo(phenylene vinylene) (OPV) rigid block and a polystyryl-type flexible block with polyoxometalate (POM) clusters as side chain pendants was developed [115]. ATRP was used to synthesize the coil block bearing pendant phthalimide protective groups while the OPV block was synthesized using Siegrist polycondensation. Subsequent conversion of the terminal aldehyde group of the PPV block to the ethynyl end group was then coupled to an azide-terminated polystyryl coil block via click chemistry. The pendent phthalimide group was then converted to the aryl amine group, which then coordinated with hexamolybdate clusters to be used in photovoltaic applications.

In solar cell applications, PPV rod–coil block copolymers can significantly influence the BHJ layer, greatly modifying the devices' performance by affecting the morphology or size of the domains of the blended donor–acceptor system. For exciton separation, a parameter necessary for efficient photovoltaic devices, the interface between the donor and the acceptor should be within 5–20 nm and the use of block copolymer domains has been found to provide good correlations. One specific PPV system, poly(2-methoxy-5-(2′-ethylhexyloxy)-1,4-phenylene vinylene) (MEH–PPV) exhibits high solubility in a variety of solvents and has been used in a substantial amount of optoelectronic device-based research. The design of MEH–PPV covalently attached to a polystyrene coil block copolymer with pendent fullerenes was accomplished and investigated as the active layer in solar cells. Its unique self-assembly to form interesting architectures resulted in a quenching of luminescent properties, thus making them attractive for photovoltaic applications [116]. In addition, polylactide (PLA)–PPV rod–coil block copolymers have also been used to influence the morphology of thin films for OSCs [117]. Depending on the volume fractions of each block, the morphology can be tailored to have lamellae perpendicular or parallel to the substrate (Figure 11.14). Selective etching of the

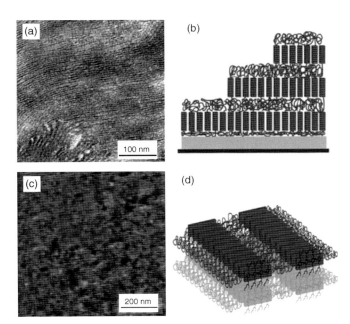

Figure 11.14 (a) TEM image of a thin film of PPV$_7$-b-PLA$_{72}$. (b) Diagram depicting lamellar structure parallel to structure. (c) AFM image of PPV$_{14}$-b-PLA$_{32}$. (d) Schematic of PPV rod alignment with parallel domains. Reproduced with permission from Ref. [117]. Copyright Wiley-VCH Verlag GmbH.

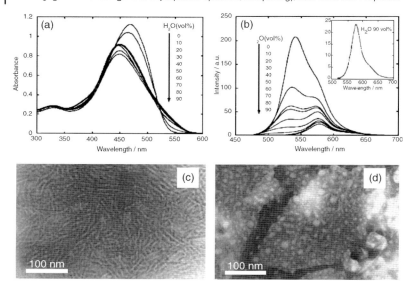

Figure 11.15 UV (a) and fluorescence (b) spectra of OPV-*b*-PEO with 62 wt% PEO with various H$_2$O concentrations. TEM images of OPV-*b*-PEO with 62 wt% PEO (c) and OPV-*b*-PEO with 79 wt% PEO (d). Reproduced with permission from Ref. [119]. Copyright Wiley-VCH Verlag GmbH.

PLA coil block followed by subsequent backfilling with fullerene derivatives for ordered donor/acceptor bulk heterojunctions can provide efficient pathways for transport in OSCs.

The remarkable self-assembling properties of rod–coil block copolymers were also shown in the synthesis of polyethylene glycol–PPV materials [118]. Yu and coworkers discovered the cylindrical micellar formation of PEG-*b*-PPV insulating–conjugated block polymers with different block lengths having a PPV core and a PEG shell. Under different solvent conditions, nanofibers can also be formed that have good crystallinity and optical and electronic properties. Tanaka and coworkers investigated the aggregation behavior of OPV-*b*-PEO polymers and concentration variations of the mixed solvent system were found to have a significant effect on the photophysical properties of the PPV block [119]. The absorption and emission changes were seen as the concentration of H$_2$O in THF was adjusted. Different morphologies such as cylindrical nanofibers versus circular micelles were formed depending on the weight fraction of each block (Figure 11.15). These materials were shown to act as a plasticizer in PPV homopolymer blends toward the design of OLEDs. As research in the areas of electrooptic devices progress, design of PPV-based materials will continue to garner interest due to its intriguing photoelectric properties.

11.4 Polyfluorenes

Polyfluorenes (PFs) are an interesting class of blue-emitting semiconducting conjugated materials that have been shown to have efficient charge transport properties and high chemical and thermal stability [120]. The fluorene monomer is made up of biphenyl units, connected by a bridged carbon atom at the C-9 position, and its appeal comes from the access to readily available monomers and its ability to be functionalized or polymerized at various positions. Some examples are listed in Figure 11.16 [2].

The synthetic strategies for substituted fluorenes involve the addition of alkyl groups (–R groups at the C-9 position), followed by bromination to give 2,7-dibromo-9,9′-dialkylfluorene monomer in high yields. The monomer for poly(3,6-fluorene) requires a more detailed route due to the fact that the 3-,6-positions cannot be directly brominated. Instead, phenanthrenequinone is used, which undergoes bromination followed by benzylic acid rearrangement and decarboxylation. Reduction gives the 3,6-dibromofluorene, which can be converted to the active monomer using standard alkylation reactions [121]. The more exotic fluorene monomers are synthesized according to the exact substitution desired. For example, for the spirofluorene monomer, the Grignard of 2-brombiphenyl is reacted with dibromofluorenone, followed by acid treatment to give 2,7-dibromo-9,9′-spirofluorene. The accessibility of these monomers has led extensive research toward the development of novel semiconducting conjugated materials. A number of synthetic methods have been utilized for the preparation of polyfluorenes. We will briefly explore some of the more common techniques, particularly focusing on end group substitution that will then lead to the formation of rod–coil block copolymers.

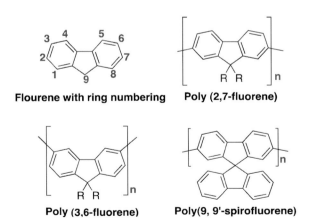

Figure 11.16 Skeleton fluorene monomer unit and selected chemical structures of polyfluorenes.

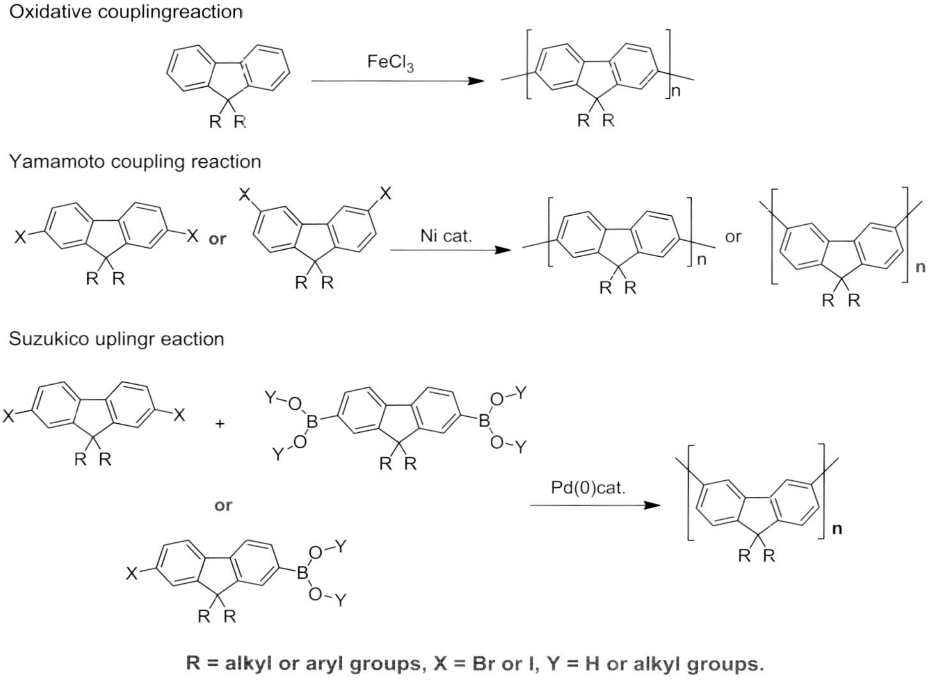

Oxidative coupling reaction

Yamamoto coupling reaction

Suzuki coupling reaction

R = alkyl or aryl groups, X = Br or I, Y = H or alkyl groups.

Figure 11.17 Examples of synthetic methods for polyfluorenes.

Typically, the polymerizations take place via carbon–carbon cross-coupling reactions, through oxidative or reductive couplings, and through transition metal-mediated condensation reactions (Figure 11.17) [122]. Oxidative coupling of fluorenes was one of the first synthetic methods and employed iron chloride as the catalyst [123]. Currently, this method is less employed due to the large number of structural defects and low molecular weight of the resultant polymers. The more common methods rely on transition metal-mediated cross-coupling and polycondensation reactions. The Yamamoto reductive coupling reaction involves the use of dihalide fluorene monomers in the presence of Ni catalysts to yield polymers with high molecular weights and low content of structural defects. The purity of the resultant polyfluorene polymers depends on the nature of the Ni catalyst and the use of more inexpensive Ni catalysts typically results in a decrease in the polymers' properties. This route is limited to the generation of homopolymers and statistical copolymers, thus making the alternative of transition metal-mediated condensation reactions preferable. For example, the Suzuki coupling reaction involves the use of substituted fluorene derivatives and precursors bearing halides and boronic acid moieties, which can be used to synthesize both homo- and copolymers with other conjugated or nonconjugated units. The Suzuki method allows for both structural modification of backbone and the introduction of other conjugated or nonconjugated blocks, giving rise to a variety of fluorene-based materials with interesting properties. Recently, the addition of boron dipyrromethene (bodipy) derivatives into the backbone of polyfluorene derivatives gave significant changes in emission wavelength as compared to homopolymers of polyfluorene [124]. The extension of the π-conjugation in the backbone caused an increase in the fluorescence and showed selectivity toward fluoride and cyanide anions.

Despite the advantageous synthetic strategies that can lead to an abundance of structurally modified polyfluorene derivatives, copolymerization with coil-like flexible segments opens up new and interesting pathways to materials with unusual morphologies and unique photophysical properties. This strategy promotes microphase separation in nanostructured materials that is typically a prerequisite in many efficient optoelectronic processes. As such, PF-containing rod–coil block copolymers have been reported and investigated for a variety of applications [125–128]. Similar to previous synthetic methods of semiconducting conjugated rod–coil block copolymers, polyfluorenes also involve end group functionalization of the PF segment, followed by subsequent polymerizations and/or coupling reactions. The most common method to generate the insulating coil block is ATRP. Chen and coworkers utilized Suzuki and ATRP polymerizations to synthesize polyfluorene-*b*-poly(stearyl acrylate) block copolymers with variable coil block lengths [129]. The crystalline coil segment was used to influence the luminescence and porosity of thin films prepared from the block copolymers. Highly ordered microporous films were formed via a breath figure process upon which the size of the pores was significantly influenced by the length of the coil block, humidity, and solution concentrations (Figure 11.18). Furthermore, removal of the upper surface of the film resulted in superhydrophobic surfaces having contact angles of $163 \pm 0.3°$, compared to the contact angles of PF and PSA, which are $95°$ and $115°$, respectively. The use of conjugated polyfluorene rod–coil block copolymers enables the development of highly optimal nanostructured materials toward a variety of broad prospect applications.

Hydrophilic coil segments of poly(*N*-isopropylacrylamide)s (PNIPAMs) were linked to polyfluorene macroinitiator rod blocks via atom transfer radical polymerization techniques (Figure 11.19a) [130]. PNIPAMs are known to be stimuli-responsive materials that undergo physical changes due to environmental stresses such as pH and temperature. Interestingly, when covalently linked to polyfluorenes, the polymeric micellar structures and fluorescence properties of the rod–coil block copolymer were significantly influenced by the solvent compositions. In water, the polymers formed micelles with the poly(*N*-isopropylacrylamide) as the corona but in mixed THF/toluene solvent systems, the reversal occurs where micelles are formed with the polyfluorene segment at the corona and the coil segment as the core. This unique ability can be used to incorporate small water-insoluble molecules via nanoencapsulation within the micelle, thus making these materials useful as carrier agents for delivery applications. The utility of the PF-*b*-PNIPAM rod–coil block copolymers was evident through successful encapsulation of the porphyrin derivatives within the core of the polymer micelles in aqueous-based solutions,

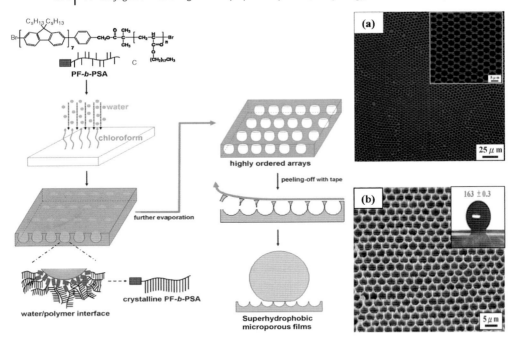

Figure 11.18 Diagram depicting the formation of ordered microporous films and delamination of upper surface layers resulting in rod-*co*-valley-like structures. (a) Microscopy image of ordered microporous films and (b) SEM of cross section after delamination of the upper layer. *Inset*: Picture of water contact angle, depicting superhydrophobicity. Reprinted (adapted) with permission from Ref. [129]. Copyright 2011, Royal Society of Chemistry.

while still maintaining fluorescence resonance energy transfer efficiency (FRET) that aids in singlet oxygen generation (Figure 11.19b).

Triblock copolymers are of increasing importance due to the capability to expand the dimension of complex morphologies available for diblock copolymers. In essence, a larger pool of interaction parameters is accessible and a range of nanostructured microdomains can be envisioned. Liu and coworkers explored the capability of coil–rod–coil block copolymers in sensor applications [131]. Self-assembled micellar nanoparticles of poly(methyl methacrylate)-*co*-4-(2-acryloyloxyethylamino)-7-nitro-2,1,3-benzooxadiazole-*block*-polyfluorene-*block*-poly(methyl methacrylate)-*co*-4-(2-acryloyloxyethylamino)-7-nitro-2,1,3-benzooxadiazole (P(MMA-*co*-NBDAE)-*b*-PF-*b*-P(MMA-*co*-NBDAE)) were used for ratiometric fluorescence detection of anions (Figure 11.20). The formation of micelles having PF cores with NBDAE-labeled PMMA coronas is observed in acetone, and in the absence of fluoride anions, the FRET processes between the PF block and the NBDAE moiety occur, resulting in strong emissions. When fluoride ions are detected, there is an 8.75-fold decrease in emission intensity and visual color changes in the solution. The micellar triblock polymers are highly sensitive to fluoride detection, resulting in

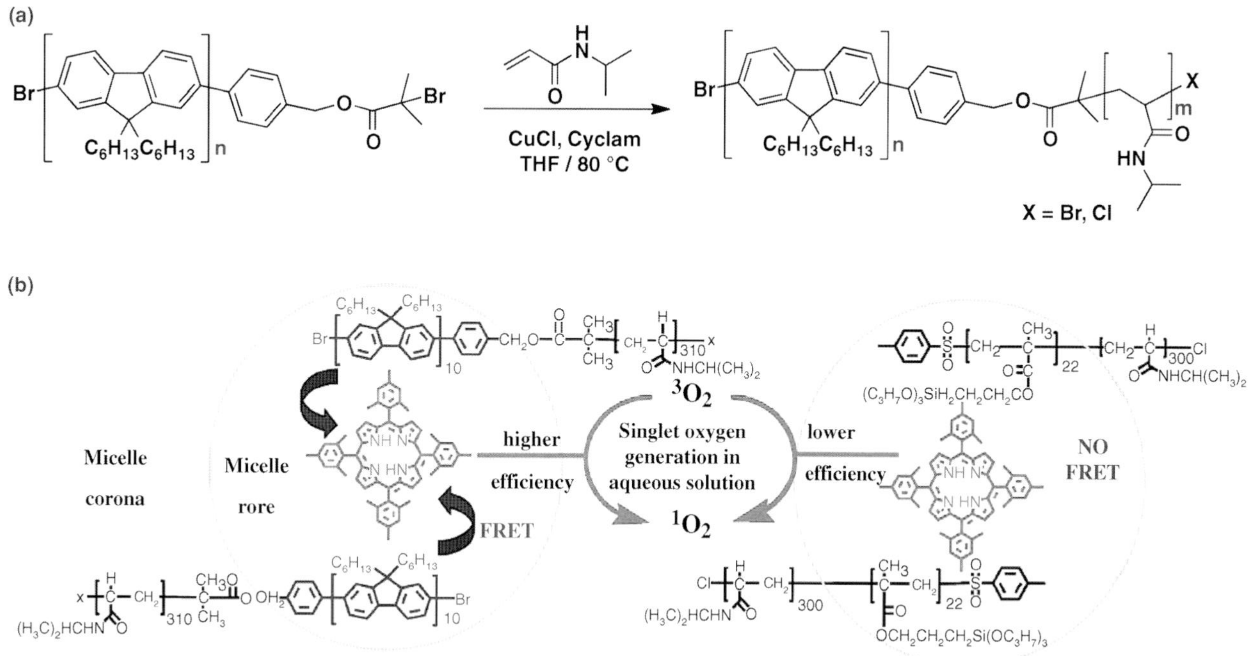

Figure 11.19 (a) Synthetic scheme for polyfluorene-*b*-poly(*N*-isopropylacrylamide) rod–coil block copolymers. (b) Schematic representation of incorporation of porphyrin derivatives in PF-*b*-PNIPA micellar core in aqueous solutions and enhancement of singlet oxygen generation via FRET. Reprinted (adapted) with permission from Ref. [130]. Copyright 2009, American Chemical Society.

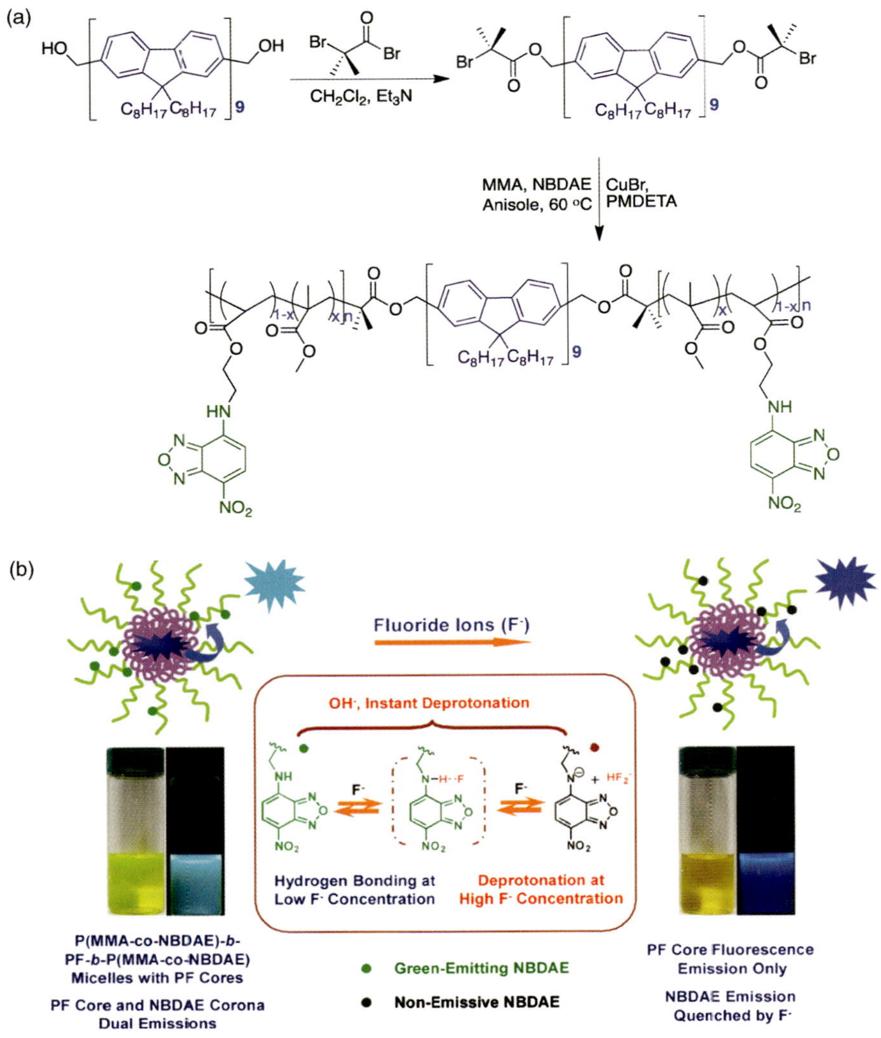

Figure 11.20 (a) Synthetic scheme for P(MMA-co-NBDAE)-b-PF-b-P(MMA-co-NBDAE) triblock copolymers. (b) Diagram depicting fluoride anion binding to NDBAE moiety in micelles and color and emission changes upon bonding and deprotonation in acetone solutions. Reproduced with permission from Ref. [131]. Copyright 2009, American Chemical Society.

visual changes occurring with as low as 0.09 ppm concentrations. The optimization of spatial distribution between the PF donor and the NBDAE acceptor created through the formation of the triblock copolymer provides high-throughput screening for fluoride ion sensing.

11.5
Other Semiconducting Rod–Coil Systems

The above sections provided a brief overview of how the utility of conjugated–insulating rod–coil polymers has been used to advance the field of conjugated materials by exploiting the interesting and unique arrays of supramolecular assemblies and morphologies. Some of the more common semiconducting conjugated materials have been discussed in relation to their supramolecular assembly and properties when covalently linked to coil-like insulating polymers, yet the large selection of semiconducting polymeric materials offers an infinite supply of building blocks that can be utilized in many applications. As such, other nontraditional semiconducting rod–coil systems have been reported and their unique electronic properties investigated. For example, Jenekhe and Chen synthesized a poly(phenylquinoline)-b-polystyrene diblock copolymer (Figure 11.21) upon which spherical, vesicular, cylindrical, and lamellar morphologies were observed upon solvent changes [132, 133]. The hollow spherical nature of the block copolymer micelles was shown to provide an ideal environment for the encapsulation of fullerene-based molecules and can thus be used in large-scale purification and extraction applications of fullerene derivatives. The ability of block copolymers to mimic self-assembled structures found in natural and living systems has been shown through the synthesis of poly(γ-benzyl-L-glutamate)-b-poly[2-(dimethylamino)ethyl methacrylate] block copolymers (Figure 11.21) [134]. This polymer combines the properties of both the rod and coil blocks, while also having external stimuli-responsive attributes, typically seen with amphiphilic materials. The PDMAEMA-b-PBLG diblock copolymers transition from vesicles to micelles upon thermal and pH changes. Also, the addition of ultrasmall superparamagnetic iron oxide

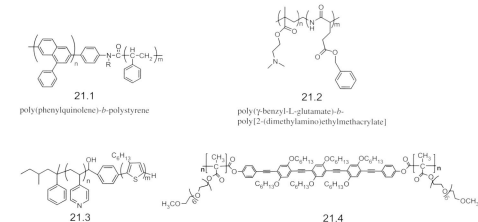

21.1 poly(phenylquinolene)-b-polystyrene

21.2 poly(γ-benzyl-L-glutamate)-b-poly[2-(dimethylamino)ethylmethacrylate]

21.3 poly(3-hexylthiophene)-b-poly(4-vinylpyridine)

21.4 poly(glycolmethacrylate)-b-oligo(phenyleneethynylene)

Figure 11.21 Examples of other conjugated–insulating systems.

nanoparticles in a block copolymer solution gave controlled and complex hybrid nanostructures via self-assembly and depletion effects, which can find applications in biolabeling and drug delivery. Mezzenga and coworkers designed a poly(3-hexylthiophene)-*b*-poly(4-vinylpyridine) (Figure 11.21) block copolymer, in which the PVP coil block had weak supramolecular interactions with the acceptor component, PCBM, in the bulk heterojunction layer of solar cell devices [135]. The PCBM would undergo preferential ordering within the coil block, providing a directed self-assembly route toward continuous pathways necessary for efficient transport in the bulk heterojunction of solar cells.

The use of 2D materials is now regarded as a way to increase the performance of semiconducting materials in optoelectronic applications. For instance, graphene has been recognized as a promising building block for nanoelectronic and spintronic devices due to its excellent physical properties such as high surface area and excellent conductivity [136–138]. As such, the functionalization of graphene is one of the leading research thrusts to improve its processability and solubility in devices requiring functional supramolecular architectures [139–142]. Zhang and coworkers used an amphiphilic rod–coil conjugated triblock copolymer to functionalize graphene oxide, thus forming interesting supramolecular nanostructures that in turn may increase its processability and biocompatibility in organic electronic devices and biological systems. A PEGylated polyphenylene ethynylene (PEG–OPE) was synthesized via ATRP with the conjugated lipophilic OPE oligomer used as the linker with the graphene by strong π–π interactions and the PEGylated chains to improve solubility (Figure 11.21). Due to the functionalization of the graphene with PEG–OPE, the solubility was greatly increased in both polar and nonpolar solvents, thus improving the applicability of 2D graphene-based materials.

11.6
Conjugated–Insulating Rod–Rod Block Copolymers

Conjugated–insulating rod–rod block copolymers are much less explored as compared to the rod–coil block copolymers. There is a potential to generate new nanostructures and optoelectronic properties by combining two rod segments. However, the inherent rigid nature of both semiconducting and insulating blocks can result in reduced solubility of the rod–rod block copolymer. Jenekhe reported the synthesis of a rod–rod triblock copolymer containing conjugated poly(9,9-dihexylfluorene-2,7-diyl) and liquid crystalline insulating poly(γ-benzyl-L-glutamate) as shown in Figure 11.22. This rod–rod triblock copolymer has been shown to form different nanostructured assemblies as a function of the copolymer composition and the secondary structure of the polypeptide block [143, 144].

The synthesis of rod–rod diblock copolymers containing rod-like conjugated P3HT and liquid crystalline insulating poly(γ-benzyl-L-glutamate) (P3HT–PBLG) was reported by Stefan's group. The H/Br-terminated P3HT was reacted with *N*-(*p*-bromobenzyl)phthalimide, followed by the deprotection of the amine group to generate the benzylamine-terminated P3HT, which was used as a macroinitiator for

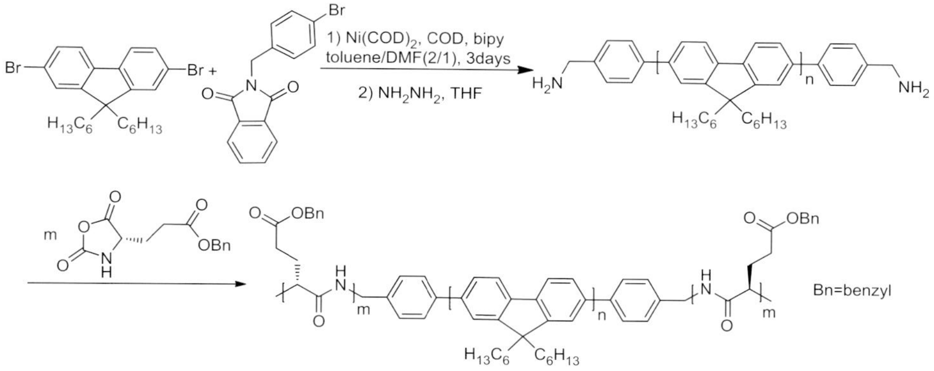

Figure 11.22 Synthesis of a rod–rod triblock copolymer containing conjugated poly(9,9-dihexylfluorene-2,7-diyl) and liquid crystalline insulating poly(γ-benzyl-L-glutamate).

the ring-opening polymerization of γ-benzyl-L-glutamate N-carboxyanhydride (Figure 11.23) [145]. The morphology of thin films of P3HT–PBLG diblock copolymer was found to depend on the casting solvents and annealing conditions (Figure 11.24) [145]. Field-effect mobilities of P3HT precursor and P3HT–PBLG diblock copolymer were measured in bottom-gate bottom-contact thin-film transistors. The P3HT precursor had a field-effect mobility of $\mu = 6.6 \times 10^{-3}$ cm^2 V^{-1} s^{-1},

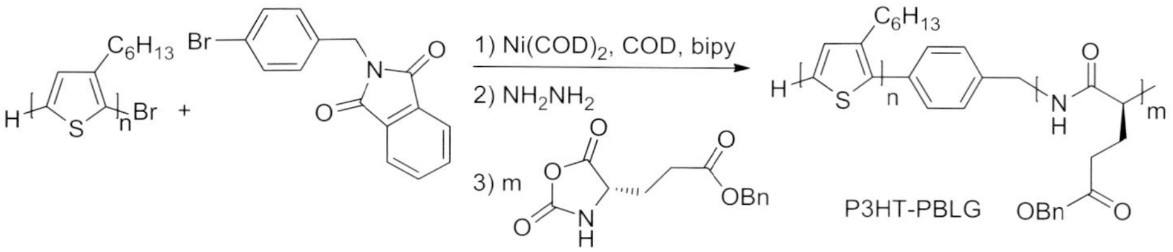

Figure 11.23 Synthesis of a rod–rod diblock copolymer containing conjugated P3HT and liquid crystalline PBLG by a combination of GRIM and ring-opening polymerization.

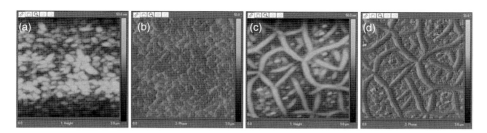

Figure 11.24 TMAFM images of poly(3-hexylthiophene)-b-poly(γ-benzyl-L-glutamate) films: (a and b) height and phase images of copolymer film deposited from chloroform; (c and d) height and phase images of copolymer deposited from trichlorobenzene. Reproduced with permission from Ref. [145]. Copyright 2011, Wiley-VCH Verlag GmbH.

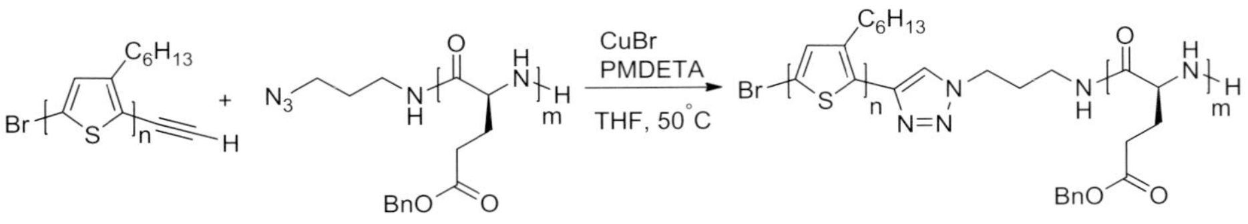

Figure 11.25 Synthesis of a rod–rod diblock copolymer containing conjugated P3HT and liquid crystalline PBLG by a combination of GRIM, copper-catalyzed [3 + 2]-cycloaddition, and ring-opening polymerization.

while the P3HT–PBLG diblock copolymer displayed a mobility of $\mu = 7.0 \times 10^{-4}$ cm^2 V^{-1} s^{-1}. The P3HT–PBLG rod–rod diblock copolymer displayed solvatochromism in THF/water mixtures.

Bielawski also reported the synthesis of P3HT-*b*-PBLG by reacting the ethynyl-terminated P3HT synthesized by GRIM with the azido-terminated PBLG via a copper-catalyzed [3 + 2]-cycloaddition to generate P3HT–PBLG diblock copolymer (Figure 11.25) [146].

11.7
Conclusions and Outlook

The above sections provided a brief and concise overview of a few main conjugated polymer systems and their role in semiconducting block copolymer materials. To date, there are many other conjugated materials that are currently being pursued and the careful design of the molecular structures can lead to an array of functional nanostructured materials. The synthetic strategies, morphological control, and processing tools are still aspects in block copolymer templates that continue to be optimized toward the use of semiconducting conjugated–insulating materials for organic electronics. Since the self-assembly of these materials depends on the nature of the monomers, crystallization of the conjugated rod moiety, the stiffness asymmetry of both blocks, and the volume fraction of the conjugated and insulating segments, it is highly likely that research in this area will continue to flourish, particularly in semiconductor devices.

References

1 Sirringhaus, H., Bird, M., Richards, T., and Zhao, N. (2010) Charge transport physics of conjugated polymer field-effect transistors. *Adv. Mater.*, **22**, 3893–3898.
2 Pang, Y. (ed.) (2010) *Design and Synthesis of Conjugated Polymers*, Wiley-VCH Verlag GmbH, Weinheim.
3 Malachoski, M.J. and Żmija, J. (2010) Organic field-effect transistors. *Opto-Electron. Rev.*, **18** (2), 121–136.
4 de Boer, B. and Facchetti, A. (2008) Semiconducting polymeric materials. *Polym. Rev.*, **48** (3), 423–431.
5 Usta, H., Faccheti, A., and Marks, T.J. (2011) High-performance n-channel

semiconductors. Materials design for efficient electron transport in OFETs. *Acc. Chem. Res.*, **44**, 501–510.

6 Horowitz, G. (1998) Organic field effect transistors. *Adv. Mater.*, **10**, 365–377.

7 Guo, Y., Yu, G., and Liu, Y. (2010) Functional organic field-effect transistors. *Adv. Materr*, **22**, 4427–4447.

8 Fachetti, A. (2011) Conjugated polymers for organic electronics and photovoltaic cell applications. *Chem. Mater.*, **23**, 733–758.

9 Pron, A. and Rannou, P. (2002) Processable conjugated polymers: from organic semiconductors to organic metals and superconductors. *Prog. Polym. Sci.*, **27**, 135–190.

10 Skotheim, T.A. and Reynolds, J.R. (eds) (2007) *Conjugated Polymers*, Taylor & Francis Group, LLC, Boca Raton, FL.

11 Park, Y. and Advincula, R.C. (2011) Hybrid semiconductor nanoparticles: π-conjugated ligands and nanostructured films. *Chem. Mater.*, **23**, 4273–4294.

12 Li, C. and Bo, Z. (2010) Three-dimensional conjugated macromolecules as light-emitting materials. *Polymer*, **51**, 4273–4294.

13 McNeill, C.R. and Greenham, N.C. (2009) Conjugated-polymer blends for optoelectronics. *Adv. Mater.*, **21**, 3840–3850.

14 Laquai, F., Park, Y.-S., Kim, J.-J., and Basche, T. (2009) Excitation energy transfer in organic materials: from fundamentals to optoelectronic devices. *Macromol. Rapid Commun.*, **30**, 1203–1231.

15 So, F., Krummacher, B., Mathai, M.K., Poplavskyy, D., Choulis, S.A., and Choong, V.-N. (2007) Recent progress in solution processable organic light emitting devices. *J. Appl. Phys.*, **102**, 091101-1–091101-21.

16 Roncali, J. (2009) Molecular bulk heterojunctions: an emerging approach to organic solar cells. *Acc. Chem. Res.*, **42**, 1719–1730.

17 Brabec, C., Dyakonov, V., and Scherf, U. (eds) (2008) *Organic Photovoltaics*, Wiley-VCH Verlag GmbH, Weinheim.

18 Weickert, J., Dunbar, R.B., Hesse, H.C., Wiedemann, W., and Schmidt-Mende, L. (2011) Nanostructured organic and hybrid solar cells. *Adv. Mater.*, **23**, 1810–1828.

19 Brabec, C.J., Heeney, M., McCulloch, I., and Nelson, J. (2011) Influence of blend microstructure on bulk heterojunction organic photovoltaic performance. *Chem. Soc. Rev.*, **40**, 1185–1199.

20 O'Neill, M. and Kelly, S.M. (2011) Ordered materials for organic electronics and photonics. *Adv. Mater.*, **23**, 566–584.

21 Zhan, X. and Zhu, D. (2010) Conjugated polymers for high-efficiency organic photovoltaics. *Polym. Chem.*, **1**, 409–419.

22 Günes, S., Neugebauer, H., and Sariciftci, N.S. (2007) Conjugated polymer-based organic solar cells. *Chem. Rev.*, **107**, 1324–1338.

23 Giridharagopal, R. and Ginger, D.S. (2010) Characterizing morphology in bulk heterojunction organic photovoltaic systems. *J. Phys. Chem. Lett.*, **1** (7), 1160–1169.

24 Thomas, S.W., III, Joly, G.D., and Swager, T.M. (2007) Chemical sensors based on amplifying conjugated polymers. *Chem. Rev.*, **107**, 1339–1386.

25 McQuade, D.T., Pullen, A.E., and Swager, T.M. (2000) Conjugated polymer-based sensory materials. *Chem. Rev.*, **100**, 2357–2574.

26 Swager, T.M. (1998) The molecular wire approach to sensory signal amplification. *Acc. Chem. Res.*, **31**, 201–207.

27 Loewe, R.S., Khersonsky, S.M., and McCullough, R.D. (1999) A simple method to prepare head-to-tail coupled, regioregular poly(3-alkylthiophenes) using Grignard metathesis. *Adv. Mater.*, **11**, 250–253.

28 Loewe, R.S., Ewbank, P.C., Liu, J., Zhai, L., and McCullough, R.D. (2001) Regioregular, head-to-tail coupled poly(3-alkylthiophenes) made easy by the GRIM method: investigation of the reaction and the origin of regioselectivity. *Macromolecules*, **34**, 4324–4333.

29 Kim, H.-C., Park, S.-M., and Hinsberg, W.D. (2010) Block copolymer based nanostructures: materials, processes, and applications to electronics. *Chem. Rev.*, **110**, 146–177.

30 Segalman, R.A., McCulloch, B., Kirmayer, S., and Urban, J.J. (2009) Block

copolymers for organic optoelectronics. *Macromolecules*, **42**, 9205–9216.

31 Liang, Y., Wang, H., Yuan, S., Lee, Y., Gan, L., and Yu, L. (2007) Conjugated block copolymers and co-oligomers: from supramolecular assembly to molecular electronics. *J. Mater. Chem.*, **17**, 2183–2194.

32 Liang, Y. and Yu, L. (2011) Conjugated block copolymers and cooligomers, in *Self-Organized Organic Semiconductors: From Materials to Device Applications*, 1st edn (ed. Q. Li), John Wiley & Sons, Ltd, Chichester, pp. 21–38.

33 Liu, C.-L., Lin, C.-H., Kuo, C.-C., Lin, S.-T., and Chen, W.-C. (2011) Conjugated rod–coil block copolymers: synthesis, morphology, photophysical properties, and stimuli-responsive applications. *Prog. Polym. Sci.*, **36**, 603–637.

34 Kuo, C.-C., Liu, C.-L., and Chen, W.-C. (2011) Morphologies and photophysical properties of conjugated rod–coil block copolymers, in *Complex Macromolecular Architecture: Synthesis, Characterization, and Self-Assembly*, 1st edn (eds A. Hirao, Y. Tezuka, and F. Du Prez), John Wiley & Sons, Ltd, Chichester, pp. 593–622.

35 de Cuendias, A., Hoirns, R.C., Cloutet, E., Vignau, L., and Cramail, H. (2010) Conjugated rod–coil block copolymers and optoelectronic applications. *Polym. Int.*, **59**, 1452–1476.

36 Olsen, B.D. and Degalman, R.A. (2008) Self-assembly of rod–coil block copolymers. *Mater. Sci. Eng. Rep. R*, **62**, 37–66.

37 Hayward, R.C. and Pochan, D.J. (2010) Tailored assemblies of block copolymers in solution: it is all about the process. *Macromolecules*, **43**, 3577–3584.

38 Lee, M., Cho, B.K., and Zin, W.C. (2001) Supramolecular structures from rod–coil block copolymers. *Chem. Rev.*, **101**, 3869–3892.

39 Radano, C.P., Scherman, O.A., Stingelin-Stutzmann, N., Müller, C., Breiby, D.W., Smith, P., Janssen, R.A.J., and Meijer, E.W. (2005) Crystalline–crystalline block copolymers of regioregular poly(3-hexylthiophene) and polyethylene by ring-opening metathesis polymerization. *J. Am. Chem. Soc.*, **127**, 12502–12503.

40 Müller, C., Goffri, S., Breiby, D.W., Andreasen, J.W., Chanzy, H.D., Janssen, R.A.J., Nielsen, M.M., Radano, C., Sirringhaus, H., Smith, P., and Stingelin-Stutzmann, N. (2007) Tough, semiconducting polyethylene–poly(3-hexylthiophene) diblock copolymers. *Adv. Funct. Mater.*, **17**, 2674–2679.

41 Mishra, A., Ma, C.-Q., and Bäuerle, P. (2009) Functional oligothiophenes: molecular design for multidimensional nanoarchitectures and their applications. *Chem. Rev.*, **109**, 1141.

42 Kumar, A., Welsh, D.M., Morvant, M.C., Piroux, F., Abboud, K.A., and Reynolds, J.R. (1998) Conducting poly(3,4-alkylenedioxythiophene) derivatives as fast electrochromics with high-contrast ratios. *Chem. Mater.*, **10**, 896–902.

43 Fichou, D. (ed.) (1999) *Handbook of Oligo- and Polythiophenes*, Wiley-VCH Verlag GmbH, Weinheim.

44 Hotta, S. (1997) Molecular conductive materials: polythiophenes and oligothiophenes, in *Handbook of Organic Conductive Molecules and Polymers*, 1st edn (ed. H.S. Nalwa), John Wiley & Sons, Ltd, Chichester, pp. 309–387.

45 Mitschke, U. and Bäuerle, P. (2000) The electroluminescence of organic materials. *J. Mater. Chem.*, **10**, 1471–1507.

46 Garnier, F., Hajlaoui, R., Yassar, A., and Srivastasa, P. (1994) All-polymer field-effect transistor realized by printing techniques. *Science*, **265**, 1684–1686.

47 Holliday, B.J. and Swager, T.M. (2005) Highly conductive metallopolymers: the role of redox matching. *Chem. Commun.*, 23–36.

48 Bäuerle, P. (1998) Sulfur-containing oligomers, in *Electronic Materials: The Oligomer Approach* (eds K. Müllen and G. Wegner), Wiley-VCH Verlag GmbH, Weinheim, pp. 105–197.

49 Perepichka, I.F. and Perepichka, D.F. (eds) (2009) *Handbook of Thiophene-Based Materials: Applications in Organic Electronics and Photonics*, John Wiley & Sons, Ltd, Chichester.

50 Kauffman, T. (1974) Oxidative coupling of organothiophenes. *Angew. Chem.*, **86**, 321–354.

51 Johno, K., Tanaka, Y., Koike, T., and Akita, M. (2011) Making organometallic molecular wires via oxidative ligand coupling. *Dalton Trans.*, **40**, 8089–8091.

52 Henssler, J.T., Zhang, X., and Matzger, A.J. (2009) Thiophene/thieno[3,2-*b*]thiophene co-oligomers: fused-ring analogs of sexithiophene. *J. Org. Chem.*, **74**, 9112–9119.

53 Tanaka, S., Tamba, S., Tanaka, D., Sugie, A., and Mori, A. (2011) Synthesis of well-defined head-to-tail-type oligothiophenes by regioselective deprotonation of 3-substituted thiophenes and nickel-catalyzed cross-coupling reaction. *J. Am. Chem. Soc.*, **133** (42), 16734–16737.

54 Wynberg, H., Logothetis, A., and VerPloeg, D. (1957) The synthesis of di- and terthienyls. *J. Am. Chem. Soc.*, **79**, 1972–1975.

55 Kagan, J. and Arora, S.K. (1983) 2,5-Di(2′-thienyl)furan and an improved synthesis of alpha-terthienyl. *Heterocycles*, **20**, 1941–1943.

56 Fanat, P.E. (1974) Ullmann synthesis of biaryls. *Synthesis*, 9–21.

57 Gronowitz, S. and Peters, D. (1990) Convenient synthesis of various terheterocyclic compounds by Pd(0)-catalyzed coupling reactions. *Heterocycles*, **30**, 645–658.

58 Crisp, G.T. (1989) Palladium mediated formation of bithiophenes. *Synth. Commun.*, **19**, 307–316.

59 Carpita, A., Rossi, R., and Veracini, C.A. (1985) Synthesis and ^{13}C NMR characterization of some π-excessive heteropolyaromatic compounds. *Tetrahedron*, **41**, 1919–1929.

60 Yamamoto, T., Sanechika, K., and Yamamoto, A. (1980) Preparation of thermostable and electric-conducting poly(2,5-thienylene). *J. Polym. Sci., Polym. Lett. Ed.*, **18**, 9–12.

61 Lin, J.W.P. and Dudek, L.P. (1980) Synthesis and properties of poly(2,5-thienylene). *J. Polym. Sci., Polym. Chem. Ed.*, **18**, 2869–2873.

62 Miller, G.G. and Elsenbaumer, R.L. (1986) Highly conducting, soluble, and environmentally-stable poly(3-alkylthiophenes). *J. Chem. Soc., Chem. Commun.*, **15**, 1346–1347.

63 Jen, K.Y., Oboodi, R., and Elsenbaumer, R.L. (1985) Processable and environmentally stable conducting polymers. *Polym. Mater. Sci. Eng.*, **53**, 79–83.

64 Elsenbaumer, R.L., Jen, K.-Y., and Oboodi, R. (1986) Processable and environmentally stable conducting polymers. *Synth. Met.*, **15**, 169–174.

65 McCullough, R.D. and Lowe, R.D. (1992) Enhanced electrical conductivity in regioselectively synthesized poly(3-alkylthiophenes). *J. Chem. Soc., Chem. Commun.*, 70–72.

66 Chen, T.A. and Rieke, R.D. (1992) The first regioregular head-to-tail poly(3-hexylthiophene-2,5-diyl) and a regiorandom isopolymer: nickel versus palladium catalysis of 2(5)-bromo-5(2)-(bromozincio)-3-hexylthiophene polymerization. *J. Am. Chem. Soc.*, **114**, 10087–10088.

67 Chen, T.A. and Rieke, R.D. (1993) Polyalkylthiophenes with the smallest bandgap and the highest intrinsic conductivity. *Synth. Met.*, **60**, 175–177.

68 Chen, T.A., O'Brien, R.A., and Rieke, R.D. (1993) Use of highly reactive zinc leads to a new, facile synthesis for polyarylenes. *Macromolecules*, **26**, 3462–3463.

69 Chen, T.A., Wu, X., and Rieke, R.D. (1995) Regiocontrolled synthesis of poly(3-alkylthiophenes) mediated by Rieke zinc: their characterization and solid-state properties. *J. Am. Chem. Soc.*, **117**, 233–244.

70 Yokoyama, A., Miyakoshi, R., and Yokozawa, T. (2004) Chain-growth polymerization for poly(3-hexylthiophene) with a defined molecular weight and a low polydispersity. *Macromolecules*, **37**, 1169–1171.

71 Miyakoshi, R., Yokoyama, A., and Yokozawa, T. (2005) Catalyst-transfer polycondensation. Mechanism of Ni-catalyzed chain-growth polymerization leading to well-defined poly(3-hexylthiophene). *J. Am. Chem. Soc.*, **127**, 17542–17547.

72 Miyakoshi, R., Yokoyama, A., and Yokozawa, T. (2004) Synthesis of poly(3-hexylthiophene) with a narrower polydispersity. *Macromol. Rapid Commun.*, **25**, 1663–1666.

73 Sheina, E.E., Liu, J., Iovu, M.C., Laird, D.W., and McCullough, R.D. (2004) Chain growth mechanism for regioregular nickel-initiated cross-coupling polymerizations. *Macromolecules*, **37**, 3526–3528.

74 Iovu, M.C., Sheina, E.E., Gil, R.R., and McCullough, R.D. (2005) Experimental evidence for the quasi-"living" nature of the Grignard metathesis method for the synthesis of regioregular poly(3-alkylthiophenes). *Macromolecules*, **38**, 8649–8656.

75 Jeffries-El, M., Sauve, G., and McCullough, R.D. (2004) In-situ end group functionalization of regioregular poly(3-alkylthiophene) using the Grignard metathesis polymerization method. *Adv. Mater.*, **16**, 1017–1019.

76 Iovu, M.C., Jeffries-EL, M., Sheina, E.E., Cooper, J.R., and McCullough, R.D. (2005) Regioregular poly(3-alkylthiophene) conducting block copolymers. *Polymer*, **46**, 8582–8586.

77 Iovu, M.C., Craley, C.R., Jeffries-EL, M., Krankowski, A.B., Zhang, R., Kowalewski, T., and McCullough, R.D. (2007) Conducting regioregular polythiophene block copolymer nanofibrils synthesized by reversible addition fragmentation chain transfer polymerization (RAFT) and nitroxide mediated polymerization (NMP). *Macromolecules*, **40**, 4733–4735.

78 Iovu, M.C., Zhang, R., Cooper, J.R., Smilgies, D.M., Javier, A.E., Sheina, E.E., Kowalewski, T., and McCullough, R.D. (2007) Conducting block copolymers of regioregular poly(3-hexylthiophene) and poly(methacrylates): electronic materials with variable conductivities and degrees of interfibrillar order. *Macromol. Rapid Commun.*, **28**, 1816–1824.

79 Liu, J., Sheina, E.E., Kowalewski, T., and McCullough, R.D. (2002) Tuning the electrical conductivity and self-assembly of regioregular polythiophene by block copolymerization: nanowire morphologies in new di- and triblock copolymers. *Angew. Chem., Int. Ed.*, **41**, 329–332.

80 Liu, J. and McCullough, R.D. (2002) End group modification of regioregular polythiophene through postpolymerization functionalization. *Macromolecules*, **35**, 9882–9889.

81 Matyjaszewski, K. and Xia, J. (2001) Atom transfer radical polymerization. *Chem. Rev.*, **101**, 2921–2990.

82 Matyjaszewski, K. (2000) *Controlled/Living Radical Polymerization: Progress in ATRP, NMP and RAFT*, ACS Symposium Series 768, American Chemical Society, Washington, DC.

83 Marsitzy, D., Klapper, M., and Mullen, K. (1999) End-functionalization of poly(2,7-fluorene): a key step toward novel luminescent rod–coil block copolymers. *Macromolecules*, **32**, 8685–8688.

84 Dai, C.A., Yen, W.C., Lee, Y.H., Ho, C.C., and Su, W.F. (2007) Facile synthesis of well defined block copolymers containing regioregular poly(3-hexyl thiophene) via anionic macroinitiation method and their self-assembly behavior. *J. Am. Chem. Soc.*, **129**, 11036–11038.

85 Chang, J.F., Sun, B., Breiby, D.W., Nielsen, M.M., Soelling, T.I., Giles, M., McCulloch, I., and Sirringhaus, H. (2004) Enhanced mobility of poly(3-hexylthiophene) transistors by spin-coating from high-boiling-point solvents. *Chem. Mater.*, **16**, 4772–4776.

86 Francois, B., Widawski, G., Rawiso, M., and Cesar, B. (1995) Block-copolymers with conjugated segments: synthesis and structural characterization. *Synth. Met.*, **69**, 463–466.

87 Yang, H., Shin, T.J., Yang, L., Cho, K., Ryu, C.Y., and Bao, Z. (2005) Effect of mesoscale crystalline structure on the field-effect mobility of regioregular poly(3-hexyl thiophene) in thin-film transistors. *Adv. Funct. Mater.*, **15**, 671–676.

88 Bazan, G.C., Miao, Y.J., Renak, M.L., and Sun, B.J. (1996) Fluorescence quantum yield of poly(*p*-phenylenevinylene) prepared via the paracyclophene route:

effect of chain length and interchain contacts. *J. Am. Chem. Soc.*, **118**, 2618–2624.

89 Greenham, N.C., Samuel, I.D.W., Hayes, G.R., Phillips, R.T., Kessener, Y.A.R., Moratti, S.C., Holmes, A.B., and Friend, R.H. (1995) Measurement of absolute photoluminescence quantum efficiencies in conjugated polymers. *Chem. Phys. Lett.*, **241**, 89–96.

90 Xu, B. and Holdcroft, S. (1993) Molecular control of luminescence from poly(3-hexylthiophenes). *Macromolecules*, **26**, 4457–4460.

91 Lim, H., Huang, K.-T., Su, W.-F., and Chao, C.Y. (2010) Facile syntheses, morphologies, and optical absorptions of P3HT coil–rod–coil triblock copolymers. *J. Polym. Sci. A*, **48**, 3311–3322.

92 Alemseghed, M.G., Gowrisanker, S., Servello, J., and Stefan, M.C. (2009) Synthesis of di-block copolymers containing regioregular poly(3-hexylthiophene) and poly(tetrahydrofuran) by a combination of Grignard metathesis and cationic polymerizations. *Macromol. Chem. Phys.*, **210**, 2007–2014.

93 Alemseghed, M.G., Servello, J., Hundt, N., Sista, P., Biewer, M.C., and Stefan, M.C. (2010) Amphiphilic block copolymers containing regioregular poly(3-hexylthiophene) and poly(2-ethyl-2-oxazoline). *Macromol. Chem. Phys.*, **211**, 1291–1297.

94 Radano, C.P., Scherman, O.A., Stingelin-Stutzmann, N., Müller, C., Breiby, D.W., Smith, P., Janssen, R.A.J., and Meijer, E.W. (2005) Crystalline–crystalline block copolymers of regioregular poly(3-hexylthiophene) and polyethylene by ring-opening metathesis polymerization. *J. Am. Chem. Soc.*, **127**, 12502–12503.

95 Huang, K.K., Fang, F.Y., Hsu, J.C., Kuo, C.C., Chang, W.H., and Chen, W.C. (2011) Synthesis, micellar structures, and multifunctional sensory properties of poly(3-hexylthiophene)-*block*-poly(2-(dimethylamino)ethyl methacrylate) rod–coil diblock copolymers. *J. Polym. Sci. A*, **49**, 147–155.

96 Shrotriya, V., Li, G., Yao, Y., Moriarty, T., Emery, K., and Yang, Y. (2006) Accurate measurement and characterization of organic solar cells. *Adv. Funct. Mater.*, **16**, 2016–2023.

97 Sariciftci, N.S., Braun, D., Zhang, C., Srdanov, V.I., Heeger, A.J., Stucky, G., and Wudl, F. (1993) Semiconducting polymer–buckminsterfullerene heterojunctions: diodes, photodiodes, and photovoltaic cells. *Appl. Phys. Lett.*, **62**, 585–587.

98 Dante, M., Yang, C., Walker, B., Wudl, F., and Nyugen, T.-Q. (2010) Self-assembly and charge-transport properties of a polythiophene–fullerene triblock copolymer. *Adv. Mater.*, **22**, 1835–1838.

99 Sauve, G., Zhang, R., Jia, S., Kowalewski, T., and McCullough, R.D. (2006) Synthesis, mobility, and conductivity of well-defined regioregular poly(3-hexylthiophene) and diblock copolymers of regioregular poly(3-hexylthiophene). *Proc. SPIE*, **6336**, 633614/1–63364/10.

100 DeLongchamp, D.M., Kline, R.J., Fischer, D.A., Richter, L.J., and Toney, F.T. (2011) Molecular characterization of organic electronic films. *Adv. Mater.*, **23**, 319–337.

101 Zhang, R., Li, B., Iovu, M.C., Jeffries-El, M., Suave, G., Cooper, J., Jia, S., Tristam-Nagle, S., Smilgies, D.M., Lambeth, D., McCullough, R.D., and Kowalewski, T. (2006) Nanostructure dependence of field-effect mobility in regioregular poly(3-hexylthiophene) thin film field effect transistors. *J. Am. Chem. Soc.*, **128**, 3480–3481.

102 Yu, X., Xiao, K., Chen, J., Lavrik, N.V., Hong, K., Sumpter, B.G., and Geohegan, D.B. (2011) High-performance field-effect transistors based on polystyrene-*b*-poly(3-hexylthiophene) diblock copolymers. *ACS Nano*, **5**, 3559–3567.

103 Suave, G. and McCullough, R.D. (2007) High field-effect mobilities for diblock copolymers of poly(3-hexylthiophene) and poly(methyl acrylate). *Adv. Mater.*, **19**, 1822–1825.

104 Dini, D. (2005) Electrochemiluminescence from organic emitters. *Chem. Mater.*, **17**, 1933–1945.

105 Kraft, A., Grimsdale, A.C., and Holmes, A.B. (1998) Electroluminescent conjugated polymers – seeing polymers

in a new light. *Angew. Chem., Int. Ed. Engl.*, **37**, 402–428.

106 Gunes, S., Neugebauer, H., and Sariciftci, N.S. (2007) Conjugated polymer-based organic solar cell. *Chem. Rev.*, **107**, 1324–1338.

107 Segura, J.L., Martin, N., and Guldi, D.M. (2005) Materials for organic solar cells: the C_{60}/π-conjugated oligomer approach. *Chem. Soc. Rev.*, **34**, 31–47.

108 Thomas, S.W., Joly, G.D., and Swager, T.M. (2007) Chemical sensors based on amplifying fluorescent conjugated polymers. *Chem. Rev.*, **107**, 1339–1386.

109 Becker, H., Spreitzer, H., Kreuder, W., Kluge, E., Schenk, H., Parker, I., and Cao, Y. (2000) Soluble PPVs with enhanced performance – a mechanistic approach. *Adv. Mater.*, **12**, 42–48.

110 Furstner, A., Guth, O., Rumbo, A., and Seidel, G. (1999) Ring closing alkyne metathesis. Comparative investigation of two different catalyst systems and application to the stereoselective synthesis of olfactory lactones, azamacrolides, and the macrocyclic perimeter of the marine alkaloid nakadomarin A. *J. Am. Chem. Soc.*, **121**, 11108–11113.

111 Furstner, A., Guth, O., Rumbo, A., and Seidel, G. (1999) Mo[N(*t*-Bu)(Ar)]$_3$ complexes as catalyst precursors: *in situ* activation and application to metathesis reactions of alkynes and diynes. *J. Am. Chem. Soc.*, **121**, 9453–9454.

112 Swager, T.M. (2008) Iptycenes in the design of high performance polymers. *Acc. Chem. Res.*, **41**, 1181–1189.

113 Zhao, D.H. and Swager, T.M. (2005) Conjugated polymers containing large soluble diethynyl iptycenes. *Org. Lett.*, **7**, 4357–4360.

114 Maddux, T.M., Li, W.J., and Yu, L.P. (1997) Stepwise synthesis of substituted oligo(phenylene-vinylene) via an orthogonal approach. *J. Am. Chem. Soc.*, **119**, 844–845.

115 Chakraborty, S., Keightley, A., Dusevich, V., Wang, Y., and Peng, Z. (2010) Synthesis and optical properties of a rod–coil diblock copolymer with polyoxometalate clusters covalently attached to the coil block. *Chem. Mater.*, **22**, 3995–4006.

116 Stalmach, U., de Boer, B., Videlot, C., and van Hutten, P.F., and Hadziioannou, G. (2000) Semiconducting diblock copolymers synthesized by means of controlled radical polymerization techniques. *J. Am. Chem. Soc.*, **122**, 5464–5472.

117 Braun, C.H., Schöpf, B., Ngov, C., Brochon, C., Hadziioannou, G., Crossland, E.J.W., and Ludwigs, S. (2011) Synthesis and thin film phase behaviour of functional rod–coil block copolymers based on poly(*para*-phenylenevinylene) and poly(lactic acid). *Macromol. Rapid Commun.*, **32**, 813–819.

118 Wang, H., Wang, H.H., Urban, V.S., Littrell, K.C., Thiyagarajan, P., and Yu, L. (2000) Syntheses of amphiphilic diblock copolymers containing a conjugated block and their self-assembling properties. *J. Am. Chem. Soc.*, **122**, 6855–6861.

119 Mori, T., Watanabe, T., Minagawa, K., and Tanaka, M. (2005) Self-assembly of oligo(*p*-phenylenevinylene)-*block*-poly(ethylene oxide) in polar media and solubilization of an oligo(*p*-phenylenevinylene) homooligomer inside the assembly. *J. Polym. Sci. A*, **43**, 1569–1578.

120 Zhao, Q., Liu, S.-J., and Huang, W. (2009) Polyfluorene-based blue-emitting materials. *Macromol. Chem. Phys.*, **210**, 1580–1590.

121 Wu, Z.L., Xiong, Y., Zou, J.H., Wang, L., Liu, J.H., Chen, Q.L., Yang, W., Peng, J.B., and Cao, Y. (2008) High-triplet-energy poly(9,9′-bis(2-ethylhexyl)-3,6-fluorene) as host for blue and green phosphorescent complexes. *Adv. Mater.*, **20**, 2359–2364.

122 Grimsdale, A.C. (ed.) (2006) *The Synthesis of Electroluminescent Polymers. Organic Light Emitting Devices*, Wiley-VCH Verlag GmbH, Weinheim.

123 Fukuda, M., Sawada, K., and Yoshino, K. (1993) Synthesis of fusible and soluble conducting polyfluorene derivatives and their characteristics. *J. Polym. Sci. A*, **31**, 2465–2471.

124 Meng, G., Velayudham, S., Smith, A., Luck, R., and Liu, H. (2009) Color tuning of polyfluorene emission with bodipy monomers. *Macromolecules*, **42**, 1995–2001.

125 Kong, X. and Jenekhe, S.A. (2004) Block copolymers containing conjugated polymer and polypeptide sequences: synthesis and self-assembly of electroactive and photoactive nanostructures. *Macromolecules*, **37**, 8180–8183.

126 Tsolakis, P.K. and Kallitsis, J.K. (2003) Synthesis and characterization of luminescent rod–coil block copolymers by atom transfer radical polymerization: utilization of novel end-functionalized terfluorenes as macroinitiators. *Chem. Eur. J.*, **9**, 936–943.

127 Lu, S., Liu, T.X., Ke, L., Ma, D.G., Chua, S.J., and Huang, W. (2005) Polyfluorene-based light-emitting rod–coil block copolymers. *Macromolecules*, **38**, 8494–8502.

128 Lin, S.T., Tung, Y.C., and Chen, W.C. (2008) Synthesis, structures and multifunctional sensory properties of poly[2,7-(9,9-dihexylfluorene)]-*block*-poly[2-(dimethylamino)ethyl methacrylate] rod–coil diblock copolymers. *J. Mater. Chem.*, **18**, 3985–3992.

129 Chiu, Y.-C., Kuo, C.-C., Lin, C.-J., and Chen, W.-C. (2011) Highly ordered luminescent microporous films prepared from crystalline conjugated rod–coil diblock copolymers of PF-*b*-PSA and their superhydrophobic characteristics. *Soft Matter*, **7**, 9350–9358.

130 Tian, Y., Chen, C.-Y., Yip, H.-L., Wu, W.-C., Chen, W.-C., and Jen, A.K.-Y. (2010) Synthesis, nanostructure, functionality, and application of polyfluorene-*block*-poly(*N*-isopropylacrylamide)s. *Macromolecules*, **43**, 282–291.

131 Hu, J., Zhang, G., Geng, Y., and Liu, S. (2011) Micellar nanoparticles of coil–rod–coil triblock copolymers for highly sensitive and ratiometric fluorescent detection of fluoride ions. *Macromolecules*, **44**, 8207–8214.

132 Jenekhe, S.A. and Chen, L.X. (1999) Self-assembled aggregates of rod–coil block copolymers and their solubilization and encapsulation of fullerenes. *Science*, **279**, 1903–1907.

133 Jenekhe, S.A. and Chen, L.X. (1999) Self-assembly of ordered microporous materials from rod–coil block copolymers. *Science*, **283**, 372–375.

134 Agut, W., Taton, D., Brulet, A., Sandre, O., and Lecommandoux, S. (2011) Depletion induced vesicle-to-micelle transition from self-assembled rod–coil diblock copolymers with spherical magnetic nanoparticles. *Soft Matter*, **7**, 9744–9750.

135 Sary, N., Richard, F., Brochon, C., Leclerc, N., Lévêque, P., Audinot, J.-N., Berson, S., Heiser, T., Hadziioannou, G., and Mezzenga, R. (2010) A new supramolecular route for using rod–coil block copolymers in photovoltaic applications. *Adv. Mater.*, **22**, 763–768.

136 Geim, A.K. and Novoselov, K.S. (2007) The rise of graphene. *Nat. Mater.*, **6**, 183–191.

137 Allen, M.J., Tung, V.C., and Kaner, R.B. (2010) Honeycomb carbon: a review of graphene. *Chem. Rev.*, **110**, 132–145.

138 Yang, W., Ratinac, K.R., Ringer, S.P., Thordarson, P., Gooding, J.J., and Braet, F. (2010) Carbon nanomaterials in biosensors: should you use nanotubes or graphene? *Angew. Chem., Int. Ed.*, **49**, 2114–2138.

139 Guo, S. and Dong, S. (2011) Graphene and its derivative-based sensing materials for analytical devices. *J. Mater. Chem.*, **21**, 18503–18516.

140 Zhang, Y.B., Tang, T.T., Girit, C., Hao, Z., Martin, M.C., Zettl, A., Crommie, M.F., Shen, Y.R., and Wang, F. (2009) Direct observation of a widely tunable bandgap in bilayer graphene. *Nature*, **459**, 820–823.

141 Watson, M., Fechtenkoetter, A., and Müllen, K. (2001) Big is beautiful – "aromaticity" revisited from the viewpoint of macromolecular and supramolecular benzene chemistry. *Chem. Rev.*, **101**, 1267–1300.

142 Qi, X., Pu, K.-Y., Li, H., Zhou, X., Wu, S., Fan, Q.-L., Liu, B., Boey, F., Huang, W., and Zhang, H. (2010) Amphiphilic

graphene composites. *Angew. Chem., Int. Ed.*, **49**, 9426–9429.
143 Kong, X. and Jenekhe, S.A. (2004) Block copolymers containing conjugated polymer and polypeptide sequences: synthesis and self-assembly of electroactive and photoactive nanostructures. *Macromolecules*, **37**, 8180–8183.
144 Rubatat, L., Kong, X., Jenekhe, S.A., Ruokolainen, J., Hojeij, M., and Mezzenga, R. (2008) Self-assembly of polypeptide/π-conjugated polymer/polypeptide triblock copolymers in rod–rod–rod and coil–rod–coil conformations. *Macromolecules*, **41**, 1846–1852.
145 Hundt, N., Hoang, Q., Nguyen, H., Sista, P., Hao, J., Servello, J., Palaniappan, K., Alemseghed, M., Biewer, M.C., and Stefan, M.C. (2011) Synthesis and characterization of a block copolymer containing regioregular poly(3-hexylthiophene) and poly(γ-benzyl-L-glutamate). *Macromol. Rapid Commun.*, **32**, 302–308.
146 Wu, Z.-Q., Ono, R.J., Chen, Z., Li, Z., and Bielawski, C.W. (2011) Polythiophene–block–poly(γ-benzyl L-glutamate): synthesis and study of a new rod–rod block copolymer. *Polym. Chem.*, **2**, 300–302.

12
Fullerene/Conjugated Polymer Composite for the State-of-the-Art Polymer Solar Cells
Wanli Ma

12.1
Introduction

In recent years, there has been an intensive search for cost-effective photovoltaics to replace the expensive silicon-based p–n junction solar cells. Among all the alternative technologies, solar cells based on conjugated polymers are the approach that could lead to the most significant cost reduction. As an advantage over the traditional inorganic semiconductors, conjugated polymers can be processed from solution at room temperature by application of spin coating or even conventional printing techniques. It is believed that a simple coating or printing process will allow a roll-to-roll manufacturing of flexible, lightweight PV modules that should enable cost-efficient production and the development of portable organic optoelectronic devices – even integrated on a single substrate. One major obstacle to a large-scale commercialization of the polymer solar cells (PSCs) is the relatively low device efficiencies. The best energy conversion efficiencies published for small-area devices have been in the range of 5–6% for many years [1–4]. In the last couple of years, however, exciting progress has been made in the synthesis of novel low-bandgap polymer materials. The power conversion efficiency (PCE) of state-of-the-art PSCs has already exceeded 8% in scientific literature [5, 6]. Companies such as Konarka and Solarmer also achieved PCEs of over 8% for their PSCs. Very recently, Heliatek, a company from Germany, announced a certified 9.8% cell efficiency for a $1.1\,cm^2$ organic tandem solar cell. These relatively high conversion efficiencies make organic solar cell more attractive and a promising candidate not only for an alternative to conventional silicon-based solar cells but also for many applications in emerging niche markets.

Semiconducting Polymer Composites: Principles, Morphologies, Properties and Applications, First Edition.
Edited by Xiaoniu Yang.
© 2012 Wiley-VCH Verlag GmbH & Co. KGaA. Published 2012 by Wiley-VCH Verlag GmbH & Co. KGaA.

12.2
Working Mechanism

12.2.1
Unique Properties of Organic Solar Cells

The most apparent distinction between organic and inorganic photovoltaic materials is the different binding energies of their exciton, which is the photoexcited state of an electron–hole pair bound together by the Coulomb attraction. In order to achieve photocurrent, these excited electron–hole pairs need to be dissociated into free charge carriers. In a classical inorganic solar cell, weakly Coulomb bound Mott–Wannier excitons can be separated by the internal electric field. In organic semiconductors, however, the screening of opposite charges is much weaker as the dielectric constant is lower. This leads to the formation of the Frenkel excitons that have a much stronger binding energy. As this binding is more difficult to overcome than that in inorganic systems, the concept of donor (D)–acceptor (A) has to be adopted in organic solar cells. The large driving force needed for the exciton dissociation derives from the difference in ionization potential and electron affinity between donor and acceptor, respectively [7]. Usually, hole-conducting small molecules or conjugated polymers have been used as donor materials. The most widely used acceptor is buckminsterfullerene (C_{60}), which is strongly electronegative and can accept as many as six electrons. Conjugated polymer/fullerene is by far the most efficient D/A composite for PSCs and it will be discussed in detail in this chapter.

Another significant difference between organic and inorganic solar cells is that organic semiconductors are amorphous and thus charge transport is more difficult than in crystals. As a result, the lower mobility of organic materials significantly limits the optimum thickness of the active layer, which is usually around a few hundred nanometers. Fortunately, because of the relatively strong absorbance coefficients (usually $\geq 10^5$ cm^{-1}) of organic materials, such a thin film can capture most of the incident photons within the absorption region. As a comparison, a conventional crystalline silicon solar cell takes about 300 μm or more of silicon depth to absorb the same amount of sunlight.

12.2.2
Understanding the Bulk Heterojunction Structures

These two fundamental differences have a huge impact on the developing history of organic solar cells. It took more than three decades of active research for people to gradually realize that the design concept and device optimization of organic solar cells are essentially different from those of inorganic solar cells. In 1986, Tang introduced the first planar donor (D), acceptor (A), heterojunction bilayer device and successfully increased the conversion efficiency from 0.1 to 1% [8]. In the beginning of the 1990s, the novel bulk heterojunction (BHJ) solar cell concept was introduced by Heeger and coworkers [9, 10], as shown in Figure 12.1. And it

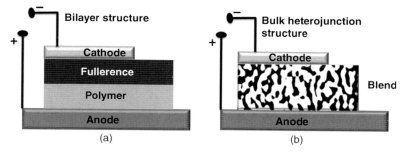

Figure 12.1 Typical bilayer structure (a) and bulk heterojunction structure (b).

still remains the most efficient device architecture for polymer solar cells up to now.

In order to understand the remarkable improvement of device efficiency introduced by BHJ structures, the working mechanism of polymer solar cells has to be clarified. The working mechanism of organic thin-film solar cells can be broken down as follows:

a) Generation of excitons in D/A molecules by light absorption.
b) Diffusion of excitons to D–A interfaces.
c) Dissociation of excitons into free carriers via charge transfer at D–A interfaces.
d) Transport of free carriers to the corresponding electrode and charge collection by an external circuit.

Each step can be evaluated by η_A (the absorption efficiency), η_{ED} (the exciton diffusion efficiency), η_{CT} (the charge transfer efficiency), and η_{CC} (the carrier collection efficiency), respectively.

The external quantum efficiency of an organic solar cell equals

$$\eta_{EQE} = \eta_A \times \eta_{ED} \times \eta_{CT} \times \eta_{CC}. \tag{12.1}$$

First, η_{CT} is considered to be almost unity since the photoinduced charge transfer takes place very rapidly within the order of femtoseconds, which is much faster than other competing charge relaxation processes (see Figure 12.2). For bilayer solar cells, charge transport happens in a single material phase, which greatly reduces the chance of recombination. In this case, η_{CC} is approaching unity and η_{EQE} is determined by the product of $\eta_A \times \eta_{ED}$. That is to say, the conversion efficiency of bilayer solar cells is limited by the exciton diffusion to the D–A interface and/or photon absorption processes. For amorphous organic material, the exciton diffusion length ($L_D \sim 100$ Å) is about one-tenth of the optical absorption length ($L_A \sim 1000$ Å). In order to separate most of the excitons and increase η_{ED}, the thickness of the active film has to be made very thin, which in turn decreases η_A. Thus, there is a trade-off between η_A and η_{ED}. In general, the external quantum efficiency of bilayer solar cells is mainly limited by the short exciton diffusion length (low η_{ED}) of organic materials.

Consequently, interpenetrating phase-separated D/A network composites, that is, bulk heterojunctions, appear to be ideal photovoltaic composites. The fundamental

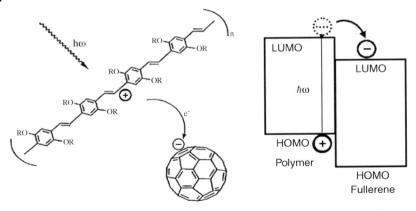

Figure 12.2 The photoinduced charge transfer from a donor (PPV) to an acceptor (fullerene).

concept of BHJ is the solution-processed active layer consisting of an interpenetrating network of a usually polymeric electron donor/hole acceptor and a usually fullerene-based electron acceptor/hole donor. The blend of a conjugated polymer as a donor and a soluble fullerene derivative PCBM as an acceptor is the most studied composite today. In the bilayer heterojunction, only the geometrical interface between the conjugated polymer and fullerene layers is the area for exciton separation but in the BHJ the entire volume of the composite layer is involved. Therefore, the distance between every generated exciton and the D/A interface lies within a distance L_D, and η_{ED} becomes almost unity. Thus, unlike bilayer solar cells η_A can be improved by increasing the thickness of the BHJ layer since η_{ED} is thickness independent. As a result, BHJ gives rise to short-circuit current orders of magnitude higher than the previously described bilayer devices due to largely improved η_{ED}.

However, note that in a BHJ solar cell $\eta_{ED} \times \eta_{CC}$ strongly depends on the network morphology of the polymer/fullerene composite. The process of carrier generation (η_{ED}) requires small D/A domain size whereas the charge transport process (η_{CC}) needs big D/A domain size to ensure percolating paths to avoid the recombination of the charge carriers due to trapping in dead ends on an isolated material's domain. The trade-off between η_{CC} and η_{ED} indicates that network morphology plays a critical role in the optimization of a polymer/fullerene BHJ photovoltaic device. An ideal morphology requires a well-connected interpenetrating network with domain size on the order of the exciton diffusion length. Once optimized, the BHJ serves as the state-of-the-art concept for polymer/fullerene solar cells, reaching power conversion efficiencies of up to 8%.

12.2.3
Device Parameters and Theoretical Efficiency

In order to further improve the performance of BHJ solar cells, it is essential to understand the nature of the device parameters that define the solar cell PCEs and

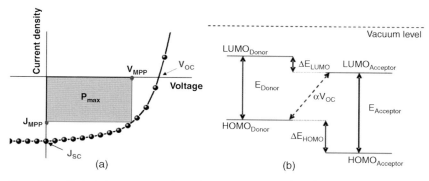

Figure 12.3 Typical photocurrent–voltage characteristics in a solar cell. J_{SC} is the short-circuit current density, V_{OC} is the open-circuit voltage, and J_{MPP} and V_{MPP} are the current and voltage at the maximum power point, respectively.

find the right strategies to optimize these parameters:

$$\eta_{POWER} = \frac{P_{OUT}}{P_{IN}} = \frac{J_{SC} V_{OC} FF}{P_{IN}}, \qquad (12.2)$$

where P_{OUT} is the output power, P_{IN} is the input power, J_{SC} is the short-circuit current density, V_{OC} is the open-circuit voltage, and the fill factor (FF) can be defined by the maximum power point (J_{MPP} and V_{MPP}) in the device J–V curve (shown in Figure 12.3):

$$FF = \frac{J_{MPP} V_{MPP}}{J_{SC} V_{OC}}. \qquad (12.3)$$

12.2.3.1 Short-Circuit Current Density

Once we obtain the external quantum efficiency of polymer solar cells, the value of J_{SC} can be calculated from the overlap between the absorption spectra of the materials and the sunlight AM 1.5 G spectrum.

$$J_{SC} = \frac{e}{hc} \int_{\lambda_1}^{\lambda_2} \eta_{EQE} \lambda P_{IN}(\lambda) \lambda \, d\lambda, \qquad (12.4)$$

where η_{EQE} is the external quantum efficiency (also known as incident photon-to-electron conversion efficiency, IPCE), h is Planck's constant, c is the speed of light in vacuum, and e is the elementary charge. λ_1 and λ_2 are the initial and final wavelengths that define the absorption band of the active materials. Although affected by η_{EQE}, the maximum values of J_{SC} are ultimately limited by the optoelectronic properties of the active materials. For example, in the case of the poly(3-hexylthiophene) (P3HT) (bandgap = 1.9 eV), if we assume an EQE of 100%, the maximum possible J_{SC} is about 18.7 mA cm^{-2}, which is significantly lower than that of crystalline silicon material (bandgap = 1.1 eV). Hence, the ultimate way of improving the J_{SC} of polymer solar cells is by designing new active materials with lower bandgaps. In a real device, the J_{SC} strongly depends on the value of η_{EQE}, which is usually lower than 70% for most of the current polymer solar cells. As we can infer from the definition of η_{EQE} in Eq. (12.1), a high η_{EQE} requires efficient photon absorption (η_A), charge separation, transport, and extraction processes ($\eta_{ED} \times \eta_{CT} \times \eta_{CC}$). Among them, η_A can be improved by increasing the active layer thickness, which, however, demands higher carrier mobilities to avoid the extra recombination induced by longer charge transport paths. As a result, the optimum thickness of polymer solar cells is mostly determined by the carrier mobilities inside the active film, while for efficient charge generation and collection in a BHJ cell, both high mobilities and long lifetime of the carriers within the interpenetrating networks are needed. The mobility of the individual components is a true material parameter, while the lifetime of the carriers in the BHJ cell is mainly determined by the quality of the D/A networks. In general, in order to improve the short-circuit current of organic solar cells, novel donor and acceptor materials will have to be synthesized, targeting on two main properties: larger carrier mobilities and wider absorption spectrum. Meanwhile, strategies leading to fine control of the nanoscale morphology have to be well developed.

12.2.3.2 Open-Circuit Voltage

For thin-film photovoltaic devices, such as polymer solar cells, the built-in electric field can influence charge dissociations, charge transport, and charge collection. The open-circuit voltage (V_{OC}) is a straightforward indication of the strength of the device internal electric field. For photodiodes using pristine conjugated polymer, such as the MIM (metal/insulator/metal) devices, V_{OC} can be estimated from the difference between the work functions of metal electrodes [11]. However, mixing fullerene with conjugated polymer into a composite active layer completely modifies the nature of the device V_{OC}. The origin of the V_{OC} for a BHJ cell has long been a topic of debate. Now it is clear that it is not defined by the work function of the electrodes but rather by the energy levels of the active materials [12, 13]. Several reports show no correlation of V_{OC} with the energy levels of the used cathode materials [14, 15]. Heeger et al. described the origin of V_{OC} in detail: under illumination, the absorption of photons leads to a generation of carriers that will occupy the available states of different energies according to a Fermi distribution. As a result, the Fermi energy in the donor domain is located near the valence band edge of the polymer, and the Fermi energy in the fullerene acceptor domain is located near the conduction band. The difference between the two quasi-Fermi levels will be the upper limit for the V_{OC} (see Figure 12.3). The built-in field is considered to be established by the accumulated electrons and holes at the polymer/fullerene interface to form a "parallel plate capacitor." And the voltage across the interface is equal to V_{OC} [16]. This microscopic picture of the origin of V_{OC} is translated to macroscopic V_{OC} by the use of two different metal contacts with different work functions and by using charge selective interlayer. A simple empirical expression was established to correlate V_{OC} in OPV devices with the energy difference between the HOMO level of the donor and the LUMO level of the acceptor [12]:

$$V_{OC} = \left|E_{HOMO}^{Donor}\right| - \left|E_{LUMO}^{Acceptor}\right| - 0.3\,V. \tag{12.5}$$

The 0.3 V is an empirical factor, mainly caused by the dark current–voltage curve of the device, which is determined by the ideality factor n and the reverse dark current.

The authors deduced the above relationship after using 26 different organic semiconductors donors, keeping PCBM as the acceptor, and correlating the value of V_{OC} to the HOMO energy level position of the donors. More recently, it was reported that the minimum energy loss from the charge transfer (CT) state energy to V_{OC} is 0.5 eV after the measurements of device V_{OC} using 18 different materials in different donor–acceptor combinations [17]. Since the minimum energy required for charge transfer is estimated as 0.1 eV, the open-circuit voltage can be directly related to the optical bandgap (E_g) of the donor material via $V_{OC} = E_g - 0.6\,eV$.

Following the discussion, in order to improve V_{OC}, the design rule of new materials is to maximize the energy difference between the HOMO level of the donor and the LUMO level of the acceptor, more specifically, to make the HOMO level of polymers deeper and LUMO level of fullerene derivatives shallower.

12.2.3.3 Fill Factor

By definition, the FF is determined by the position of the maximum power point in the fourth quadrant *J–V* curve of the illuminated solar cells. A large FF requires that the photocurrent rises abruptly as the applied voltage approaches V_{OC}. This optimum condition can only be met when the photogenerated carriers are mostly extracted without significant loss from recombination even at extremely low electric field. Hence, the FF strongly depends on charge transport and therefore on charge carrier mobility [18]. In other words, the FF is limited by the carrier drift length, L_d, which is defined as

$$L_d = \mu \tau E, \tag{12.6}$$

where μ is the carrier mobility, τ is the carrier recombination lifetime, and E is the electric field. In order to prevent significant loss by recombination, L_d must be longer than the active layer thickness L, which is usually around 100 nm for most optimized polymer solar cells [19]. Hence, a high mobility, long lifetime, or a thin film is necessary for efficient charge carrier extraction. If we assume a typical value for the internal electric field of $10^5\,V\,cm^{-1}$ and charge carrier lifetimes of 10^2 ns, the carrier mobility has to be larger than $10^{-3}\,cm^2\,V^{-1}\,s^{-1}$ to make sure L_d exceeds 100 nm at short-circuit conditions [12].

The FF also indicates the quality of the photodiode. Counter diode will lead to a negative curvature of the *J–V* curve in the fourth quadrant and reduce the FF below 25%. Large serial resistance and small parallel resistance can also reduce the FF. In addition, high FF value requires a photovoltaic device with a strict selection principle for the separation of positive and negative carriers [20]. In general, in order to improve the FF, the network structure of the D/A composites and the interface morphology have to be well controlled for reduced serial resistance and shunt. The introduction of the ultrathin interlayer between the active layer and the electrodes is usually necessary to improve the charge selectivity.

12.2.3.4 Theoretical Efficiency

Based on the discussion of the device parameters, it is possible to predict the realistic maximum achievable efficiency of polymer solar cells. Power conversion efficiency of nearly 11% for single-junction OPVs can be achieved by assuming the FF = 65%, EQE = 65%, and an idea difference of 0.3 eV between the LUMOs of polymer and fullerene. This efficiency is expected for a donor polymer having an ideal bandgap energy of 1.5 eV [21]. Further efficiency improvement can be achieved by adding antireflection layers or light trapping structures or by using tandem architectures. For current state-of-the-art polymer solar cells, FF over 70% and EQE over 80% have been achieved [22, 23]. And there is still potential to improve the performance of single-junction cells by further fine-tuning the EQE and FF values. If we assume PSCs with an FF of 70% and an EQE of 90%, which are challenging but realistic values for fully optimized organic solar cells, the maximum theoretical efficiency can reach a remarkable 20% for the scenario with a total loss of 0.5 eV for V_{OC} and charge transfer.

Discovering the potential of plastic solar cells and understanding the nature of device parameters are of tremendous importance when considering strategies for optimizing the morphology of BHJ polymer solar cells. Meanwhile, these investigations also give a fundamental insight into the loss mechanisms of BHJ solar cells and clearly outline the new material design rules for achieving 10% efficiency: a new polymer material should have an HOMO energy level lower than −5.2 eV to increase both device V_{OC} and air stability together with a bandgap of around 1.5 eV to achieve high J_{SC} and maintain large V_{OC}. Meanwhile, the hole mobility of the pristine polymer should be greater than $10^{-3}\,cm^2\,V^{-1}\,s^{-1}$ to avoid low FF.

12.3
Optimization of Fullerene/Polymer Solar Cells

After the discovery of BHJ cells, researchers have successfully developed a number of optimization methods that can roughly be divided into two categories: new materials design and device optimization. The first one involves the synthesis of novel donor and acceptor materials with improved charge mobility and wider absorption spectrum. The second one aims for a more favorable inner structure with regard to exciton dissociation and charge transport by means of using thermal or solvent annealing, solvent additives, interlayer and new device architectures, and so on. In the following chapters, we will review some of the most important and recent developments in BHJ polymer/fullerene solar cells from the material and device perspective, respectively.

12.3.1
Design of New Materials

Conjugated polymers are chosen for photovoltaic application because they are conducing and can absorb light in the UV–visible or even near-infrared part of the solar spectrum. In addition, with solubility in common organic solvents governed through side chains, they can be processed from solution by applying spin coating, doctor blading, or even conventional printing techniques for a controlled film deposition.

During the early stage of BHJ PSCs, many common conjugated polymers have been used as active materials, but two classes have attracted most interest. The soluble poly(phenylene vinylene)s (PPVs) exemplified by MEH-PPV and MDMO-PPV were synthesized by introduction of side chains to the conjugated and insoluble core. They were widely used in the early PSCs and devices were prepared from MDMO-PPV/PCBM blends reaching 2.5% by using proper solvent [20]. Subsequent optimization of electrodes as well as side chains yielded efficiencies of about 3% [24, 25]. Nevertheless, the relatively large bandgap of the PPV polymers together with their low carrier mobility limited the further improvement of device power conversion efficiencies. More recently, highly regioregular P3HT has become a more promising choice of active material because of its lower bandgap and higher hole mobility. By tuning several parameters such as solvent, annealing conditions, and interface morphology, efficiencies of approximately 5% have been achieved [1, 2]. Due to its excellent photovoltaic properties, P3HT has long been the most popular and extensively studied active material for PSCs. However, the highest V_{OC} reported for well-performing P3HT/PCBM solar cells was around 0.66 V [26], which indicates a serious V_{OC} loss due to the mismatch of LUMO energy levels between P3HT and fullerene.

In order to achieve device efficiency of over 5%, new active materials have to be synthesized complying with the following design rules:

1) A low bandgap to capture photons in visible and near-infrared region.
2) Optimized LUMO and HOMO energy levels for both efficient exciton dissociation and high V_{OC}.
3) A high charge carrier mobility of over $10^{-3}\,cm^2\,V^{-1}\,s^{-1}$ is essential for efficient charge extraction and a good fill factor.
4) Good miscibility of donor polymers with PCBM and excellent solubility in common organic solvents.

Since these material properties are essential for high-performance PSCs, the recent progress in molecular design of photovoltaic materials will be reviewed according to these properties.

12.3.1.1 Absorption Enhancement
The widely used strategies for reducing polymer bandgaps are as follows:

1) Increasing the effective conjugation length by using fused ring system [27–29], conjugated side chain [30–32], or intermolecular interactions [33, 34].
2) Incorporation of electron-donating or electron-withdrawing substituent onto the main chain or side chain [35, 36].
3) Synthesis of an alternating copolymer from electron-rich (donor) and electron-deficient (acceptor) units in its backbone [37, 38].

The donor–acceptor copolymerization is one of the most effective approaches to reduce bandgap and tune the electronic energy levels of the conjugated polymers. The push–pull driving forces of D–A structure facilitate the formation of quinoid mesomeric structure (D^+–A^-) and therefore reduce the bond length alternation. Photoinduced intramolecular charge transfer correlated with the high-lying HOMO of the donor unit and the low-lying LUMO of the acceptor unit can also account for the reduced bandgap. The degree of bandgap reduction strongly depends on the strength of donor and acceptor units. More specifically, the HOMO level of the polymers mainly depends on the donor unit, and their LUMO level is mainly determined by the acceptor unit. Hence, fine-tuning of the electronic energy levels of the copolymers can be easily realized by selecting suitable donor and acceptor units.

Among the promising donating groups, there are benzo[1,2-b;3,4-b]dithiophene (BDT), cyclopentadithiophene, carbazole, fluorene, dibenzo(thieno)silole, and dialkoxybenzodithiophene, whereas the acceptors are benzothiadiazole (BT), 4,7-dithien-5-yl-2,1,3-benzodiathiazole (DTBT), quinoxaline, thienopyrazine, and diketopyrrolopyrrole (DPP). Most of them have been extensively discussed in the several recent review articles [39–44]. Herein, our review only focuses on BDT and fluorene-like donor units that are the two mostly used building blocks for current state-of-the-art PSCs with efficiencies over 5%.

Fluorene-Like Donor Units Fluorene and its derivatives are rigid, planar molecules that are usually associated with relatively large bandgaps and low-lying HOMO energy levels, which give them high photodegradation and thermal stability, large hole mobility, and potentially high open-circuit voltages. However, their device performance is usually limited by the relatively low photocurrent due to their large bandgaps. Hence, the highest power conversion efficiency of a polyfluorene-based solar cell is only 4.5%, reported by Yang and coworkers [45]. In order to further reduce the polymer bandgap, a variety of D–A copolymers based on fluorene-like donor units were synthesized and used in PSCs with efficiencies of over 5%.

An interesting variation of polymer P1 in Figure 12.4 is obtained by replacing the fluorene unit by dibenzosilole. Leclerc and coworkers were the first ones to report this polymer with a molecular weight of 15 kDa, an optical bandgap of 1.86 eV, and a PCE of 1.6% [46]. Cao and coworkers were able to obtain the same polymer with a V_{OC} of 0.90 V, a J_{SC} of 9.5 mA cm^{-2}, and a PCE of 5.4% [47]. The much higher molecular weight of 79 kDa may play an important role in the device performance

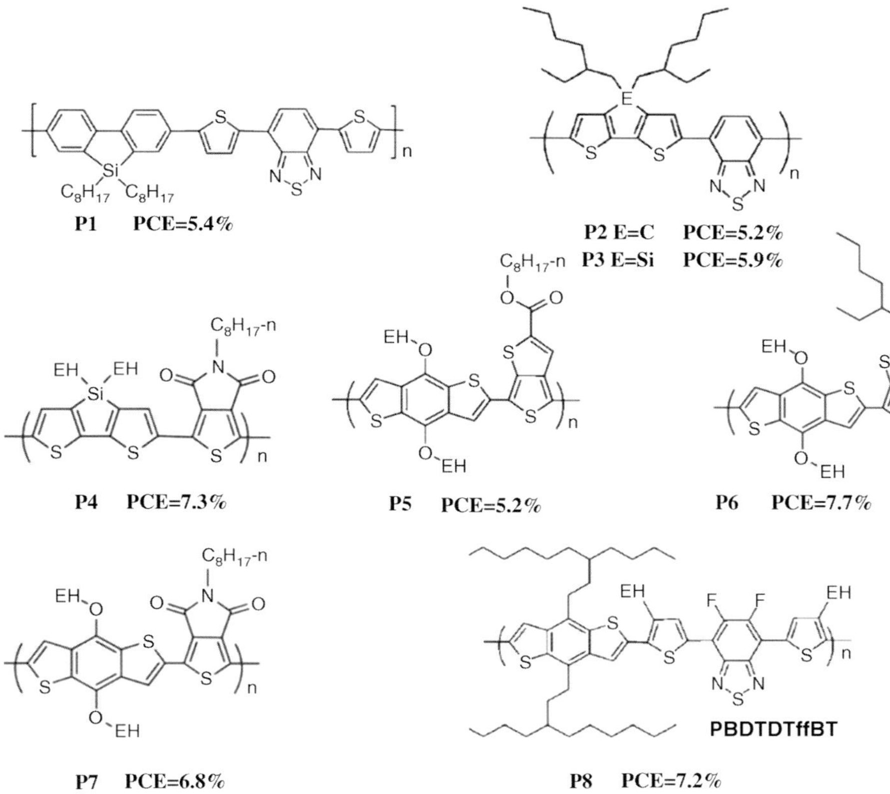

P1 PCE=5.4%

P2 E=C PCE=5.2%
P3 E=Si PCE=5.9%

P4 PCE=7.3%

P5 PCE=5.2%

P6 PCE=7.7%

P7 PCE=6.8%

P8 PCE=7.2% (PBDTDTffBT)

Figure 12.4 Chemical structures of some popular low-bandgap polymers.

improvement. Another two fluorene-like donor units based on dithiophene have been more extensively studied over the years, namely, cyclopentadithiophene and dithienosilole (DTS). Poly[2,6-(4,4-bis-(2-ethylhexyl)-4H-cyclopenta[2,1-b;3,4-b2]-dithiophene)-*alt*-4,7-(2,1,3-benzothiadiazole)] (PCPDTBT, polymer P2, Figure 12.4) is the most prominent candidate of cyclopentadithiophene-based copolymers [48]. This polymer is a real low-bandgap material ($E_g = 1.45$ eV), as well as an excellent charge transporter, with hole mobility as high as 1.4 cm^2 V^{-1} s^{-1} under optimized conditions [49, 50]. Heeger and coworkers improved the PCPDTBT/PCBM device efficiency to beyond 5% by using solvent additives to form a more coarse nanomorphology. The enhanced PCE was attributed to increased carrier generation efficiency and carrier lifetime. One of the drawbacks of this polymer is the instability of the carbon atom on top of the five-membered ring that could undergo an oxidation into a ketone. Therefore, a new derivative P3, that is, dithienosilole, was synthesized in order to prevent this type of oxidation. It is showed that the C—Si bond is longer than the C—C bond in the fluorine core [51]. Consequently, the less steric hindrance created in the dibenzosilole core leads to a better π–π stacking, that is, an improved hole mobility for P3. Yang and coworkers were the first ones to use this unit for OPV applications by adding ethylhexyl side chains on the silicon atom to achieve good solubility [52]. They demonstrated a PCE of 5.1%, which was further improved to 5.9% by replacing the 2-ethylhexyl chains on the silicon atom with long n-dodecyl side chains [53]. Recently, by choosing a new acceptor unit thienopyrroledione (TPD) to fine-tune the energy levels, Tao and coworkers synthesized a new copolymer of DTS-TPD (P4) and demonstrated a high PCE of 7.3% [54].

Benzo[1,2-*b*;3,4-*b*]dithiophene In general, copolymers with donor units based on fluorene and carbazole show a large V_{OC} and a low photocurrent. Whereas polymers containing cyclopentadithiophene and dithienosilole usually exhibit a low V_{OC} and a large photocurrent, polymers based on BDT have an apparently better balance between V_{OC} and J_{SC} because they can simultaneously possess small bandgap, deep HOMO level, and high mobility, which are the keys to realize large J_{SC}, high V_{OC}, and good FF, respectively. Consequently, current state-of-the-art PSCs with PCE over 6.5% are mostly using polymers based on BDT comonomer.

BDT-based polymers were first used in organic field-effect transistors (OFETs) because of their symmetric and planar conjugated structure, which can easily realize ordered π–π stacking in a large domain size. In 2007, a BDT-thiophene-based polymer was reported to have a hole mobility of $0.25\,\mathrm{cm^2\,V^{-1}\,s^{-1}}$, which was one of the highest values for polymer-based OFET [55]. In 2008, Hou et al. first introduced BDT-based conjugated polymer for photovoltaic applications and several BDT-based photovoltaic polymers were designed and synthesized with a highest PCE of only 1.6% [56]. Yu and coworkers used a thiophene derivative, namely thieno[3,4-*b*]thiophene, as the other comonomer and developed numerous copolymers based on BDT [57]. The new material has a bandgap of 1.62 eV and a hole mobility of approximately $10^{-4}\,\mathrm{cm^2\,V^{-1}\,s^{-1}}$. Devices using this polymer achieved a large J_{SC} of $15.6\,\mathrm{mA\,cm^{-2}}$ and a good PCE of 5.6%. However, the high-lying position of the HOMO energy level (−4.90 eV) resulted in low V_{OC} and air instability. In order to achieve higher V_{OC}, further optimizations on the structure and the processing of this class of polymers have been carried out by Yang and coworkers. Their research focused on fine-tuning the side chain on the thienothiophene comonomer [58]. The original side chain on the thienothiophene unit of P5 contained an ester function with a long alkyl side chain. A shorter side chain (n-octyl) was used and the HOMO energy level of the polymer was reduced to −5.01 eV and the V_{OC} was increased to 0.62 V (polymer P5). After that, they changed the ester group to a ketone and replaced the n-octyl by a 2-ethylhexyl moiety, which moved the HOMO further down to −5.12 eV and increased the V_{OC} to 0.7 V. Finally, for the polymer with the best performance (P6), the functional group on the thienothiophene was still a ketone with an n-octyl attached to it and a fluorine atom was added at the other free position of the thienothiophene. The strong electron affinity of the fluorine atom was reported to have effect on reducing the HOMO energy level of polymers [59]. As anticipated, the HOMO energy level was further reduced to −5.22 eV and a large V_{OC} of 0.76 V was obtained. Despite the reduced HOMO, P6 still has a low bandgap of 1.6 eV and the best device fabricated with this polymer reached a PCE of 7.7%.

Another widely reported acceptor unit copolymerized with BDT is thieno[3,4-c]pyrrole-4,6-dione. The synthesis of TPD unit has been well studied and it possesses excellent tunability because the nitrogen atom can be functionalized by a wide variety of groups. It is very interesting that four groups have reported similar copolymers based on BDT-TPD during a very short period of time [60–63]. The best performances were obtained by Frechet and coworkers (polymer P7). A PCE up to 6.8% was achieved with the addition of 1,8-diiodooctane during the processing of the film [61]. The device parameters are $J_{SC} = 11.5$ mA cm^{-2}, $V_{OC} = 0.85$ V, and FF = 0.70. BT and DTBT are probably the mostly widely used electron-deficient acceptor units in conjugated polymers and many important materials have been achieved by using these units. You and coworkers reported a new polymer based on BDT and fluorine atoms modified DTBT comonomers. The target polymer, PDBTDTffBT (P8), showed a bandgap of 1.7 eV, a HOMO level of -5.54 eV, a high V_{OC} of 0.91 V, and an excellent PCE of 7.2% [64].

Conjugated Side Chains The effective conjugation length of the one-dimensional main chain has a significant impact on the polymer bandgap and carrier mobilities. The concept of two-dimensional molecular structure involves the introduction of conjugated side chains to the polymer main chain to enhance the coplanarity of the polymer and hence improve the carrier delocalization over the entire 2D domains. Consequently, a broader absorption, higher charge mobilities, and deeper HOMO may be realized by the improved planarity and intermolecular π–π stacking. Pioneering works done by Li and coworkers showed that the incorporation of conjugated side chains into polythiophene or poly(thienylene vinylene) (PTV) significantly broadened the absorption spectrum and enhanced the power conversion efficiencies of photovoltaic cells [30, 31].

Despite the improved device performance, the 2D polymer still showed a PCE below 5% due to the limited potential of PT and PTV polymers. As we discussed above, BDT comonomer has been the current most effective building blocks for PSCs and it shows great potential for further optimization. The symmetric structure of BDT has an advantage over asymmetric PT or PTV for expanding to 2D structure by the introduction of conjugated side chains. Huo *et al.* introduced two thienyl conjugated side chains to the BDT unit (BDT-T) and successfully synthesized a serials of high-performance polymers by copolymerizing with different acceptor units, such as DTBT [65], thiazolo[5,4-d]thiazole (TTZ) [66], naphtho[1,2-c:5,6-c]bis[1,2,5]-thiadiazole (NT) [67], and thieno[3,4-b]thiophene [68]. All the synthesized 2D polymers showed broadened absorption, improved carrier mobilities, and enhanced performance with PCEs from 5.66 to 7.59%.

In general, proper side chains have to be carefully chosen to keep the coplanarity of the entire 2D molecular structure. If the side chains are too short, the improvement of conjugation length is limited; if the side chains are too long or too bulky, the intermolecular packing will the hindered and the twisted side chains can break the conjugation of the structure. By fine-tuning the conjugated side chain, the 2D

12.3.1.2 Fine-Tuning of HOMO and LUMO Energy Levels

The positions of the HOMO and LUMO energy levels have direct impact on the V_{OC} and J_{SC} of the polymer solar cells. Since fullerene is the mostly used electron acceptor with a fixed LUMO position, the V_{OC} is mainly determined by the position of the polymer HOMO energy level, whereas the J_{SC} depends on the polymer bandgap, which is the energy difference between its HOMO and LUMO levels. In order to achieve a high J_{SC} and meanwhile maintain a large V_{OC}, strategies have to be taken to well tune the polymer HOMO and LUMO positions. Currently, there are two approaches demonstrated to be mostly effective.

1) A weak donor–strong acceptor strategy.

    Alternating donor and acceptor units in copolymers has been proven to be an effective approach to lower the bandgap via internal charge transfer. The interaction between the D–A units implies that the HOMO level of the polymer mainly depends on the donor unit, and the LUMO level is mainly determined by the acceptor unit. Therefore, we envisioned that incorporating a less electron-rich donor (weak donor) would lead to a low-lying HOMO energy level and introducing a more electron-deficient acceptor (strong acceptor) would lead to a smaller bandgap and maintain the low HOMO energy level in the newly designed materials.

    You and coworkers proposed this weak donor–strong acceptor concept and successfully applied polycyclic aromatics-based weak donors such as naphtho[2,1-b:3,4-b′]dithiophene (NDT), quadrathienonaphthalene (QTN), and dithieno-[3,2-f:2′,3′-h]quinoxaline (QDT) in PSCs yielding V_{OC} over 0.7 V [69, 70]. Another good example of weak donor is BDT comonomer, which has been widely demonstrated in PSCs with both high V_{OC} and J_{SC} [32]. You and coworkers also reported that the use of strong acceptors based on BT, DTBT, and thiadiazolo[3,4-c]pyridine (PyT) can greatly reduce the bandgap and simultaneously maintain a high V_{OC}. The PCE of PSCs based on PyT reached 6.32% [71–74].

2) Incorporation of electron-withdrawing or electron-donating side groups.

    In general, the effect of "weaker donor–strong acceptor" strategy can be further enhanced by using less electron-donating side groups on the donor comonomers or using electron-withdrawing side groups on the acceptor units.

    Tao and coworkers and Ferraris *et al.* demonstrated that the replacement of the alkoxy chains on BDT comonomer with less electron-donating alkyl or dialkylthio chains makes BDT an even weaker donor, thereby lowering the HOMO energy levels of this class of polymers without increasing their bandgap energy [75, 76].

    Yang and coworkers tuned the HOMO level of the copolymer based on BDT and thieno[3,4-b]thiophene (TT) [58]. By changing the ester group on the acceptor unit to a more strongly electron-withdrawing ketone group, the

LUMO and HOMO levels of the polymer dropped by 0.11 eV. By attaching a strong electron-withdrawing fluorine atom on the TT unit, the HOMO level was shifted downward further by 0.1 eV. The decrease of the HOMO level with keeping the same bandgap of the polymers leads to higher V_{OC} and high PCE of over 7%.

12.3.1.3 Mobility and Solubility Improvement

In order to enhance the π–π stacking and then extend the conjugation length, it is necessary to avoid the twisting of the main chain and maintain the coplanarity. Copolymers of pentacyclic fused thiophene–phenylene–thiophene (TPT) unit with different acceptor units were reported by Chen and Li and coworkers [77–79]. The rigid planar structure of TPT facilitates effective π-conjugation and strong intermolecular interactions. High hole mobility and PCE beyond 6% were demonstrated. As discussed earlier, the introduction of 2D structure by attachment of conjugated side chains has also proved to be effective for the improvement of carrier mobility [65, 66].

The solubility of polymers can directly influence the solution processability, nanoscale morphology, and device efficiency. Polymers based on DPP and DTBT comonomers usually exhibit poor solubility in common organic solvents because of their rigid planar structure. Janssen and coworkers successfully improved the solubility of DPP-based polymers by using extended side chains (i.e. 2-hexyldecyl (HD)) on DPP and achieved a high PCE of 5.5% [80, 81]. Qin *et al.* reported a PCE of 6.1% for a copolymer based on carbazole and octyloxy chains-substituted DTBT [82]. The flexible side chains improve the solubility of the polymer and meanwhile help maintain a planar conformation of the polymer chain.

12.3.1.4 New Fullerene Derivative

PCBM was firstly introduced in PSCs in 1995 [9] and since then no other acceptors can outperform it. However, the photovoltaic properties of the C_{60} derivatives can be further enhanced by fine-tuning their LUMO energy levels and improving their absorbance in the visible region.

The potential to increase efficiency comes from tuning the LUMO levels of the fullerenes relative to the LUMO of the donor to ensure efficient charge transfer and a high open-circuit voltage simultaneously. So far, one successful strategy to modify fullerene LUMO is attaching electron-donating groups to the carbon cage [83], which might make the fullerene less electron deficient and hence shift its LUMO energy level up. Another breakthrough came when Blom and coworkers successfully demonstrated that fullerene multiadducts give a 100–200 mV higher lying LUMO compared to pristine C_{60} [84]. The attachment of multiple side groups increased the fullerene bandgap by reducing the conjugation of carbon cage, thereby shifting the HOMO level up. Combining the effects of the electron-donating indene moiety and the effect of bisadduct on the upshift of the LUMO energy level of PCBM, Li and coworkers synthesized indene-C_{60} bisadduct (ICBA) for raising its LUMO level. The PSCs based on P3HT/ICBA reached a PCE of 6.48% with a high V_{OC} of 0.84 V compared to a V_{OC} of 0.58 V for PC60BM-based devices [22].

However, despite this remarkably high efficiency, ICBA has not shown good performance when used with other low-bandgap polymers. An alternative path to modify the LUMO level of fullerenes for OPV applications was using expensive endohedral fullerenes introduced by Drees and coworkers [85].

PC60BM exhibits weak absorption in the visible region due to its highly symmetric structure, whereas PC70BM and PC84BM show much stronger absorbance than that of PC60BM in the UV–visible region from 200 to 700 nm. However, their cost is much higher than that of PC60BM, which could be a problem for future commercial applications in PSCs.

In general, the material cost for endohedral fullerenes, PC70BM, and PC84BM is high whereas bis-PCBM has not shown good compatibility with polymers other than P3HT. Thus, the search for new fullerene derivatives with strong visible absorption and high-lying LUMO energy levels still remains a challenge. Recently, an interesting fullerene derivative was reported by Mikroyannidis et al. [86], showing an improved absorption in UV–Vis region and a high LUMO level. The PSCs using this new fullerene together with P3HT show a high V_{OC} of 0.86 V and a PCE of 5.25%. The shift of the LUMO level is attributed to the presence of the strong electron-withdrawing nitro and cyano groups attached on the side chain.

12.3.2
Optimization of Polymer Solar Cell Devices

Since the invention of the BHJ structure, there has been great success in the development of high-efficiency polymer solar cells. Although recently PCEs of over 8% have been achieved for lab-scale PSCs [5, 6], many challenges remain to approach the theoretical efficiency of OPV, that is, 15–20% [21]. Further optimization of PSCs includes the morphology improvement of the D/A blend nanostructure and the use of new device architectures to minimize energy losses. We will review recent advances in these two fields, emphasizing on processing parameters, postprocessing treatments, and tandem structure.

12.3.2.1 Morphology Control

The active layer of a BHJ cell is typically a binary blend of a conjugated hole transporting polymer and a soluble fullerene, codeposited from solution. These two components have different surface energies and entropies of mixing. Hence, nanoscale phase separation readily occurs upon film formation from solution cast. In general, three domains will be formed in the blend film, including pure crystalline and amorphous domains of each component and amorphous domains containing mixed components. High-purity, crystalline domains possess efficient charge transport but with probability of less efficient exciton dissociation, whereas mixed domains have the opposite properties. The size of the demixed domain is determined by the degree of phase segregation governed by different processing conditions or postprocessing treatments. A minimum degree of phase separation results in well-mixed film with small domain size and low conductivity while a maximum degree of demixing results in large domain size and reduced D/A interfaces, thereby resulting in poor charge dissociation. By well control of the phase segregation, the demixed domain can form an interpenetrating bicontinuous network for efficient charge transport. Meanwhile, the domain sizes can be kept comparable to the exciton diffusion length to ensure excellent charge dissociation. In order to probe the film morphology, transmission electron microscopy (TEM), scanning electron microscope (SEM), and 2D grazing incidence X-ray scattering (2D GISAXS or GIWAXS) have been widely used to gather the information.

In general, phase segregation has to be introduced and the aggregation sizes have to be well controlled. The trade-off between network connectivity and optimum domain sizes becomes the key factor for achieving the ideal morphology. The manipulation of the size and connectivity of demixed phases strongly depends on the fabrication parameters such as blending ratio, solvent selection, and annealing treatment, which have been discussed in detail in some recent reviews [87–92].

Blending Ratio For BHJ solar cells, it has long been observed that the optimum device performance strongly depends on the ratio of the components involved. For the system of P3HT/PCBM, an optimum blending ratio of around 1 : 1 was widely reported. For MDMO-PPV/PCBM composites, however, the optimum blending ratio is 1 : 4. Little is known about what material properties are responsible for determining the optimal blend ratio.

The chemical structure of blend materials, such as the type, length, and regioregularity of the polymer side chains, was reported to strongly influence the morphology and performance of PSCs [93–95]. Moreover, recently the density of the polymer side chains has been suggested to be the origin of the blend ratios required to achieve optimal performance in some BHJs [96, 97]. A well-ordered semicrystalline polymer poly(2,5-bis(3-tetradecyllthiophen-2-yl)thieno[3,2-b]thiophene) (pBTTT) was used together with fullerene to make BHJ cells. They found that 1 : 4 blends had a cell efficiency of 2.35%, while the 1 : 1 blends had a low efficiency of only 0.16%. They calculated that there is enough space between the side chains along the polymer backbone to incorporate a single fullerene molecule per repeat unit (see Figure 12.5). However, such a 1 : 1 blending ratio can only fill all the free volume between side chains with fullerenes, without providing an effective percolation path for electron transport. As a result of unbalanced charge transport, the device efficiency of the 1 : 1 blend is low. For higher loadings, excess PCBM leads to the formation of electron percolation pathway outside the mixed crystals and sets the optimal blend ratio of 1 : 4 PBTTT/PCBM for higher PCE.

The length and density of polymer side chains determine whether intercalation of fullerene occurs or not. For materials with short and high-density side chains, such as P3HT, there is insufficient free volume for intercalation to occur. Therefore, a 1 : 1 blending ratio can provide enough PCBM to form the electron-transporting network and achieve optimum device performance.

Solvent Effects The essence of choosing different solvent types, mixed solvents, and solvent additives is to better control the time and degree of phase segregation of the blending film. The spin casting is a short and dynamic process that, in most

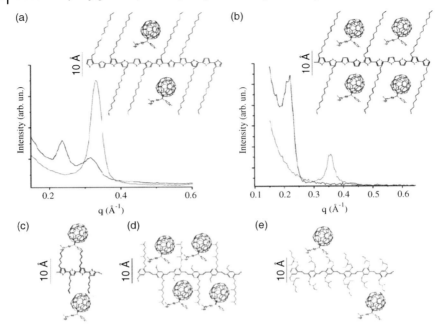

Figure 12.5 Fullerene intercalation in different polymer/fullerene systems. (a) The expansion of the d-spacing of the pTT upon the addition of PC60BM and the inset shows how the PC60BM fits between the side chains. (b) The same situation exists for PQT. (c) There is insufficient room between the side chains of P3HT to allow for intercalation. (d) There is sufficient room for PC60BM intercalation between the side chains in amorphous MDMO-PPV. (e) BisOC10-PPV, however, does not have sufficient room. Reproduced with permission from Ref. [96].

cases, does not provide enough time and intermolecular drive force for the development of phase separation and molecular self-organization. Hence, various strategies have been developed to solve this problem.

The phase segregation of PCBM was reported to be too big if "bad solvent" such as toluene was used [20]. Due to the poor solubility of PCBM in toluene, the average size of PCBM-rich domains reaches ~100 nm, which causes inefficient exciton dissociation. The use of "good solvent" such as chlorobenzene can greatly reduce the size of PCBM clusters and hence improve the device performance. However, using mixture of "bad solvent" and "good solvent" provides a new strategy to develop phase segregation earlier in solution phase rather than in spin coating process. For polymers, the adding of bad solvent can reduce the polymer–solvent interaction, and the polymer chains adopt a tight and contracted conformation with more interchain interactions, resulting in highly crystalline aggregations in both solutions and film [26, 98]. For fullerenes, in the presence of a bad solvent, their molecules crystallize to well-distributed clusters in the host solvent [99]. Therefore, by carefully adjusting the ratio of the solvent mixture, the size of the aggregates in solution can be well controlled so that the as-cast BHJ film has a finely demixed nanomorphology that does not need further postprocessing treatment. Several groups reported enhancements in the performance of polymer:PCBM BHJs by forming aggregated polymer whiskers in solution prior to spin casting [26, 100–102]. Very recently, Tao and coworkers reported a PCE improvement from 6.0 to 7.1% by adding bad solvents such as dimethyl sulfoxide or dimethylformamide in orthodichlorobenzene solution. The enhancement of the PSCs was attributed to the improved domain structure and hole mobility in the active layer [98].

In general, longer drying time during spin casting can help the formation of phase segregation and highly ordered polymer domains. High boiling point solvents such as chlorobenzene and dichlorobenzene are often used due to their low evaporation rate. Recently, the use of solvent additives has been widely reported as an effective approach to form desired nanonetwork morphology [3, 103, 104]. The additives are typically higher boiling point solvents, such as alkylthiols with the ability to selectively dissolve the fullerene component, whereas the polymer is less soluble. As the host solvent evaporates during spin coating, high boiling point additive stays in the film to keep the fullerene molecules in dissolved form for a longer time. The mobile fullerene can diffuse out of the polymer matrix during the self-organization of polymer chains, therefore allowing for a better demixed binary network structure, as shown in Figure 12.6. So far, the mostly studied and understood processing additives are alkylthiol-based compounds. However, current

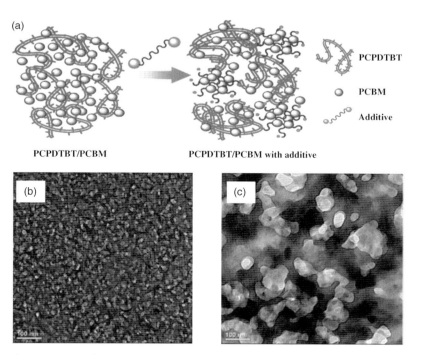

Figure 12.6 (a) Schematic depiction of the role of the processing additive in the self-assembly of bulk heterojunction blend materials. TEM images of exposed PCPDTBT networks after removal of PCBM without (b) and with (c) 1,8-octanedithiol. Reproduced with permission from Ref. [103].

state-of-the-art PSCs were mostly fabricated by using 1,8-diiodooctane [103] as solvent additive [5, 58, 98, 105].

Solvent annealing is another widely used approach to effectively extend the time of film formation by slowing the evaporation of solvent in an enclosed atmosphere. The elongated drying time allows polymer chains to keep mobile and facilitating self-organization. Yang and coworkers successfully introduced solvent annealing step by slowing the evaporation of 1,2-dichlorobenzene and 1,2,4-trichlorobenzene during casting [1, 106]. Zhao *et al.* reported a solvent vapor annealing approach with similar steps [107]. The improved packing of polymer domain was reflected in higher hole mobility, higher fill factors, reduced serial resistance, and overall improved device efficiency.

Thermal Annealing Since thermal annealing is carried out after formation of the solid film, elevated temperature is needed to keep blend molecules in a mobile state. Thermal annealing has the effect of allowing the morphology to evolve toward the most energetically favorable demixed state. The polymers tend to form crystalline domains, allowing the molecular diffusion of fullerene out of the polymer matrix upon annealing. The further crystallization of the PCBM clusters leads to the formation of bicontinuous percolation path with enhanced optical absorption and charge carrier mobilities. Annealing temperature and time are the most critical parameters to control the domain size and achieve the optimized performance.

In 2003, Padinger and coworkers achieved a high PCE of 3.5% by applying thermal annealing on a P3HT/PCBM-based BHJ device [108]. In 2005, Ma *et al.* reported a remarkable PCE of 5% for a fully optimized P3HT:PCBM cell by carefully adjusting the parameters of thermal annealing [2]. The improved current is attributed to the enhanced light absorbance and increased polymer crystallinity. The large fill factor is due to the decreased series resistance of both the film and the interface. The morphology was also investigated by TEM. The evolution of phase segregation under elevated temperature was observed and the optimum domain size was measured to be around 20 nm (see Figure 12.7).

Although thermal annealing is a powerful tool in improving the morphology and performance of P3HT: PCBM based devices, it has not proved advantageous in BHJs with many newly designed donor–acceptor polymers.

Summary of Morphology Control Postprocessing treatments such as thermal and solvent annealing have some disadvantages in comparison to approaches used during the device fabrication processing. In a dry film, the demixing of polymer and fullerene is strongly affected by their intermolecular interactions, which is determined by their material chemical structures, physical properties, and the characteristics of the as-cast film. Therefore, the optimum conditions of thermal annealing strongly depend on the used materials and processing parameters. Moreover, thermal annealing usually has negative effects on BHJ system with high PCBM loading because the fast crystallization of small fullerene molecules can lead to the overgrowth of PCBM-rich domains [2]. The solvent annealing still works for

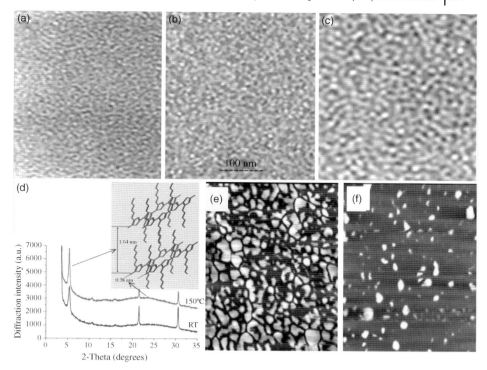

Figure 12.7 TEM images of P3HT/PCBM film bulk morphology before thermal annealing (a), after thermal annealing at 150 °C for 30 min (b) and after 2 h (c). XRD spectra of a P3HT/PCBM film drop-cast onto a PEDOT/ITO substrate with and without thermal annealing at 150 °C for 30 min (d). The inset shows the P3HT crystal structure. AFM phase images of the Al/P3HT interface annealed before and after Al deposition are shown in (e) and (f), respectively. Reproduced with permission from Ref. [2].

P3HT/PCBM BHJ cell with high PCBM loading [109]. But it has not proved to be a universal strategy for optimization of BHJ devices with different polymer/fullerene systems.

The use of mixed solvents, including additives, can develop the materials' aggregation in a much simpler solution phase [100]. The forming of high-purity aggregates is readily realized in solution due to the high mobility of dissolved material molecules and their relatively large spacing distance. The domain size of aggregates can be tuned by choosing the proper type and volume of bad solvent. So far, the use of mixed solvents is the most widely applied strategy for the optimization of BHJ polymer solar cells.

12.3.2.2 Device Architectures

The morphology control has proved to be extremely important for the optimization of PSCs. Recent developments in device architectures have also shown promising potential for further increasing the performance of PSCs.

Inverted Device Structures The conventional device structure for PSCs is indium tin oxide (ITO)/PEDOT:PSS/polymer blend/Al, where a conductive high-work-function PEDOT:PSS layer is used for anode contact, and a low-work-function metal as the cathode. Both the PEDOT:PSS layer and the low-work-function metal cathode can cause the degradation of PSCs [110–112]. The acidic PEDOT:PSS was reported to etch the ITO and cause interface instability through indium diffusion into the polymer active layer. Low-work-function metals, such as calcium and aluminum, are easily oxidized when exposed to air, increasing the series resistance at the metal/BHJ interface and degrading device performance.

In an inverted structure, the ITO substrate is covered with a low-work-function compound to create an effective electron-selective contact. On the contrary, the original cathode is substituted with a high-work-function interlayer covered by a stable metal electrode, such as Au or Ag, to collect holes. By using the inverted structure, the potential interface instability is successfully overcome. Moreover, in the inverted cell, the anode is a high-work-function metal such as Ag, which can be coated using printing technology to simplify and reduce the cost of manufacturing [113]. The challenges are the search of interlayers to improve the selectivity of charge collection at both electrodes and to introduce minimum electrical and optical losses.

The most widely used interlayers for electron extraction are based on solution-processed transition metal oxide such as titanium oxide and zinc oxide that possesses suitable energy levels for efficient electron collecting and hole blocking. Brabec and coworkers presented 3.1% PCE PSCs based on inverted device architecture using a low-temperature solution-processed TiO_x interlayer as electron-selective contact [114]. The better FF of inverted devices compared to conventional ones indicates that the ITO:TiO_x contact has better hole-blocking capability than LiF/Al. Similarly, solution-processed ZnO as the cathode buffer layer has also been reported [23, 115]. Recently, Heeger and coworkers demonstrated air-stable inverted BHJ solar cells using a low-temperature annealed sol–gel-derived ZnO film as an electron transport layer. The low annealing temperature of the ZnO film is compatible with fabrication on flexible substrates [116]. Cs_2CO_3 was also reported as an effective electron-collecting layer for PSCs by Yang and coworkers [117]. The interface dipole formed between Cs_2CO_3 and the ITO surface was considered to be the reason for reduction of the ITO work function [118]. Further optimization of the electron-collecting layer can be realized by modifying its interface with ITO/or the active organic materials. Inserting an ultrathin organic layer between the electron-selective TiO_x layer and the ITO substrate was reported to significantly improve the wetting and work function of the TiO_x interlayer [119–121].

At the anode side, thermally evaporated layers of a transition metal oxide such as MoO_3 [122], WO_3 [123], and V_2O_5 [117] have been used for efficient hole extraction. Very recently, solution-processed vanadium oxides as a hole-collecting interlayer for high-performance PSCs have been reported by several groups [124–126]. The PSCs using this sol–gel fabricated VO_x demonstrated improved device performance and durability compared to devices using thermally evaporated V_2O_5 [125].

12.3 Optimization of Fullerene/Polymer Solar Cells

Tandem Structure One of the reasons for low performance of PSCs is the narrow absorption range of polymer materials. Photons with energies smaller than bandgap will not be absorbed. Another drawback of polymer materials is their low carrier mobility, which limits the thickness of the active layer in PSCs. This restriction on the film thickness leads to insufficient absorption even in the absorption range of the active materials. In addition, when photons with energy greater than bandgap are absorbed, thermalization of hot carriers becomes a major loss for single-junction PSCs.

A typical organic tandem cell consists of two subcells stacked on top of each other, each of which has different absorption range (Figure 12.8). The two subcells are connected in series via a thin recombination layer. Hence, better coverage of the solar spectrum and reduction of the thermalization loss can be realized due to the use of materials with different bandgaps. The total thickness of the active films in

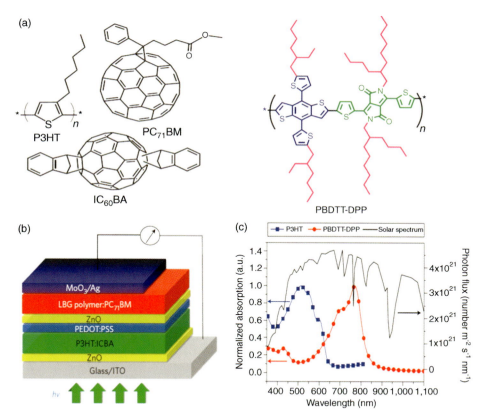

Figure 12.8 (a) Chemical structures of the materials used in tandem solar cells. (b) Device structure of the high-efficiency tandem solar cell. (c) UV–visible absorption spectra of PBDTT–DPP and P3HT films and the solar radiation spectrum. The UV–visible absorption profiles of PBDTT–DPP and P3HT show that the two materials cover the solar spectrum from 350 to 850 nm in a complementary manner. Reproduced with permission from Ref. [6].

tandem cells is also much larger than that of single cells, leading to more efficient light absorption. As a result, the theoretical efficiency of polymer tandem solar cells reaches 15% compared to 11% for single-junction PSCs [21, 127]. The optimization of tandem cells requires carefully choosing active materials in order to achieve balanced photocurrent and minimal absorption overlap between the two subcells. Another challenge is the fabrication of a solution-processed recombination layer that introduces minimal electrical and optical losses.

Kawano et al. demonstrated one of the first polymer tandem cells using MDMO-PPV/PCBM BHJ subcells that were connected by sputter deposited ITO [128]. Blom and coworkers for the first time fabricated a tandem cell using a low-bandgap and large-bandgap polymers [129]. Janssen and coworkers reported the earliest all-solution-processed tandem cells by using ZnO nanoparticle and PEDOT as the recombination layer [130]. Due to the limited choice of low-bandgap materials, the performance of tandem cells was relatively low. A major breakthrough in the area of polymer tandem cells was achieved by Heeger and coworkers in 2007 [131]. Efficient tandem cells with a PCE of 6.5% were fabricated by employing solution-processed TiO_x/PEDOT as the recombination layer. The use of a novel low-bandgap polymer and P3HT allowed a reduction of the overlap between the respective absorption spectrum of the active blends. Since then, little progress has been made [132] in demonstrating higher efficiency tandem cells mainly due to the lack of a suitable low-bandgap polymer. In 2012, Yang and coworkers reported a certified PCE of 8.62% for an inverted tandem PSC, which is the highest certified efficiency for PSCs to date [6]. The remarkably high efficiency is attributed to the use of a novel polymer (PBDTT–DPP) that has a small optical bandgap, deep HOMO level, and high hole mobility. Single-junction BHJ cells using PBDTT–DPP/C_{70} showed PCEs of more than 6%.

12.4
Outlook

Polymer/fullerene solar cells have advanced rapidly in recent years, making the technology a promising candidate for future low-cost photovoltaic applications. Recent developments in materials synthesis have improved the efficiencies of PSCs above the 7% level [58, 59, 64]. Based on these new materials, PSCs with efficiencies over 8% have also been demonstrated by using efficient interlayer and tandem structures [5, 6]. To achieve PCEs over 10%, it requires the use of novel active materials and tandem structures. Alternating donor and acceptor units in copolymers has been the most effective approach to obtain highly efficient polymers. And it shows promising potential to be used to develop future low-bandgap polymers with further improved photovoltaic properties. In parallel, recent improvements in the fabrication technology of polymer tandem cells have ensured better tandem cell efficiencies compared to the constituent subcells. Based on current advances in both material and device fields, PSCs with PCEs beyond 10% can be expected in the next few years.

References

1 Li, G. et al. (2005) High-efficiency solution processable polymer photovoltaic cells by self-organization of polymer blends. *Nat. Mater.*, **4** (11), 864–868.

2 Ma, W.L. et al. (2005) Thermally stable, efficient polymer solar cells with nanoscale control of the interpenetrating network morphology. *Adv. Funct. Mater.*, **15** (10), 1617–1622.

3 Peet, J. et al. (2007) Efficiency enhancement in low-bandgap polymer solar cells by processing with alkane dithiols. *Nat. Mater.*, **6** (7), 497–500.

4 Gaudiana, R. and Brabec, C. (2008) Organic materials – fantastic plastic. *Nat. Photon.*, **2** (5), 287–289.

5 He, Z.C. et al. (2011) Simultaneous enhancement of open-circuit voltage, short-circuit current density, and fill factor in polymer solar cells. *Adv. Mater.*, **23** (40), 4636–4640.

6 Dou, L. et al. (2012) Tandem polymer solar cells featuring a spectrally matched low-bandgap polymer. *Nat. Photon.* doi: 10.1038/NPHOTON.2011.356.

7 Sariciftci, N.S. et al. (1992) Photoinduced electron-transfer from a conducting polymer to buckminsterfullerene. *Science*, **258** (5087), 1474–1476.

8 Tang, C.W. (1986) Two-layer organic photovoltaic cell. *Appl. Phys. Lett.*, **48** (2), 183–185.

9 Yu, G. et al. (1995) Polymer photovoltaic cells – enhanced efficiencies via a network of internal donor–acceptor heterojunctions. *Science*, **270** (5243), 1789–1791.

10 Halls, J.J.M. et al. (1995) Efficient photodiodes from interpenetrating polymer networks. *Nature*, **376** (6540), 498–500.

11 Parker, I.D. (1994) Carrier tunneling and device characteristics in polymer light-emitting diodes. *J. Appl. Phys.*, **75** (3), 1656–1666.

12 Scharber, M.C. et al. (2006) Design rules for donors in bulk-heterojunction solar cells – towards 10% energy-conversion efficiency. *Adv. Mater.*, **18** (6), 789.

13 Gadisa, A. et al. (2004) Correlation between oxidation potential and open-circuit voltage of composite solar cells based on blends of polythiophenes/fullerene derivative. *Appl. Phys. Lett.*, **84** (9), 1609–1611.

14 Barker, J.A., Ramsdale, C.M., and Greenham, N.C. (2003) Modeling the current-voltage characteristics of bilayer polymer photovoltaic devices. *Phys. Rev. B*, **67** (7), 75–205.

15 Cheyns, D. et al. (2008) Analytical model for the open-circuit voltage and its associated resistance in organic planar heterojunction solar cells. *Phys. Rev. B.*, **77** (16), 165–332.

16 Heeger, A.J., Sariciftci, N.S., and Namdas, E.B. (2010) *Semiconducting and Metallic Polymers*, Oxford University Press.

17 Veldman, D., Meskers, S.C.J., and Janssen, R.A.J. (2009) The energy of charge-transfer states in electron donor–acceptor blends: insight into the energy losses in organic solar cells. *Adv. Funct. Mater.*, **19** (12), 1939–1948.

18 Schilinsky, P. et al. (2004) Simulation of light intensity dependent current characteristics of polymer solar cells. *J. Appl. Phys.*, **95** (5), 2816–2819.

19 Riedel, I. and Dyakonov, V. (2004) Influence of electronic transport properties of polymer–fullerene blends on the performance of bulk heterojunction photovoltaic devices. *Phys. Status Solidi (a)*, **201** (6), 1332–1341.

20 Shaheen, S.E. et al. (2001) 2.5% efficient organic plastic solar cells. *Appl. Phys. Lett.*, **78** (6), 841–843.

21 Ameri, T. et al. (2009) Organic tandem solar cells: a review. *Energy Environ. Sci.*, **2** (4), 347–363.

22 Zhao, G.J., He, Y.J., and Li, Y.F. (2010) 6.5% efficiency of polymer solar cells based on poly(3-hexylthiophene) and indene-C_{60} bisadduct by device optimization. *Adv. Mater.*, **22** (39), 4355–+.

23 White, M.S. et al. (2006) Inverted bulk-heterojunction organic photovoltaic device using a solution-derived ZnO underlayer. *Appl. Phys. Lett.*, **89** (14), 143–517.

24. Brabec, C.J. et al. (2002) Effect of LiF/metal electrodes on the performance of plastic solar cells. *Appl. Phys. Lett.*, **80** (7), 1288–1290.
25. Blom, P.W.M. et al. (2007) Device physics of polymer:fullerene bulk heterojunction solar cells. *Adv. Mater.*, **19** (12), 1551–1566.
26. Moule, A.J. and Meerholz, K. (2008) Controlling morphology in polymer–fullerene mixtures. *Adv. Mater.*, **20** (2), 240.
27. Jespersen, K.G. et al. (2004) The electronic states of polyfluorene copolymers with alternating donor–acceptor units. *J. Chem. Phys.*, **121** (24), 12613–12617.
28. Persson, N.K. et al. (2005) Optical properties of low band gap alternating copolyfluorenes for photovoltaic devices. *J. Chem. Phys.*, **123** (20), 204–718.
29. Yang, L., Feng, J.K., and Ren, A.M. (2005) Theoretical studies on the electronic and optical properties of two thiophene-fluorene based pi-conjugated copolymers. *Polymer*, **46** (24), 10970–10981.
30. Hou, J.H. et al. (2006) Synthesis and absorption spectra of poly(3-(phenylenevinyl)thiophene)s with conjugated side chains. *Macromolecules*, **39** (2), 594–603.
31. Hou, J.H. et al. (2006) Synthesis and photovoltaic properties of two-dimensional conjugated polythiophenes with bi(thienylenevinylene) side chains. *J. Am. Chem. Soc.*, **128** (14), 4911–4916.
32. Chang, Y.T. et al. (2009) Intramolecular donor–acceptor regioregular poly(hexylphenanthrenyl-imidazole thiophene) exhibits enhanced hole mobility for heterojunction solar cell applications. *Adv. Mater.*, **21** (20), 2093–2097.
33. van Mullekom, H.A.M., Vekemans, J.A.J.M., and Meijer, E.W. (1996) Alternating copolymer of pyrrole and 2,1,3-benzothiadiazole. *Chem. Commun.*, (18), 2163–2164.
34. Aasmundtveit, K.E. et al. (2000) Structural anisotropy of poly(alkylthiophene) films. *Macromolecules*, **33** (8), 3120–3127.
35. Pei, Q.B. et al. (1994) Electrochromic and highly stable poly(3,4-ethylenedioxythiophene) switches between opaque blue-black and transparent sky blue. *Polymer*, **35** (7), 1347–1351.
36. Zhang, Q.T. and Tour, J.M. (1998) Alternating donor/acceptor repeat units in polythiophenes. Intramolecular charge transfer for reducing band gaps in fully substituted conjugated polymers. *J. Am. Chem. Soc.*, **120** (22), 5355–5362.
37. Kitamura, C., Tanaka, S., and Yamashita, Y. (1996) Design of narrow-bandgap polymers. Syntheses and properties of monomers and polymers containing aromatic-donor and o-quinoid-acceptor units. *Chem. Mater.*, **8** (2), 570–578.
38. Brocks, G. and Tol, A. (1996) Small band gap semiconducting polymers made from dye molecules: polysquaraines. *J. Phys. Chem.*, **100** (5), 1838–1846.
39. Cheng, Y.J., Yang, S.H., and Hsu, C.S. (2009) Synthesis of conjugated polymers for organic solar cell applications. *Chem. Rev.*, **109** (11), 5868–5923.
40. Boudreault, P.L.T., Najari, A., and Leclerc, M. (2011) Processable low-bandgap polymers for photovoltaic applications. *Chem. Mater.*, **23** (3), 456–469.
41. Facchetti, A. (2011) π-Conjugated polymers for organic electronics and photovoltaic cell applications. *Chem. Mater.*, **23** (3), 733–758.
42. Chen, J.W. and Cao, Y. (2009) Development of novel conjugated donor polymers for high-efficiency bulk-heterojunction photovoltaic devices. *Acc. Chem. Res.*, **42** (11), 1709–1718.
43. Huo, L.J. and Hou, J.H. (2011) Benzo[1,2-b:4,5-b']dithiophene-based conjugated polymers: band gap and energy level control and their application in polymer solar cells. *Polym. Chem.*, **2** (11), 2453–2461.
44. Li, Y. (2011) Molecular design of photovoltaic materials for polymer solar cells: toward suitable electronic energy levels and broad absorption. *Acc. Chem. Res.* doi: 10.1021/ar2002446.
45. Chen, M.H. et al. (2009) Efficient polymer solar cells with thin active layers

based on alternating polyfluorene copolymer/fullerene bulk heterojunctions. *Adv. Mater.*, **21** (42), 4238–+.

46 Boudreault, P.L.T., Michaud, A., and Leclerc, M. (2007) A new poly(2,7-dibenzosilole) derivative in polymer solar cells. *Macromol. Rapid. Comm.*, **28** (22), 2176–2179.

47 Wang, E.G. *et al.* (2008) High-performance polymer heterojunction solar cells of a polysilafluorene derivative. *Appl. Phys. Lett.*, **92** (3), 33–307.

48 Muhlbacher, D. *et al.* (2006) High photovoltaic performance of a low-bandgap polymer. *Adv. Mater.*, **18** (21), 2884.

49 Zhang, M. *et al.* (2007) Field-effect transistors based on a benzothiadiazole–cyclopentadithiophene copolymer. *J. Am. Chem. Soc.*, **129** (12), 3472.

50 Tsao, H.N. *et al.* (2009) The influence of morphology on high-performance polymer field-effect transistors. *Adv. Mater.*, **21** (2), 209.

51 Chen, H.Y. *et al.* (2010) Silicon atom substitution enhances interchain packing in a thiophene-based polymer system. *Adv. Mater.*, **22** (3), 371.

52 Hou, J.H. *et al.* (2008) Synthesis, characterization, and photovoltaic properties of a low band gap polymer based on silole-containing polythiophenes and 2,1,3-benzothiadiazole. *J. Am. Chem. Soc.*, **130** (48), 16144.

53 Coffin, R.C. *et al.* (2009) Streamlined microwave-assisted preparation of narrow-bandgap conjugated polymers for high-performance bulk heterojunction solar cells. *Nat. Chem.*, **1** (8), 657–661.

54 Chu, T.Y. *et al.* (2011) Bulk heterojunction solar cells using thieno[3,4-c]pyrrole-4,6-dione and dithieno[3,2-b:2′,3′-d]silole copolymer with a power conversion efficiency of 7.3%. *J. Am. Chem. Soc.*, **133** (12), 4250–4253.

55 Pan, H.L. *et al.* (2007) Low-temperature, solution-processed, high-mobility polymer semiconductors for thin-film transistors. *J. Am. Chem. Soc.*, **129** (14), 4112.

56 Hou, J.H. *et al.* (2008) Bandgap and molecular energy level control of conjugated polymer photovoltaic materials based on benzo[1,2-b:4,5-b′]dithiophene. *Macromolecules*, **41** (16), 6012–6018.

57 Liang, Y.Y. *et al.* (2009) Development of new semiconducting polymers for high performance solar cells. *J. Am. Chem. Soc.*, **131** (1), 56.

58 Chen, H.Y. *et al.* (2009) Polymer solar cells with enhanced open-circuit voltage and efficiency. *Nat. Photon.*, **3** (11), 649–653.

59 Liang, Y.Y. *et al.* (2009) Highly efficient solar cell polymers developed via fine-tuning of structural and electronic properties. *J. Am. Chem. Soc.*, **131** (22), 7792–7799.

60 Zou, Y.P. *et al.* (2010) A thieno[3,4-c]pyrrole-4,6-dione-based copolymer for efficient solar cells. *J. Am. Chem. Soc.*, **132** (15), 5330.

61 Piliego, C. *et al.* (2010) Synthetic control of structural order in N-alkylthieno[3,4-c]pyrrole-4,6-dione-based polymers for efficient solar cells. *J. Am. Chem. Soc.*, **132** (22), 7595.

62 Zhang, Y. *et al.* (2010) Efficient polymer solar cells based on the copolymers of benzodithiophene and thienopyrroledione. *Chem. Mater.*, **22** (9), 2696–2698.

63 Zhang, G.B. *et al.* (2010) Benzo[1,2-b:4,5-b′]dithiophene-dioxopyrrolothiophen copolymers for high performance solar cells. *Chem. Commun.*, **46** (27), 4997–4999.

64 Zhou, H.X. *et al.* (2011) Development of fluorinated benzothiadiazole as a structural unit for a polymer solar cell of 7% efficiency. *Angew. Chem., Int. Ed.*, **50** (13), 2995–2998.

65 Huo, L.J. *et al.* (2010) A polybenzo[1,2-b:4,5-b′]dithiophene derivative with deep HOMO level and its application in high-performance polymer solar cells. *Angew. Chem., Int. Ed.*, **49** (8), 1500–1503.

66 Huo, L.J. *et al.* (2011) PBDTTTZ: a broad band gap conjugated polymer with high photovoltaic performance in polymer solar cells. *Macromolecules*, **44** (11), 4035–4037.

67 Wang, M. et al. (2011) Donor–acceptor conjugated polymer based on naphtho [1,2-c:5,6-c]bis[1,2,5]thiadiazole for high-performance polymer solar cells. *J. Am. Chem. Soc.*, **133** (25), 9638–9641.

68 Huo, L.J. et al. (2011) Replacing alkoxy groups with alkylthienyl groups: a feasible approach to improve the properties of photovoltaic polymers. *Angew. Chem., Int. Ed.*, **50** (41), 9697–9702.

69 Xiao, S.Q. et al. (2009) Conjugated polymers based on benzo[2,1-b:3,4-b'] dithiophene with low-lying highest occupied molecular orbital energy levels for organic photovoltaics. *ACS Appl. Mater. Interfaces*, **1** (7), 1613–1621.

70 Xiao, S.Q. et al. (2010) Conjugated polymer based on polycyclic aromatics for bulk heterojunction organic solar cells: a case study of quadrathienonaphthalene polymers with 2% efficiency. *Adv. Funct. Mater.*, **20** (4), 635–643.

71 Price, S.C., Stuart, A.C., and You, W. (2010) Low band gap polymers based on benzo[1,2-b:4,5-b']dithiophene: rational design of polymers leads to high photovoltaic performance. *Macromolecules*, **43** (10), 4609–4612.

72 Yang, L.Q., Zhou, H.X., and You, W. (2010) Quantitatively analyzing the influence of side chains on photovoltaic properties of polymer–fullerene solar cells. *J. Phys. Chem. C*, **114** (39), 16793–16800.

73 Zhou, H.X. et al. (2010) A weak donor–strong acceptor strategy to design ideal polymers for organic solar cells. *ACS Appl. Mater. Interfaces*, **2** (5), 1377–1383.

74 Zhou, H.X. et al. (2010) Enhanced photovoltaic performance of low-bandgap polymers with deep LUMO levels. *Angew. Chem., Int. Ed.*, **49** (43), 7992–7995.

75 Wakim, S. et al. (2011) New low band gap thieno[3,4-b]thiophene-based polymers with deep HOMO levels for organic solar cells. *J. Mater. Chem.*, **21** (29), 10920–10928.

76 Lee, D., Stone, S.W., and Ferraris, J.P. (2011) A novel dialkylthio benzo[1,2-b:4,5-b']dithiophene derivative for high open-circuit voltage in polymer solar cells. *Chem. Commun.*, **47** (39), 10987–10989.

77 Chen, Y.C. et al. (2010) Low-bandgap conjugated polymer for high efficient photovoltaic applications. *Chem. Commun.*, **46** (35), 6503–6505.

78 Zhang, Y. et al. (2011) Synthesis, characterization, charge transport, and photovoltaic properties of dithienobenzoquinoxaline- and dithienobenzopyridopyrazine-based conjugated polymers. *Macromolecules*, **44** (12), 4752–4758.

79 Zhang, M.J. et al. (2011) Synthesis and photovoltaic properties of D–A copolymers based on alkyl-substituted indacenodithiophene donor unit. *Chem. Mater.*, **23** (18), 4264–4270.

80 Bijleveld, J.C. et al. (2009) Poly (diketopyrrolopyrrole-terthiophene) for ambipolar logic and photovoltaics. *J. Am. Chem. Soc.*, **131** (46), 16616–+.

81 Bijleveld, J.C. et al. (2010) Efficient solar cells based on an easily accessible diketopyrrolopyrrole polymer. *Adv. Mater.*, **22** (35), E242.

82 Qin, R.P. et al. (2009) A planar copolymer for high efficiency polymer solar cells. *J. Am. Chem. Soc.*, **131** (41), 14612.

83 Kooistra, F.B. et al. (2007) Increasing the open circuit voltage of bulk-heterojunction solar cells by raising the LUMO level of the acceptor. *Org. Lett.*, **9** (4), 551–554.

84 Lenes, M. et al. (2008) Fullerene bisadducts for enhanced open-circuit voltages and efficiencies in polymer solar cells. *Adv. Mater.*, **20** (11), 2116.

85 Ross, R.B. et al. (2009) Endohedral fullerenes for organic photovoltaic devices. *Nat. Mater.*, **8** (3), 208–212.

86 Mikroyannidis, J.A. et al. (2011) A simple and effective modification of PCBM for use as an electron acceptor in efficient bulk heterojunction solar cells. *Adv. Funct. Mater.*, **21** (4), 746–755.

87 Brabec, C.J. et al. (2010) Polymer–fullerene bulk-heterojunction solar cells. *Adv. Mater.*, **22** (34), 3839–3856.

88 Chen, L.M. et al. (2009) Recent progress in polymer solar cells: manipulation of polymer:fullerene morphology and the

formation of efficient inverted polymer solar cells. *Adv. Mater.*, **21** (14–15), 1434–1449.

89 Ruderer, M.A. and Muller-Buschbaum, P. (2011) Morphology of polymer-based bulk heterojunction films for organic photovoltaics. *Soft Matter*, **7** (12), 5482–5493.

90 Pivrikas, A., Neugebauer, H., and Sariciftci, N.S. (2011) Influence of processing additives to nano-morphology and efficiency of bulk-heterojunction solar cells: a comparative review. *Sol. Energy*, **85** (6), 1226–1237.

91 Brabec, C.J. et al. (2011) Influence of blend microstructure on bulk heterojunction organic photovoltaic performance. *Chem. Soc. Rev.*, **40** (3), 1185–1199.

92 Slota, J.E., He, X.M., and Huck, W.T.S. (2010) Controlling nanoscale morphology in polymer photovoltaic devices. *Nano Today*, **5** (3), 231–242.

93 Nguyen, L.H. et al. (2007) Effects of annealing on the nanomorphology and performance of poly(alkylthiophene): fullerene bulk-heterojunction solar cells. *Adv. Funct. Mater.*, **17** (7), 1071–1078.

94 Kim, Y. et al. (2006) A strong regioregularity effect in self-organizing conjugated polymer films and high-efficiency polythiophene:fullerene solar cells. *Nat. Mater.*, **5** (3), 197–203.

95 Szarko, J.M. et al. (2010) When function follows form: effects of donor copolymer side chains on film morphology and BHJ solar cell performance. *Adv. Mater.*, **22** (48), 5468–5472.

96 Mayer, A.C. et al. (2009) Bimolecular crystals of fullerenes in conjugated polymers and the implications of molecular mixing for solar cells. *Adv. Funct. Mater.*, **19** (8), 1173–1179.

97 Koppe, M. et al. (2007) Polyterthiophenes as donors for polymer solar cells. *Adv. Funct. Mater.*, **17** (8), 1371–1376.

98 Chu, T.Y. et al. (2011) Morphology control in polycarbazole based bulk heterojunction solar cells and its impact on device performance. *Appl. Phys. Lett.*, **98** (25), 253–301.

99 Li, L.G., Lu, G.H., and Yang, X.N. (2008) Improving performance of polymer photovoltaic devices using an annealing-free approach via construction of ordered aggregates in solution. *J. Mater. Chem.*, **18** (17), 1984–1990.

100 Xin, H., Kim, F.S., and Jenekhe, S.A. (2008) Highly efficient solar cells based on poly(3-butylthiophene) nanowires. *J. Am. Chem. Soc.*, **130** (16), 5424.

101 Chen, F.C., Tseng, H.C., and Ko, C.J. (2008) Solvent mixtures for improving device efficiency of polymer photovoltaic devices. *Appl. Phys. Lett.*, **92** (10), 103–316.

102 Yao, Y. et al. (2008) Effect of solvent mixture on the nanoscale phase separation in polymer solar cells. *Adv. Funct. Mater.*, **18** (12), 1783–1789.

103 Lee, J.K. et al. (2008) Processing additives for improved efficiency from bulk heterojunction solar cells. *J. Am. Chem. Soc.*, **130** (11), 3619–3623.

104 Zhang, F.L. et al. (2006) Influence of solvent mixing on the morphology and performance of solar cells based on polyfluorene copolymer/fullerene blends. *Adv. Funct. Mater.*, **16** (5), 667–674.

105 Liang, Y.Y. et al. (2010) For the bright future – bulk heterojunction polymer solar cells with power conversion efficiency of 7.4%. *Adv. Mater.*, **22** (20), E135.

106 Chu, C.W. et al. (2008) Control of the nanoscale crystallinity and phase separation in polymer solar cells. *Appl. Phys. Lett.*, **92** (10), 103–306.

107 Zhao, Y. et al. (2007) Solvent-vapor treatment induced performance enhancement of poly(3-hexylthiophene): methanofullerene bulk-heterojunction photovoltaic cells. *Appl. Phys. Lett.*, **90** (4), 43–504.

108 Padinger, F., Rittberger, R.S., and Sariciftci, N.S. (2003) Effects of postproduction treatment on plastic solar cells. *Adv. Funct. Mater.*, **13** (1), 85–88.

109 Li, G. et al. (2007) "Solvent annealing" effect in polymer solar cells based on poly(3-hexylthiophene) and methanofullerenes. *Adv. Funct. Mater.*, **17** (10), 1636–1644.

110 Watanabe, A. and Kasuya, A. (2005) Effect of atmospheres on the open-circuit

photovoltage of nanoporous TiO$_2$/poly(3-hexylthiophene) heterojunction solar cell. *Thin Solid Films*, **483** (1–2), 358–366.

111 de Jong, M.P., van Ijzendoorn, L.J., and de Voigt, M.J.A. (2000) Stability of the interface between indium-tin-oxide and poly(3,4-ethylenedioxythiophene)/poly(styrenesulfonate) in polymer light-emitting diodes. *Appl. Phys. Lett.*, **77** (14), 2255–2257.

112 Sahin, Y. *et al.* (2005) Development of air stable polymer solar cells using an inverted gold on top anode structure. *Thin Solid Films*, **476** (2), 340–343.

113 Krebs, F.C. (2009) Polymer solar cell modules prepared using roll-to-roll methods: knife-over-edge coating, slot-die coating and screen printing. *Sol. Energy Mater. Sol. Cells*, **93** (4), 465–475.

114 Waldauf, C. *et al.* (2006) Highly efficient inverted organic photovoltaics using solution based titanium oxide as electron selective contact. *Appl. Phys. Lett.*, **89** (23), 233–517.

115 Hau, S.K. *et al.* (2008) Air-stable inverted flexible polymer solar cells using zinc oxide nanoparticles as an electron selective layer. *Appl. Phys. Lett.*, **92** (25), 253–301.

116 Sun, Y.M. *et al.* (2011) Inverted polymer solar cells integrated with a low-temperature-annealed sol-gel-derived ZnO film as an electron transport layer. *Adv. Mater.*, **23** (14), 1679.

117 Li, G. *et al.* (2006) Efficient inverted polymer solar cells. *Appl. Phys. Lett.*, **88** (25), 253–503.

118 Huang, J.S., Li, G., and Yang, Y. (2008) A semi-transparent plastic solar cell fabricated by a lamination process. *Adv. Mater.*, **20** (3), 415.

119 Steim, R. *et al.* (2008) Interface modification for highly efficient organic photovoltaics. *Appl. Phys. Lett.*, **92** (9), 93–303.

120 Tang, Z. *et al.* (2012) Interlayer for modified cathode in highly efficient inverted ITO-free organic solar cells. *Adv. Mater.*, **24** (4), 554.

121 Choi, H. *et al.* (2011) Combination of titanium oxide and a conjugated polyelectrolyte for high-performance inverted-type organic optoelectronic devices. *Adv. Mater.*, **23** (24), 2759–2763.

122 Zhao, D.W. *et al.* (2008) Efficient tandem organic solar cells with an Al/MoO$_3$ intermediate layer. *Appl. Phys. Lett.*, **93** (8), 83–305.

123 Tao, C. *et al.* (2009) Role of tungsten oxide in inverted polymer solar cells. *Appl. Phys. Lett.*, **94** (4), 43–311.

124 Chang, C.Y. *et al.* (2012) Combination of molecular, morphological, and interfacial engineering to achieve highly efficient and stable plastic solar cells. *Adv. Mater.*, **24** (4), 549.

125 Chen, C.P., Chen, Y.D., and Chuang, S.C. (2011) High-performance and highly durable inverted organic photovoltaics embedding solution-processable vanadium oxides as an interfacial hole-transporting layer. *Adv. Mater.*, **23** (33), 3859.

126 Zilberberg, K. *et al.* (2011) Inverted organic solar cells with sol-gel processed high work-function vanadium oxide hole-extraction layers. *Adv. Funct. Mater.*, **21** (24), 4776–4783.

127 Dennler, G. *et al.* (2008) Design rules for donors in bulk-heterojunction tandem solar cells – towards 15% energy-conversion efficiency. *Adv. Mater.*, **20** (3), 579.

128 Kawano, K. *et al.* (2006) Open circuit voltage of stacked bulk heterojunction organic solar cells. *Appl. Phys. Lett.*, **88** (7), 73–514.

129 Hadipour, A. *et al.* (2006) Solution-processed organic tandem solar cells. *Adv. Funct. Mater.*, **16** (14), 1897–1903.

130 Gilot, J., Wienk, M.M., and Janssen, R.A.J. (2007) Double and triple junction polymer solar cells processed from solution. *Appl. Phys. Lett.*, **90** (14), 143–512.

131 Kim, J.Y. *et al.* (2007) Efficient tandem polymer solar cells fabricated by all-solution processing. *Science*, **317** (5835), 222–225.

132 Yang, J. *et al.* (2011) A robust inter-connecting layer for achieving high performance tandem polymer solar cells. *Adv. Mater.*, **23** (30), 3465.

13
Semiconducting Nanocrystal/Conjugated Polymer Composites for Applications in Hybrid Polymer Solar Cells

Michael Krueger, Michael Eck, Yunfei Zhou, and Frank-Stefan Riehle

13.1
Introduction

Polymer–nanoparticle composite materials have attracted increasing attention during the past decade since the properties of polymers and nanoparticles (NPs) can be combined and new properties based on the two material components can be utilized in functional films. Examples are the integration of carbon nanotubes and nanofibers into polymers to enhance significantly the mechanical strength [1], the incorporation of luminescent nanoparticles into a polymer host to generate light conversion layers [2], and the integration of semiconductor nanoparticles into conjugated polymers for LED [3–6] and photovoltaic applications [7]. In this chapter, we present the state of the art of hybrid solar cells based on conjugated polymers and semiconductor nanocrystals. Materials, working principle, and solar cell device structures and performances are summarized as well as novel concepts and results for improving the device performance of these hybrid film-based systems are illustrated. In a perspective viewpoint, the chances for this technology in comparison to other photovoltaic technologies are discussed. Since the number of review articles and book chapters addressing this topic has been increasing recently, we try to provide a comprehensive overview, relating to recently published reviews for further details, and focus here in more detail on novel aspects, concepts, and recent developments.

13.2
Composite Materials

13.2.1
Colloidal Semiconductor Nanocrystals

Semiconductor nanocrystals (NCs) are nanosized crystals of semiconductor materials such as ZnO, ZnSe, TiO_2, CdS, CdSe, CdTe, PbS, PbSe, InP, GaAs, and Si. Their usual sizes range between 2 and 20 nm, which correspond to a number of

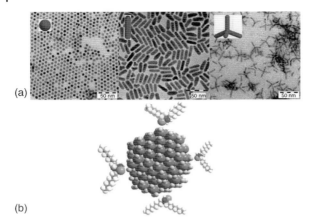

Figure 13.1 (a) TEM images of different shaped semiconductor NCs. From left to the right: CdSe quantum dots, CdSe nanorods, and CdTe tetrapods. (b) Schematic illustration of a colloidal semiconductor QD capped with surface ligands.

atoms between 100 and 10 000. They are available in various shapes resulting in spherical NCs called quantum dots (QDs) or elongated or branched crystals such as rods, wires, and multipods. Figure 13.1a illustrates examples of various available shapes of CdX NCs. One-dimensional NC wires can have a larger length scale, exceeding 100 nm in one direction. In the easiest case, colloidal NCs usually consist of crystalline cores and organic ligand molecules attached to surface atoms of the crystals, enabling the solubility and colloidal stability in solution, which is schematically illustrated in Figure 13.1b. Details of various synthesis approaches for colloidal NCs are reviewed elsewhere in detail [8]. In brief, there are two distinct routes to produce nanoparticles: by physical approaches where they can be grown by lithographic methods, ion implantation, and molecular beam deposition; or by chemical approaches where they are often synthesized by colloidal chemistry in a solvent medium. Here, metastable precursor molecules such as organometallic reagents are decomposed thermally in the presence of surface-active molecules, so-called ligands. Typical organic ligands for semiconducting NPs often contain metal coordinating head groups, which have a high affinity to the electron-poor metal atoms on the NP surface [9], avoiding the further growth and aggregation of NPs. They also usually have longer aliphatic tails, which provide a hydrophobic surface to the outside environment and steric stabilization of the nanoparticles in solution [10] ensuring colloidal stability. Examples of coordinating ligands are long-chain fatty acids, alkyl phosphonic acids, alkyl phosphines, and alkyl phosphine oxides (e.g., strearic acid, trioctylphosphine (TOP), or trioctylphosphine oxide (TOPO)) as well as alkylamines (e.g., octylamine or hexadecylamine) [9, 10]. Colloidal synthetic methods are widely used and are promising for large batch production and commercial applications because of their scalability and the relative simplicity of the process involved. In academic research, the so-called hot injection method is still dominant where one precursor is injected to the rest of the reaction mixture at

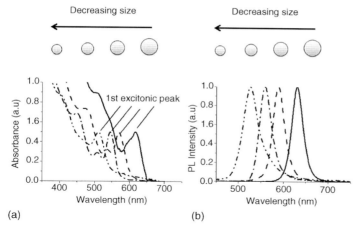

Figure 13.2 (a) Absorption spectra of different sized CdSe QDs. (b) Corresponding PL spectra representing red, orange, yellow, and green (from right to left) emitting CdSe QDs under UV excitation at 380 nm.

high temperature to induce a fast nucleation followed by a moderately slow growth of the nanoparticles [10]. In recent years, novel one-pot synthesis methods have been established by utilizing, for example, a microwave reactor [11] or a microfluidic reactor [12, 13] up to fully automated synthesis setups for the synthesis of nanoparticles and the development of novel reactions [14, 15].

Semiconductor NCs usually have broad absorption spectra starting to absorb photons from the UV until the occurrence of the first excitonic absorption peak in the visible or near-infrared region. With decreasing particle size, the first excitonic peaks of the UV–Vis spectra are blueshifted due to the widening of the bandgap (Figure 13.2a). The corresponding photoluminescence (PL) graphs for CdSe QDs are shown in Figure 13.2b. The PL emission wavelengths are slightly redshifted compared to the respective first excitonic absorption peak. The physical and chemical properties of NCs can differ significantly from those of the corresponding bulk materials due to their sizes down to the nanometer scale. The wavelengths of the electrons and holes are more and more spatially confined with decreasing size on the nanoscale, resulting in an increased Coulomb energy of the electron–hole pair compared to bulk material of the same composition where the electrons and holes can often move freely within the semiconductor bulk crystals. The resulting increased electron–hole binding energy, also called exciton binding energy, is mainly responsible for the bandgap widening of NCs compared to bulk material of the same composition affecting the absorption and luminescence properties significantly.

In Figure 13.3a a schematic energy level diagram for semiconductor NCs in comparison to bulk material is illustrated. The number of energy levels is significantly reduced compared to the situation in bulk materials where continuous band structures are apparent. After light absorption a confined strongly bound exciton is generated in the NC (Fig. 13.3b). In principle three processes can then occur: light emission, exciton transfer or charge transfer (Fig. 13.3c). The energy levels in NCs

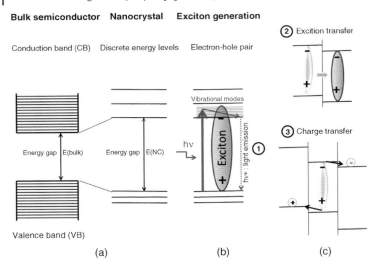

Figure 13.3 (a) Schematic comparison of the electronic states of bulk semiconductors and semiconductor NCs. A characteristic bandgap widening is observed for NCs. (b) After exciton generation by light absorption, three basic responses are in principle possible: after relaxation via vibrational modes light emission at higher wavelengths can occur (1). (c) In the presence of an acceptor that can be a molecule with matching HOMO–LUMO energy levels, another semiconducting NC or simply an electrode, exciton transfer (2) or charge transfer (3) might occur, respectively.

can also split up into different energetic levels due to the influence of an internal crystal field that is mainly influenced by the shape of NCs and the electron–hole interactions in the crystals. As shown in Figure 13.3, a bulk semiconductor has continuous conduction and valence energy bands separated by a fixed energy gap, whereas a semiconductor NC is characterized by discrete atomic-like states and a size-dependent energy gap. In general, the bandgap energy increases with decreasing particle size.

In a first simple effective mass approximation, the relation between the bandgap energy of the bulk material $E_{g,bulk}$ and the bandgap energy $E_{g,NC}$ of a spherical nanocrystal with a radius r is given by the following formula [16]:

$$E_{g,NC}(r) = E_{g,bulk} + \frac{h^2}{8r^2}\left[\frac{1}{m_e^*} + \frac{1}{m_h^*}\right],$$

where h represents the Planck constant, r is the particle radius, and m_e^* and m_h^* are the effective masses of the electron and the hole, respectively, which depend on the semiconducting material. More detailed theories and approximations exist, taking into account additional effects such as Coulomb interactions between electron and hole [16], influencing effects by the crystal field [17], or the deviation from particle sphericity [18].

The schematic shown in Figure 13.4 illustrates the development of the energy levels and the bandgap as a function of particle size in semiconductor CdSe QDs

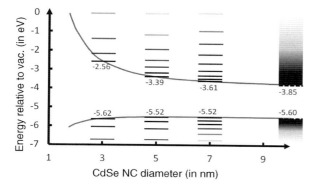

Figure 13.4 Schematic development of the energy band diagram from bulk values of CdSe QDs with decreasing particle size from right to the left. The values for the energy levels of the first excited states of CdSe QDs are calculated for various sizes and included in the schematics. The bandgap increases and the energy levels of the VB and CB split into discrete levels.

based on calculated values for the energy levels as a function of particle size utilizing the above-mentioned formula based on the effective mass approximation.

The increase in the optical bandgap for decreasing particle size is caused by the confinement of the electron and hole wavefunctions of the generated exciton (electron–hole pair). Due to the lower effective mass of the electron, its wavefunction is more sensitive to the confinement than for the hole. This results in an unsymmetrical bandgap widening shown in Figure 13.4.

In conclusion, the optical bandgap and the energy levels of semiconductor NCs can be tuned by adjusting the size and/or the material composition. In Table 13.1,

Table 13.1 Measured and calculated energy values for selected bulk and NC-type semiconductors.

Material	Bulk CB/VB (eV)	Reference	NC diameter (nm)	Measured LUMO/ HOMO (eV)	Reference	Calculated LUMO/HOMO (eV)
CdSe	−3.85/−5.6	[117]	4.8	−3.32/−5.36	[20]	−3.35/−5.52
CdS	−3.74/−6.21	[35]	4	−3.6/−6.6	[21]	−3.35/−6.10
CdTe	−3.52/−5.13	[35]	4	−3.06/−5.01	[22]	−2.67/−5.13
ZnO	−4.4/−7.9	[149]	55	−4.19/−7.39	[24]	−4.40/−7.75
CIS (CuInS$_2$)	−4.1/−5.6	[26]	7	−3.1/−6.0	[26]	−3.91/−5.55
TiO$_2$	−4.2/−7.4	[118]	5	−3.7/−6.9	[28]	−4.19/−7.37
Si	−4.0/−5.12	[30]	4	−3.9/−5.3	[30]	−4.02/−5.34
PbSe	−4.56/−4.84	[36]	6	−4.20/−5.02	[32]	−4.43/−4.93
PbS	−4.35/−4.76	[36]	4.8	−4.00/−5.00	[33]	−4.09/−4.96

HOMO and LUMO levels of the NCs are calculated based on reported bulk values, effective electron and hole mobilities of the bulk material, and the relative permittivities following the formula for the effective mass approximation from Brus [34].

measured and/or calculated energy levels in electron volt (eV) relative to vacuum level for different semiconductor nanocrystals at certain diameters are summarized and compared with respective bulk values.

For the calculation of energy levels $E_{HOMO}(r)$ and $E_{LUMO}(r)$ of the NCs with a certain size and radius r, the following formulas developed by Brus [34], using the effective mass approximation and considering Coulomb interaction of the electron hole pair, have been applied:

$$E_{HOMO}(r) = E_{VB} - \left(\frac{h^2}{8m_h^* r^2} - \frac{1.8e^2}{4\pi r \varepsilon_r \varepsilon_0} \right),$$

$$E_{LUMO}(r) = E_{CB} + \frac{h^2}{8m_e^* r^2},$$

where E_{VB} and E_{CB} are the energy levels of the bulk material respective to vacuum level, h is the Planck constant, m_h^* is the effective hole mass, m_e^* is the effective electron mass, e is the elementary electric charge, ε_r is the relative permittivity, and ε_0 is the vacuum permittivity.

Often bulk values are obtained as well by approximations [35]. The experimental results can vary due to different applied measurement methods for extracting the respective values of the lowest unoccupied molecular orbital (LUMO) and highest occupied molecular orbital (HOMO) levels. There are several experimental methods for determining the absolute values of the HOMO and LUMO levels, such as cyclic voltammetry (CV), ultraviolet photoelectron spectroscopy (UPS), and X-ray photoelectron spectroscopy (XPS). These methods are often combined with UV–Vis spectroscopy, with which optical bandgaps are determined, by adding this value to the HOMO level determined by, for example, CV to obtain the LUMO values, in order to increase the accuracy [36]. Since the bandgap and the absolute values for the energy levels are size dependent for NCs, the respective diameters are given additionally in Table 13.1.

To sum up, the optical bandgap of NCs can be engineered by material and size composition over a wide range from UV to the NIR regime. However, it has to be kept in mind that for photovoltaic application not only the bandgap but also the absolute values for the energy levels are crucial to enable effective charge transfer at the interface between an electron-donating and an electron-accepting material as will be described later in more detail.

13.2.2
Conjugated Polymers

In photoactive composite films of hybrid solar cells, conjugated polymers usually are the main light absorbing material and act as electron donors and hole acceptors. They form the hole conductive phase and usually have decent mobilities ranging from 5×10^{-5} cm^2/(Vs) for MEH-PPV [37] to 2×10^{-2} cm^2/(Vs) for PCPDTBT [38]. Similarly as in organic photovoltaics (OPVs), the trend to utilize low-bandgap polymers with improved absorption capabilities toward the NIR led to improved solar cell performance. In Figure 13.5, the AM 1.5 G solar spectrum together with the

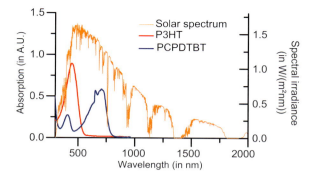

Figure 13.5 Comparison between the solar spectrum (AM 1.5G) and the absorption spectra of the conjugated polymers P3HT and PCPDTBT.

absorption spectra of P3HT and PCPDTBT is displayed showing clearly the better utilization of the solar spectrum by low-bandgap polymers. Usually, the energy levels of common conjugated polymers are optimized for PCBM as acceptors and hybrid solar cell technology just utilized the new developed polymers as well. Table 13.2 gives an overview of the most commonly used polymers in hybrid solar cells as well as their measured HOMO–LUMO levels. The energy levels can be tuned by introducing chemical functionalities. References [39–41] give a more detailed insight into the principal families of conjugated polymers for organic electronics. In Table 13.2, the most common conjugated polymers used in hybrid solar cells are summarized, including their structural formulas, their short and full names, and their HOMO–LUMO energy levels.

13.3 Device Structure

13.3.1 Photoactive Layer

In hybrid solar cells, the photoactive layer is typically formed by simple physical mixing of organic polymer with inorganic nanoparticles. NPs from CdS, CdSe, CdTe, ZnO, SnO_2, TiO_2, Si, PbS, PbSe, and $CuInS_2$ (CIS) have been used for hybrid solar cells, which combine the advantages of both the individual organic and inorganic components. Here, the solubility in a common organic solvent or solvent mixture for both components, the NCs and the conjugated polymers, has to match. In academic research, the photoactive layer is usually deposited on a substrate by spin coating, but similarly as in pure OPV other more large-scale compatible deposition processes such as printing, roll-to-roll processing, inkjet printing, or other coating techniques from solution are in principle applicable [49, 50]. A common device structure for a hybrid solar cell with three different realizations for the photoactive NC–polymer hybrid film is schematically illustrated in Figure 13.6. As a

Table 13.2 Examples of conjugated polymers utilized so far in hybrid solar cells with respective HOMO–LUMO levels and references from literature.

Structural formula	Short name	Full name	LUMO/HOMO (eV)	Reference
	PCPDT-BT	Poly[2,6-(4,4-bis-(2-ethylhexyl)-4H-cyclopenta[2,1-b;3,4-b']-dithiophene)-alt-4,7-(2,1,3-benzothiadiazole]]	−3.57/ − 5.3	[38]
	PDTP-BT	Poly[2,6-(N-(1-octylnonyl)dithieno[3,2-b:2',3'-d]pyrrole)-alt-4,7-(2,1,3-benzothiadiazole)	−3.08/ − 4.89	[42]
	PSiF-DBT	Poly[2,1,3-benzothiadiazole-4,7-diyl-2,5-thiophenediyl(9,9-dioctyl-9H-9-silafluorene-2,7-diyl)-2,5-thiophenediyl]	−3.57/ − 5.39	[43]
	P3EBT	Poly[3-(ethyl-4-butanoate)thiophene-2,5-diyl]	−3.5/ − 5.4	[44]

P3HT	Poly(3-hexylthiophene-2,5-diyl)	−3.2/−5.2	[45]
OC₁C₁₀PPV (MDMO-PPV)	Poly(2-methoxy-5-(3,7-dimethyloctoxy)-p-phenylene vinylene)	−2.8/−5.0	[37]
MEH-PPV	Poly(2-methoxy-5-(2-ethylhexoxy)-1,4-phenylene vinylene)	−2.8/−5.0	[37]
APFO-3	Poly(2,7-(9,9-dioctyl-fluorene)-alt-5,5-(4′,7′-di-2-thienyl-2′,1′,3′-benzothiadiazole))	−3.5/−5.8	[46]

(continued)

Table 13.2 (Continued)

Structural formula	Short name	Full name	LUMO/HOMO (eV)	Reference
(structure with CH₃ octyl chain on thiophene)	P3OT	Poly(3-octylthiophene-2,5-diyl)	−2.85/ −5.25	[47]
(PEDOT and PSS structures with SO₃⁻)	PEDOT:PSS	Poly(3,4-ethylenedioxythiophene):poly(styrenesulfonic acid)	−2.2/ −5.2	[48]

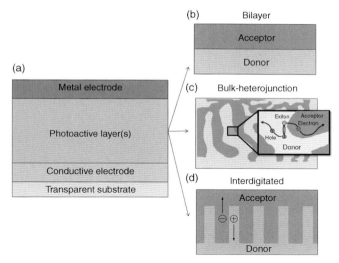

Figure 13.6 (a) Schematic illustration of typical device structures for hybrid solar cells. The hybrid film consists usually of a conjugated polymer as donor and NCs as acceptor materials and can be structured as bilayer (b), bulk heterojunction (c), or in an interdigitated manner (d).

typical thin film device, it usually consists of photoactive layer(s) with a thickness ranging between 100 and 200 nm, sandwiched between two electrodes of different work functions. Indium tin oxide (ITO) on a flexible plastic or glass substrate is often used as the anode where the light comes in because of its high work function, conductive behavior, and high transparency. The photoactive light absorbing layer consists of conjugated polymer as organic part and semiconducting NPs as inorganic part. A top metal electrode (e.g., Al or Ag) is usually vacuum deposited onto the photoactive layer. Between the active layer and the anode, a PEDOT:PSS interlayer is used as hole extraction and electron blocking layer. Between the cathode and the photoactive film, an additional organic semiconducting hole blocking layer, for example, bathocuproine [21], is deposited in some cases. In addition, in many cases a 10–20 nm thin layer of, for example, LiF, Ca, or Mg is deposited before the top metal electrode for hindering the metal diffusion inside the active layer and to act as hole blocking layer [51, 52].

13.3.2
Device Principle

Similar to organic solar cells, photocurrent generation is a multistep process in NC–polymer hybrid bulk heterojunction solar cells, as demonstrated in Figure 13.7. Briefly, when a photon is absorbed by the absorbing material, electrons are exited from the valance band (VB) to the conduction band (CB) to form excitons. The excitons diffuse to the donor/acceptor interface where charge transfer occurs, leading to the dissociation of the excitons into free electrons and holes. Driven by the

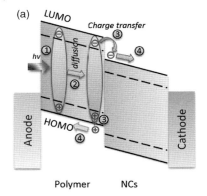 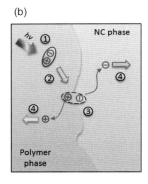

Figure 13.7 Schematic diagram of the photocurrent generation mechanism in bulk heterojunction hybrid solar cells: (a) schematics of the different energy levels of the donor and acceptor materials and the individual processes taking place: exciton generation (1), exciton diffusion (2), charge transfer (3), charge carrier transport, and collection (4). (b) Schematic of a bulk heterojunction contact at the polymer–NC interface. The same processes mentioned in (a) are illustrated directly at the donor–acceptor interface. (Reproduced with permission from Ref. [54].)

internal electric field, these carriers are transported through the respective donor or acceptor material domains and are finally collected at the respective electrodes. To sum up, there are four main steps: photon absorption (1), exciton diffusion (2), charge separation (3), and charge carrier transport and collection (4). For a detailed physical description of the underlying processes in organic/hybrid solar cells, the readers may refer to Refs [39, 53].

The overall solar cell efficiency depends on the efficiency of each single step. The absorption of light or respective energy $E = h\nu$ is mainly done by the conjugated polymers, which usually have large absorption coefficients. In the case of hybrid solar cells, the NCs additionally contribute to the light absorption. A number of excitons are created after light absorption, which represent electron–hole (e–h) pairs with a certain binding energy in the range of 200–500 meV [55, 56]. These excitons have a certain lifetime before recombination or energy relaxation processes might take place and energy is released as heat or light emission. In order to avoid these energy losses, charge separation has to take place within the exciton lifetime. Excitons in conjugated polymers are not locally defined; they can diffuse along a polymer chain or between adjacent polymer chains. The diffusion length characterizes the distance an exciton can travel before recombination takes place and is in the range of 10–20 nm [57–59]. Since charge separation takes place at the polymer–NC interface, the internal structure on the nanoscale of the hybrid photoactive film is of utmost importance for the efficiency of the charge separation. Figure 13.6b illustrates a bilayer heterojunction structure of a hybrid film where two layers of donor and acceptor materials are stacked to form a planar donor/acceptor interface. However, the exciton dissociation efficiency is limited due to the limiting interfacial area. Since the exciton diffusion lengths in conjugated polymers are typically around 10–20 nm, the optimum distance of the exciton to the donor/acceptor (D/A) interface, where charge transfer takes place and excitons dissociate into free charge carriers, should be in the same length range. This requirement limits the part of the active layer that contributes to the photocurrent to a very small region near the D/A interface. In other words, excitons generated in the remaining area of the device are lost for power conversion. Therefore, for efficient exciton dissociation at the heterojunction, the donor and the acceptor materials have to be at a suitable distance. In order to overcome the problem that not all excitons are able to reach the D/A interface, the so-called bulk heterojunction structure was introduced [60, 61], which is schematically shown in Figure 13.6c. The bulk heterojunction composite is made by mixing both the electron donor and electron acceptor intimately together as shown in Figure 13.6c; thus, the interfacial area is dramatically increased and the distance that excitons have to travel to reach the interface is reduced compared to a bilayer structure. After exciton dissociation into free charge carriers, holes and electrons are transported via polymer and NP percolation pathways toward the respective electrodes. Compared to the bilayer heterojunction structure where donor and acceptor phases contact the respective anode and cathode selectively, the bulk heterojunction requires percolated pathways for the charge carrier transporting phases to the respective electrodes. Thus, the donor and acceptor phases should form a bicontinuous and interpenetrating network for efficient charge transport after exciton dissociation takes place to enable efficient charge carrier transport. Therefore, nanoscale morphology control within the blend is very important for obtaining suitable bulk heterojunction devices. Ideally, an interdigitated donor–acceptor configuration would be a perfect structure for efficient exciton dissociation and charge transport as depicted in Figure 13.6d. In such a structure, the distance from exciton generation sites, either in the donor or in the acceptor phase, to the D/A interface would be in the range of the exciton diffusion length. After exciton dissociation, both holes and electrons will be transported within their prestructured donor or acceptor phases along a direct percolation pathway to the respective electrodes. This interdigitated structure can be realized by various nanostructuring approaches, which will be discussed in detail later in chapter 13.5.4.

13.3.3
Band Alignment and Choice of Donor/Acceptor Pairs

As we discussed previously, by decreasing the size of semiconductor NCs, the energy levels turn from continuous states to discrete ones, resulting in a widening of the bandgap apparent as a blueshift in optical properties. As a result, the bandgap as well as the energy levels of NPs can be tuned easily by controlling the size of NPs in order to match the various energy levels of their polymer donors for efficient exciton dissociation at the D/A interface (Figure 13.8). Figure 13.9 shows the energy levels (in eV) of commonly used conjugated polymers as donors, and the energy levels of NPs as acceptors for hybrid solar cells. The Fermi levels of the electrodes and the energy levels of PCBM ([6,6]-phenyl-C_{61}-butyric acid methyl

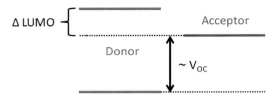

Figure 13.8 Schematic illustrating the energy level matching of a donor and an acceptor. In order to allow an efficient electron extraction from the donor, the LUMO difference (ΔLUMO) between donor and acceptor must be at least 0.3 eV. The Voc is considered proportional to the energy gap of the HOMO level of the donor and the LUMO level of the acceptor.

ester, a typical acceptor in organic solar cells) are shown as well. The variation in the values for the energy levels is derived from different references based on different experimental measurements that are summarized in a recently published review article [54]. For efficient charge separation and transfer at the D/A interface, the band energy levels have to match (see Figure 13.8). An optimum energy gap of about 0.3 eV was proposed for the LUMO level of the donor and the LUMO level of the acceptor in order to enable an effective charge separation [62] while granting a maximum possible open-circuit voltage (V_{OC}) in the device [63]. The V_{OC} is considered proportional to the energy gap of the HOMO level of the donor and the LUMO level of the acceptor.

For efficient charge extraction, the Fermi level of the electrodes has to match as well with the respective HOMO and LUMO levels of the donor and acceptor, respectively. It has to be kept in mind that the band energy values shown in Figure 13.9 can give a rough guidance if D/A pairs are suitable for forming a photoactive layer with efficient charge separation. If a donor and an acceptor material are brought in contact with each other or with an electrode, energy level alignment takes place. Other effects such as energy barriers deriving from organic ligand molecules or defect states occur additionally influencing the charge separation process.

13.4
State of the Art of Hybrid Solar Cells

Greenham *et al.* first reported on the incorporation of spherical CdSe NCs, also called quantum dots (QDs), into MEH-PPV polymer in 1996 [64]. Since then, hybrid solar cells based on conjugated polymers and semiconducting NPs have been attracting increasing attention. As the first type of NPs being incorporated into hybrid solar cells, CdSe NP-based devices are still one of the most efficient hybrid solar cells and are under extensive investigations because of their absorption at a useful spectral range for harvesting solar emission, good electron acceptor properties in combination with conjugated polymers, and well-established synthesis methods for reproducible syntheses. After different elongated NC shapes were

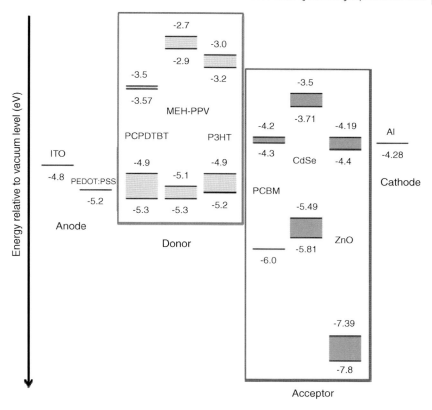

Figure 13.9 Energy levels (in eV) of selected conjugated polymers and NCs. [6,6]-Phenyl-C_{61}-butyric acid methyl ester (PCBM) as a typical acceptor in organic solar cells is also shown as a reference as well as the Fermi levels of ITO and Al. (Reproduced with permission from Ref. [54].)

synthetically available [65], numerous approaches were published regarding the synthesis of various morphologies and structures of CdSe NPs such as QDs, nanorods (NRs), and tetrapods (TPs) and their integration into hybrid solar cells. A significant advancement was reported in 2002 by Huynh *et al.*, who demonstrated efficient hybrid solar cells based on P3HT and elongated CdSe NRs, which were used for providing extended pathways for the formation of NP networks and effective electron transport [7]. It is reported that electron transport can be improved by increasing the NR length, resulting in an improvement in the external quantum efficiency (EQE) of a solar cell device. The optimized devices consisting of 90 wt% pyridine-treated nanorods (7 nm in diameter and 60 nm in length) and P3HT exhibited an EQE over 54% and a power conversion efficiency (PCE) of 1.7%. Later on, high boiling point 1,2,4-trichlorobenzene (TCB) was used instead of chlorobenzene as solvent for the polymers and NCs influencing the resulting hybrid film morphology. Fibrillar morphology of P3HT was formed and provided extended pathways for hole transport, resulting in improved device PCEs utilizing CdSe NRs up to 2.6% [66]. Further improvement was achieved by using CdSe TPs since TPs always have an extension perpendicular to the electrode for more efficient electron transport in comparison to NRs, which are preferentially oriented more parallel to the electrode. Devices based on CdSe TPs and poly(2-methoxy-5-(3′,7′-dimethyloctyloxy)-*p*-phenylene vinylene) (OC_1C_{10}-PPV) as the donor polymer processed with high boiling point solvent TCB exhibited PCEs up to 2.8% [67].

Elongated or branched NPs can in principle provide more extended and directed electrical conductive pathways, thus reducing the number of interparticle hopping events for extracting electrons toward the electrode. However, despite the relatively high intrinsic conductivity within the individual NPs, the electron mobility through the NP network in hybrid solar cells is quite low due to the electrical insulating organic ligands on the NC surface. For example, very low electron mobilities on the order of 10^{-5} cm^2 V^{-1} s^{-1} were measured in the thin films of spherical CdSe NPs with TOPO ligand sandwiched between two metal electrodes [68], compared to the electron mobility in bulk CdSe of about 10^2 cm^2 V^{-1} s^{-1} [69]. In most cases, the ligands used for preventing aggregation during the growth of the NPs contain long alkyl chains, such as oleic acid (OA), TOPO, or hexadecylamine (HDA). Such ligand shell around NP core forms an electrically insulating layer, preventing an efficient charge transfer between NPs and polymer, as well as electron transport between the individual NPs [64, 70]. In order to overcome this problem, post-synthetic treatment on the NPs is very important and has been investigated extensively. Three general strategies of post-synthetic treatment do exist in general: (a) ligand exchange from original long alkyl ligand to shorter molecule, for example, pyridine; (b) chemical surface treatment and washing for reducing the ligand shell; and (c) a combination of (a) and (b) where ligand shell reduction is applied prior to ligand exchange.

Pyridine ligand exchange is one of the most effective post-synthetic procedures that was till recently the method of choice to obtain efficient solar cell devices [7]. For all previously described results, pyridine ligand exchange was applied to the NCs prior to the formation of the photoactive hybrid film. Generally, as-synthesized NPs are washed by methanol several times and consequently refluxed in pure pyridine at the boiling point of pyridine under inert atmosphere overnight. This pyridine treatment is proved to replace the synthetic insulating ligand with shorter and more conductive pyridine molecules. Other small molecules containing functional chemical groups such as chloride [71], amine [40], and thiols [29, 72] were also investigated as potential ligand exchange ligands. Aldakov *et al.* systematically investigated CdSe NPs modified by various small ligand molecules with nuclear magnetic resonance (NMR), optical spectroscopy, and electrochemistry, although their hybrid devices exhibited low efficiencies [72]. Olson *et al.* reported on CdSe/P3HT blended devices exhibiting PCEs up to 1.77% when butylamine was used as a shorter capping ligand for the NPs [40]. In an alternative approach, shortening of the insulating ligands by thermal decomposition was demonstrated and led to a relative improvement in the PCEs of the CdSe/P3HT-based solar cells [73].

However, redispersed NPs after ligand exchange from bulky molecules to molecules with shorter alkyl chains, usually suffer from colloidal instability resulting in agglomeration and precipitation out of the organic solvent. Therefore it is difficult to obtain stable blends of NPs and polymer. In contrast to the organic PCBM nanoparticle, which has a well-defined molecular structure, colloidal semiconductor NCs are composed of a size-tunable inorganic core and a diffuse organic ligand shell that protects the NCs from coalescence and bulk formation. The organic shell comprises an inner layer or monolayer of ligands that are strongly bound (chemisorption) to the crystal surface and a diffuse outer layer containing one or more sublayers of weakly bound (physisorption) ligands. The inner layer is essential for the stabilization of the NCs during their synthesis at high temperatures whereas the outer layer is assembled after the synthesis during the cooling-down and its formation can be seen as directed crystallization of the organic ligands around the NCs. The size of the outer layer can vary individually and strongly depends on the nature of the crystal surface. The weakly bound ligands that can desorb from the NC surface are in a dynamic equilibrium with free ligands in solution. This equilibrium depends on the kind of solvent, ligand, temperature, and NC concentration (Figure 13.10). The quality of the ligand shell strongly influences the chemical and physical behavior of the NCs thus their suitability for possible applications. Because of its complex and dynamic structure, the proper surface investigation of NCs is challenging. The size of the ligand shell can be determined by e.g. dynamic light scattering (DLS) or directly visualized by TEM using a special technique superposing the dark-field (organic material) and the bright-field (inorganic material) image [74]. A new and innovative method for analyzing the exact composition

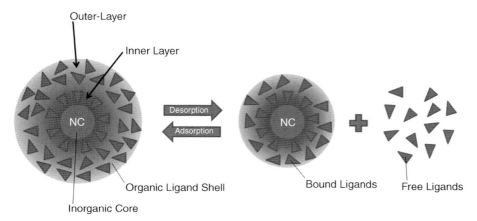

Figure 13.10 Dynamic equilibrium between bound ligands on the nanocrystal surface and free ligands in solution.

of the ligand shell is using mass spectrometry coupled with TGA. According to their different surface affinities, free, weakly, and strongly bound ligands can be separated by their different evaporation temperatures in TGA and afterwards detected by their mass in the mass spectrometer [75]. A new strategy for post-synthetic treatment has been demonstrated for spherical CdSe QDs using hexanoic acid to reduce the initial ligand sphere formed around the NCs after synthesis [74]. One advantage of this treatment is that the NPs retain a good dispersibility allowing a high loading of CdSe NCs in the blend, which is preferable for an efficient percolation network formation during the annealing step of the photoactive composite film. Devices with optimized ratios of QDs to P3HT exhibited reproducible PCEs up to 2.1% [74]. A simple reduced ligand sphere model was proposed to explain the possible reason for improved photovoltaic device efficiencies after acid treatment based on the illustration in Figure 13.10. Hexanoic acid, as a chemically active solvent, can effectively remove the ligand sphere by e.g. protonation of weakly bound surface ligands forming a salt which can be dissolved and separated from the NCs at elevated temperatures by centrifugation.

Absorption of a large fraction of the incident photons is required for harvesting the maximum possible amount of the solar energy. Generally, incident photons are mainly absorbed by the donor polymer materials and partially also from the inorganic NPs. For example, in blends containing 90 wt% CdSe nanoparticles in P3HT, about 60% of the total absorbed light energy can be attributed to P3HT due to its strong absorption coefficient [76]. Therefore, using low-bandgap polymer that can absorb a larger fraction of the solar emission at longer wavelength is another important strategy for obtaining efficient hybrid solar cells. Most low-bandgap polymers are from the material classes of thiophene-, fluorene-, carbazole-, and cylopentadithiophene-based polymers, which are reviewed in detail in several articles [62, 41]. Among those low-bandgap polymers, PCPDTBT (chemical structure shown in Table 13.2) with a bandgap of ~1.4 eV and a relatively high hole mobility of up to $1.5-2 \times 10^{-2}$ cm^2 V^{-1} s^{-1} [38, 77] appears to be an excellent candidate as a photon absorbing and electron-donating material [78]. Devices based on PCPDTBT:PC$_{70}$BM system achieved already efficiencies up to 5.5% [79] and 6.1% [80]. Dayal *et al.* reported a hybrid solar cell based on pyridine-treated CdSe tetrapods and PCPDTBT with an efficiency of 3.19% [81]. Zhou *et al.* demonstrated a direct comparison study of using PCPDTBT and P3HT as donor polymer for hexanoic acid-treated spherical CdSe QD-based hybrid solar cells [19]. The PCPDTBT-based device showed a considerable enhancement of PCE to 2.7% compared to the P3HT-based device mainly due to the increase in photocurrent. Recently, devices based on hexanoic acid-treated large-sized QDs (7.1 ± 0.4 nm) and PCPDTBT with PCEs up to 3.1% are reported [82]. The I–V characteristics and the EQE spectrum are shown in Figure 13.11a and b, respectively. In the EQE spectrum, there is a broad peak between 300 and 800 nm with four pronounced shoulders at around 400, 500, 630, and 720 nm. Compared to the absorption spectrum of the PCPDTBT thin film [82], the photocurrent around 500 and 630 nm region is mainly due to the absorption from the QD part, revealing that both polymer and QD components are contributing to the photocurrent generation.

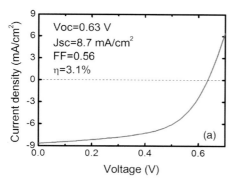
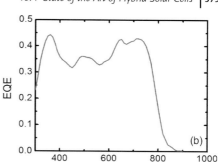

Figure 13.11 (a) Current density–voltage characteristic of hybrid solar cell fabricated from a blend of PCPDTBT:CdSe QDs with a PCE of 3.1%. (b) Corresponding EQE spectrum. (Image taken with permission from Ref. [63].)

A new approach by using mixtures of CdSe QDs and NRs as the NP part was demonstrated recently [82]. As mentioned before, although elongated NRs can naturally provide improved conductive pathways for electron transport, they show the tendency of being mostly oriented close to parallel to the plane of the substrate [83]. Using branched NCs such as tetrapods that always have an extension perpendicular to the electrode could be a solution for more efficient electron transport, although they are more difficult to synthesize and lack reproducibility on a larger scale. Therefore, an alternative way for improving the NC network connectivity by simply blending hexanoic acid washed QDs and NRs together at certain weight ratios was explored. Promising first results showed that devices with a mixture of QD:NR exhibited better PCE compared to either QD-only or NR-only devices (Figure 13.12). The addition of QDs into NRs is supposed to be helpful in reducing the NR–NR horizontal aggregation, thus improving the electron transport in the vertical direction. Recently, this result was confirmed with pyridine ligand exchanged mixtures of CdSe QDs and NRs by Jeltsch *et al.* leading to device efficiencies up to 3.5% [23].

A combination of post-synthetic washing and pyridine ligand exchange on CdSe NRs and their integration in low-bandgap polymer PCPDTBT led to a significant improvement in hybrid solar cells with PCEs approaching 3.5%. Here, the successful sequential removal of the ligand sphere active as soft ligand shell could be detected by TGA-MS, highlighting the importance of optimizing the interface between the polymer chains and the NCs [75].

On the other hand, extension of the NP absorption range could also lead to overall enhancement of hybrid solar cell performance. In this sense, low-bandgap NPs such as CdTe, PbS, and PbSe NPs are promising acceptor materials due to their ability of absorbing light at longer wavelengths. For instance, CdTe NPs have a smaller bandgap compared to CdSe NPs, while their synthesis routes including shape control are similar to that of CdSe NPs [84]. However, suitable CdTe/polymer systems have not yet been found, and reported PCEs of 0.1% based on CdTe/MEH-PPV are quite low [25]. This could be attributed to

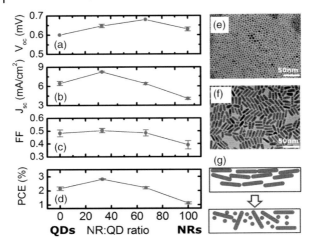

Figure 13.12 Variation of values for (a) V_{OC}, (b) J_{sc}, (c) FF, and (d) PCEs of PCPDTBT-based devices with various CdSe QD:NR weight ratios in the blend. The TEM images of QDs (e) and NRs (f) used in the blend and the proposed model for improvement in NR alignment by addition of QDs (g) are also shown. (Image taken from Ref. [82].)

the possibility that energy transfer rather than charge transfer could occur from the polymer to CdTe NPs in CdTe/polymer blends, resulting in an insufficient generation of free charge carriers [22, 27]. Further lowering of the NC bandgap could be achieved by using semiconductors such as PbS or PbSe. Zhang et al. have reported hybrid solar cells based on blends of MEH-PPV and PbS NPs, but unfortunately the PCEs were very low [85]. Watt et al. have developed a novel surfactant-free synthetic route where PbS NPs were synthesized *in situ* within a MEH-PPV film [31, 86]. In general, using low-bandgap NPs as electron acceptor in polymer/NPs system has been for a long time not very successful as yet, because energy transfer from polymer to low-bandgap NPs is the most likely outcome, resulting in inefficient exciton dissociation. Only very recently, Seo et al. reported on high-efficiency hybrid solar cells based on PbS QDs in combination with the low-bandgap polymer PDTPBT reaching PCE values above 2% [87]. Further introduction of a hole blocking TiO_2 layer between the photoactive film and the LiF/Al cathode and optimization of the PbS QD size led to a further efficiency increase to 3.78%, which is quite remarkable. Also here postsynthetic treatment of the hybrid film with the short ligand ethanedithiol (EDT) leading to ligand exchange of the original oleic acid ligands was one key factor for improving the overall device performance together with matching of the PbS NC LUMO level to the LUMO level of the polymer. The contribution of the PbS QDs to the photocurrent could be seen in the EQE spectrum with contribution between 900 and 1100 nm where only the PbS QDs and not the respective donor polymer show light absorption.

13.5
Novel Approaches in Hybrid Solar Cell Development

13.5.1
Utilization of Less Toxic Semiconductor NCs

For future applications, there is a basic need to develop hybrid solar cells utilizing nontoxic Cd-free NPs. Si NPs are one promising acceptor material for hybrid solar cells due to the abundance of Si compounds, no apparent toxicity, and strong UV absorption. Hybrid solar cells based on Si NPs and P3HT have recently been tested showing promising results and PCEs above 1% [30]. The Si NPs were synthesized by radio frequency plasma via dissociation of silane, and the size can be tuned between 2 and 20 nm by changing the chamber pressure, precursor flow rate, and radio frequency power. Devices made out of optimized Si NPs with a size of 3–5 nm and 50 wt% loading ratio exhibited a PCE of 1.47% [88]. In a similar manner, colloidal copper indium disulfide (CIS) nanoparticles were integrated into P3EBT to form hybrid films leading to hybrid solar cells with a PCE of 0.4%. Oosterhout *et al.* synthesized ZnO NC directly in P3HT polymer by mixing diethylzinc precursor with P3HT and by heating the spin-coated film. Devices of hybrid solar cells with PCEs up to 2% were obtained [89]. TEM tomography was utilized for imaging the three-dimensional nanomorphology and the distribution of the ZnO NC phase in the polymer phase (Figure 13.13). A quantitative correlation among solar cell performance, photophysical data, and the three-dimensional morphology was obtained. These results confirm that TEM tomography is a powerful method for imaging the nanomorphology in hybrid films enabling direct correlations between structure and function in real solar cell devices, for example, by imaging the connectivity of the inorganic percolation network in three dimensions.

13.5.2
In Situ **Synthesis of Ligand-Free Semiconductors in Conjugated Polymers**

Besides the described *in situ* synthesis of ZnO NCs in P3HT, there have been recently additional approaches demonstrating the generation of semiconductor metal sulfide NPs directly in the conjugated polymer matrix. These *in situ* synthesis approaches overcome on the one hand the problems with surface attached ligands that hinder charge separation and charge transport within hybrid solar cells and on the other hand the solubility problem that occurs without stable surface attached ligands since the nanoparticles are *in situ* generated into the already formed polymer. The polymer can act as capping agent preventing extensive particle growth. Nanocomposite films based on ZnS, CdS [90], and CIS [91] have been fabricated starting from the corresponding metal salts and thiourea as soluble precursors. In order to avoid nonvolatile by-products remaining in the active layer, metal xanthates were introduced as single source precursors for the *in situ* generation of metal

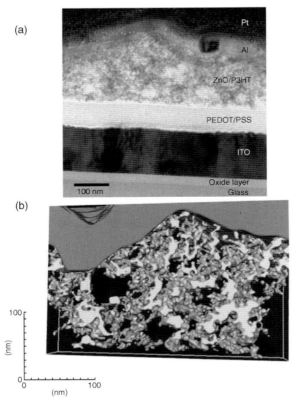

Figure 13.13 (a) TEM micrograph of a cross section of a ZnO–P3HT hybrid solar cell device. (b) Reconstructed image from TEM tomography results revealing the distribution of the ZnO and polymer volumes in the hybrid film of a ZnO–P3HT hybrid solar cell device. (Reconstructed with permission from Ref. [89].)

sulfides in conjugated polymers where only volatile decomposition products are formed leading to the fabrication of high-efficiency CdS/P3HT [92, 93] and CIS/-PSiF-DBT [94] nanocomposite-based hybrid solar cells with PCEs of 2.17 and 2.8%, respectively. Figure 13.14 illustrates the principle of the *in situ* formation of CIS NP in a conjugated polymer to form a donor–acceptor nanocomposite film by a simple thermal conversion step.

13.5.3
Utilization of One-Dimensional Structured Donor–Acceptor Nanostructures for Hybrid Film Formation

Liao *et al.* formed photoactive CdS CdS NR/P3HT hybrid films by utilizing the P3HT as a soft molecular template for the growth of CdS NRs. Improved PL quenching occurring in such nanocomposites indicates a close contact of the CdS

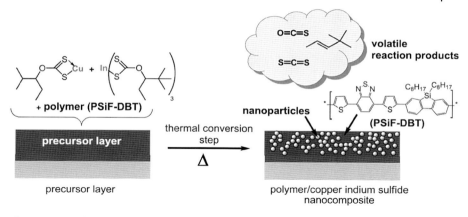

Figure 13.14 Schematic illustration of the formation of CIS NP from two single source Cu and In xanthate precursors mixed in the polymer film. NPs are directly formed after a thermal annealing step releasing additionally volatile reaction products. The resulting nanocomposite hybrid film can be utilized as an active layer for the generation of high efficient hybrid solar cells. (Reproduced with permission from Ref. [94].)

NRs to the polymer. Hybrid solar cells based on these nanocomposites showed impressively high PCE values up to 2.9% [95].

Yang et al. reported recently on the controlled self-assembly of CdTe and polymer QDs to one-dimensional nanofibers by a directed host–guest assembly of the NPs into the polymer matrix [96]. This directed assembly can be, for example, driven by solvent effects minimizing the interfacial energy for both components forming one-dimensional superstructures. Ren et al. very recently utilized the so-called chemical grafting method to generate P3HT nanofibers decorated with CdS QDs [21]. The nondecorated P3HT nanowires already showed an improved absorption characteristic compared to pristine P3HT (Figure 13.15) indicating a higher structural order leading to a broader absorption signal. Additional ligand exchange on the CdS QDs replacing n-butylamine-capped QDs by the significant shorter capping agent ethanedithiol improved the QD–polymer interfacial contact significantly, which was proven by XPS, transient PL measurements detecting a shorter lifetime, and PL quenching experiments. In Figure 13.15b, the absorbance of the hybrid film as well as the PL signal of CdS/P3HT hybrid films based on physical mixing (nongrafted) and directed CdS attachment (grafted) is shown. A schematic of the device architecture and the corresponding flat band diagram is shown in Figure 13.15a. Improved hybrid solar cells with a loading of about 80% CdS (wt%) exhibited PCEs up to 4.1% under AM 1.5 solar illumination that is to date the highest value for polymer–nanocrystal hybrid solar cells (Figure 13.15b). Improved charge separation and optimized charge transport caused by the CdS decorated P3HT nanowires together with the high open-circuit voltage V_{OC} of 1.0 V seem to be key factors for this significant performance improvement in hybrid solar cells. Therefore, the approach of Ren et al. has a high potential for generally improving the efficiency of nanocrystal hybrid solar cells. This performance of CdS/P3HT-based

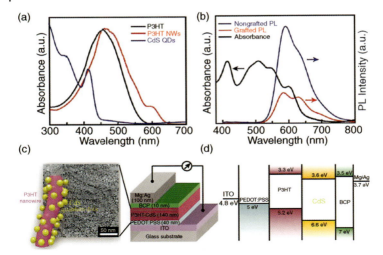

Figure 13.15 (a) Absorption spectra of solutions of pristine P3HT, P3HT NWs, and CdS QDs. (b) Absorption spectrum of hybrid films based on grafted CdS–P3HT NWs and PL spectra of hybrid films based on grafted and nongrafted CdS–P3HT NW hybrid films. (c) Illustration of the device architecture of CdS–P3HT NW-based hybrid solar cells. (d) Flat band energy diagram of the hybrid solar cell device. (Reproduced with permission from Ref. [21].)

hybrid solar cells is reaching already close to the highest reported values for PCBM–P3HT-based organic solar cells that reach optimized PCEs of about 5% [97, 98].

In Figure 13.15c, the device structure of the CdS/P3HT NW-based hybrid solar cell device is schematically illustrated together with the corresponding flat band diagram (Figure 13.15d).

Before summarizing recent approaches for nanostructuring electrodes and donor acceptor phases in a directed way toward interdigitated device structures as depicted in Figure 13.6d, the results of various NC–polymer-based hybrid solar cells are summarized in Table 13.3.

13.5.4
Toward Nanostructured Donor–Acceptor Phases

In Figure 13.6d, an ideal schematic structure of a bulk heterojunction solar cell was introduced where the donor–acceptor phases are arranged in an interdigitated manner with spacings of about 10–20 nm, which is in the size range of the exciton diffusion length. The direct pathways of the donor and acceptor phases toward the respective electrodes would ensure a direct carrier transport and the pure acceptor phase at the electron collecting electrode as well as the pure donor phase at the hole collecting electrode would minimize charge recombination processes at the respective electrode interfaces. There have been several methods applied for the

13.5 Novel Approaches in Hybrid Solar Cell Development | 385

Table 13.3 Summary of high efficient hybrid solar cell results with details about the material composition, treatment and hybrid film formation as well as best cell parameters obtained.

NC type	NC shape	Synthesis ligand	NC treatment	Polymer	Solvent (NC/polymer)	PCE (%)	V_{OC} (V)	FF	Reference
CdSe	NR	TDPA	Prewashing + pyridine ligand exchange	PCPDTBT	CB/TCB	3.42	0.63	0.45	[75]
CdSe	QD/NR	TDPA	Pyridine ligand exchange	PCPDTBT		3.5			[23]
CdSe	QD/NR	HDA/TDPA	Hexanoic acid	PCPDTBT	CB/TCB	2.82	0.65	0.5	[82]
CdSe	TP	OPA	Pyridine ligand exchange	PCPDTBT	$CHCl_3$ + pyridine/TCB	3.19	0.69	0.51	[81]
CdSe	QD	HDA	Hexanoic acid	PCPDTBT	CB/TCB	3.1	0.63	0.56	[82]
CdSe	TP	OPA	Pyridine ligand exchange	OC_1C_{10}-PPV	$CHCl_3$ + pyridine/TCB	2.8 (2.4^a)	0.76^a	0.44^a	[67]
CdSe	NR	TDPA	Pyridine ligand exchange	P3HT	$CHCl_3$/TCB	2.65	0.55	0.49	[99]
CdSe	NR	OPA	Pyridine ligand exchange	P3HT	$CHCl_3$ + pyridine/TCB	2.6 (2.9^b)	0.62^b	0.5^b	[66]
CdSe	TP	OPA	Pyridine ligand exchange	APFO-3	$CHCl_3$ + pyridine/p-xylene	2.4 (2.6^c)	0.95^c	0.38^c	[100]
CdSe	Hyperbranched	TDPA + CEPA	—	P3HT	$CHCl_3$	2.18	0.6	0.5	[101]
CdSe	QD	HDA	Hexanoic acid	P3HT	DCB	2.0	0.62	0.56	[74]
CdSe	QD	TDPA	Pyridine + octylamine ligand exchange	P3HT	$CHCl_3$	1.8	0.55	0.47	[40]
CdSe	NR	—	—	P3HT	$CHCl_3$ + pyridine	1.7	0.7	0.4	[7]
ZnO	—	In situ	None	P3HT	Toluene + THF/CB	2.0	0.75	0.52	[89]
ZnO	NR	MeOH	None	P3HT	Methanol + CB + $CHCl_3$ + DCM	1.4^d	0.84^d	0.53^d	[102]
CdS	In situ	None	None	P3HT	CB	2.17	0.842	0.53	[93]
CdS	NR	In situ	None	P3HT	DCB	2.9	0.68	0.48	[95]
CdS	QD	OA	Prewashing + n-butylamine ligand exchange + EDT ligand	P3HT fibrils	Octane/DCB + $(CH_2)_5CO$	4.1	1.1	0.35	[21]

(continued)

Table 13.3 (Continued)

NC type	NC shape	Synthesis ligand	NC treatment	Polymer	Solvent (NC/polymer)	PCE (%)	V_{OC} (V)	FF	Reference
CdTe	NR	TDPA	exchange inside hybrid film	MEH-PPV	$CHCl_3$ + pyridine/CB	0.05	0.37	0.27	[25]
CdTe	NR	Electrochem. dep. in AAO	Pyridine ligand exchange None	P3OT	$CHCl_3$	1.06	0.71	0.48	[103]
PbS	QD	In situ	None	MEH-PPV	Toluene	0.7	1.0	0.28	[86]
PbS	QD	OA	Ethanedithiol ligand exchange on deposited hybrid blend	PDTP-BT	$CHCl_3$	3.78	0.57	0.51	[87]
PbSe	QD	Oleate	None	P3HT	C_2Cl_4	0.14	0.35	0.37	[32]
Si	QD	Hydro-gene terminated	None	P3HT	DCB	1.47	0.8	0.47	[30]
CIS	NR	TPP	None	PCBM	Toluene	0.086	0.71	0.44	[26]
CIS	—	In situ	None	P3EBT	Pyridine	0.4	0.66	0.27	[91]
CIS	—	In situ	None	PSiF-DBT	CB	2.8	0.54	0.5	[94]

TDPA: tetradecylphosphonic acid; HDA: hexadecylamine; OPA: octylphosphonic acid; CEPA: carboxyethylphosphonic acid; OA: oleic acid; TPP: triphenyl phosphite. *Solvents*: CB, chlorobenzene; DCB, dichlorobenzene; TCB, trichlorobenzene; THF, tetrahydrofuran; DCM, dichloromethane.

a) Average value.
b) Measured at 0.92 sun.
c) No spectral mismatch.
d) Measured at 1.7 sun.

realization of such interdigitated nanostructured electrodes. However, no significant breakthrough occurred compared to the existing randomly oriented bulk heterojunction devices so far. Device performances based on directed nanostructured donor–acceptor phases are still lagging behind. In the following, some examples of nanostructuring technologies are highlighted that might be promising in the future, pushing the performance of hybrid and organic PV towards new frontiers.

Snaith *et al.* utilized the so-called polymer brushes as hole conductors in hybrid solar cell devices. CdSe NCs have been attached to the brushes by infiltration to form a thin hybrid film [104]. Devices showed an improved hole conduction compared to spin-coated films of the same polymer [105]. However, device efficiencies remained quite low due to limited absorption of the solar light.

The formation of nanostructured arrays of conjugated polymers by the utilization of nanoporous templates has been reported. The deposition of the polymer inside the pores can be achieved by filling the pores with a solution of polymer and evaporation of the solvent or by the direct synthesis of conjugated polymer inside the pores by chemical or electrochemical approaches. Porous templates were based on track-etched polycarbonate membranes [106–108] or alumina that is obtained by anodic aluminum oxidation (AAO) [109–111]. Thus, periodic vertical channels with diameters between 20 and 120 nm are formed by first electrochemical oxidation and etching and then subsequent etching for pore widening (Figure 13.16).

There are several reported methods for creating polymeric nanopillars with diameters down to 50 nm inside porous alumina. One method is spin casting of the conjugated polymer on the AAO template followed by a heating step [112, 113]. In order to avoid air inclusions during the filling process, vacuum can be applied additionally to heat to support capillary forces [102]. Another approach for facilitating the process is the filling of the pores with monomers and a subsequent start of the polymerization *in situ* within the pores. One reported example for this approach is the use of the divinyltriphenylamine (DVTPA) monomer [110]. After the filling, the AAO template can be removed by a KOH or NaOH solution. The resulting nanopillars exhibit an

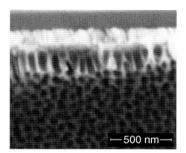

Figure 13.16 SEM image of a porous AAO membrane formed on ITO, manufactured by anodic oxidation of aluminum at 40 V in the presence of 0.3 M oxalic acid and a subsequent following pore widening for 60 min in phosphoric acid.

increased hole mobility due to an improved vertical alignment of the polymeric chains within the AAO template [113, 114]. For example, the hole mobility of P3HT rose from 3×10^{-4} cm^2 V^{-1} s^{-1} for a flat polymer layer in diode configuration to 1×10^{-1} cm^2 V^{-1} s^{-1} for the aligned polymer inside the AAO pores [113]. For the F8T2 polymer, it was shown by thin-film transistor (TFT) measurements that the hole mobility in direction of the polymer backbone can reach 0.2 cm^2 V^{-1} s^{-1} in comparison to 1×10^{-3} cm^2 V^{-1} s^{-1} for an unaligned polymer solution [115].

There is one report of a PCE of 1% using vertically aligned CdTe nanorods obtained by electrodeposition into an AAO template. After the AAO template was removed, poly(3-octylthiophene) (P3OT) was filled in by spin coating. This indicates that CdTe NCs may be useful for hybrid solar cells when their energy levels match the energy levels of the polymer [103].

One notable example for the integration of a nanostructuring method into solar cell device fabrication is the use of AAO templates for the deposition of CdS leading to aligned nanopillars utilizing a vapor liquid solid (VLS) process. Additional VLS deposition of CdTe resulted in a nanostructured all inorganic solar cell with an impressive PCE of about 6% [116]. A few attempts to combine AAO templates to obtain nanostructured hybrid solar cells also exist. These resulted so far in devices with significantly lower efficiencies compared to state-of-the-art hybrid solar cells without additional nanostructuring step. One example for the utilization of an AAO template for a nanostructured hybrid solar cell was published by Kuo et al. [118] and is schematically illustrated in Figure 13.17. A direct comparison between a nanostructured bulk heterojunction hybrid solar cell and a bilayer-based hybrid solar cell was performed. First, freestanding nanopillars of TiO$_2$ were formed by spin coating of a TiO$_2$ dispersion onto the AAO template. After sintering at 450 °C for 1 h and the subsequent removal of the 300 nm thick AAO template by NaOH, the TiO$_2$ nanopillars were obtained. By covering the TiO$_2$ structure with P3HT by spin coating and subsequent evaporation of Au contacts, hybrid solar cells were manufactured with a PCE of 0.512% in comparison to 0.12% for the bilayer structure of the same donor–acceptor material composition. By this method an inverted solar cell was created, using gold as top electrode. A drawback in this

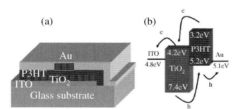

Figure 13.17 (a) Schematic illustration of an inverted TiO$_2$/P3HT hybrid solar cell manufactured by Kuo et al. using an AAO template for the formation of parallel aligned TiO$_2$ nanopillars subsequently filled by P3HT. (b) Schematic illustration of the energy level diagram of the fabricated hybrid solar cell. (Reproduced with permission from Ref. [118].)

design is that donor and acceptor materials are in direct contact with the ITO substrate, where both holes and electrons could be extracted and recombination processes can lower the device performance.

Another method that was successfully applied for the formation of a nanostructured bulk heterojunction organic solar cell is nanoimprint lithography (NIL). In direct comparison with a flat bilayer organic solar cell design, the nanostructured version exhibited nearly a double increase in solar PCE [119]. An AAO template was also used as a mask for etching a Si substrate using a two-step inductively coupled plasma (ICP) etching process [119]. Thus, a silicon mold as shown in Figure 13.18a is formed. This mold is then used for creating nanorods in a film of a conjugated polymer (e.g., regioregular P3HT) (Figure 13.18b). The created polymeric rods (Figure 13.18c) show an increased crystallinity and preferential alignment of the polymer molecules in the vertical direction [114] as well. The space between the polymer rods could in principle be filled with an acceptor material. After the evaporation of a top electrode, the hybrid solar cell would be complete.

Kim *et al.* used NIL to create nanostructured organic solar cells combining the molded polydithiophene derivative TDPDT with PCBM leading to a PCE of 0.8% compared to 0.25% for a bilayer structure [150]. In a similar manner the formation of inorganic nanopillars is reported. Cathodic electrodeposition of CdS, CdSe, CdTe [103, 120–123], and ZnSe [123], or the sol–gel method to create TiO_2 [124] in an AAO template was reported. For most reported organic and inorganic nanopillars, an aggregation of the pillars is observed into several bundles, or even bending is observed for longer pillars.

Vertically aligned TiO_2 nanotubes [109] with pore diameters down to 10 nm [125] can be obtained, by an anodization process, similar to AAO, that can be utilized not only as template for material deposition but also as electron-accepting semiconductor providing continuous pathways for charge extraction (Figure 13.19).

Also for TiO_2 nanotubes the filling with polymer by melt infiltration is reported [112], but in order to improve the charge transfer from polymer to the acceptor, the polymer should be chemically attached to the TiO_2 surface. Therefore,

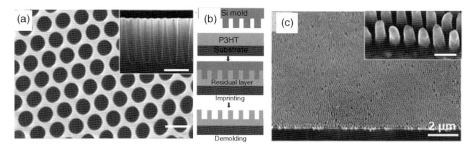

Figure 13.18 (a) Silicon mold created by ICP etching of silicon using an AAO template as mask (*inset*: side view of the mold); (b) illustration of the molding process applied to P3HT; (c) molded parallel aligned P3HT nanopillars [114]. (Reproduced with permission from Ref. [114].)

Figure 13.19 Side view SEM image of porous TiO$_2$ nanotubes [126] fabricated by anodization of titanium.

an *in situ* synthesis was proposed, where the first step is the attachment of a polymerization initiator to the surface of the TiO$_2$ pore surface followed by filling the pores with monomers [127]. The resulting chemically bound polymers after the *in situ* polymerization exhibited a better charge transfer behavior to the TiO$_2$ than mechanically attached ones [127].

Lim *et al.* demonstrated the successful infiltration of P3HT into TiO$_2$ nanotubes of diameters of 60–80 nm [128]. However, the diameters of the filled pores are above the desired diameters for an efficient charge extraction so here the reproducible and complete filling of the TiO$_2$ nanotubes is still one of the main challenges to be solved for photovoltaic cells [112, 128–131].

13.6
Outlook and Perspectives

13.6.1
Hybrid Solar Cells Versus Pure OPVs

Research in the field of OPV based on small molecules or polymer/PCBM systems has developed tremendously within the past decade. Champion laboratory cells reaching efficiencies of 10% have been reported recently [132]. However, average solar cell efficiencies of optimized laboratory cells ranging in between 5 and 6% and PCE values of solar cell modules already commercially available on the market are usually lowered by a factor of about 2.

The performance of NC–polymer hybrid solar cells increased constantly as well during the past years, but PCE values are still by a factor of 2 lower than that for pure OPV. Here, hybrid solar cell technologies benefit from the development of low-bandgap polymers from OPV, leading to an increased utilization of the incident solar radiation [81, 19]. However, the development of suitable low-bandgap polymers has been optimized for their utilization with fullerene-based acceptors such as PCPM and its derivatives, and the tailored design of polymers for specific

NC systems has still not been developed yet. Xu and Qiao predicted the highest achievable cell efficiencies obtainable for NC–polymer hybrid solar cells as a function of polymer bandgap and polymer energy levels for TiO_2, ZnO, and CdSe NCs [133]. Under the assumption that all photons are absorbed by the respective hybrid film and the V_{OC} equals the optimized energy offset between the polymer HOMO and the NC LUMO levels, power conversion efficiencies beyond 10% are achievable in all three NC polymer hybrid systems. Besides the optimized matching of the polymer and NC energy levels, the design and introduction of functional chemical groups into the polymer substituting surface ligands by directly attaching to the NC surface might improve the charge transfer at the donor–acceptor interphase significantly. For P3HT-based hybrid solar cells, the efficiency has reached nearly values of P3HT based OPV cells [21], highlighting the bright potential of hybrid solar cells to catch up with OPV on the long term. Research on hybrid solar cells is up to now performed on a more basic level without too intensive efforts of optimization of individual parameters such as additives, or device structure such as the introduction of additional hole and/or electron blocking layers that have been extensively studied for OPV leading to improved device performances. Here additional gain in performance improvement can be expected. One increasing important aspect will be the control of the nanomorphology by specially designed interdigitated donor–acceptor nanophases. Semiconductor NCs can form continuous wires while the optical properties such as energy levels and bandgap are maintained by keeping the wire width in the nanometer dimension. They can be attached and directed to electrodes. This is not achievable in a comparable manner with fullerenes so far since "elongated fullerenes" such as carbon nanotubes have significantly different optical and electronic properties. Further improvements in hybrid solar cell technology would be beneficial as well for the development of various polymer–NC-based applications such as LEDs, photodetectors, and so on.

13.6.2
Hybrid PV and OPV Versus Other PV Technologies

The efficiency of solar cells is generally often an overestimated criterion for the performance of solar cells but also for their potential impact on the solar cell market. Third-generation solar cell technologies that include OPV and hybrid solar cells as well as dye-sensitized solar cells (DSSCs) are still in the development phase even if first products are commercially available. Parameters such as long-term stability, module price, installation costs, and so on play a much more important role for the future fate of a technology being a key technology in PV or a temporarily appearance. Module costs of traditional solar cells of the first and second generations have been reduced significantly recently, making it much harder for the third-generation technologies to enter into the solar cell market. However, OPV or hybrid solar cells might at least enter into niche markets such as indoor PV or their integration into facades of buildings for harvesting indirect sunlight and having additional functionalities such as light

Table 13.4 Comparison of best and average PCE values of single solar cells and modules of different PV technologies.

PV technology	Best cell PCEs	Reference	Average cell PCEs	Best module PCEs	Reference	Average module PCEs
Si (bulk)	25.0% (monocryst.)	[134]	—	22.9% (monocryst.)	[135]	14–20% (monocryst.)
	20.4% (polycryst.)	[136]		19.5% (polycryst.)	[137]	12–17% (polycryst.)
	10.1% (amorphous)	[138]				7–9% (amorphous)
CIGS (thin film)	20.3%	[139]	—	15.7%	[140]	12–13%
CdTe (thin film)	17.3%	[141]	—	14.4%	[142]	10–12%
DSSC	11.4%	[143]	5–9%	5.38%	[144]	—
OPV (thin film)	9.0% (Konarka, 2012)	[145]	3–5%	4.2%	[146]	1–3%
	10.7% (Heliatek, 2012)	[147]				
	10.1% (Mitsubishi, 2011)	[148]				

protection as well as miniaturized energy harvesters for sensors with a limited sensor lifetime where traditional solar cells have disadvantages in cost. The increasing need for energy autarkic miniaturized systems taking its energy from the environment and avoiding expensive wiring or batteries such as sensors and actuators will certainly open new growing market opportunities for energy harvesting application including photovoltaics. In Table 13.4, PCE values of different PV technologies are compared on the single cell and module level. Values for best cells as well as average values are given. Especially for the emerging third-generation PV technologies, the average efficiencies are significantly lower than the results obtained for best champion cells or modules.

Acknowledgments

The contributing researchers thank the German Federal Ministry of Education and Research (BMBF) within the project "Nanopolysol" under the contract No. 03X3517E and the German Research Foundation (DFG) graduate school GRK 1322 "Micro Energy Harvesting" for their financial report for performing research in the field of hybrid solar cells enabling us to provide this insight into the recent developments of hybrid polymer solar cells.

References

1. Liu, J., Yue, Z., and Fong, H. (2009) *Small*, **5** (5), 536–542.
2. Lee, J., Sundar, V.C., Heine, J.R., Bawendi, M.G., and Jensen, K.F. (2000) *Adv. Mater.*, **12** (15), 1102–1105.
3. Colvin, V.L., Schlamp, M.C., and Alivisatos, A.P. (1994) *Nature*, **370** (6488), 354–357.
4. Gao, M., Richter, B., and Kirstein, S. (1997) *Adv. Mater.*, **9** (10), 802–805.
5. Tessler, N., Medvedev, V., Kazes, M., Kan, S., and Banin, U. (2002) *Science*, **295** (5559), 1506–1508.
6. Anikeeva, P.O., Halpert, J.E., Bawendi, M.G., and Bulović, V. (2009) *Nano Lett.*, **9** (7), 2532–2536.
7. Huynh, W.U., Dittmer, J.J., and Alivisatos, A.P. (2002) *Science*, **295** (5564), 2425–2427.
8. Rogach, A.L., *Semiconductor Nanocrystal Quantum Dots*, 1st edition 2008, Springer Wien, New York.
9. Talapin, D.V., Lee, J.-S., Kovalenko, M.V., and Shevchenko, E.V. (2009) *Chem. Rev.*, **110** (1), 389–458.
10. Yin, Y. and Alivisatos, A.P. (2005) *Nature*, **437** (7059), 664–670.
11. Washington, A.L., II and Strouse, G.F. (2008) *J. Am. Chem. Soc.*, **130**, 8916–8922.
12. Chan, E.M., Mathies, R.A., and Alivisatos, A.P. (2003) *Nano Lett.*, **3** (2), 199–201.
13. Yen, B.K., Stott, N.E., Jensen, K.F., and Bawendi, M.G. (2003) *Adv. Mater.*, **15** (21), 1858–1862.
14. Chan, E.M., Xu, C., Mao, A.W., Han, G., Owen, J.S., Cohen, B.E., and Milliron, D.J. (2010) *Nano Lett.*, **10** (5), 1874–1885.
15. Einwächter, S. and Krüger, M. (2011) MRS Fall Meeting Proceedings 2010, p. 1284.
16. Brus, L.E. (1984) *J. Chem. Phys.*, **80**, 4403.
17. Efros, A.L. (1992) *Phys. Rev. B*, **46** (12), 7448–7458.
18. Efros, A.L. and Rodina, A.V. (1993) *Phys. Rev. B*, **47** (15), 10005–10007.
19. Zhou, Y., Eck, M., Veit, C., Zimmermann, B., Rauscher, F., Niyamakom, P., Yilmaz, S., Dumsch, I., Allard, S., Scherf, U., and Krüger, M. (2011) *Sol. Energy Mater. Sol. Cells*, **95** (4), 1232–1237.
20. Kuçur, E., Bücking, W., Giernoth, R., and Nann, T. (2005) *J. Phys. Chem. B*, **109** (43), 20355–20360.
21. Ren, S., Chang, L.-Y., Lim, S.-K., Zhao, J., Smith, M., Zhao, N., Bulović, V., Bawendi, M., and Gradečak, S. (2011) *Nano Lett.*, **11** (9), 3998–4002.
22. Zhou, Y., Li, Y., Zhong, H., Hou, J., Ding, Y., Yang, C., and Li, Y. (2006) *Nanotechnology*, **17**, 4041–4047.
23. Jeltsch, K.F., Schädel, M., Bonekamp, J.B., Niyamakom, P., Rauscher, F., Lademann, H.W.A., Dumsch, I., Allard, S., Scherf, U., Meerholz, K. (2012) *Adv. Funct. Mater.*, **22** (2), 397–404.
24. Lin, J.M., Lin, H.Y., Cheng, C.L., and Chen, Y.F. (2006) *Nanotechnology*, **17**, 4391–4394.
25. Kumar, S. and Nann, T. (2004) *J. Mater. Res.*, **19** (07), 1990–1994.
26. Arici, E., Sariciftci, N.S., and Meissner, D. (2003) *Adv. Funct. Mater.*, **13** (2), 165–171.
27. van Beek, R., Zoombelt, A.P., Jenneskens, L.W., van Walree, C.A., de Mello Donegá, C., Veldman, D., and Janssen, R.A.J. (2006) *Chem. Eur. J.*, **12** (31), 8075–8083.
28. Petrella, A., Tamborra, M., Curri, M.L., Cosma, P., Striccoli, M., Cozzoli, P.D., and Agostiano, A. (2005) *J. Phys. Chem. B*, **109** (4), 1554–1562.
29. Sih, B.C. and Wolf, M.O. (2007) *J. Phys. Chem. C*, **111** (46), 17184–17192.
30. Liu, C.-Y., Holman, Z.C., and Kortshagen, U.R. (2008) *Nano Lett.*, **9** (1), 449–452.
31. Watt, A., Thomsen, E., Meredith, P., and Rubinsztein-Dunlop, H. (2004) *Chem. Commun.*, (20), 2334–2335.
32. Cui, D., Xu, J., Zhu, T., Paradee, G., Ashok, S., and Gerhold, M. (2006) *Appl. Phys. Lett.*, **88**, 183111.
33. Hyun, B.-R., Zhong, Y.-W., Bartnik, A.C., Sun, L., Abruña, H.D., Wise, F.W., Goodreau, J.D., Matthews, J.R., Leslie, T.M., and Borrelli, N.F. (2008) *ACS Nano*, **2** (11), 2206–2212.

34 Brus, L. (1986) *J. Phys. Chem.*, **90** (12), 2555–2560.
35 Van de Walle, C.G. and Neugebauer, J. (2003) *Nature*, **423** (6940), 626–628.
36 Jasieniak, J., Califano, M., and Watkins, S.E. (2011) *ACS Nano*, **5** (7), 5888–5902.
37 Chua, L.-L., Zaumseil, J., Chang, J.-F., Ou, E.C.-W., Ho, P.K.-H., Sirringhaus, H., and Friend, R.H. (2005) *Nature*, **434** (7030), 194–199.
38 Mühlbacher, D., Scharber, M., Morana, M., Zhu, Z., Waller, D., Gaudiana, R., and Brabec, C. (2006) *Adv. Mater.*, **18** (21), 2884–2889.
39 Reiss, P., Couderc, E., Girolamo, J.D., and Pron, A. (2010) *Nanoscale*, **3** (2), 446–489.
40 Olson, J.D., Gray, G.P., and Carter, S.A. (2009) *Sol. Energy Mater. Sol. Cells*, **93** (4), 519–523.
41 Riede, M., Mueller, T., Tress, W., Schueppel, R., and Leo, K. (2008) *Nanotechnology*, **19**, 424001.
42 Yue, W., Zhao, Y., Shao, S., Tian, H., Xie, Z., Geng, Y., and Wang, F. (2009) *J. Mater. Chem.*, **19** (15), 2199–2206.
43 Wang, E., Wang, L., Lan, L., Luo, C., Zhuang, W., Peng, J., and Cao, Y. (2008) *Appl. Phys. Lett.*, **92**, 033307.
44 Gruber, M., Stickler, B.A., Trimmel, G., Schürrer, F., and Zojer, K. (2010) *Org. Electron.*, **11** (12), 1999–2011.
45 Thompson, B.C. and Fréchet, J.M.J. (2008) *Angew. Chem., Int. Ed.*, **47** (1), 58–77.
46 Andersson, B.V., Huang, D.M., Moulé, A.J., and Inganäs, O. (2009) *Appl. Phys. Lett.*, **94**, 043302.
47 Kymakis, E. and Amaratunga, G.A.J. (2002) *Appl. Phys. Lett.*, **80**, 112.
48 Othman, M.K., Salleh, M.M., and Mat, A.F. (2006) IEEE International Conference on Semiconductor Electronics (ICSE '06), pp. 134–137.
49 Krebs, F.C. (2009) *Sol. Energy Mater. Sol. Cells*, **93** (4), 394–412.
50 Wengeler, L., Schmidt-Hansberg, B., Peters, K., Scharfer, P., and Schabel, W. (2011) *Chem. Eng. Process.*, **50** (5–6), 478–482.
51 Brabec, C.J., Shaheen, S.E., Winder, C., Sariciftci, N.S., and Denk, P. (2002) *Appl. Phys. Lett.*, **80**, 1288.
52 Reese, M.O., White, M.S., Rumbles, G., Ginley, D.S., and Shaheen, S.E. (2008) *Appl. Phys. Lett.*, **92**, 053307.
53 Kippelen, B. and Brédas, J.-L. (2009) *Energy Environ. Sci.*, **2** (3), 251–261.
54 Zhou, Y., Eck, M., and Krüger, M. (2010) *Energy Environ. Sci.*, **3** (12), 1851–1864.
55 Marks, R.N., Halls, J.J.M., Bradley, D.D.C., Friend, R.H., and Holmes, A.B. (1994) *J. Phys.: Condens. Matter*, **6**, 1379–1394.
56 Barth, S. and Bässler, H. (1997) *Phys. Rev. Lett.*, **79** (22), 4445–4448.
57 Halls, J.J.M., Pichler, K., Friend, R.H., Moratti, S.C., and Holmes, A.B. (1996) *Appl. Phys. Lett.*, **68**, 3120.
58 Markov, D.E., Amsterdam, E., Blom, P.W.M., Sieval, A.B., and Hummelen, J.C. (2005) *J. Phys. Chem. A*, **109** (24), 5266–5274.
59 Markov, D.E., Tanase, C., Blom, P.W.M., and Wildeman, J. (2005) *Phys. Rev. B*, **72** (4), 045217.
60 Yu, G., Gao, J., Hummelen, J.C., Wudl, F., and Heeger, A.J. (1995) *Science*, **270** (5243), 1789–1791.
61 Halls, J.J.M., Walsh, C.A., Greenham, N.C., Marseglia, E.A., Friend, R.H., Moratti, S.C., and Holmes, A.B. (1995) *Nature*, **376** (6540), 498–500.
62 Scharber, M.C., Mühlbacher, D., Koppe, M., Denk, P., Waldauf, C., Heeger, A.J., and Brabec, C.J. (2006) *Adv. Mater.*, **18** (6), 789–794.
63 Brabec, C.J., Cravino, A., Meissner, D., Sariciftci, N.S., Fromherz, T., Rispens, M.T., Sanchez, L., and Hummelen, J.C. (2001) *Adv. Funct. Mater.*, **11** (5), 374–380.
64 Greenham, N.C., Peng, X., and Alivisatos, A.P. (1996) *Phys. Rev. B*, **54** (24), 17628–17637.
65 Peng, X., Manna, L., Yang, W., Wickham, J., Scher, E., Kadavanich, A., and Alivisatos, A.P. (2000) *Nature*, **404** (6773), 59–61.
66 Sun, B. and Greenham, N.C. (2006) *Phys. Chem. Chem. Phys.*, **8**, 3557.
67 Sun, B., Snaith, H.J., Dhoot, A.S., Westenhoff, S., and Greenham, N.C. (2005) *J. Appl. Phys.*, **97**, 014914.
68 Ginger, D.S. and Greenham, N.C. (2000) *J. Appl. Phys.*, **87**, 1361.

69 Rode, D.L. (1970) *Phys. Rev. B*, **2** (10), 4036.

70 Huynh, W.U., Dittmer, J.J., Libby, W.C., Whiting, G.L., and Alivisatos, A.P. (2003) *Adv. Funct. Mater.*, **13** (1), 73–79.

71 Owen, J.S., Park, J., Trudeau, P.-E., and Alivisatos, A.P. (2008) *J. Am. Chem. Soc.*, **130** (37), 12279–12281.

72 Aldakov, D., Chandezon, F., De Bettignies, R., Firon, M., Reiss, P., and Pron, A. (2007) *Eur. Phys. J.*, **36**, 261–265.

73 Seo, J., Kim, W.J., Kim, S.J., Lee, K.-S., Cartwright, A.N., and Prasad, P.N. (2009) *Appl. Phys. Lett.*, **94**, 133302.

74 Zhou, Y., Riehle, F.S., Yuan, Y., Schleiermacher, H.-F., Niggemann, M., Urban, G.A., and Krüger, M. (2010) *Appl. Phys. Lett.*, **96**, 013304.

75 Celik, D., Krüger, M., Veit, C., Schleiermacher, H.-F., Zimmermann, B., Allard, S., Dumsch, I., Scherf, U., Rauscher, F., and Nyamakom, P. (2012) *Sol. Energy Mater. Sol. Cells*, **98**, 433–440.

76 Dayal, S., Reese, M.O., Ferguson, A.J., Ginley, D.S., Rumbles, G., and Kopidakis, N. (2010) *Adv. Funct. Mater.*, **20** (16), 2629–2635.

77 Morana, M., Wegscheider, M., Bonanni, A., Kopidakis, N., Shaheen, S., Scharber, M., Zhu, Z., Waller, D., Gaudiana, R., and Brabec, C. (2008) *Adv. Funct. Mater.*, **18** (12), 1757–1766.

78 Soci, C., Hwang, I.-W., Moses, D., Zhu, Z., Waller, D., Gaudiana, R., Brabec, C.J., and Heeger, A.J. (2007) *Adv. Funct. Mater.*, **17** (4), 632–636.

79 Peet, J., Kim, J.Y., Coates, N.E., Ma, W.L., Moses, D., Heeger, A.J., and Bazan, G.C. (2007) *Nat. Mater.*, **6** (7), 497–500.

80 Park, S.H., Roy, A., Beaupre, S., Cho, S., Coates, N., Moon, J.S., Moses, D., Leclerc, M., Lee, K., and Heeger, A.J. (2009) *Nat. Photon.*, **3** (5), 297–302.

81 Dayal, S., Kopidakis, N., Olson, D.C., Ginley, D.S., and Rumbles, G. (2011) *Nano Lett.*, **10** (1), 239–242.

82 Zhou, Y., Eck, M., Men, C., Rauscher, F., Niyamakom, P., Yilmaz, S., Dumsch, I., Allard, S., Scherf, U., and Krüger, M. (2011) *Sol. Energy Mater. Sol. Cells*, **95** (12), 3227–3232.

83 Hindson, J.C., Saghi, Z., Hernandez-Garrido, J.-C., Midgley, P.A., and Greenham, N.C. (2011) *Nano Lett.*, **11** (2), 904–909.

84 Peng, Z.A. and Peng, X. (2000) *J. Am. Chem. Soc.*, **123** (1), 183–184.

85 Zhang, S., Cyr, P.W., McDonald, S.A., Konstantatos, G., and Sargent, E.H. (2005) *Appl. Phys. Lett.*, **87**, 233101.

86 Watt, A.A.R., Blake, D., Warner, J.H., Thomsen, E.A., Tavenner, E.L., Rubinsztein-Dunlop, H., and Meredith, P. (2005) *J. Phys. D*, **38**, 2006–2012.

87 Seo, J., Cho, M.J., Lee, D., Cartwright, A. N., and Prasad, P.N. (2011) *Adv. Mater.*, **23** (34), 3984–3988.

88 Liu, C., Holman, Z.C., and Kortshagen, U.R. (2010) *Adv. Funct. Mater.*, **20** (13), 2157–2164.

89 Oosterhout, S.D., Wienk, M.M., van Bavel, S.S., Thiedmann, R., Jan Anton Koster, L., Gilot, J., Loos, J., Schmidt, V., and Janssen, R.A.J. (2009) *Nat. Mater.*, **8** (10), 818–824.

90 Maier, E., Fischereder, A., Haas, W., Mauthner, G., Albering, J., Rath, T., Hofer, F., List, E.J.W., and Trimmel, G. (2011) *Thin Solid Films*, **519** (13), 4201–4206.

91 Maier, E., Rath, T., Haas, W., Werzer, O., Saf, R., Hofer, F., Meissner, D., Volobujeva, O., Bereznev, S., Mellikov, E., Amenitsch, H., Resel, R., and Trimmel, G. (2011) *Sol. Energy Mater. Sol. Cells*, **95** (5), 1354–1361.

92 Leventis, H.C., King, S.P., Sudlow, A., Hill, M.S., Molloy, K.C., and Haque, S.A. (2010) *Nano Lett.*, **10** (4), 1253–1258.

93 Dowland, S., Lutz, T., Ward, A., King, S. P., Sudlow, A., Hill, M.S., Molloy, K.C., and Haque, S.A. (2011) *Adv. Mater.*, **23** (24), 2739–2744.

94 Rath, T., Edler, M., Haas, W., Fischereder, A., Moscher, S., Schenk, A., Trattnig, R., Sezen, M., Mauthner, G., Pein, A., Meischler, D., Bartl, K., Saf, R., Bansal, N., Haque, S.A., Hofer, F., List, E.J.W., and Trimmel, G. (2011) *Adv. Energy Mater.*, **1**, 1046–1050.

95 Liao, H.-C., and Chen, S.-Y., and Liu, D.-M. (2009) *Macromolecules*, **42** (17), 6558–6563.

96 Yang, S., Wang, C.-F., and Chen, S. (2011) *J. Am. Chem. Soc.*, **133** (22), 8412–8415.

97 Reyes-Reyes, M., Kim, K., and Carroll, D.L. (2005) *Appl. Phys. Lett.*, **87**, 083506.
98 Ma, W., Yang, C., Gong, X., Lee, K., and Heeger, A.J. (2005) *Adv. Funct. Mater.*, **15** (10), 1617–1622.
99 Wu, Y. and Zhang, G. (2010) *Nano Lett.*, **10** (5), 1628–1631.
100 Wang, P., Abrusci, A., Wong, H.M.P., Svensson, M., Andersson, M.R., and Greenham, N.C. (2006) *Nano Lett.*, **6** (8), 1789–1793.
101 Gur, I., Fromer, N.A., Chen, C.-P., Kanaras, A.G., and Alivisatos, A.P. (2006) *Nano Lett.*, **7** (2), 409–414.
102 Baek, S., Park, J.B., Lee, W., Han, S.-H., Lee, J., and Lee, S.-H. (2009) *New J. Chem.*, **33**, 986.
103 Kang, Y., Park, N.-G., and Kim, D. (2005) *Appl. Phys. Lett.*, **86**, 113101.
104 Snaith, H.J., Whiting, G.L., Sun, B., Greenham, N.C., Huck, W.T.S., and Friend, R.H. (2005) *Nano Lett.*, **5** (9), 1653–1657.
105 Whiting, G.L., Snaith, H.J., Khodabakhsh, S., Andreasen, J.W., Breiby, D.W., Nielsen, M.M., Greenham, N.C., Friend, R.H., and Huck, W.T.S. (2006) *Nano Lett.*, **6** (3), 573–578.
106 Schönenberger, C., van der Zande, B.M.I., Fokkink, L.G.J., Henny, M., Schmid, C., Krüger, M., Bachtold, A., Huber, R., Birk, H., and Staufer, U. (1997) *J. Phys. Chem. B*, **101** (28), 5497–5505.
107 Dauginet-De Pra, L., Ferain, E., Legras, R., and Demoustier-Champagne, S. (2002) *Nucl. Instrum. Methods B*, **196** (1–2), 81–88.
108 Cornelius, T.W., Schiedt, B., Severin, D., Pépy, G., Toulemonde, M., Apel, P.Y., Boesecke, P., and Trautmann, C. (2010) *Nanotechnology*, **21**, 155702.
109 Jessensky, O., Muller, F., and Gosele, U. (1998) *Appl. Phys. Lett.*, **72** (10), 1173–1175.
110 Haberkorn, N., Gutmann, J.S., and Theato, P. (2009) *ACS Nano*, **3** (6), 1415–1422.
111 Liu, P., Singh, V.P., and Rajaputra, S. (2010) *Nanotechnology*, **21**, 115303.
112 Coakley, K.M. and McGehee, M.D. (2003) *Appl. Phys. Lett.*, **83** (16), 3380–3382.
113 Coakley, K.M., Srinivasan, B.S., Ziebarth, J.M., Goh, C., Liu, Y., and McGehee, M.D. (2005) *Adv. Funct. Mater.*, **15** (12), 1927–1932.
114 Aryal, M., Trivedi, K., and Hu, W. (Walter) (2009) *ACS Nano*, **3** (10), 3085–3090.
115 Stolz Roman, L., Inganäs, O., Granlund, T., Nyberg, T., Svensson, M., Andersson, M.R., and Hummelen, J.C. (2000) *Adv. Mater.*, **12** (3), 189–195.
116 Fan, Z., Razavi, H., Do, J.-won, Moriwaki, A., Ergen, O., Chueh, Y.-L., Leu, P.W., Ho, J.C., Takahashi, T., Reichertz, L.A., Neale, S., Yu, K., Wu, M., Ager, J.W., and Javey, A. (2009) *Nat. Mater.*, **8** (8), 648–653.
117 Markus, T.Z., Wu, M., Waldeck, D.H., Orin, D., Naaman, R. (2009) *J. Phys. Chem. C*, **113**, 14200–14206.
118 Kuo, C.Y., Tang, W.C., Gau, C., Guo, T.F., and Jeng, D.Z. (2008) *Appl. Phys. Lett.*, **93** (3), 033307-033307-3.
119 Aryal, M., Buyukserin, F., Mielczarek, K., Zhao, X.-M., Gao, J., Zakhidov, A., and Hu, W. (2008) *J. Vac. Sci. Technol. B*, **26**, 2562.
120 Berding, M.A. (1999) *Appl. Phys. Lett.*, **74**, 552.
121 Xu, D., Chen, D., Xu, Y., Shi, X., Guo, G., Gui, L., and Tang, Y. (2000) *Pure Appl. Chem.*, **72** (1–2), 127–135.
122 Kouklin, N., Menon, L., Wong, A.Z., Thompson, D.W., Woollam, J.A., Williams, P.F., and Bandyopadhyay, S. (2001) *Appl. Phys. Lett.*, **79**, 4423.
123 Peña, D.J., Mbindyo, J.K.N., Carado, A.J., Mallouk, T.E., Keating, C.D., Razavi, B., and Mayer, T.S. (2002) *J. Phys. Chem. B*, **106** (30), 7458–7462.
124 Lei, Y., Zhang, L.D., Meng, G.W., Li, G.H., Zhang, X.Y., Liang, C.H., Chen, W., and Wang, S.X. (2001) *Appl. Phys. Lett.*, **78**, 1125.
125 Chen, X., Schriver, M., Suen, T., and Mao, S.S. (2007) *Thin Solid Films*, **515** (24), 8511–8514.
126 Macák, J.M., Tsuchiya, H., and Schmuki, P. (2005) *Angew. Chem., Int. Ed.*, **44** (14), 2100–2102.
127 Zhang, Y., Wang, C., Rothberg, L., and Ng, M.-K. (2006) *J. Mater. Chem.*, **16**, 3721.
128 Lim, S.L., Liu, Y., Liu, G., Xu, S.Y., Pan, H.Y., Kang, E., and Ong, C.K. (2011) *Phys. Status Solidi (a)*, **208** (3), 658–663.

129 Wei, Q., Hirota, K., Tajima, K., and Hashimoto, K. (2006) *Chem. Mater.*, **18** (21), 5080–5087.
130 Lee, J. and Jho, J.Y. (2011) *Sol. Energy Mater. Sol. Cells*, **95** (11), 3152–3156.
131 Foong, T.R.B., Shen, Y., Hu, X., and Sellinger, A. (2010) *Adv. Funct. Mater.*, **20** (9), 1390–1396.
132 Service, R.F. (2011) *Science*, **332** (6027), 293.
133 Xu, T. and Qiao, Q. (2011) *Energy Environ. Sci.*, **4** (8), 2700–2720.
134 Zhao, J., Wang, A., Green, M.A., and Ferrazza, F. (1998) *Appl. Phys. Lett.*, **73**, 1991.
135 Zhao, J., Wang, A., Yun, F., Zhang, G., Roche, D.M., Wenham, S.R., and Green, M.A. (1997) *Prog. Photovolt.: Res. Appl.*, **5** (4), 269–276.
136 Schultz, O., Glunz, S.W., and Willeke, G.P. (2004) *Prog. Photovolt.: Res. Appl.*, **12** (7), 553–558.
137 Engelhart, P., Wendt, J., Schulze, A., Klenke, C., Mohr, A., Petter, K., Stenzel, F., Hörnlein, S., Kauert, M., Junghänel, M., Barkenfelt, B., Schmidt, S., Rychtarik, D., Fischer, M., Müller, J.W., and Wawer, P. (2011) *Energy Procedia*, **8** (0), 313–317.
138 Benagli, S. (2009) 24th European Photovoltaic Solar Energy Conference, September 21–25, 2009, Hamburg, Germany, pp. 2293–2298.
139 Jackson, P., Hariskos, D., Lotter, E. *et al.* (2011) *Prog. Photovolt. Res. Appl.*, **19**, 894–897.
140 MiaSolé Achieves 15.7% Efficiency with Commercial-Scale CIGS Thin Film Solar Modules press release December 02, 2010 http://www.miasole.com/sites/default/files/MiaSole_release_Dec_02_2010.pdf
141 First Solar Sets World Record for CdTe Solar PV Efficiency: 17.3 Percent Efficiency Confirmed by NREL (press release July 26, 2011) http://investor.firstsolar.com/releasedetail.cfm?ReleaseID=593994
142 First Solar Sets Another World Record for CdTe Solar PV Efficiency,14.4 Percent Total Area Efficiency Module Confirmed by NREL (press release January 16, 2012) http://investor.firstsolar.com/releasedetail.cfm?ReleaseID=639463
143 NIMS Sets a New World Record for the Highest Conversion Efficiency in Dye-Sensitized Solar Cells (press release 30.08.2011) http://www.nims.go.jp/eng/news/press/2011/08/p201108250.html
144 Goldstein, J., Yakupov, I., and Breen, B. (2010) *Sol. Energy Mater. Sol. Cells*, **94** (4), 638–641.
145 Konarka Technologies Advances Award Winning Power Plastic Solar Cell Efficiency with 9% Certification (press release Feb. 28, 2012) http://www.konarka.com/index.php/site/pressreleasedetail/konarka_technologies_advances_award_winning_power_plastic_solar_cell_effici
146 Green, M.A., Emery, K., Hishikawa, Y., Warta, W., Dunlop, E.D. (2012) *Prog. Photovolt. Res. Appl.*, **20**, 12–20.
147 Heliatek sets new world record efficiency of 10.7% for its organic tandem cell (press release April 27, 2012) http://www.heliatek.com/?p=1923&lang=en
148 Achieving the world's highest photoelectric conversion efficiency of more than 10% (press release June 2011 from Mitsubishi Chemical) http://www.m-kagaku.co.jp/english/aboutmcc/RC/special/feature1.html
149 Kuwabara, T., Kawahara, Y., Yamaguchi, T., Takahashi, K. (2009) *ACS Appl. Mater. Interfaces*, **1**, 2107–2110.
150 Kim, M.S., Kim, J.S., Cho, J.C., Shtein, M., Guo, L.J., Kim, J. (2007) *Appl. Phys. Lett.*, **90** (12), 123113.

14
Conjugated Polymer Blends: Toward All-Polymer Solar Cells
Christopher R. McNeill

14.1
Introduction

An alternative to fullerene derivatives and inorganic nanocrystals as electron acceptors in polymer solar cells are electron-accepting polymers. The pairing of two conjugated polymers as donor and acceptor in so-called all-polymer solar cells has several potential advantages. First, there is greater flexibility in the synthetic chemistry of conjugated polymers compared to fullerene derivatives and inorganic nanocrystals allowing for greater tuning of the optoelectronic properties of the acceptor material. Second, as conjugated polymers have higher extinction coefficients than fullerenes or inorganic nanocrystals, the acceptor phase in polymer/polymer blends contributes as much to photocurrent as the donor phase, with a higher overall absorption cross section of the blend film. These first two advantages therefore offer greater potential for tuning and optimizing of the optoelectronic properties of donor and acceptor, with donor and acceptor covering complementary regions of the solar spectrum without the need for a tandem architecture. Finally, by covalently linking donor and acceptor polymer, there is great potential for exact control of nanomorphology through the use of block copolymers.

The efficiency of all-polymer blends has lagged behind that of polymer/fullerene blends and polymer/nanocrystal blends in recent years, with all-polymer blends recording record efficiencies of ~2% compared to over 3% [1] for polymer/nanocrystal blends and over 9% [2] for polymer/fullerene blends. A number of reasons have been given for the comparatively low efficiency of all-polymer blends including the lower electron mobility of acceptor polymers compared to fullerene derivatives [3], poor interfacial charge separation [4], and lack of morphological control [5]. Another reason for the lower efficiency of all-polymer blends is that less synthetic effort has been dedicated to the synthesis of electron-accepting polymers with the majority of new conjugated polymers for organic photovoltaics synthesized for use with fullerene derivatives. For polymer solar cells in general, the link between chemical structure and device performance is not well established, with efficiency improvements gained largely through synthetic trial and error. Given the nature of materials development in this field, the lack of a synthetic effort directed toward the

Semiconducting Polymer Composites: Principles, Morphologies, Properties and Applications, First Edition.
Edited by Xiaoniu Yang.
© 2012 Wiley-VCH Verlag GmbH & Co. KGaA. Published 2012 by Wiley-VCH Verlag GmbH & Co. KGaA.

development of new acceptor polymers has likely also contributed to the lower efficiency of all-polymer cells compared to other approaches.

This chapter aims to provide an overview of the development of all-polymer solar cells, reviewing progress made to date and discussing key issues affecting cell operation. Perspectives for future work will also be given in light of the challenges facing the field.

14.2
Review of Polymer Photophysics and Device Operation

Due to the fact that excitons (tightly bound electron–hole pairs) are the primary product of photoexcitation in conjugated polymer films, heterojunctions between electron-donating and electron-accepting materials are required to facilitate charge generation [6]. Coupled with the low exciton diffusion length of conjugated polymer films, ~10 nm [7], a blended or bulk heterojunction architecture is used to optimize light absorption, exciton dissociation, and charge collection (see Figure 14.1a). An active layer thickness of ~100–200 nm is sufficient to absorb the majority of photons with energy within the absorption band of each polymer in

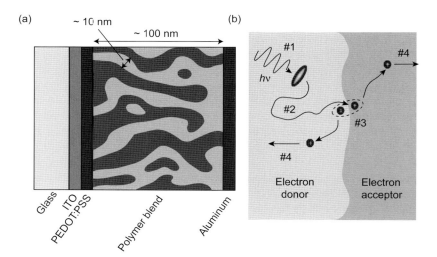

Figure 14.1 (a) Schematic of the device structure of an all-polymer solar cell. (b) Diagram representing the operating mechanisms of an organic solar cell: A photon of energy $h\nu$ excites the electron-donor phase creating an exciton with binding energy of ~0.4 eV, shown in step 1. The exciton then diffuses (step 2) to a donor/acceptor interface where it dissociates into an interfacial electron–hole pair. Step 3 represents the separation of interfacial electron–hole pairs with step 4 the bulk transport of free charges to the electrodes for collection. (Reprinted with permission from Ref. [9]. Copyright 2009, Wiley-VCH Verlag GmbH.)

the blend. Phase-separated domains of donor and acceptor with domain size on the order of the exciton diffusion length ensure that the majority of photoexcited excitons reach and dissociate at a donor/acceptor interface. Interpenetrating networks of donor and acceptor with percolation paths to the electrodes ensure that separated charges are collected at the relevant electrodes. As will be seen, this "ideal" morphology is generally not realized in actual solution-processed blends, with impure phases and a hierarchy of domains with a broad distribution of domain sizes observed. Sharp interfaces between donor and acceptor phases as portrayed in Figure 14.1 may also not be realized in solution-processed blends [8].

The device structure of all-polymer blends is generally the same as employed for polymer/fullerene and polymer/nanocrystal blends (see Figure 14.1a). Specifically, indium tin oxide (ITO)-coated glass is used as the semitransparent electrode with the conducting polymer PEDOT:PSS (poly(3,4-ethylenedioxythiophene):poly(styrene sulfonate)) used as an interfacial layer between ITO and the active semiconductor layer. The active layer is deposited from a blended solution on top of the PEDOT:PSS layer (typically via spin coating) and the device completed by the evaporation of a low-work-function metal such as aluminum. Due to the low lateral conductivity of the cells, the active area is generally defined by the overlap between the top and bottom electrodes in the sandwich-style configuration. Flexible substrates may be used instead of glass for industry-relevant processing, and other device geometries are possible such as so-called inverted cells [10].

With reference to Figure 14.1b, device operation can be described as follows: An incident photon is absorbed by donor or acceptor polymer producing an exciton. This exciton then diffuses to a donor/acceptor interface where it is dissociated, producing an interfacial electron–hole pair. Dissociation of this interfacial electron–hole pair can also be problematic (see Section 14.5.1) with their so-called geminate recombination thought to be a dominant loss mechanism in all-polymer cells. Under favorable conditions, the electron and hole are separated from the interface and are collected at the relevant electrode (electrons at the top, low-work-function electrode and holes at the bottom, high-work-function electrode) avoiding the bimolecular recombination of charges.

14.3
Material Considerations

In order to facilitate efficient device action, donor and acceptor polymers must possess compatible optical, electronic, and physical properties. Efficient exciton dissociation requires sufficient differences in the electron affinity and ionization potential of the paired polymers. As a rule of thumb, a 0.4 eV difference in the lowest unoccupied molecular orbitals (LUMOs) is typically required to drive dissociation of excitons generated in the donor phase via electron transfer (see Figure 14.2 for a schematic energy level diagram). A similar offset in the highest occupied molecular orbitals (HOMOs) is required to drive dissociation of excitons generated in the acceptor phase via hole transfer. Note that the assignment of donor and

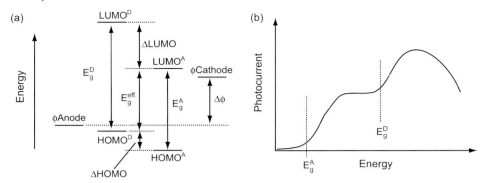

Figure 14.2 (a) Schematic energy level diagram depicting the energy levels and bandgaps (E_g) of donor and acceptor, work functions (ϕ) of the electrodes, and energy level offsets. The effective bandgap, E_g^{eff}, is the offset between the LUMO of the acceptor and HOMO of the donor and places an upper limit on the V_{OC} of the system. (b) Schematic photocurrent action spectrum of an all-polymer solar cell with onsets of absorption by donor and acceptor marked.

acceptor is relative, since a polymer that acts as an electron acceptor in one combination may work as an electron donor in another combination, as determined by the relative position of HOMO and LUMO [11]. Ideally, donor and acceptor will possess complementary optical absorption profiles to cover different regions of the solar spectrum, as depicted schematically in Figure 14.2. However, the open-circuit voltage (V_{OC}) is also affected by the energy levels of donor and acceptor, with the effective bandgap, E_g^{eff}, defined as the difference between the HOMO of the donor and LUMO of the acceptor (see Figure 14.2), placing an upper limit on the V_{OC} of the system. Thus, device optimization requires a consideration of both the effective bandgap and the covering of as much of the solar spectrum as possible. The work functions of the electrodes may also influence V_{OC}; for the case of polymer/fullerene cells, V_{OC} is largely governed by E_g^{eff} for the case of ohmic contacts, while the work function difference of the electrodes determines V_{OC} for the case of non-ohmic contacts [12]. While less study has been devoted to the origin of V_{OC} in all-polymer systems, it is likely to reflect the findings of studies into polymer/fullerene cells. Another consideration is the charge transport properties of the materials; donor and acceptor should possess high hole and electron mobilities, respectively, to facilitate charge extraction and minimize charge recombination. Finally, it is important that donor and acceptor possess favorable mixing properties in order to facilitate the creation of nanoscale, phase-separated morphologies required for optimum device operation. Thus, both polymers should be soluble in the same solvent and interact in a way to favor phase-separated domains. In addition, where the favorable properties (e.g., mobility, optical absorption) of either or both components depend on the material being in a crystalline or semicrystalline state, it is essential that crystallization is not hindered in the blend.

Synthetically, conjugated polymers with deep LUMOs suitable for electron acceptors are generally achieved through the use of electron-withdrawing groups, either as side-chain substituents or as units in the main chain of donor–acceptor polymers (see Table 14.1). Cyano (CN) groups have been commonly used as side-chain substituents [13–21], while benzothiadiazole is a common main-chain electron-withdrawing group used in donor–acceptor polymers [11, 22, 23]. The ladder polymer BBL has also been used to good effect as an electron donor in bilayer solar cells [24]; however, due to processing constraints, it has not been adopted in blends. More recently, perylene diimide [25, 26] and naphthalene diimide [27–29] units have been used as acceptor units in deep-LUMO donor–acceptor polymers. Other novel electron acceptors developed recently include organobromo polymers [30], diketopyrrolopyrrole-based acceptors [31], and polymers using novel electron-deficient units such as quinoxaline [32], though these novel electron-accepting polymers are yet to show promising efficiencies. For donor polymers, polyphenylene vinylene (PPV) and polythiophene derivatives have typically been used.

14.4
Device Achievements to Date

Table 14.1 presents a selection of donor/acceptor combinations that have shown notable solar cell efficiencies. Power conversion efficiencies of ∼1.5–2.2% have been demonstrated for a number of systems, with most blends utilizing either a CN-PPV derivative or the fluorene–dithienylbenzothiadiazole copolymer F8TBT as the acceptor. Despite their high electron mobilities, perylene diimide- and naphthalene diimide-based acceptors have shown disappointing efficiencies, though with optimization Zhou *et al.* have recently been able to achieve a power conversion efficiency of 2.2% with a perylene diimide acceptor [25].

High open-circuit voltages are often demonstrated by all-polymer cells, with a V_{OC} of 1.4 V demonstrated for MDMO-PPV:PF1CVTP blends (see Table 14.1). In comparison, the open-circuit voltages of polymer/fullerene blends and polymer/nanocrystal blends are typically less than 1 V. For acceptors with deep LUMOs, where the LUMO/LUMO offset is significantly greater than the exciton binding energy, this excess energy is not necessarily "lost" during the charge transfer process, since it may play a critical role in assisting the separation of interfacial electron–hole pairs [34]. However, for perylene diimide- and naphthalene diimide-based acceptors that possess LUMOs as deep as fullerene derivatives, an increase in short-circuit current relative to higher LUMO acceptors is not observed. All-polymer cells in general possess lower fill factors compared to other polymer-based photovoltaics. Of the cells that have shown the highest efficiencies to date, fill factors as low as 0.35 have been exhibited. Recent research has shown that high fill factors can be achieved in all-polymer solar cells, with a fill factor of 0.67 demonstrated for P3HT:P(NDI2OD-T2) solar cells achieved by tuning film morphology. However, these high fill factors have not been combined with high short-circuit currents meaning that overall efficiency of such devices has remained low. Example current–voltage

Table 14.1 Donor/acceptor combinations used in efficient all-polymer cells.

Electron donor	Electron acceptor	J_{SC} (mA cm^{-2})	V_{OC} (V)	FF	PCE (%)	References
POPT	MEH-CN-PPV	5.4	1.06	0.35	2.0	[13, 29]
M3EH-PPV	CN-ether-PPV	3.6	1.36	0.35	1.7	[16]

14.4 Device Achievements to Date | 405

MDMO-PPV	3.0	1.4	0.37	1.5	[17]
PF1CVTP					
P3HT	4.0	1.25	0.45	1.8	[11, 33]
F8TBT					

(continued)

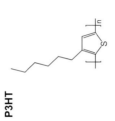

Table 14.1 (Continued)

Electron donor	Electron acceptor	J_{SC} (mA cm^{-2})	V_{OC} (V)	FF	PCE (%)	References
PT1	PC-PDI	6.35	0.70	0.50	2.23	[25]
P3HT	P(NDI2OD-T2)	2.39	0.48	0.54	0.62	[27–29]

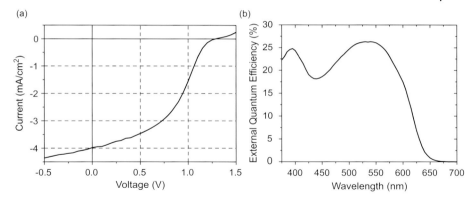

Figure 14.3 Current–voltage characteristics (a) and external quantum efficiency spectrum of a P3HT:F8TBT blend device (b). Note that the "kink" in the current–voltage curve about V_{OC} is likely due to one of the contacts blocking charge extraction under forward bias. (Reprinted with permission from Ref. [11]. Copyright 2007, American Institute of Physics.)

characteristics and EQE spectrum of an efficient P3HT:F8TBT cell are presented in Figure 14.3.

14.5 Key Issues Affecting All-Polymer Solar Cells

14.5.1 Interfacial Charge Separation

The low fill factor of all-polymer cells can be taken as being symptomatic of low charge separation efficiencies. All-polymer cells typically exhibit a linear dependence of short-circuit current versus light intensity despite showing a significant slope of the current–voltage curve as it goes through short-circuit conditions [16, 35–38]. This strong voltage dependence of photocurrent in the absence of a nonlinear light intensity dependence of photocurrent (that should accompany bimolecular recombination associated with charge recombination after charge separation) has been explained by a voltage dependence of charge separation. The optimized active layer thickness of many efficient all-polymer cells is also low (55–70 nm [16, 37]), which is consistent with high internal electric fields being necessary to drive device action. Direct evidence for a link between applied voltage and charge separation efficiency has been shown by Gonzalez-Rabade *et al.* who correlated quenching of exciplex emission and photocurrent collection in blends of the polyfluorene polymers PFB and F8BT (see Figure 14.4a). Figure 14.4a presents a schematic diagram of the charge generation and recombination processes. After exciton dissociation at a PFB/F8BT

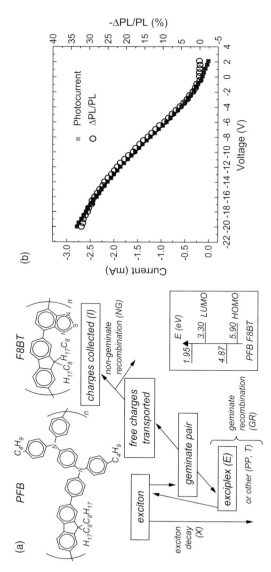

Figure 14.4 (a) Chemical structures of F8BT and PFB and relative HOMO and LUMO energy levels as well as a schematic of charge generation and recombination pathways. (b) Simultaneous measurement of current and relative electric field quenching of the exciplex as a function of applied voltage for a PFB:F8BT blend device. (Reprinted with permission from Ref. [40]. Copyright 2009, Wiley-VCH Verlag GmbH.)

interface, an interfacial electron–hole pair (or "geminate pair") is formed. This electron–hole pair can either be separated into free charges or recombine. In the PFB/F8BT system, a proportion of electron–hole pairs recombine through an emissive, exciplex state that has redshifted emission allowing for exciplex recombination to be distinguished from PFB and F8BT fluorescence. As the voltage applied to a PFB/F8BT cell is varied, the change in exciplex emission is compared to the voltage dependence of photocurrent. Strikingly, a direct correlation between reduced exciplex emission and increased photocurrent collection is observed (Figure 14.4b), indicating that an increased internal electric field aids the separation of interfacial electron–hole pairs that otherwise would recombine. A similar anticorrelation between relative exciplex emission and photovoltaic efficiency has been observed by Yin *et al.* who studied M3EH-PPV:CN-ether-PPV blends [35]. A strong electric field dependence of charge transfer exciton emission in polyfluorene copolymer/fullerene blends has also been observed by Veldman *et al.*, indicating that this phenomenon is not exclusive to all-polymer systems [39].

The poor charge separation efficiency of all-polymer blends is attributed (in part) to the low dielectric constant of conjugated polymers. Charge separation efficiency has also been related to charge transport mobility [39, 41], film morphology [38, 42], and the interfacial conformation and energetic structure of the donor/acceptor interface [43, 44]. The reason why polymer/fullerene blends exhibit superior charge separation efficiencies is not entirely clear. Fullerene derivatives possess a higher dielectric constant ($\varepsilon_r \sim 4$) compared to conjugated polymers ($\varepsilon_r \sim 2-3$) [36, 45], though this difference alone is not likely to explain the superior performance of polymer/fullerene devices. The low electron mobility of many acceptor polymers compared to fullerene derivatives is likely a strong contributing factor to the poor charge separation efficiency of all-polymer cells [35–37]. However, recent results using P(NDI2OD-T2) as the acceptor in all-polymer cells have demonstrated that low electron mobility is not the only factor. P(NDI2OD-T2) has been shown to exhibit superior electron transport properties in both transistor and diode configurations with an electron mobility comparable to fullerene derivatives [46, 47]. However, polythiophene:P(NDI2OD-T2) solar cells [27–29] have not been able match the efficiency of other all-polymer cells that exhibit electron mobilities many orders of magnitude lower.

Ultrafast pump-probe experiments on all-polymer blends confirm the fast geminate recombination of interfacial electron–hole pairs that competes with charge separation [4, 28, 48]. For PFB/F8BT blends, rapid geminate recombination is facilitated by the intersystem crossing of electron–hole pairs that terminally recombine into F8BT triplet excitons [48] demonstrating the importance of engineering the energy levels of the system such that the triplet energy is higher than that of interfacial charge pair states [49]. For more efficient P3HT:F8TBT and P3HT:P(NDI2OD-T2) blends, a significant proportion of charge pairs recombine in the first few nanoseconds with the remaining charge pairs exhibiting similar lifetimes to efficient polymer/fullerene blends [4, 28]. This observation of different populations of charge pairs with distinct lifetimes

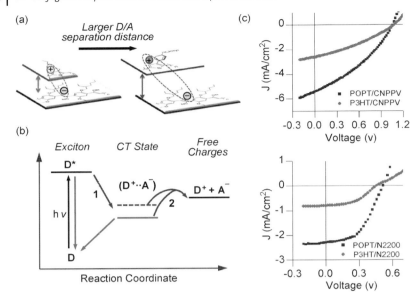

Figure 14.5 (a and b) Schematic diagrams demonstrating how steric interactions can lead to an increase in backbone spacing and destabilization of the geminate pair due to a different energy landscape. (c) Example current–voltage curves showing the enhanced performance of POPT-based devices compared to P3HT-based devices. (Reprinted with permission from Ref. [29]. Copyright 2011, American Chemical Society.)

suggests a range of interfacial donor/acceptor conformations, some favorable and others unfavorable for charge separation. It is likely that differences between the conformational and associated energetic nature of the donor/acceptor interfaces in all-polymer blends and polymer/fullerene blends largely explain the differences in efficiency and device behavior. Indeed, recent measurements have found evidence for an interfacial dipole of ∼0.2 V nm^{-1} in P3HT:PCBM blends that may explain the high fill factor and charge separation efficiency in this system [44, 50]. Recent measurements on all-polymer systems have also demonstrated the importance of donor/acceptor spacing in influencing interfacial energy levels and device efficiency [29]. By introducing a bulky substituent onto the polymer backbone, the physical separation of donor and acceptor chains was altered. With increased donor/acceptor separation, the energy of relaxed interfacial electron–hole pairs was raised, decreasing the barrier to charge separation and resulting in increased photocurrent generation (see Figure 14.5). Thus, the interfacial nature of donor/acceptor systems plays a key role in determining charge separation, and an improved understanding of interfacial processes will facilitate the design of donor and acceptor polymers for improved charge separation efficiencies [51].

14.5.2
Morphology

The morphology of the bulk heterojunction film also plays a key role in determining overall device efficiency. Films with domains that are too coarse will result in losses due to excitons not reaching donor/acceptor interfaces, while blends that are too well mixed may result in low charge separation and charge collection efficiencies. Percolating charge collection pathways are also required as "dead ends" lead to reduced charge collection. The morphology of solution-processed bulk heterojunction films, however, is hard to control. The nature of the thin film morphology produced will depend on many parameters such as polymer/polymer and polymer/solvent interactions, polymer solubility, solvent boiling point, molecular weight, substrate surface energy, and so on. A number of experimental handles have been used to influence film morphology including solvent choice, donor/acceptor mixing ratio, thermal and solvent annealing of cast films, and the use of cosolvents or solvent additives.

Associated with the challenge of controlling film morphology is the challenge of characterizing film morphology. The characterization of the three-dimensional nanostructure of polymer blends is challenging due to the similar chemical composition and mass density of donor and acceptor polymers. Atomic force microscopy (AFM) is a convenient and widely employed method for characterizing film microstructure; however, it has poor chemical contrast and images surface topography only. Thus, features observed with AFM are hard to interpret and often only reflect the structure of enriched surface layers that bear little correlation with the underlying bulk morphology. Transmission electron microscopy (TEM) probes bulk structure with high spatial resolution; however, conventional modes have limited chemical sensitivity. For polymer/fullerene blends contrast in bright-field TEM experiments is afforded by density differences with fullerene derivatives such as PCBM having a higher mass density compared to conjugated polymers ($\sim 1.5\,\mathrm{g\,cm^{-3}}$ compared to $\sim 1\,\mathrm{g\,cm^{-3}}$ [52]). For polymer/polymer blends, chemical staining or the use of energy filtering of transmitted electrons is often required in order to produce contrast when imaging with TEM. Figure 14.6 presents

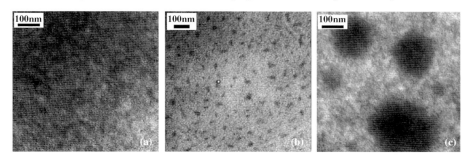

Figure 14.6 Zero-loss energy-filtered transmission electron microscopy images of MDMO–PPV: PCNEPV blends with low (a), medium (b), and high (c) molecular weights of PCNEPV. (Reprinted with permission from Ref. [53]. Copyright 2007, Wiley-VCH Verlag GmbH.)

energy-filtered TEM images of MDMO-PPV:PCNEPV blends with different PCNEPV molecular weight [53]. Zero-loss images are presented that provide enhanced contrast relative to standard bright-field images due to a reduction in chromatic blurring. Nitrogen elemental maps confirm that the features seen in Figure 14.6 are the result of chemical contrast rather than thickness variations with the dark regions corresponding to PCNEPV-rich domains [54]. These MDMO-PPV: PCNEPV films were prepared with a weight ratio of 1 : 1; however, examining Figure 14.6, it is clear that the PCNEPV domains do not fill 50% of the images, providing a clear indication of intermixing with in phases, confirmed by fluorescence quenching measurements. Strikingly, the device performance of blends based on such films shows little difference, suggesting that device performance is more sensitive to the fine-scale intermixing than the large features observed in these TEM images.

The submicron hierarchical domain structure of conjugated polymer blends (with phase separation characterized by multiple characteristic length scales) has also been observed with soft X-ray microscopy. Soft X-ray microscopy utilizes differences in the unoccupied electronic structure for contrast, typically through the use of photons with energies around the carbon K-edge ($\sim$285 eV). By tuning into specific molecular resonances of each material (corresponding to transitions from a 1s core state to an antibonding molecular orbital), chemical contrast is achieved. Indeed, it is also possible to derive quantitative chemical maps based on images taken at multiple energies with knowledge of the thickness-calibrated X-ray absorption spectra of neat films. Figure 14.7 presents chemical composition maps of PFB: F8BT blends spin coated from chloroform. Films were annealed subsequent to film deposition for 10 min followed by quenching to room temperature. Chloroform was chosen as the solvent with the aim of producing molecularly intermixed blends, since chloroform is a good solvent for both polymers and possesses a low boiling point. Annealing was then used to coarsen the morphology in a pseudo-controlled way. However, as revealed in the image of the as-spun film, structure is already present in this film with a length scale of $\sim$80 nm, much larger than the exciton diffusion length. Photoluminescence is strongly quenched in the blend, however, indicating intermixing within the 80 nm features. With annealing, no change in the structure of the film is observed with soft X-ray microscopy until an annealing temperature of 160 °C, where enhanced contrast is observed followed by coarsening of the morphology at higher temperatures. Photoluminescence intensity does increase with annealing to 140 °C indicating evolution of morphology on the length scale of $\sim$10 nm. Device efficiency is optimized at an annealing temperature of 140 °C, indicating that although coarsening of the morphology results in a reduction in exciton dissociation yield, it improves efficiency through an improvement in charge separation efficiency.

Since the resolution of soft X-ray microscopy is limited to $\sim$30 nm, other approaches are required for providing information on finer length scales. Resonant soft X-ray scattering (R-SoXS) has been developed to utilize the chemical contrast afforded by soft X-rays with improved spatial resolution enabling structures as fine as the wavelength of the corresponding soft X-rays ($\sim$4 nm at the carbon K-edge). Figure 14.8 presents the raw scattering profiles and the corresponding pair

14.5 Key Issues Affecting All-Polymer Solar Cells

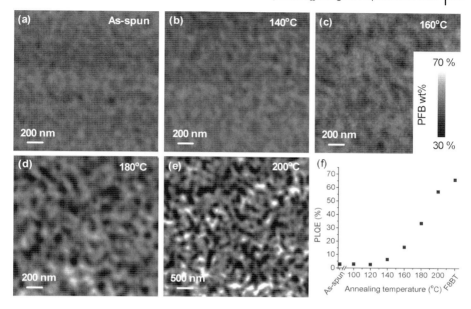

Figure 14.7 (a–e) Chemical composition maps of PFB:F8BT blends spin coated from chloroform with subsequent thermal annealing (temperature indicated). (f) Evolution of photoluminescence quantum efficiency of PFB:F8BT blends with annealing. (Reproduced with permission from Ref. [55]. Copyright 2008, Institute of Physics Publishing Ltd.)

distribution function. The pair distribution function, $P(r)$, assists in the interpretation of the scattering data providing ensemble-averaged information about the distances between the phases in the sample and the shape of the domains. The first zero crossing is an indication of the mean domain size, while the magnitude of $P(r)$

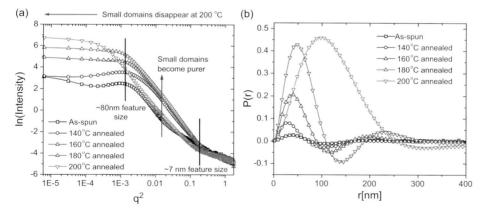

Figure 14.8 Resonant soft X-ray scattering profiles of PFB:F8BT blends (a) and the corresponding pair distribution function (b). (Reprinted with permission from Ref. [5]. Copyright 2010, American Chemical Society.)

is a measure of domain purity. For the as-cast sample, the first crossing of $P(r)$ is at ∼80 nm matching the domain sizes imaged with X-ray microscopy. With annealing, the magnitude of $P(r)$ increases with annealing at 140 °C with the zero crossing shifting to larger values only for annealing above 160 °C. Thus, the morphology evolves initially through a purifying of the 80 nm domains before an evolution in their size. Soft X-ray investigation of higher efficiency systems such as P3HT:F8TBT and P3HT:P(NDI2OD-T2) blends [28] has found similar results indicating that this hierarchical nature of nanoscale phase separation is likely to be a common feature of conjugated polymer blends.

The results of these morphological studies indicate that our ability to control nanostructure using conventional processing methods is not as refined as desired. Ideally, pure, interpenetrating domains with a characteristic length scale on the order of the exciton diffusion length are required. Even through the use of high boiling point solvents, domains significantly larger than the exciton diffusion length are produced. While these domains possess intermixing on a length scale sufficient to quench exciton generation, the local blend ratio deviates from the specified mixing ratio. Thus, there is no direct control of the resulting domain composition that may be suboptimal for device performance. It is not clear exactly how suboptimal these structures are; however, improved device efficiencies are likely to result from improved morphological control.

A number of novel approaches have been developed to provide improved morphological control. One approach has been to create polymer nanoparticles of well-defined size via a miniemulsion process (see Figure 14.9) [56, 57]. Films are prepared from an aqueous dispersion containing polymer nanoparticles. Separate donor and acceptor nanoparticles can be prepared and then blended, or nanoparticles can be prepared where both donor and acceptor polymers are contained in each nanoparticle. In both cases, the upper limit for the dimension of phase separation is determined by the size of the individual nanoparticles. A limitation of this approach is that the smallest size nanoparticles that can be produced (a few tens of nanometers [57]) already have sizes larger than the exciton diffusion length, and the morphology of blended nanospheres tends to adopt a core–shell structure that may not be ideal for device operation [58].

Another approach is to blend the donor polymer with a sacrificial, insulating polymer that acts as a phase-directing agent (PDA) (see Figure 14.10) [59]. This phase-directing polymer is chosen to be soluble in a solvent that the donor is not that allows it to be washed away once the film has been deposited, leaving a porous nanostructured matrix. The donor matrix is then cross-linked and backfilled with the acceptor phase. The use of the PDA offers greater flexibility in tuning the morphology of the blend through choice of the PDA and tuning of PDA molecular weight. This approach avoids intermixing of donor and acceptor phases, though a minority component of the PDA is likely to remain in the donor phase. Superior device efficiencies have been demonstrated for PFB/F8BT structures prepared in this way compared to blends [59]; however, the demonstration of efficiencies beyond state of the art are yet to be demonstrated. There are also likely to be issues associated with attaining phase-separated structures on the length scale of

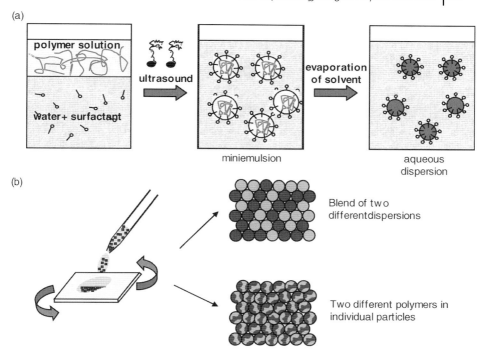

Figure 14.9 Schematic of the process of preparing polymer nanoparticles and films via a miniemulsion route. (a) First, a solution of the polymer in an organic solvent is mixed with water containing an appropriate surfactant. A miniemulsion is then formed on stirring and ultrasonication. Finally, the solvent is evaporated, resulting in solid polymer nanoparticles dispersed in water. (b) Phase-separated structures at the nanometer scale can be prepared either by coating a layer from a dispersion containing nanoparticles of two different polymers or by using dispersions that contain both polymers in each individual nanoparticle. (Reprinted with permission from Ref. [57]. Copyright 2003, Macmillan Publishers Ltd.)

10–20 nm with this approach associated with the difficulty of washing away and backfilling into nanoporous structures on such a length scale (50 nm was the smallest lateral feature demonstrated [59]).

Nanoimprint lithography has also recently been used to create films with nanostructured donor/acceptor interfaces. A nanopatterned silicon master (produced by electron beam lithography) is used to imprint a predefined structure into a polymer film, assisted by either heating of the polymer film or swelling with a solvent-saturated atmosphere. Once formed, this first nanoimprinted polymer film can be used to imprint another polymer film (that has a lower glass transition temperature) creating an interpenetrating network of donor and acceptor polymers. A device structure is created by preparing the polymer films on substrates with predeposited electrodes (see Figure 14.11). The structures fabricated by the NIL technique also ensure complete coverage of the anode by the electron-donating polymer and of

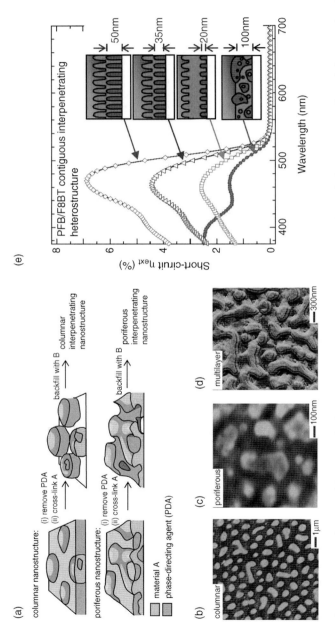

Figure 14.10 (a) Schematic of the process for forming nanostructured donor/acceptor blends through the use of a sacrificial PDA. The donor polymer is mixed with the PDA to form a phase-separated blend structure. The PDA is then dissolved, the donor polymer cross-linked, and the structure backfilled with the acceptor polymer. (b–d) AFM images of morphologies produced. (e) Comparison of the performance of PFB:F8BT devices fabricated through conventional blending and through the use of a PDA. (Adapted and reprinted with permission from Ref. [59]. Copyright 2009, Macmillan Publishers Ltd.)

14.5 Key Issues Affecting All-Polymer Solar Cells | 417

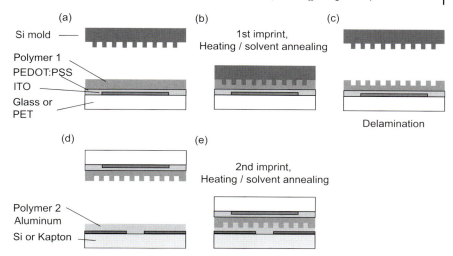

Figure 14.11 Schematic of the nanoimprint lithography approach for creating nanostructured heterojunction devices.

the cathode by the electron-accepting polymer. P3HT/F8TBT nanoimprinted devices have been fabricated with a power conversion efficiency twice that of a control blend device (using a 25 nm diameter columns with 50 nm pitch), [33] and equal to the best previously reported efficiency reported for blend devices based on the donor/acceptor system used. A strong correlation between the feature size of the nanoimprinted structure and device performance was also observed (Figure 14.12), consistent with the increased interfacial area. Structures finer than 25 nm were not able to be produced due to the instability of high aspect ratio imprinted structures. Given the slope of the efficiency versus feature size plot

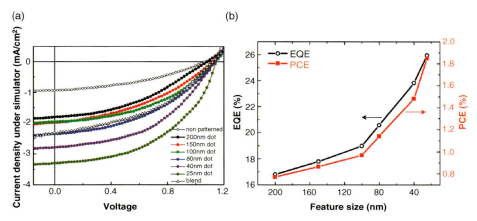

Figure 14.12 Device performance of nanoimprinted polymer solar cells as a function of nanoimprinted feature size. (Reprinted with permission from Ref. [33]. Copyright 2010, American Chemical Society.)

Figure 14.12, significantly higher efficiencies could be expected with feature sizes below 20 nm. These results are encouraging and demonstrate that improved control of film morphology is indeed a promising route to engineering efficiency improvements.

Ultimate control of film morphology may be achieved utilizing block copolymers [60–62]. Through the covalent linking of donor and acceptor polymers, severe restrictions are placed on the phase separation of donor and acceptor blocks. Furthermore, by controlling the molecular weight and polydispersity of the two blocks, the self-assembly of well-ordered nanoscopic morphologies including gyroidal, cylindrical, and lamellar phases with controlled length scales can be achieved [63]. The opportunity to use synthesis to engineer controlled and well-defined nanostructures through self-assembly therefore holds great appeal. However, synthesis of conjugated block copolymers is challenging compared to traditional insulating block copolymers. Furthermore, the different physical properties of rod-like conjugated polymers compared to insulating coil-like polymers traditionally used for block copolymers present new challenges for understanding and controlling phase behavior. While several donor–acceptor block copolymers have been demonstrated, well-organized nanostructures have been slow to appear [61]. Some success has been achieved with block copolymers composed of a P3HT donor with an acceptor consisting of a saturated backbone but with pendant perylene diimide units [64], encouraging further work in this area, though achieving efficiencies beyond state of the art still seems some way away.

14.5.3
Charge Transport

Given that the low mobility of many acceptor polymers (compared to fullerene derivatives such as PCBM) is often cited as a reason for the poor performance of all-polymer solar cells [23, 32, 36], it is worth spending some time discussing the issue of charge transport. As discussed in Section 14.5.1, poor charge separation can be regarded as the main limiting step in photocurrent generation in all-polymer solar cells. Though the efficiency of charge separation can be related to charge mobility [36], it relates more to the local mobility of charges near interfaces rather than bulk mobility. As demonstrated for P3HT:P(NDI2OD-T2) blends, a high bulk charge carrier mobility of the acceptor is not sufficient in itself to ensure improved power conversion efficiencies. P3HT:F8TBT blends [11] exhibit a threefold higher power conversion efficiency than P3HT:P(NDI2OD-T2) blends [27] despite F8TBT exhibiting an electron mobility nearly three orders of magnitude lower ($\sim 9 \times 10^{-6}$ cm^2 V^{-1} s^{-1} for F8TBT [65] compared to 5×10^{-3} cm^2 V^{-1} s^{-1} for P(NDI2OD-T2) [46]). Thus, understanding and overcoming the nature of rapid geminate recombination in all-polymer solar cells appear to be a higher priority than the development of new electron-accepting polymers with high mobilities *per se*.

That said, it is worth examining the potential influence of low electron mobility on the collection of charge subsequent to charge separation. For most all-polymer systems charge transport is imbalanced, with the electron mobility of the acceptor phase at least one order of magnitude lower than the hole mobility of the donor phase [65, 66]. The electron mobility of conjugated polymers used as electron acceptors is often found to be limited by traps [66, 67]. The origin of these traps is not clear, but it is worth remembering that the electron mobility of conjugated polymers in general is not necessarily lower than the hole mobility [47, 68]. One possibility is that the use of electron-withdrawing groups such as cyano substituents or benzothiadiazole units used to lower the LUMO results in localized states [69]. Thus, the method in which the LUMO of a conjugated polymer is lowered in order to impart an electron-accepting character may be responsible for the poor electron transporting properties observed.

The trapping of charge in a device provides a number of mechanisms for enhanced charge recombination. Trap-assisted recombination, where a free carrier recombines with a trapped charge, is one such mechanism, though device simulations have indicated that under solar intensities the Langevin (or bimolecular) recombination of mobile carriers dominates over trap-assisted recombination [36]. The buildup of trapped charge in the device can, however, lead to a redistribution of the internal electric field [10, 70]. This electric field redistribution slows the extraction of mobile charges resulting in increased and overlapping electron/hole densities fostering enhanced bimolecular recombination. Furthermore, since most of the light is absorbed (and hence most of the geminate charge pairs produced) close to the transparent bottom electrode, a reduction in the internal electric field here can actually result in a net lowering of the charge separation efficiency in the device [70]. Thus, trap-assisted recombination (as described by the Shockley–Read–Hall mechanism) may not dominate at high intensities, with the buildup of charge caused by charge trapping leading to increased recombination *via* space-charge effects. The effect of charge trapping on all-polymer solar cell operation, in particular its ability to enhance recombination at high light intensities, has been demonstrated with photocurrent transient measurements on P3HT:F8TBT blend solar cells [67] (see Figure 14.13a). At low light intensities, a monotonic rise in photocurrent is observed after the turn-on of the square light pulse at $0\,\mu s$. With increasing light intensity, a transient peak in the photocurrent turn-on dynamics evolves. The timescale of this dynamic is related to the time for the trap population within the device to reach equilibrium, as determined by the trapping/detrapping rates and trap density. This feature has been reproduced with time-dependent drift diffusion modeling incorporating electron trapping (see Figure 14.13b), confirming the origin of this feature as being due to trap-mediated recombination through a combination of increased geminate and bimolecular recombination [70]. Thus, charge trapping in all-polymer solar cells is an issue affecting the collection efficiency of separated charges, and improved electron transport in the acceptor phase is likely to improve charge collection as well as charge separation. It is worth noting that charge trapping has also been observed in a number of high-efficiency polymer/fullerene systems [71, 72] demonstrating that charge trapping is not a problem exclusive to all-polymer solar cells and a greater knowledge of the origin of charge tapping may assist in engineering future efficiency improvements of organic photovoltaic devices in general.

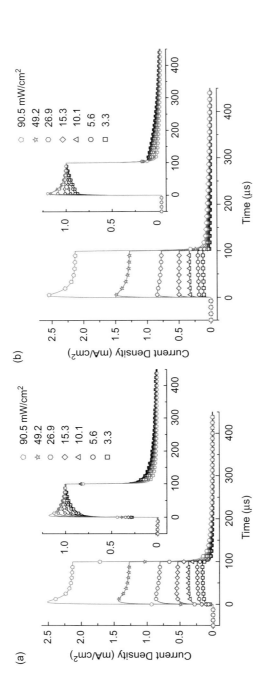

Figure 14.13 Response of the short-circuit photocurrent of P3HT:F8TBT solar cells to a square pulse illumination of varying intensity. Turn-on of the source is at 0 μs and turn-off at 100 μs. (a) The experimental photocurrent transients; (b) the photocurrent transients simulated using a time-dependent drift diffusion model incorporating charge trapping. Insets show normalized photocurrent traces. (Reprinted with permission from Ref. [67]. Copyright 2009, American Institute of Physics.)

14.6
Summary and Outlook

The use of conjugated polymers as electron acceptors in all-polymer solar cells is attractive due to their ability to complement the absorption of the donor polymer. The flexibility in the synthetic chemistry of conjugated polymers compared to fullerene derivatives and inorganic nanocrystals also allows for superior tuning of their optoelectronic properties. Despite these potential advantages, efficiencies of all-polymer cells have remained low, largely due to the problem of interfacial charge separation. Our understanding of interfacial processes is improving, however, with the tuning of interfacial structure through tuning of polymer structure, a promising route to create interfacial energy profiles that promote charge separation. A renewed synthetic effort toward the development of new acceptor polymers will also benefit the field, with conjugated polymers with deep but delocalized LUMOs an attractive area for future research. Improved control of blend morphology is also likely to impart efficiency improvements, with exact control over film morphology through the use of block copolymers a tantalizing prospect.

References

1 Dayal, S., Kopidakis, N., Olson, D.C., Ginley, D.S., and Rumbles, G. (2009) Photovoltaic devices with a low band gap polymer and CdSe nanostructures exceeding 3% efficiency. *Nano Lett.*, **10**, 239–242.

2 Service, R.F. (2011) Outlook brightens for plastic solar cells. *Science*, **332**, 293.

3 Kim, Y., Cook, S., Choulis, S.A., Nelson, J., Durrant, J.R., and Bradley, D.D.C. (2004) Organic photovoltaic devices based on blends of regioregular poly(3-hexylthiophene) and poly(9,9-dioctylfluorene-*co*-benzothiadiazole). *Chem. Mater.*, **16**, 4812–4818.

4 Hodgkiss, J.M., Campbell, A.R., Marsh, R.A., Rao, A., Albert-Seifried, S., and Friend, R.H. (2010) Subnanosecond geminate charge recombination in polymer–polymer photovoltaic devices. *Phys. Rev. Lett.*, **104**, 177701.

5 Swaraj, S., Wang, C., Yan, H., Watts, B., Lüning, J., McNeill, C.R., and Ade, H. (2010) Nanomorphology of bulk heterojunction photovoltaic thin films probed with resonant soft X-ray scattering. *Nano Lett.*, **10**, 2863–2869.

6 Blom, P.W.M., Mihailetchi, V.D., Koster, L.J.A., and Markov, D.E. (2007) Device physics of polymer:fullerene bulk heterojunction solar cells. *Adv. Mater.*, **19**, 1551–1566.

7 Markov, D.E., Amsterdam, E., Blom, P.W.M., Sieval, A.B., and Hummelen, J.C. (2005) Accurate measurement of the exciton diffusion length in a conjugated polymer using a heterostructure with a side-chain cross-linked fullerene layer. *J. Phys. Chem. A*, **109**, 5266–5274.

8 Yan, H., Swaraj, S., Wang, C., Hwang, I., Greenham, N.C., Groves, C., Ade, H., and McNeill, C.R. (2010) Influence of annealing and interfacial roughness on the performance of bilayer donor/acceptor polymer photovoltaic devices. *Adv. Funct. Mater.*, **20**, 4329–2337.

9 McNeill, C.R. and Greenham, N.C. (2009) Conjugated-polymer blends for optoelectronics. *Adv. Mater.*, **21**, 3840–3850.

10 Brenner, T.J.K., Hwang, I., Greenham, N.C., and McNeill, C.R. (2010) Device physics of inverted all-polymer solar cells. *J. Appl. Phys.*, **107**, 114501.

11 McNeill, C.R., Abrusci, A., Zaumseil, J., Wilson, R., McKiernan, M.J., Halls, J.J.M., Greenham, N.C., and Friend, R.H. (2007) Dual electron donor/electron acceptor character of a conjugated polymer in efficient photovoltaic diodes. *Appl. Phys. Lett.*, **90**, 193506.

12 Mihailetchi, V.D., Blom, P.W.M., Hummelen, J.C., and Rispens, M.T. (2003) Cathode dependence of the open-circuit voltage of polymer:fullerene bulk heterojunction solar cells. *J. Appl. Phys.*, **94**, 6849–6854.

13 Granström, M., Petritsch, K., Arias, A.C., Lux, A., Andersson, M.R., and Friend, R.H. (1998) Laminated fabrication of polymeric photovoltaic diodes. *Nature*, **395**, 257.

14 Halls, J.J.M., Walsh, C.A., Greenham, N.C., Marseglia, E.A., Friend, R.H., Moratti, S.C., and Holmes, A.B. (1995) Efficient photodiodes from interpenetrating polymer networks. *Nature*, **376**, 498–500.

15 Holcombe, T.W., Woo, C.H., Kavulak, D.F.J., Thompson, B.C., and Frechet, J.M.J. (2009) All-polymer photovoltaic devices of poly(3-(4-n-octyl)-phenylthiophene) from Grignard metathesis (GRIM) polymerization. *J. Am. Chem. Soc.*, **131**, 14160–14161.

16 Kietzke, T., Hörhold, H.-H., and Neher, D. (2005) Efficient polymer solar cells based on M3EH-PPV. *Chem. Mater.*, **17**, 6532–6537.

17 Koetse, M.M., Sweelssen, J., Hoekerd, K.T., Schoo, H.F.M., Veenstra, S.C., Kroon, J.M., Yang, X., and Loos, J. (2006) Efficient polymer:polymer bulk heterojunction solar cells. *Appl. Phys. Lett.*, **88**, 083504.

18 Yu, G. and Heeger, A.J. (1995) Charge separation and photovoltaic conversion in polymer composites with internal donor/acceptor heterojunctions. *J. Appl. Phys.*, **78**, 4510.

19 Sang, G., Zou, Y., Huang, Y., Zhao, G., Yang, Y., and Li, Y. (2009) All-polymer solar cells based on a blend of poly[3-(10-n-octyl-3-phenothiazine-vinylene)thiophene-co-2,5-thiophene] and poly[1,4-dioctyloxyl-p-2,5-dicyanophenylenevinylene]. *Appl. Phys. Lett.*, **94**, 193302.

20 Gupta, D., Kabra, D., Kolishetti, N., Ramakrishnan, S., and Narayan, K.S. (2007) An efficient bulk-heterojunction photovoltaic cell based on energy transfer in graded-bandgap polymers. *Adv. Funct. Mater.*, **17**, 226–232.

21 Roman, L.S., Arias, A.C., Theander, M., Andersson, M.R., and Inganäs, O. (2003) Photovoltaic devices based on photo induced charge transfer in polythiophene: CN-PPV blends. *Braz. J. Phys.*, **33**, 376–381.

22 Halls, J.J.M., Arias, A.C., MacKenzie, J.D., Wu, W., Inbasekaran, M., Woo, E.P., and Friend, R.H. (2000) Photodiodes based on polyfluorene composites: influence of morphology. *Adv. Mater.*, **12**, 498–502.

23 Kim, J.S., Ho, P.K.H., Murphy, C.E., and Friend, R.H. (2004) Phase separation in polyfluorene-based conjugated polymer blends: lateral and vertical analysis of blend spin-cast thin films. *Macromolecules*, **37**, 2861–2871.

24 Jenekhe, S.A. and Yi, S. (2000) Efficient photovoltaic cells from semiconducting polymer heterostructures. *Appl. Phys. Lett.*, **77**, 2635.

25 Zhou, E., Cong, J., Wei, Q., Tajima, K., Yang, C., and Hashimoto, K. (2011) All-polymer solar cells from perylene diimide based copolymers: material design and phase separation control. *Angew. Chem., Int. Ed.*, **50**, 2799–2803.

26 Zhan, X., Tan, Z., Domercq, B., An, Z., Zhang, X., Barlow, S., Li, Y., Zhu, D., Kippelen, B., and Marder, S.R. (2007) A high-mobility electron-transport polymer with broad absorption and its use in field-effect transistors and all-polymer solar cells. *J. Am. Chem. Soc.*, **129**, 7246–7247.

27 Fabiano, S., Chen, Z., Vahedi, S., Facchetti, A., Pignataro, B., and Loi, M.A. (2011) Role of photoactive layer morphology in high fill factor all-polymer bulk heterojunction solar cells. *J. Mater. Chem.*, **21**, 5891–5896.

28 Moore, J.R., Albert-Seifried, S., Rao, A., Massip, S., Watts, B., Morgan, D.J., Friend, R.H., McNeill, C.R., and Sirringhaus, H. (2011) Polymer blend solar cells based on a high-mobility naphthalenediimide-based polymer acceptor: device physics,

29 Holcombe, T.W., Norton, J.E., Rivnay, J., Woo, C.H., Goris, L., Piliego, C., Griffini, G., Sellinger, A., Brédas, J.-L., Salleo, A., and Fréchet, J.M.J. (2011) Steric control of the donor/acceptor interface: implications in organic photovoltaic charge generation. *J. Am. Chem. Soc.*, **133**, 12106–12114.

30 Cataldo, S., Fabiano, S., Ferrante, F., Previti, F., Patané, S., and Pignataro, B. (2011) Organoboron polymers for photovoltaic bulk heterojunctions. *Macromol. Rapid. Commun.*, **31**, 1281–1286.

31 Falzon, M.-F., Zoombelt, A.P., Wienk, M.M., and Janssen, R.A.J. (2011) Diketopyrrolopyrrole-based acceptor polymers for photovoltaic application. *Phys. Chem. Chem. Phys.*, **13**, 8931–8939.

32 Falzon, M.-F., Wienk, M.M., and Janssen, R.A.J. (2011) Designing acceptor polymers for organic photovoltaic devices. *J. Phys. Chem. C*, **115**, 3178–3187.

33 He, X., Gao, F., Tu, G., Hasko, D., Hüttner, S., Steiner, U., Greenham, N.C., Friend, R.H., and Huck, W.T.S. (2010) Formation of nanopatterned polymer blends in photovoltaic devices. *Nano Lett.*, **10**, 1302–1307.

34 Ohkita, H., Cook, S., Astuti, Y., Duffy, W., Tierney, S., Zhang, W., Heeney, M., McCulloch, L., Nelson, J., Bradley, D.D.C., and Durrant, J.R. (2008) Charge carrier formation in polythiophene/fullerene blend films studied by transient absorption spectroscopy. *J. Am. Chem. Soc.*, **130**, 3030–3042.

35 Yin, C., Kietzke, T., Neher, D., and Hörhold, H.H. (2007) Photovoltaic properties and exciplex emission of polyphenylenevinylene-based blend solar cells. *Appl. Phys. Lett.*, **90**, 092116.

36 Mandoc, M.M., Veurman, W., Koster, L.J.A., de Boer, B., and Blom, P.W.M. (2007) Origin of the reduced fill factor and photocurrent in MDMO-PPV:PCNEPV all-polymer solar cells. *Adv. Funct. Mater*, **17** 2167–2173.

37 McNeill, C.R., Halls, J.J.M., Wilson, R., Whiting, G.L., Berkebile, S., Ramsey, M.G., Friend, R.H., and Greenham, N.C. (2008) Efficient polythiophene/polyfluorene co-polymer bulk heterojunction photovoltaic devices: device physics and annealing effects. *Adv. Funct. Mater.*, **18**, 2309–2321.

38 McNeill, C.R., Westenhoff, S., Groves, C., Friend, R.H., and Greenham, N.C. (2007) Influence of nanoscale phase separation on the charge generation dynamics and photovoltaic performance of conjugated polymer blends – balancing charge generation and separation. *J. Phys. Chem. C*, **111**, 19153–19160.

39 Veldman, D., Ipek, O., Meskers, S.C.J., Sweelssen, J., Koetse, M.M., Veenstra, S.C., Kroon, J.M., van Bavel, S.S., Loos, J., and Janssen, R.A.J. (2008) Compositional and electric field dependence of the dissociation of charge transfer excitons in alternating polyfluorene copolymer/fullerene blends. *J. Am. Chem. Soc.*, **130**, 7721–7735.

40 Gonzalez-Rabade, A., Morteani, A.C., and Friend, R.H. (2009) Correlation of heterojunction luminescence quenching and photocurrent in polymer-blend photovoltaic diodes. *Adv. Mater.*, **21**, 3924–3927.

41 Marsh, R.A., McNeill, C.R., Abrusci, A., Campbell, A.R., and Friend, R.H. (2008) A unified description of current–voltage characteristics in organic and hybrid photovoltaics under low light intensity. *Nano Lett.*, **8**, 1393–1398.

42 Groves, C., Marsh, R.A., and Greenham, N.C. (2008) Monte Carlo modeling of geminate recombination in polymer–polymer photovoltaic devices. *J. Chem. Phys.*, **129**, 114903.

43 Huang, Y.-S., Westenhoff, S., Avilov, I., Sreearunothai, P., Hodgkiss, J.M., Deleener, C., Friend, R.H., and Beljonne, D. (2008) Electronic structures of interfacial states formed at polymeric semiconductor heterojunctions. *Nat. Mater.*, **7**, 483–489.

44 Aarnio, H., Sehati, P., Braun, S., Nyman, M., de Jong, M.P., Fahlman, M., and Österbacka, R. (2011) Spontaneous charge transfer and dipole formation at the interface between P3HT and PCBM. *Adv. Energy Mater.*, **1**, 792–797.

45 Mihailetchi, V.D., Koster, L.J.A., Blom, P.W.M., Melzer, C., de Boer, B.,

van Duren, J.K.L., and Janssen, R.A.J. (2005) Compositional dependence of the performance of poly(p-phenylene vinylene):methanofullerene bulk-heterojunction solar cells. *Adv. Funct. Mater.*, **15**, 795–801.

46. Steyrleuthner, R., Schubert, M., Jaiser, F., Blakesley, J.C., Chen, Z., Facchetti, A., and Neher, D. (2010) Bulk electron transport and charge injection in a high mobility n-type semiconducting polymer. *Adv. Mater.*, **22**, 2799–2803.

47. Yan, H., Chen, Z., Zheng, Y., Newman, C., Quinn, J.R., Dötz, F., Kastler, M., and Facchetti, A. (2009) A high-mobility electron-transporting polymer for printed transistors. *Nature*, **457**, 679–687.

48. Westenhoff, S., Howard, I.A., Hodgkiss, J.M., Kirov, K.R., Bronstein, H.A., Williams, C.K., Greenham, N.C., and Friend, R.H. (2008) Charge recombination in organic photovoltaic devices with high open-circuit voltages. *J. Am. Chem. Soc.*, **130**, 13653–13658.

49. Veldman, D., Meskers, S.C.J., and Janssen, R.A.J. (2009) The energy of charge-transfer states in electron donor–acceptor blends: insight into the energy losses in organic solar cells. *Adv. Funct. Mater.*, **19**, 1939–1948.

50. Gearba, R.I., Mills, T., Morris, J., Pindak, R., Black, C.T., and Zhu, X. (2011) Quantifying interfacial electric fields and local crystallinity in polymer–fullerene bulk-heterojunction solar cells. *Adv. Funct. Mater.*, **21**, 2666–2673.

51. Brédas, J.-L., Norton, J.E., Cornil, J., and Coropceanu, V. (2009) Molecular understanding of organic solar cells: the challenges. *Acc. Chem. Res.*, **42**, 1691–1699.

52. Bullie-Lieuwma, C.W.T., van Gennip, W.J.H., van Duren, J.K.J., Jonkheijm, P., Janssen, R.A.J., and Niemantsverdriet, J.W. (2003) Characterization of polymer solar cells by TOF-SIMS depth profiling. *Appl. Surf. Sci.*, **203–204**, 547–550.

53. Veenstra, S.C., Loos, J., and Kroon, J.M. (2007) Nanoscale structure of solar cells based on pure conjugated polymer blends. *Prog. Photovolt.: Res. Appl.*, **15**, 727–740.

54. Loos, J., Yang, X., Koetse, M.M., Sweelssen, J., Schoo, H.F.M., Veenstra, S.C., Grogger, W., Kothleitner, G., and Hofer, F. (2005) Morphology determination of functional poly[2-methoxy-5-(3,7-dimethyloctyloxy)-1,4-phenylenevinylene]/poly[oxa-1,4-phenylene-1,2-(1-cyanovinylene)-2-methoxy,5-(3,7-dimethyloctyloxy)-1,4-phenylene-1,2-(2-cyanovinylene)-1,4-phenylene] blends as used for all-polymer solar cells. *J. Appl. Polym. Sci.*, **97**, 1001–1007.

55. McNeill, C.R., Watts, B., Swaraj, S., Ade, H., Thomsen, L., Belcher, W.J., and Dastoor, P.C. (2008) Evolution of the nanomorphology of photovoltaic polyfluorene blends: sub-100nm resolution with X-ray spectromicroscopy. *Nanotechnology*, **19**, 424015.

56. Kietzke, T., Neher, D., Kumke, M., Montenegro, R., Landfester, K., and Scherf, U. (2004) A nanoparticle approach to control the phase separation in polyfluorene photovoltaic devices. *Macromolecules*, **37**, 4882–4890.

57. Kietzke, T., Neher, D., Landfester, K., Montenegro, R., Guntner, R., and Scherf, U. (2003) Novel approaches to polymer blends based on polymer nanoparticles. *Nat. Mater.*, **2**, 408–412.

58. Burke, K.B., Stapleton, A.J., Vaughan, B., Zhou, X., Kilcoyne, A.L.D., Belcher, W.J., and Dastoor, P.C. (2011) Scanning transmission X-ray microscopy of polymer nanoparticles: probing morphology on sub-10nm length scales. *Nanotechnology*, **22**, 265710.

59. Png, R.Q., Chia, P.J., Tang, J.C., Liu, B., Sivaramakrishnan, S., Zhou, M., Khong, S.H., Chan, H.S.O., Burroughes, J.H., Chua, L.L., Friend, R.H., and Ho, P.K.H. (2009) High-performance polymer semiconducting heterostructure devices by nitrene-mediated photocrosslinking of alkyl side chains. *Nat. Mater.*, **9**, 152–158.

60. Darling, S.B. (2009) Block copolymers for photovoltaics. *Energy Environ. Sci.*, **2**, 1266–1273.

61. Segalman, R.A., McCulloch, B., Kirmayer, S., and Urban, J.J. (2009) Block copolymers for organic optoelectronics. *Macromolecules*, **42**, 9205–9216.

62. Sommer, M., Huettner, S., and Thelakkat, M. (2010) Donor–acceptor

63 Bates, F.S. (1991) Polymer–polymer phase behaviour. *Science*, **251**, 898–905.
64 Sommer, M., Huttner, S., Steiner, U., and Thelakkat, M. (2009) Influence of molecular weight on the solar cell performance of double-crystalline donor–acceptor block copolymers. *Appl. Phys. Lett.*, **95**, 183308.
65 McNeill, C.R. and Greenham, N.C. (2008) Charge transport dynamics of polymer solar cells under operating conditions: influence of trap filling. *Appl. Phys. Lett.*, **93**, 203310.
66 Mandoc, M.M., Veurman, W., Koster, L.J.A., Koetse, M.M., Sweelssen, J., and de Boer, B., and Blom, P.W.M. (2007) Charge transport in MDMO-PPV:PCNEPV all-polymer solar cells. *J. Appl. Phys.*, **101**, 104512.
67 McNeill, C.R., Hwang, I., and Greenham, N.C. (2009) Photocurrent transients in all-polymer solar cells: trapping and detrapping effects. *J. Appl. Phys.*, **106**, 024507.
68 Chua, L.-L., Zaumseil, J., Chang, J.F., Ou, E.C.-W., Ho, P.K.-H., Sirringhaus, H., and Friend, R.H. (2005) General observation of n-type field-effect behaviour in organic semiconductors. *Nature*, **434**, 194–199.
69 Winfield, J.M., Van Vooren, A., Park, M.-J., Hwang, D.-H., Cornil, J., Kim, J.-S., and Friend, R.H. (2009) Charge-transfer character of excitons in poly[2,7-(9,9-di-n-octylfluorene)$_{(1-x)}$-co-4,7-(2,1,3-benzothiadiazole)$_{(x)}$]. *J. Chem. Phys.*, **131**, 035104.
70 Hwang, I., McNeill, C.R., and Greenham, N.C. (2009) Modelling the photocurrent transients currents in bulk heterojunction solar cells. *J. Appl. Phys.*, **106**, 094506.
71 Li, Z. and McNeill, C.R. (2011) Transient photocurrent measurements of PCDTBT:PC$_{70}$BM and PCPDTBT:PC$_{70}$BM solar cells: evidence for charge trapping in efficient polymer/fullerene blends. *J. Appl. Phys.*, **109**, 074513.
72 Beiley, Z.M., Hoke, E.T., Noriega, R., Dacuña, J., Burkhard, G.F., Bartelt, J.A., Salleo, A., Toney, M.F., and McGehee, M.D. (2011) Morphology-dependent trap formation in high performance polymer bulk heterojunction solar cells. *Adv. Energy Mater.*, **1**, 954–962.

15
Conjugated Polymer Composites and Copolymers for Light-Emitting Diodes and Laser

Thien Phap Nguyen and Pascale Jolinat

15.1
Introduction

Hybrid composites are materials made of at least two components of different chemical nature, which are mixed together to develop specific interactions between molecules or atoms, forming a solid structure and having properties that could not be obtained by a single constituent. Hybrid composites have been used since a long time. The oldest hybrid materials used even today are composites seen in Maya paintings, which were first discovered in around eighth century. In the electronic field, hybrid composites are interesting because they well combine the electrical, optical, and mechanical properties of the components, thus providing unique materials that can be used in several electronic applications, which was not possible with conventional semiconductors. These applications are essentially the so-called organic electronic devices, which include organic light-emitting diodes (OLEDs) and organic photovoltaic cells (OPVs). Other organic devices have also been studied, but their applications are still hypothetical because of their low performance compared to those of inorganic counterparts.

The hybrid materials used in organic electronics are based on conjugated polymers or small molecules, to which are added inorganic components of different nature. The organic component provides flexibility to hybrid devices, which is useful for specific applications. For most of the devices treated here, the principal electronic functions are ensured through the use of organic materials, although in some structures the inorganic semiconductors play a key role, as important or even more important than its organic counterpart. In general, as organic material is sensitive to light and oxygen or water, the stability of devices is poor. Incorporation of inorganic component compensates this drawback and provides a better environmental resistance to the material, with or without changes in the desired physical properties.

In this chapter, we shall review the properties and the use of composites in OLEDs and lasers. Applications of OLEDs are nowadays available in everyday electronic devices such as mp4 players, cell phones, and TV screens, while those of organic lasers are still not available. This chapter is organized as follows. In the first

---

Semiconducting Polymer Composites: Principles, Morphologies, Properties and Applications, First Edition.
Edited by Xiaoniu Yang.
© 2012 Wiley-VCH Verlag GmbH & Co. KGaA. Published 2012 by Wiley-VCH Verlag GmbH & Co. KGaA.

section, the principal characteristics of organic materials are reviewed with emphasis on their optical properties. Next, polymer-based composites and the photophysical processes in these materials are described. The working principle of devices including OLEDs and lasers, using the composites as active components, will be then given and different structures are also examined and analyzed. Finally, the advantages as well as the disadvantages of the devices are discussed to foresee their use in future practical applications.

15.2
Properties of Organic Semiconductors

The properties of conventional semiconductors can be described by classical quantum theory on the basis of the crystalline structure of these materials. According to this description, the band diagram of a semiconductor is defined by two energy levels representing the limits of the conduction band (E_C) and the valence band (E_V). The bandgap (E_G) defined as the difference of energy between E_C and E_V plays a key role in electrical and optical properties of the semiconductor. Indeed, electrical conduction in a semiconducting material is determined by electrons in the conduction band and holes in the valence band. These charge carriers are created in their corresponding bands either by external excitation (electrical, thermal, or optical processes) or by doping (and a small amount of energy). Under an external applied field, they can move freely in their bands. Also, the bandgap will define the energy of photons that can be absorbed by the material and create electron–hole pairs whose energy will define the wavelength of the emitted light when they recombine.

Organic materials having physical behavior similar to that of conventional semiconductors are also ranged in semiconductor category from the conduction point of view. However, because of their noncrystalline structure, the energy states for electrons and holes are localized on molecules and do not form a band structure. The conductivity of organic materials is based on the presence of delocalized π-electrons in unsaturated hydrocarbon molecules. This delocalization occurs when the system contains alternating single and double carbon bonds. The π-electrons can no longer be attributed to one specific C—C bond, and the corresponding wavefunction is delocalized over the whole organic structure. The delocalized electrons are weakly bound and can move along the molecules giving a semiconducting material. Because of the amorphous character of the materials with no specific molecular ordering, the orbital overlap between molecules is small and the charge carrier movement is limited and consequently their mobility is low.

Two energy levels are characteristics of the structure: the highest occupied molecular orbital (HOMO) and the lowest unoccupied molecular orbital (LUMO), which are equivalent to the conduction and the valence bands in crystalline semiconductors, respectively. It should be noted that because of their amorphous character, a spread of the density of states into the bandgap through HOMO and LUMO levels of organic semiconductors is usually obtained, with a distribution of these states within the gap.

When an electron and a hole are formed on neighboring molecules within the organic material, they can recombine on a single molecule, forming either an exciton or a phonon. An exciton is a electron–hole pair bound together, characterized by a binding energy and a total spin. If the total spin is 0 (in units of $\hbar$), the exciton is referred to as a *singlet*, and if it is 1, the exciton is referred to as a *triplet*. Excitons can be created by the association of a free electron and a free hole, which are created by external excitation (either electrical or optical). With optical excitation, only singlet excitons are created, while electrical excitation creates both singlet and triplet excited states [1]. They can also dissociate by charge transfer to a neighboring molecule, yielding either two free charges of opposite signs or an emitted photon.

An important optical process in organic materials concerns the energy transfer between molecules via excitons. Two mechanisms can be distinguished: Förster and Dexter transfer. In the Förster mechanism, the energy released from the exciton upon its dissociation is nonradiatively transferred to a molecule, which in turn creates another exciton by a process previously described. This mechanism occurs between the donor and acceptor molecules and is a long-range distance process. Several conditions should be fulfilled to expect an efficient energy transfer by the Förster mechanism:

- An overlap between the emission spectrum of the donor molecule and the absorption spectrum of the acceptor molecule.
- A small average distance between the donor and the acceptor.
- A mutual alignment between the donor and the acceptor.
- A high quantum yield of the donor and the acceptor.

Dexter mechanism is a charge transfer between neighboring molecules, during which an electron is given to the acceptor by the donor. Both energy transfer processes can be used in host–guest systems to increase the triplet states, which in turn increase the light emission in the materials. The different transfer possibilities are the following [1]:

- Excitons are formed in the host and recombine on the guest.
- Förster transfer of singlet excitons from the host to the guest.
- Dexter transfer of triplet excitons from the host to the guest.

Both Förster and Dexter transfers can also be used to gather the excited states for obtaining an optimized emission from devices [2]. These processes are essential in the photophysics of composites and will be discussed in detail in the following sections.

15.3
Polymer-Based Composites

Polymer-based composites designate materials whose one component at least is a polymer. In this chapter, besides the materials, we also consider devices using several components, one of them at least being made of a polymer.

Polymer-based composites can be prepared by two different ways. The first preparation consists in adding inorganic fillers, which are usually metals, semiconductors, or oxide nanoparticles (NC) such as silica (SiO_2) or titania (TiO_2), to the polymer matrix. The nature and amount of nanoparticles are determined by the desired properties of the composites. The composition of the prepared composites can be precisely controlled by adjusting the amount of inorganic particles. However, it is difficult to obtain a uniform distribution of particles because they tend to form aggregates in the polymer matrix. In order to better control their dispersion, the particles can be coated with a polymer shell, which significantly decreases the surface energy of the bare particles [3]. Another approach consists in using polymeric surfactants, which adsorb strongly onto the particle surface. These processes allow a uniform distribution of the particles, but can modify the active surface contact between the particles and the polymer, and, consequently, the expected effects introduced by the nanocomposites (NCs). Description of the synthesis, properties, and transport in NCs has been recently reviewed [4, 5].

The second preparation consists in introducing the polymer into a porous (organized or not) inorganic template such as zinc oxide (ZnO) or TiO_2. The choice of polymers as well as the nature of the template depends on the expected properties of the composites. The advantage of such preparation is the uniformity of composite when depositing the polymer in a well-designed template. The results depend strongly on the ability of the polymer solution to penetrate the template structure (highly soluble). Zinc oxide is a particularly interesting wide bandgap (~3.3 eV) material for optoelectronic applications. This oxide can be prepared by the hydrothermal technique for obtaining vertically aligned nanorods, forming a template into which polymer can be incorporated to yield nanostructured composites [6]. The preparation of such template consists in coating a substrate with ZnO particles (seeds), which act as nucleation sites for the formation of vertical nanorods in aqueous solution heated at 90 °C [7]. The rod size and length can be controlled by synthesis parameters. However, reproducibility of the nanostructure to identical templates is still an issue.

15.4
Use of Polymer Composites in Photonic Applications

We will focus on the light emission of composite materials for their use in devices including organic light-emitting diodes and organic semiconductor lasers (OLs).

15.4.1
Organic Light-Emitting Diodes

The first OLEDs working at low applied voltages were proposed by Tang and VanSlyke [8]. They used a tris(8-hydroxy-quinoline) aluminum (Alq_3) thin film, which is an organic material for the emitting layer, and sandwiched between two electrodes. The devices emitted a yellow green light with an applied voltage of ~10 V.

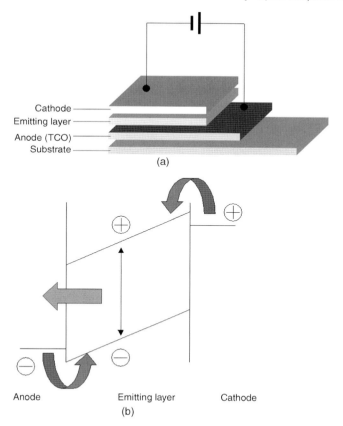

Figure 15.1 Structure and working principle of OLEDs.

The working principle of a basic OLED is shown in Figure 15.1. The device is composed of three layers deposited on a substrate made of glass or plastic. The anode is a transparent conducting oxide (TCO) film, which allows the emitted light in the active layer to go out to the external medium. Its ionization energy is high, enabling hole injection into the HOMO of the organic film. The organic emitter film of thickness ∼100 nm is deposited over the anode by spin-coating (of polymer) or vacuum deposition (of organic materials or small molecules). The cathode is a metal of low work function (Ca, Mg), which allows a favorable injection of electrons into the LUMO of the organic film.

Under an applied voltage, holes from the anode and electrons from the cathode are injected into the bands of the active layer forming excitons, which give rise to a light emission after their dissociation as explained previously. The color of the emitted light depends on the bandgap of the organic layer.

In practice, the structure of an OLED is more complex and may contain several additional layers, which are organic semiconductors and are introduced to enhance the efficiency of the diode. These layers comprise electron transporting layer (ETL), hole blocking layer (HBL), hole transporting layer (HTL), and electron blocking layer (EBL). The basic OLED structure usually comprises four layers, namely, the anode, the HTL, the active organic emitter, and the cathode.

15.4.1.1 Efficiency of OLEDs

Generally speaking, to describe the performance of OLEDs, several parameters can be used. The internal quantum efficiency η_I is defined as the ratio of emitted photons from the active layer to injected electrons. This parameter indicates the ability of the material to convert electrical energy into light. However, only a fraction of the emitted photons will be collected in the external medium (air) because they have to cross the interface between the emitter and the air, and a certain number of photons will be reflected at the interface. The optical efficiency η_{op} is defined as the ratio of the collected photons in the external medium to those emitted from the active layer. This parameter represents the ability of the emitter/air interface to transmit the generated light to the air. The external quantum efficiency η_e is defined as the ratio of collected photons in the external medium to injected electrons. This parameter represents the true efficiency of the diode, that is, its ability to convert electrical energy into usable light. The relationship between the different efficiencies is as follows:

$$\eta_e = \eta_I \times \eta_{op}. \tag{15.1}$$

The diode efficiency can be improved by improving both η_{op} and η_I. As the external medium is generally air, the optical efficiency depends on the refractive index of the active layer, which is almost constant for usual polymers or organic materials. Therefore, η_{op} can only be improved by an appropriate design of the diode surface to avoid subsequent reflection of the emitted light. On the contrary, η_I depends only on the quality of the materials, which are involved in the injection, transport, and recombination processes. The use of nanocomposites can favor these processes by improving the internal efficiency of diodes as follows.

Injection of Charge Carriers When applying an electrical field to a typical basic diode, electrons are injected from the cathode into the active layer and holes from the anode are injected into the HTL. The injection mechanism is well known and several examples have been given in the literature concerning diodes using known polymers or organic materials as emitting materials. The number of injected carriers depends on the potential barrier existing at the contact between the electrode and the organic layer and on the contact surface. When using a conjugated polymer as an emitter and a metal as a cathode material in an OLED, the injection of electrons will be normally defined by the choice of the materials. It is, however, possible to increase the charge injection by using a composite material made of the same polymer, which is mixed with NCs. It is indeed shown that the incorporated particles tend to agglomerate at the interface between the materials and increase the contact surface, allowing more charge carriers to enter the emitter. OLEDs using composites of poly(2-methoxy-5-(2'-ethyl-hexyloxy)-1,4-phenylene vinylene

(MEH–PPV)/TiO$_2$ or MEH–PPV/SiO$_2$ show a higher current density and a better efficiency than diodes using pristine polymer as an active layer [9]. Figure 15.2 shows the current density versus the applied field in ITO-(PPV–NCs)–MgAg devices [10]. As in the case of MEH–PPV composite based-diodes, the current is increased when nanoparticles are incorporated into the polymer. However, the electrical characteristics also depend on the nature and size of the particles and suggest that the transport in the bulk of the polymer layer is also affected by these particles. In the same way, hole injection from the anode side into the HTL can also be improved by replacing the PEDOT:PSS by a composite material made of PEDOT:PSS mixed with NCs [11–13]. The limiting factor of the use of composites is a possible alteration of the light emission by quenching of created photons due to the NCs when their concentration becomes high enough. For the HTL layer, the incorporation of NCs in the PEDOT:PSS film may alter the light transmission to the external medium, and thus decrease the overall efficiency.

Transport and Recombination of Charge Carriers As previously mentioned, the mobilities of electrons and holes are different in organic materials. A direct consequence of this difference is an inefficient recombination process in most organic devices because there will be little chance for an electron to recombine with a hole when they are moving under an applied field. Indeed, holes are usually more mobile than electrons in organic materials, and they can quickly reach the cathode and go out of the device into the external circuit without meeting electrons to form electron–hole pairs. If they meet electrons, the recombination will occur near the cathode and the created excitons can be quenched in the interface region [14]. Imbalance of charge will also occur if the mobility of electrons is higher than that of holes, and quenching of excitons will take place at the anode in that case. It is then necessary to slow down their motion in the organic material in order to facilitate the formation of electron–hole pairs and thus to enhance the light emission efficiency. This operation can be realized by adding layers such as HBL or EBL to the device structure in order to block the movement of the charge carriers and to allow them to recombine with those of opposite charge. The use of a composite material can play the role of such layers and increase the recombination rate. For instance, the efficiency of OLEDs using a poly[2-phenyl-3-(9,9-dihexyl-fluoren-2-yl) phenylene vinylene]-co-[2-methoxy-5-(2′-ethylhexyloxy) phenylene vinylene] (FP-PPV-co-MEH–PPV) copolymer as a photoactive layer is enhanced on incorporating a small amount of cadmium selenide CdSe(ZnS) quantum dots (QDs) to the polymer film [15]. The increase in luminance of devices was assigned to an improved balance of the charge carriers in the composite, which was enabled by the nanoparticles by blocking the charge carriers during their transport.

15.4.1.2 Color Emission

One of the main advantages of OLEDs over inorganic devices consists in controlling the emitted color by modifying the bandgap of the materials by simple chemistry [16]. Basic colors RGB (red, green, and blue) can be obtained with either polymers or small molecules, and the diodes emitting RGB are used in flat panels

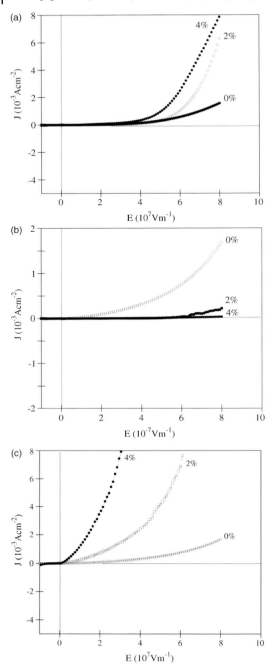

Figure 15.2 Current density versus applied field in ITO composite–MgAg diodes: (a) PPV/SiO$_2$ (100 nm); (b) PPV/SiO$_2$ (20 nm); (c) PPV/TiO$_2$ (20 nm). (Reprinted with permission from Ref. [10]. Copyright 2005, Elsevier B.V.)

display technology. Future and promising applications of OLEDs concern lighting, which requires a combination of different emitted colors to obtain a controlled white light with high luminance [17, 18].

There are two approaches for realizing a white OLED. The first approach consists in using a multilayer structure where the resulting light is composed of three basic RGB components (Figure 15.3).

Each color is emitted by a single organic layer having an adequate bandgap for providing the desired wavelength when it is excited. In order to obtain a pure white light, the intensity of each light component should be the same, and this is a technical difficulty to be solved. This technique can also be adapted to a single-layer structure by using an electroluminescent polymer and by adding blue, green, and red chromophores to achieve the white light emission [19]. The resulting spectrum can be controlled by adjusting the chromophore concentrations. The fabrication of the photoactive layer can be simplified by using a blue emitting (co)polymer with green and red chromophores [20].

The second approach consists in using the emission from different materials for obtaining a spectrum covering the wavelengths corresponding to a white light. The most interesting materials are quantum dots that are incorporated into a conjugated polymer matrix. Quantum dots are semiconductor nanoparticles exhibiting quantum confinement effects due to their size. The energy bandgap of QDs is

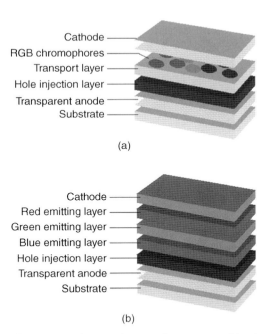

Figure 15.3 Schematic diagram of two different approaches to obtain white light: (a) color mixing from red, green, and blue light-emitting layers in multilayer structures; (b) color mixing from red, green, and blue nanoparticles embedded in a conjugated polymer matrix.

determined by the size of the particles, which offers the possibility of obtaining a variety of emitted colors by using a unique QD material but with different sizes. Several semiconductor QDs can be used in OLEDs, and the most popular material is spherical cadmium selenide (CdSe) [21]. These QDs yield high photoluminescence efficiency and the size can be controlled giving a narrow size distribution. For further improvement, these QDs can be coated with a higher bandgap material (e.g., ZnS) to confine the exciton to the CdSe core and to reduce the defects forming on the surface of the particles. The emitted light from the composites will depend on the structure of the diodes. Indeed, white light spectrum is composed of RGB components that can be generated by QDs of different sizes. The first type of diodes can be fabricated by forming an emissive monolayer with multiple QD colors inside a standard device structure [22]. The process is simple but the control of the emitted light is complicated. For instance, to fabricate white OLEDs, QDs RGB chromophores of CdSe in a monolayer are sandwiched between two organic layers of a diode structure, as shown in Figure 15.4 [23]. One of them is a hole transporting layer (TPD or N,N'-bis(3-methylphenyl)-N,N'-bis(phenylbenzidine)) and the other one is a hole blocking layer (TAZ or 3,4,5-triphenyl-1,2,4-triazole). These layers facilitate the transport of holes in the diodes and prevent the formation of excitons at the QD sites. The obtained electroluminescence spectrum shows that the red, green, and blue components from the QDs are dominant, and residual emissions from organic layers (TPD and Alq$_3$) are also observed, resulting from energy transfer in QDs (Figure 15.5). Besides, Förster energy transfer can occur between QDs from blue particles to green and red ones, and from green QDs to red ones (down conversion). Consequently, the QD concentrations should be carefully chosen to obtain desired color coordinates. The second type of diodes consists of several layers of polymer containing different proportions of QDs that yield a white light emission. The deposition is more complex, but the control of the emitted light can be carried out by the PL measurement of the composite films, prior to the realization of the devices. The third type of diodes makes use of the Förster energy transfer between the QDs and luminescent materials to produce a white

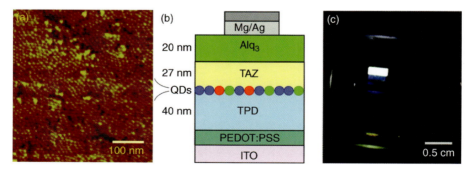

Figure 15.4 (a) Atomic force microscope phase image of blue QDs formed on a 40 nm thick TPD film. (b) Device cross section of a white QD LED. (c) Photograph of a white QD LED in operation with 10 V of applied voltage. (Reprinted with permission from Ref. [23]. Copyright 2007, American Chemical Society.)

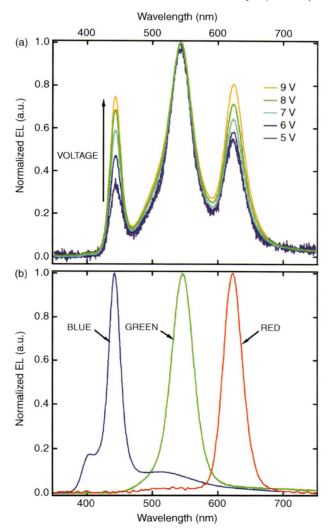

Figure 15.5 (a) Normalized electroluminescence spectra of a white QD LED for a set of increasing applied voltages. (b) Normalized electroluminescence spectra of red, green, and blue QD LEDs. (Reprinted with permission from Ref. [23]. Copyright 2007, American Chemical Society.)

light. The emitter is obtained by mixing an emitting organic material and the QDs. The formed excitons in the organic layer undergo Förster energy transfer to the lower energy QD sites where they recombine. White light OLEDs can be obtained by using RGB CdSe QDs of different sizes in a CBP organic matrix [24], with a high brightness (1500 cd m^{-2} at 62 mA cm^{-2}), which is promising for lighting applications. In these devices, the light emission originates from the inorganic materials and not from the organic ones. White light emitter can also be fabricated by blending a blue–green emitting polymer (BADF or 2,7-bis[2-(4-diphenylamino-phenyl)-1,3,4-oxydazol-5-yl]-9,9-dihexylfluorene) with red CdSe QDs [25]. In this case, energy transfer between the QDs and the polymer matrix was evidenced by separating the organic materials and the QDs. The diode was fabricated by using BADF as an active layer on a glass substrate and the QDs were incorporated into a polymethacrylate matrix, which was deposited on the backside of the glass substrate. When the diode is switched on, the emitted light is white, proving that an energy transfer process has occurred.

ZnO is a wide bandgap material with good optical and electro-optical properties that can be exploited in optoelectronic devices. White OLEDs using hybrid polyvinylcarbazole (PVK) and QDs of ZnO emit a white light of CIE (Commission Internationale de l'Eclairage) chromaticity coordinates of (0.39, 0.41) with a good stabilization [26]. An important advantage of ZnO is that it can be obtained in various nanostructures on different types of substrates, making it very attractive for fabricating OLEDs. Nanowires and nanorods have been used with luminescent conjugated polymers to fabricate white light-emitting OLEDs [27–31]. Indeed, the size of the nanostructured ZnO wires and rods being much bigger than that of QDs, they contain a large number of radiative intrinsic and extrinsic defects, which emit blue, green, yellow, and red colors. They can hence be used as a complementary emitter to conjugated polymers for obtaining a white light emission. A simple process particularly interesting for fabricating hybrid light-emitting diodes is the use of freestanding films of ZnO nanowires, which are deposited onto the organic or polymer layer by a roller-pressing process [29]. Figure 15.6 shows a ZnO nanowire sheet and its deposition process in the hybrid light-emitting diodes, whose electroluminescence spectra show emission from ZnO and its defects. Furthermore, as the size and the concentration of ZnO nanorods can be controlled by the growth parameters,

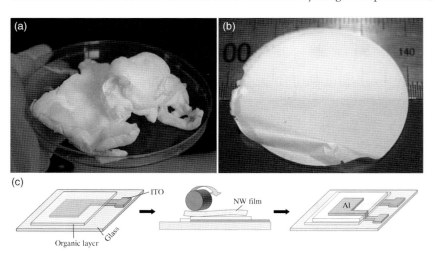

Figure 15.6 (a) A cotton-like product consisting of ZnO nanowires. (b) A freestanding film made of ZnO nanowires. (c) Schematic of the process to fabricate ZnO/organic hybrid LEDs using the nanowire film. (Reprinted with permission from Ref. [28]. Copyright 2009, IOP Publishing.)

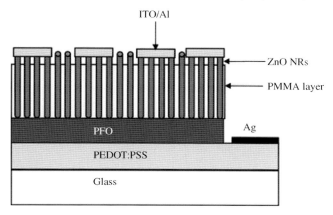

Figure 15.7 Schematic diagram showing the LED device structure using ZnO nanorods polyfluorene–PEDOT:PSS heterostructure. (Reprinted with permission from Ref. [31]. Copyright 2011, Springer.)

the ZnO/polymer composite materials are more homogeneous than those using QDs and their characteristics are also more reproducible to provide reliable devices. Figure 15.7 shows a schematic diagram of a diode using a hybrid composite with ZnO nanorods. The nanorods are embedded in an insulating matrix in order to prevent the individual ZnO from short circuit and the emitting polymer from a direct contact with the top electrode. Diodes of structure ITO/polyfluorene (PF)/ZnO nanorods:polysilane (SOG)/Al have been fabricated with ZnO nanorod composite and polyfluorene, a blue emitting polymer. An electroluminescence spectrum is obtained covering a broad emission range from 400 to 800 nm, which results from a blue emission of the polyfluorene and green and red emissions of the defects in ZnO nanorods [28] (Figure 15.8).

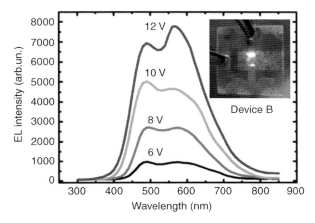

Figure 15.8 Electroluminescence spectra of device B (ITO/PF/ZnO nanorods –SOG/Al) as a function of the applied voltage. *Inset*: Light emission from device B under a bias of 10 V. (Reprinted with permission from Ref. [28]. Copyright 2009, IOP Publishing.)

15.4.1.3 Stability

Stability of organic devices is a key issue for their commercialization. Generally speaking, organic materials are more susceptible to chemical degradation than inorganic materials. The mechanisms that affect the lifetime of organic devices can be classified into two groups: intrinsic and extrinsic degradations [32].

Extrinsic degradation is attributed to chemical reactions of the active materials with its environment. The most common cause of degradation is the contact with oxygen and water in the presence of light, which leads to a rapid decrease of performance of the diodes, resulting from a modification of the structure of the active material, which is furthermore accelerated by electrical stress. The degradation onset corresponds to a formation of dark spots [33], which are nonemissive areas of the device surface. Most of the devices should therefore be protected by an encapsulation to prevent the contact with ambient air. A properly protected encapsulated diode will not develop chemical reactions that affect its lifetime. An alternative approach is the use of metal oxide layers combined with high work function and air-stable metals such as Al or Au to replace the usual transport layers (PEDOT:PSS or LiF) to fabricate diodes that can be used without encapsulation [34].

Intrinsic degradation refers to mechanisms occurring inside the devices that cover all the processes that affect the loss of efficiency over time during operation. Several causes of degradation have been identified. In the active layer, the formation of defects or traps is likely responsible for a decrease of the device performance. Indeed, charge carriers can be captured by traps, resulting in a decrease in the mobility and an increase in the operating voltage. Moreover, traps also act as quenching centers and diminish the radiative recombinations and hence the light emission from the emitting layer. Whether the traps are responsible for the degradation [35] or they are produced by it [36, 37], there is a strong correlation between the defects and the intrinsic failure of devices. At the contact between the emitter and the transport layers or the electrodes, interface interactions [38] and diffusion of metal or oxygen from the electrodes can occur [39]. A diffusion of indium into the active polymer layer of devices using a PEDOT:PSS as a HTL has been observed, even in the absence of an applied voltage [40]. The modifications of the emissive layer induced by these mechanisms are the culprit of degradation as observed in several organic or polymer-based diodes.

Many studies have reported that the lifetime of OLEDs using a composite material is strongly improved compared to that of devices using pristine organic or polymer material as an emitter [9, 41]. One of the reasons for the increased stability is that the inorganic components are chemically and thermally stable and are not affected by the physical changes of the device during the operation. On the contrary, in some systems, it has been shown that the incorporation of nanoparticles in a polymer matrix significantly reduced the amount of defects and improved the stability of the material and devices [42, 43]. Indeed, the incorporated particles may neutralize the polymer defect sites (chain ends, bonding defects) and prevent the chemical reactions with the diffused species (oxygen, metal ions) between them from occurring. Furthermore, some oxide nanoparticles, for instance, TiO_2, by their specific optical properties can absorb energetic radiations and spare the organic material heating. In electron or hole transport layers, the incorporation of nanoparticles into the polymer film not only enhances the injection of charge carriers into the active layer but also prevents the diffusion of species from the electrodes. Therefore, the stability of the devices is also improved.

15.4.2
Organic Semiconductor Lasers

15.4.2.1 Working Principle

In 1960, the first demonstration of lasing from ruby was reported [44], which led to considerable developments in science and technology and a wide range of applications, from medicine to telecommunications and consumer electronics. Basically, the acronym LASER derives from the phrase "light amplification by stimulated emission of radiation." However, from a technological point of view, a laser is more precisely defined as the combination of an optical amplifier (the so-called gain medium) and a feedback structure (the so-called resonator), giving the characteristic light output properties:

- spectral narrow line emission coinciding with resonant cavity modes
- lasing threshold
- spatial coherence
- strongly polarized output

Schematically, excitation of a two-level system (Figure 15.9) is likely to give rise to light emission either by spontaneous or by stimulated emission (SE). In order to achieve significant contribution from stimulated rather than from spontaneous emission, the population of the upper state should exceed that of the lower state,

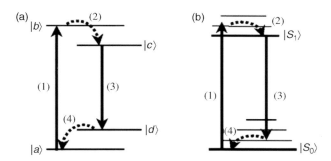

Figure 15.9 Energy levels and transitions (a) in a generic four-level laser material and (b) corresponding to the lowest two singlet states in an organic semiconductor. Transitions 1 and 3 are optical absorption and emission, and transitions 2 and 4 are thermal relaxations. (Reprinted with permission from Ref. [63]. Copyright 2007, American Chemical Society.)

giving the so-called population inversion. This can be achieved by either optical or electrical pumping, the latter process being involved in laser diodes, the ultimate interest of this chapter. In the former process, an incident photon stimulates a transition between the excited state and ground state of the medium, generating further photons. Inversion cannot, however, be practically achieved in a system with just two energy levels. A three- or four-level system is more appropriate since it allows very low rate of excitation and lasing threshold. As light travels through a gain medium, optical amplification may be achieved by stimulated emission. The intensity is then expected to increase exponentially with the distance z:

$$I(z) = I(0) \exp(gz), \tag{15.2}$$

where g is the wavelength-dependent gain coefficient of the medium.

The maximum net gain achievable is affected by gain losses that contribute negatively to the amplification process. They stem from linear absorption or excited-state absorption and may also result from light scattering within the laser medium. These photophysical processes are generally investigated and quantified [45, 46] by pump–probe experiments (Figure 15.10).

As mentioned above, the gain medium of every kind of laser is associated with an optical resonator, whose structure will be detailed below. The simplest Fabry–Perot-type resonator comprises only two mirrors, between which the gain medium is situated. The feedback structure imposes two basic properties upon the oscillating laser field. It defines not only the allowed resonant frequencies within the gain medium's emission spectrum but also the spatial characteristics of the laser beam.

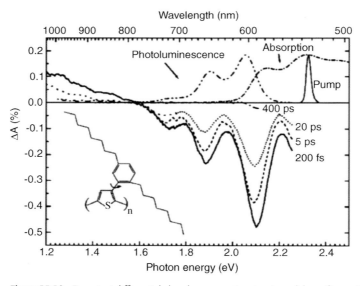

Figure 15.10 Transient differential absorbance spectra at various delays after excitation for poly(3-(2,5-dioctylphenyl)thiophene) film. The ground-state absorption and photoluminescence spectra are given for reference. (Reprinted with permission from Ref. [46]. Copyright 1998, Elsevier B.V.)

To achieve a sustained oscillation in a laser, amplification in the gain medium must at least balance out with the optical loss during each round-trip of the cavity. Therefore, when the pump rate increases beyond a threshold value, an intense coherent laser beam is generated whose power rises linearly with the excess pump rate. At low pumping rates, the excitations in the gain medium are radiated in all directions as spontaneous emission.

15.4.2.2 Materials

Organic laser dyes are widely used in commercial UV-Vis infrared tunable lasers employing solutions of these molecules. Nevertheless, the area of applications of such liquid solution-based systems is restricted due to the need of complex and bulky design, as well as the need to employ large volumes of organic solvents. Achieving similar systems with a solid-state organic gain medium is highly desirable. However, a direct employment of the dye materials in neat films is not feasible due to the significant luminescence quenching resulting from a strong interaction of excited-state molecules. A strategy to circumvent this concentration quenching is to incorporate organic dyes as guests in host materials, for example, polymers [47]. Unlike laser dyes, organic semiconductors can have high photoluminescence quantum yield even as neat films in the solid state, which should result in stronger pump absorption and gain. Furthermore, their semiconducting properties open up the possibility of electrical pumping, the objective that is still pursued. Intrinsic emissive properties of organic laser materials should be combined with photothermal and photooxidative stability, which is one of the most important issues to be addressed.

Lasing from a semiconducting polymer was first reported in 1992 by Moses [48], who used MEH–PPV dissolved in xylene to replace dye solution in a laser cavity. The author concluded that the quantum yield of this laser was comparable to that of those using rhodamine 6G laser dye. The following results were both important and original since they concerned laser emission from solutions and films containing titanium dioxide nanocrystals. In 1994, Lawandy *et al.* [49] performed experiments on colloidal solutions containing rhodamine 640 perchlorate dye in methanol and TiO_2 nanoparticles, constituting strongly scattering media. By optical pumping, they evidenced emission characteristic of a multimode laser oscillator, although the system contained no external cavity. The authors suggested a possible interpretation in terms of a feedback mechanism supplied by photon diffusion, which would make the system a "random laser." According to Wiersma *et al.* [50], however, this observation had to be related to amplified spontaneous emission (ASE). Within a pumped region, ASE is observed in the direction(s) of highest gain, that is, generally in the direction(s) in which this region is most extended. Lawandy *et al.* replied that although ASE was a well-known phenomenon occurring without any scattering particles, their observations were closely related to the presence of particles and occurred in both side and front directions. They also reported similar results with a poly(methyl methacrylate) (PMMA) polymer sheet containing the same laser dye and TiO_2 nanoparticles [51]. Based on these results, Hide *et al.* [52] incorporated MEH–PPV in cyclohexanone solutions or polystyrene dilute blend films since ultrafast spectroscopy showed that stimulated emission was inhibited in pure MEH–PPV. In the presence of scattering particles, they observed dramatic narrowing of the emission spectrum (the so-called gain narrowing) in both media, which they attributed to laser action. As a conclusion, the authors evaluated transient current densities that would be necessary to achieve the same excitation densities as in optically pumped films, which should be in excess of $10^3\,A\,cm^{-2}$. The same research group demonstrated gain narrowing in neat films of submicrometer thickness for poly(p-phenylenevinylene) (PPV), poly(p-phenylene) (PPP), and polyfluorene (PF) derivatives [53]. In previous works, ultrafast spectroscopy had shown that the SE in neat films either did not exist or decayed in at most a few picoseconds. The general explanation is that interchain interactions give rise to photoinduced absorption, which is strong enough to overwhelm the SE. However, the authors showed that by a judicious choice of functionalized side chains on the polymer backbone, long SE lifetimes could be obtained. The thickness dependence of gain narrowing was studied for neat films of the synthesized polymers. In all cases, a well-defined thickness cutoff was observed below which gain narrowing did not occur. The authors attributed this phenomenon to some waveguiding effects. They also observed that the threshold pump energy for gain narrowing and the final linewidth depend on the choice of solvent, which suggested a role of film morphology and chain conformation. Heliotis et al. [54] obtained similar results (Figure 15.11) by using a high-gain PF polymer. Denton et al. [55] investigated more precisely gain narrowing in PPV films and found no evidence of a threshold in output intensity or increase in directionality accompanying the narrowing, the effects that are expected for lasing. Therefore, the authors proposed that fluorescence was amplified by stimulated emission and waveguided

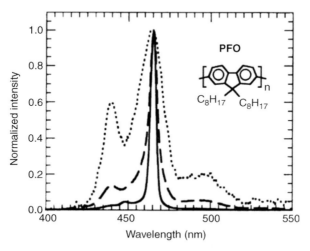

Figure 15.11 Normalized emission spectra from a poly(9,9-dioctylfluorene) (PFO) waveguide excited at 390 nm with pump energy densities of $10\,\mu J\,cm^{-2}$ (dotted line), $37.5\,\mu J\,cm^{-2}$ (dashed line), and $13.3\,\mu J\,cm^{-2}$ (solid line). (Reprinted with permission from Ref. [54]. Copyright 2002, American Institute of Physics.)

Figure 15.12 Optically pumped waveguide structure based on a ladder-type poly(paraphenylene) (m-LPPP) polymer film on glass, using pulsed excitation at 444 nm wavelength. (Reprinted with permission from Ref. [56]. Copyright 1997, American Institute of Physics.)

through the excited region. Zenz et al. [56] realized similar experiments by using the high-gain ladder-type poly(p-phenylene) (m-LPPP) as the active material (Figure 15.12). They observed a highly directional output whose polarization was parallel to the polymer film and independent of the pump beam polarization. The same phenomenon was reported for a glass/ITO/m-LPPP (130 nm) structure that could in principle be applicable for electrical pumping. By comparing the threshold values for optically pumped GaN and m-LPPP devices, the authors concluded that the threshold current for m-LPPP laser diodes in this structure should be about two orders of magnitude lower.

Although solid films were thought to be unusable due to strongly competing induced absorptions, Tessler et al. [57] argued against this perceived wisdom by first reporting optically driven laser activity in devices based on films of PPV. In this work, PPV was inserted in a microcavity structure with highly reflecting mirrors, the bottom one being a distributed Bragg reflector (DBR). A dramatic change in the photoluminescence spectrum was observed, with the mode at the PPV gain peak wavelength (545 nm) dominating the emission at high output powers. While microcavity devices have the appropriate structure for laser diode operation, the most important question to be answered is whether electrical injection can generate the required excitation densities for lasing to occur. Indeed, the estimated current density of $10^3 \, A \, cm^{-2}$ that would be needed is much higher than is normally sustainable within a light-emitting diode. By using tunneling injection from a scanning tunneling microscope (STM), Lidzey and Bradley [58] showed, however, that a PPV derivative could sustain a current density as high as $10^4 \, A \, cm^{-2}$. Electroluminescence was observed in these conditions, although the emission was highly nonuniform on length scales of 10 nm and greater. Tessler et al. [59] studied conventional polymer light-emitting diodes based on PPV derivatives under pulsed electrical excitation. They confirmed that the devices could support current densities up to $10^3\,A\,cm^{-2}$. After considering that these results were unexpected due to very low charge carrier mobility in PPV, they evidenced a sharp increase of mobility, by a factor of 10^5, at large current densities. Their conclusion was that device efficiency, rather than peak current density, would limit laser action.

In 2000, Schön et al. [60] first reported on electrically driven amplified spontaneous emission and lasing tetracene single crystals. Field-effect device structures were prepared on freshly cleaved crystal surfaces, providing reflections and feedback for laser action. By increasing injected current density, an abrupt spectral narrowing was first observed at $\sim30\,A\,cm^{-2}$, which was typical of amplified spontaneous emission along with gain guiding. Then, a second decrease of the linewidth at higher current densities ($>500\,A\,cm^{-2}$) indicated the onset of laser action. According to the authors, the high electron and hole mobilities and the balanced charge carrier injection, as well as the reduced charge-induced absorption in these high-quality single crystals, were key parameters in device performance. Baldo et al. [61] studied the tetracene device in detail and were not convinced by the demonstration of electrically pumped lasing. In particular, they were surprised by the observation of spectral narrowing at the reported threshold current density, due to the low optical confinement in this structure. The publication from Schön and coworkers was retracted in 2002 [62]. More generally, Baldo et al. concluded that while OLEDs could be highly efficient at low current densities, increases in non-radiative losses with excitation strength were a serious obstacle to obtaining an electrically pumped laser. They found that the performance of potential lasing materials could be quantified by the product of external quantum efficiency and current density, which should be $\sim5\,A\,cm^{-2}$. To date, the best amorphous devices possess $\sim0.3\,A\,cm^{-2}$. We can conclude that the development of organic laser diodes is a very challenging goal that will require further innovation. At the present time, an alternative approach involves indirect electrical pumping, in which an efficient electrically driven light source is used to pump an organic semiconductor laser optically.

15.4.2.3 Types of Resonators

Optically pumped organic semiconductor lasers have been demonstrated by using various resonant structures (Figure 15.13), leading to a variety of spectral, spatial, and power properties. We have already seen that the simplest Fabry–Perot-type resonator comprises two mirrors, between which the gain medium is situated. This structure was used in initial studies on dye lasers with liquid or solid solutions of conjugated polymers, showing that MEH–PPV performed as well as a rhodamine dye. Practically, a cuvette or a polymer block is sandwiched between a pair of mirrors, or one of them may be replaced by a diffraction grating. Such macroscopic lasers tend to operate at quite high energies, with threshold of tens to hundreds of microjoules [63]. We have seen that Tessler et al. used a similar structure for the first demonstration of lasing from a PPV film in a planar microcavity. In this case, however, the length of the gain medium

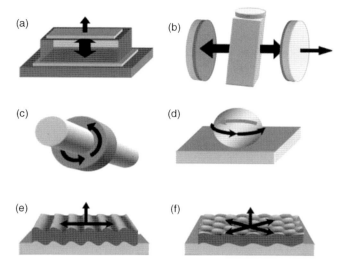

Figure 15.13 Schematic resonators used for organic semiconductor lasers showing propagation directions of the resonant laser field. (a) Planar microcavity. (b) Fabry–Perot dye laser cavity. (c) Microring resonator, coated around an optical fiber. (d) Spherical microcavity. (e) Distributed feedback resonator. (f) Two-dimensional DFB/photonic crystal resonator. (Reprinted with permission from Ref. [63]. Copyright 1997, American Institute of Physics.)

was only 100 nm, proving the enormous gain available from conjugated polymers, with lasing above a pump density of $\sim 200\, nJ\, cm^{-2}$. Such structures are comparable to the vertical cavity surface emitting lasers (VCSELs) that have been widely studied in inorganic semiconductors. An alternative to this conventional Fabry–Perot structure is obtained in slab waveguide lasers, where mirrors are formed parallel to the plane of the film. This is the typical low-cost configuration for inorganic semiconductor lasers, where the cleaved semiconductor crystal forms very flat facets of typically 30% reflectivity, due to the very high refractive index of the semiconductor. In the case of organic semiconductors, however, this structure is less attractive due to their rather low refractive index and the difficulty to form good-quality edges with polymer films. Kozlov *et al.* [64] demonstrated such a laser based on the low molecular weight material Alq_3 doped with DCM2, with a 1 mm long cavity. Due to the dimensions of the system, a highly divergent output emission was observed perpendicular to the plane of the film at very low excitation density ($\sim 1\, \mu J\, cm^{-2}$). Easy processing of organic semiconductors has allowed the demonstration of less conventional structures, the so-called whispering-gallery mode lasers [65], which are based on dielectric resonators with circular symmetry. Microring lasers, for example, consist of a thin polymer waveguide deposited around a dielectric or metallic core. Alternative geometries are the microdisk and microsphere cavities. In these structures, reflections of the light beam occur by total internal reflection at the interfaces between polymer and surrounding media. In addition to the often-complicated spectral output, all these resonators are distinctive in that they do not emit a well-defined directional output beam.

A final important class of resonators for organic semiconductor lasers comprise diffractive systems based on periodic, wavelength-scale microstructures. They can be incorporated into planar organic semiconductor waveguides, thus avoiding the need for good-quality end facets to provide optical feedback. The angle through which light is diffracted, or Bragg scattered, is highly wavelength dependent. For a given period of the corrugation, a particular set of wavelengths λ is likely to be diffracted from a propagating mode of the waveguide, according to the Bragg condition:

$$m\lambda = 2n_{\text{eff}}\Lambda, \qquad (15.3)$$

where Λ is the period of the structure and n_{eff} is the so-called effective refractive index of the waveguide.

This simple description does not take into account the so-called photonic stopband effect, which forbids propagation at wavelength exactly satisfying the Bragg condition. According to the coupled mode theory [66, 67], distributed feedback (DFB) lasers normally oscillate on the edge of this photonic stopband (Figure 15.14).

Generally speaking, DFB structures have an advantage over the Fabry–Perot lasers in that they provide both a long resonator length in which light can interact with the gain medium (giving a low oscillation threshold) and a strong spectral selection of the resonant light. One-dimensional (1D) DFB lasers have been demonstrated with a variety of organic semiconductors, and are commonly used with their inorganic counterparts in telecommunication applications. Alternative 2D DFB structures have been proposed, such as photonic crystals with square, hexagonal, or honeycomb lattices. Finally, 3D structures may act as DFB resonators, one of note being the "random laser" structure obtained by blending TiO_2 nanoparticles with a conjugated polymer. While silica or inert polymer gratings have most often been used in DFB organic lasers, metallic gratings, arrays of metal nanodisks, or titania and alumina gratings have also been reported. All resonators require submicrometer periodic structures that are difficult to produce by conventional photolithography. Alternative techniques include holography, electron beam lithography, or, most interestingly, direct patterning of a polymer layer (Figure 15.15) by nanoimprint techniques [68, 69].

15.4.2.4 Applications and Developments

We have just seen that a large variety of resonators have been proposed for the realization of organic semiconductor lasers. In particular, diffractive feedback structures have attracted much interest since they combine the advantages of lowthreshold surface emission and good spectral selection. As recent works have shown, fine design of the resonator structure allows tuning the laser output properties to a given application. Generally speaking, organic semiconductor lasers offer a range of attractive features, such as simple low-cost processing and broadly tunable

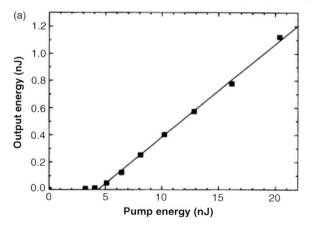

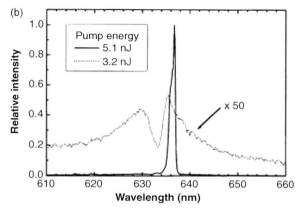

Figure 15.14 Output characteristics of a surface-emitting two-dimensional distributed feedback laser based on the conjugated polymer MEH–PPV. Microchip laser is used as the pump source. (a) Output energy as a function of pump energy. (b) Emission spectra for different pump energies. (Reprinted with permission from Ref. [67]. Copyright 2003, American Institute of Physics.)

emission throughout the visible spectrum. Two main fields of application have been developed in the last years. The first one concerns spectroscopy and chemical sensing, the spectral range of these emitters being particularly adapted to organic molecules, including biological systems. Woggon et al. [70], for instance, developed a compact optical transmission spectrometer based on a continuously tunable organic DFB laser. Tuning of the laser wavelength was accomplished by pumping at different spatial positions of a monomode waveguide with varying thickness. Alternatively, organic semiconductor lasers may be integrated in combination with passive photonic components and microfluidic structures to realize plastic lab-on-a-chip systems (Figure 15.16), as

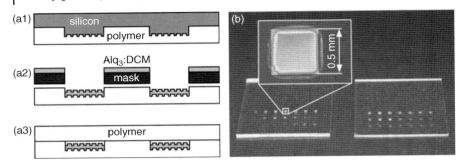

Figure 15.15 All-polymer encapsulated organic semiconductor laser chips. (a1)–(a3) Scheme of the main fabrication steps, including thermal nanoimprint. (b) Photograph of two chips and a microscope image of a laser (*inset*). (Reprinted with permission from Ref. [69]. Copyright 2010, Optical Society of America.)

demonstrated by Vannahme et al. [71]. More basically, optical amplification in organic semiconductors may find applications in telecommunication systems, where organic gain media could be coupled with polymer optical fibers. Whatever the application may be, the issue of stability of organic semiconductors to

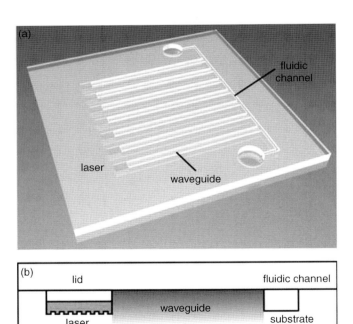

Figure 15.16 (a) Scheme of an exemplary plastic lab-on-a-chip system with integrated lasers as used for fluorescence excitation. (b) Sectional view of one detection unit (not to scale). (Reprinted with permission from Ref. [71]. Copyright 2011, Optical Society of America.)

photooxidation will have to be considered in optically pumped lasers, while still pursuing the goal of electrically pumped organic lasers [72].

15.5 Conclusions

Composites consisting of organic and inorganic materials offer advantages that enable their use in high-tech applications such as organic light-emitting diodes, lasers, and solar cells. The use of these materials as active layers of devices provides an enhancement of the performance in terms of efficiency and stability, which are important for the fabrication of commercialized products. Therefore, organic–inorganic hybrid materials are certainly the materials for the future development of organic electronics. For evident economical reasons, the organic devices will be more and more produced with solution-processable materials on large surface substrates by using techniques such as ink-jet screen or roll-to-roll printing. Composites can be adapted to such techniques as the size of particles can be controlled at the nanoscale in the form of quantum dots. The chemistry of nanocrystals has greatly improved the quality of the inorganic particles used in composites. However, despite the progress made in diverse fields for improving the composite materials and their use, several problems are still to be solved. One of the key issues for the reproducibility of nanostructured materials is the aggregation of nanoparticles that is not easy to control. Several approaches have been proposed for synthesizing the nanoparticles in the presence of polymer for an efficient spatial control of the particle dispersion. However, the synthesis of such composites is hard to carry out. Modification of the nanoparticles by surface chemistry (application of surfactants and ligands) may improve the dispersability and organization in bulk polymer thin films. Another means of ensuring that the dispersion of the inorganic fillers is uniform in a polymer matrix is to employ templates such as nanorods to host the polymer solution. This approach has given promising results and further progress in the future for the fabrication of reliable devices will require the use of this technique. In addition to these basic problems, the quality and the size control of nanoparticles are important factors for obtaining the uniformity of the physical properties of composite films and, consequently, highly efficient devices using them as photoactive layers [4]. In addition, the morphology of the emitting layer in contact with the other layers of the diode structure has impacts on the device performance [73] and needs to be controlled by the material processing and the film surface treatments.

Many technical challenges remain for the use of nanocomposite materials in future optical applications. These challenges cover the fundamental and experimental aspects of the research in materials, designs, and devices. The progress made in nanomaterials and nanotechnology has been encouraging and has allowed the fabrication and use of new nanostructures such as nanoparticles, nanowires, nanorods, quantum dots in numerous new applications. It is expected that the parallel development of these sciences will provide efficient solutions to enable the commercialization of nanocomposite-based devices in the near future.

References

1 Köhler, A. and Bässler, H. (2009) Triplet states in organic semiconductors. *Mater. Sci. Eng.*, **R66**, 71–109.
2 Sun, Y.R., Giebink, N.C., Kanno, H., Ma, B.W., Thompson, M.E., and Forrest, S.R. (2006) Management of singlet and triplet excitons for efficient white organic light-emitting devices. *Nature*, **440**, 908–912.
3 Rozenberg, B.A. and Tenne, R. (2008) Polymer assisted fabrication of nanoparticles and nanocomposites. *Prog. Polym. Sci.*, **33**, 40–112.
4 Talapin, D.V., Lee, J.S., Kovalenko, M.V., and Shevchenko, E.V. (2008) Prospects of colloidal nanocrystals for electronic and optoelectronic applications. *Chem. Rev.*, **110**, 389–458.
5 Reiss, P., Couderc, E., De Girolamo, J., and Pron, A. (2011) Conjugated polymers/semiconductor nanocrystals hybrid materials: preparation, electrical transport properties and applications. *Nanoscale*, **3**, 446–489.
6 Gonzalez-Valls, I. and Lira-Cantu, M. (2009) Vertically aligned nanostructures of ZnO for excitonic solar cells: a review. *Energ. Environ. Sci.*, **2**, 19–34.
7 Greene, L.E., Law, M., Tan, D.H., Montano, M., Goldberger, J., Somorjai, G., and Yang, P. (2005) General route to vertical ZnO nanowire arrays using textured ZnO seeds. *Nano Lett.*, **5**, 1231–1236.
8 Tang, C.W. and VanSlyke, S.A. (1987) Organic electroluminescent diodes. *Appl. Phys. Lett.*, **51**, 913–915.
9 Carter, S.A., Scott, J.C., and Brock, P.J. (1997) Enhanced luminance in polymer composite light-emitting diodes. *Appl. Phys. Lett.*, **71**, 1145–1147.
10 Yang, S.H., Nguyen, T.P., Le Rendu, P., and Hsu, C.S. (2005) Optical and electrical investigations of poly(p-phenylene vinylene)/silicon oxide and poly(p-phenylene vinylene)/titanium oxide nanocomposites. *Thin Solid Films*, **471**, 230–235.
11 Xu, X., Yu, G., Liu, Y., and Zhu, D. (2006) Electrode modification in organic light-emitting diodes. *Displays*, **27**, 24–34.
12 Wang, G.F., Tao, X.M., and Wang, R.X. (2008) Fabrication and characterization of OLEDs using PEDOT:PSS and MWCNT nanocomposites. *Compos. Sci. Technol.*, **68**, 2837–2841.
13 Inigo, A.R., Underwood, J.M., and Silva, S.R.P. (2011) Carbon nanotube modified electrodes for enhanced brightness in organic light-emitting diodes. *Carbon*, **49**, 4211–4217.
14 Krummacher, B., Mathai, M.K., Choong, V.E., Choulis, S.A., So, F., and Winnacker, A. (2006) Influence of charge balance and microcavity effects on resultant efficiency of organic-light-emitting devices. *Org. Electron.*, **7**, 313–318.
15 Lee, C.W., Chou, C.H., Huang, J.H., Hsu, C.S., and Nguyen, T.P. (2008) Investigations of organic light-emitting diodes with CdSe(ZnS) quantum dots. *Mater. Sci. Eng. B*, **147**, 307–311.
16 Friend, R.H. (2001) Conjugated polymers: new materials for optoelectronic devices. *Pure Appl. Chem.*, **73**, 425–430.
17 Kamtekar, K.T., Monkman, A.P., and Bryce, M.R. (2010) Recent advances in white organic light-emitting materials and devices. *Adv. Mater.*, **22**, 572–582.
18 D'Andrade, B.W. and Forrest, S.R. (2004) White organic light-emitting devices for solid state lighting. *Adv. Mater.*, **16**, 1585–1595.
19 Gather, M.C., Köber, S., Heun, S., and Meerholz, K. (2009) Improving the lifetime of white polymeric organic light-emitting diodes. *J. Appl. Phys.*, **106**, 024506-1–024506-10.
20 Renaud, C. and Nguyen, T.P. (2008) Study of trap states in polyspirobifluorene based devices: influence of chromophore

addition. *J. Appl. Phys.*, **104**, 113705-1–113705-8.
21 Colvin, V.L., Schlamp, M.C., and Alivisatos, A.P. (1994) Light-emitting diodes made from cadmium selenide nanocrystals and a semiconducting polymer. *Nature*, **370**, 354–357.
22 Coe, S., Woo, W.K., Bawendi, M.G., and Bulović, V. (2002) Electroluminescence from single monolayers of nanocrystals in molecular organic devices. *Nature*, **420**, 800–803.
23 Anikeeva, P.O., Halpert, J.E., Bawendi, M.G., and Bulovic, V. (2007) Electroluminescnce from a mixed red-green-blue colloidal quantum dot monolayer. *Nano Lett.*, **7**, 2196–2200.
24 Li, Y.Q., Rizzo, A., Cingolani, R., and Gigli, G. (2006) Bright white light-emitting devices from ternary nanocrystal composites. *Adv. Mater.*, **18**, 2545–2548.
25 Ahn, J.H., Bertoni, C., Dunn, S., Wang, C., Talapin, D.V., Gaponik, N., Eychmüller, A., Hua, Y., Bryce, M.R., and Petty, M.C. (2007) White organic light-emitting devices incorporating nanoparticles of II–VI semiconductors. *Nanotechnology*, **18**, 335202-1–335202-7.
26 Son, D.I., You, C.H., Kim, W.T., and Kim, T.W. (2009) White light-emitting diodes fabricated utilizing hybrid polymer-colloidal ZnO quantum dots. *Nanotechnology*, **20**, 365206.
27 Zhang, T., Xu, Z., Qian, L., Tao, D.L., Teng, F., and Xu, X.R. (2006) Influence of ZnO nanorod on the luminescent and electrical properties of fluorescent dye-doped polymer nanocomposites. *Opt. Mater.*, **29**, 216–219.
28 Lee, C.Y., Wang, J.Y., Chou, Y., Cheng, C.L., Chao, C.H., Chiu, S.C., Hung, S.C., Chao, J.J., Liu, M.Y., Su, W.F., Chen, Y.F., and Liu, C.F. (2009) White-light electroluminescence from ZnO nanorods/polyfluorene by solution based growth. *Nanotechnology*, **20**, 425202-1–425202-5.
29 Liu, J., Ahn, Y.H., Park, J.Y., Koh, K.H., and Lee, S. (2009) Hybrid light-emitting diodes based on flexible sheets of mass produced ZnO nanowires. *Nanotechnology*, **20**, 445203-1–445203-6.
30 Wilander, M., Nur, O., Bano, N., and Sultana, K. (2009) Zinc oxide nanorod based heterostructures on solid and soft substrates for white emitting diodes. *New J. Phys.*, **11**, 125020-1–125020-16.
31 Hussain, I., Bano, N., Hussain, S., Nur, O., and Willander, M. (2011) Study of intrinsic white light emission and its components from ZnO nanorods/p-polymer hybrid junctions grown on glass substrates. *J. Mater. Sci.*, **46**, 7437–7442.
32 So, F. and Kondakov, D. (2010) Degradation mechanisms in small-molecule and polymer organic light-emitting diodes. *Adv. Mater.*, **22**, 3762–3777.
33 Sheats, J.R., Antoniadis, H., Hueschen, M., Leonard, W., Miller, J., Moon, R., Roitman, D., and Stocking, A. (1996) Organic electroluminescent devices. *Science*, **273**, 884–888.
34 Sessolo, M. and Bolink, H.J. (2011) Hybrid organic–inorganic light-emitting diodes. *Adv. Mater.*, **23**, 1829–1845.
35 Silvestre, G.C.M., Johnson, M.T., Giraldo, A., and Shannon, J.M. (2001) Light degradation and voltage drift in polymer light-emitting diodes. *Appl. Phys. Lett.*, **78**, 1619–1621.
36 Wang, Q., Luo, Y., and Aziz, H. (2010) Evidence of intermolecular species formation with electrical aging in anthracene-based blue organic light-emitting devices. *J. Appl. Phys.*, **107**, 084506-1–084506-6.
37 Renaud, C. and Nguyen, T.P. (2009) Study of trap states in spirobifluorene-based devices: influence of aging by electrical stress. *J. Appl. Phys.*, **106**, 053707-1–053707-11.
38 Papadimitratos, A., Fong, H.H., and Malliaras, G.G. (2007) Degradation of hole injection at the contact between a conducting polymer and a fluorene copolymer. *Appl. Phys. Lett.*, **91**, 042116-1–042116-3.
39 Lee, T.W., Kim, M.G., Kim, S.Y., Park, S.H., Kwon, O., Noh, T., and Oh, T.S. (2006) Hole-transporting interlayers for improving the device lifetime in the polymer light-emitting diodes. *Appl. Phys. Lett.*, **89**, 123505-1–123505-3.

40 Nguyen, T.P. and de Vos, S.A. (2004) An investigation into the effect of chemical and thermal treatments on the structural changes of poly(3,4-ethylenedioxythiophene)/polystyrenesulfonate and consequences on its use on indium tin oxide substrate. *Appl. Surf. Sci.*, **221**, 330–339.

41 Park, J.H., Kim, T.H., Yu, J.W., Kim, J.K., Kim, Y.C., and Park, O.O. (2005) Enhanced color purity and stability from polymer/nanoporous silica nanocomposite blue light-emitting diodes. *Synth. Met.*, **154**, 145–148.

42 Zou, J.P., Le Rendu, P., Musa, I., Yang, S.H., Dan, Y., Ton That, C., and Nguyen, T.P. (2011) Investigation of the optical properties of polyfluorene/ZnO nanocomposites. *Thin Solid Films*, **519**, 3997–4003.

43 Lee, C.W., Renaud, C., Hsu, C.S., and Nguyen, T.P. (2008) Traps and performance of MEH–PPV/CdSe(ZnS) nanocomposite based organic light-emitting diodes. *Nanotechnology*, **19**, 455202-1–455202-7.

44 Mainman, T.H. (1960) Stimulated optical radiation in ruby. *Nature*, **187**, 493–494.

45 Kranzelbinder, G. and Leising, G. (2000) Organic solid-state lasers. *Rep. Prog. Phys.*, **63**, 729–762.

46 Ruseckas, A., Theander, M., Valkunas, L., Andersson, M.R., Inganäs, O., and Sundström, V. (1998) Energy transfer in a conjugated polymer with reduced inter-chain coupling. *J. Lumin.*, **76–77**, 474–477.

47 Soffer, B.H. and McFarland, B.B. (1967) Continuously tunable, narrow-band organic dye lasers. *Appl. Phys. Lett.*, **10**, 266–267.

48 Moses, D. (1992) High quantum efficiency luminescence from a conducting polymer in solution: a novel polymer laser dye. *Appl. Phys. Lett.*, **60**, 3215–3216.

49 Lawandy, N.M., Balachandran, R.M., Gomes, A.S.L., and Sauvain, E. (1994) Laser action in strongly scattering media. *Nature*, **368**, 436–438.

50 Wiersma, D.S., van Albada, M.P., and Lagendijk, A. (1995) Random laser? *Nature*, **373**, 203–204.

51 Balachandran, R.M., Pacheco, D.P., and Lawandy, N.M. (1996) Laser action in polymeric gain media containing scattering particles. *Appl. Opt.*, **35**, 640–643.

52 Hide, F., Schwartz, B.J., Diaz-Garcia, M.A., and Heeger, A.J. (1996) Laser emission from solutions and films containing semiconducting polymer and titanium dioxide nanocrystals. *Chem. Phys. Lett.*, **256**, 424–430.

53 Hide, F., Diaz-Garcia, M.A., Schwartz, B.J., Andersson, M.R., Pei, Q., and Heeger, A.J. (1996) Semiconducting polymers: a new class of solid-state laser materials. *Science*, **273**, 1833–1836.

54 Heliotis, G., Bradley, D.D.C., Turnbull, G.A., and Samuel, I.D.W. (2002) Light amplification and gain in polyfluorene waveguides. *Appl. Phys. Lett.*, **81**, 415–417.

55 Denton, G.J., Tessler, N., Stevens, M.A., and Friend, R.H. (1997) Spectral narrowing in optically pumped poly(p-phenylenevinylene) films. *Adv. Mater.*, **9**, 547–551.

56 Zenz, C., Graupner, W., Tasch, S., Leising, G., Müllen, K., and Scherf, U. (1997) Blue green stimulated emission from a high gain conjugated polymer. *Appl. Phys. Lett.*, **71**, 2566–2568.

57 Tessler, N., Denton, G.J., and Friend, R.H. (1996) Lasing from conjugated-polymer microcavities. *Nature*, **382**, 695–697.

58 Lidzey, D. and Bradley, D.D.C. (1997) Electroluminescence in polymer films. *Nature*, **386**, 135.

59 Tessler, N., Harrison, N.T., and Friend, R.H. (1998) High peak brightness polymer light-emitting diodes. *Adv. Mater.*, **10**, 64–68.

60 Schön, J.H., Kloc, C., Dodabalapur, A., and Batlogg, B. (2000) An organic solid state injection laser. *Science*, **289**, 599–601.

61 Baldo, M.A., Holmes, R.J., and Forrest, S.R. (2002) Prospects for electrically pumped organic lasers. *Phys. Rev. B*, **66**, 035321-1–035321-15.

62 Bao, Z., Batlogg, B., Berg, S., Dodabalapur, A., Haddon, R.C., Hwang, H., Kloc, C., Meng, H., and Schön, J.H. (2002) Retraction. *Science*, **298** (5595), 961.

63 Samuel, I.D.W. and Turnbull, G.A. (2007) Organic semiconductor lasers. *Chem. Rev.*, **107**, 1272–1295.

64 Koslov, V.G., Bulovic, V., Burrows, P.E., and Forrest, S.R. (1997) Laser action in organic semiconductor waveguide and double-heterostructure devices. *Nature*, **389**, 362–364.

65 Dodabalapur, A., Berggren, M., Slusher, R.E., Bao, Z., Timko, A., Schiortino, P., Laskowski, E., Katz, H.E., and Nalamasu, O. (1998) Resonators and materials for organic lasers based on energy transfer. *IEEE J. Sel. Top. Quant. Electron.*, **4**, 67–74.

66 Kogelnik, H. and Shank, C.V. (1972) Coupled-wave theory of distributed feedback lasers. *J. Appl. Phys.*, **43**, 2327–2336.

67 Turnbull, G.A., Andrew, P., Barnes, W.L., and Samuel, I.D.W. (2003) Operating characteristics of a semiconducting polymer laser pumped by a microchip laser. *Appl. Phys. Lett.*, **82**, 313–315.

68 Ge, C., Lu, M., Jian, X., Tan, Y., and Cunningham, B.T. (2010) Large-area organic distributed feedback laser fabricated by nanoreplica molding and horizontal dipping. *Opt. Express*, **18**, 12980–12991.

69 Vannahme, C., Klinkhammer, S., Christiansen, M.B., Kolew, A., Kristensen, A., Lemmer, U., and Mappes, T. (2010) All-polymer organic semiconductor laser chips: parallel fabrication and encapsulation. *Opt. Express*, **18**, 24881–24887.

70 Woggon, T., Klinkhammer, S., and Lemmer, U. (2010) Compact spectroscopy system based on tunable organic semiconductor lasers. *Appl. Phys. B*, **99**, 47–51.

71 Vannahme, C., Klinkhammer, S., Lemmer, U., and Mappes, T. (2011) Plastic lab-on-a-chip for fluorescence excitation with integrated organic semiconductor lasers. *Opt. Express*, **19**, 8179–8186.

72 Samuel, I.D.W., Namdas, E.B., and Turnbull, G.A. (2009) How to recognize lasing. *Nat. Photonics*, **3**, 546–549.

73 Kabra, D., Song, M.H., Wenger, B., Friend, R.H., and Snaith, H.J. (2008) High efficiency composite metal oxide-polymer electroluminescent devices: a morphological and material based investigation. *Adv. Mater.*, **20**, 3447–3452.

16
Semiconducting Polymer Composite Based Bipolar Transistors
Claudia Piliego, Krisztina Szendrei, and Maria Antonietta Loi

16.1
Introduction

Since the realization of the first field-effect transistor (FET) in 1928 [1], the electronic devices have been optimized to the point that they have become essential building blocks of today's electronics. The efficient electronic circuit design and their mass production are based on the integration of tiny FETs in single chips with improved performance and reliability. The manufacturing of electronic devices has been dominated by the silicon technology; however, in the past 10 years [2], a new class of semiconductors, based on organic compounds, has also entered into the world of FETs. Plastic transistors, fabricated using organic materials for the active layer, could serve as key components in cheap and flexible electronic devices [3] such as radio frequency identification (RF-ID) tags [4] and flexible displays [5, 6]. Unlike most of the inorganic semiconductors, organic materials can be processed onto various flexible plastic substrates, thanks to their mechanical properties. Moreover, the simplification of the manufacturing processes, by introducing solution processing, and the vapor deposition at low temperature, opens the way toward large area, low-cost fabrication of electronic devices.

The performance of organic semiconductor field-effect transistor has been improved tremendously in the last few years, focusing on increasing the environmental stability and charge carrier mobility of p- and n-type polymers and small molecules. In particular, for polymers the record mobility values reported till date are approaching $2\,cm^2\,(V^{-1}\,s^{-1})$ for p-type polymers [7] and $1\,cm^2\,(V^{-1}\,s^{-1})$ for n-type polymers [8]. The p-type conduction has been till last few years much superior than n-type one, due to the oxygen and water sensitivity of many organic anions [9]. Only recently, n-type polymers with comparable performance have been reported.

Complementary metal oxide–semiconductor (CMOS) technology is the fundamental technology for integrated circuit fabrication. Typically in silicon technology complementary and symmetrical pairs of p- and n-type transistors are used for logic functions. In the case of organic semiconductors several different strategies have been investigated to fabricate a single FET having the functionality of the

Semiconducting Polymer Composites: Principles, Morphologies, Properties and Applications, First Edition.
Edited by Xiaoniu Yang.
© 2012 Wiley-VCH Verlag GmbH & Co. KGaA. Published 2012 by Wiley-VCH Verlag GmbH & Co. KGaA.

complementary symmetrical pair of n- and p-type transistors [10]. Bipolar (ambipolar) transistors are able to provide p- and n-channel functionality by using single components [11], bilayers [12], and blends [13–15] as active semiconducting layer. One of the major challenges in achieving good bipolar transport is the efficient injection of holes and electrons from the same metal electrode into the HOMO (highest occupied molecular orbital) and the LUMO (lowest unoccupied molecular orbital) of the semiconductor(s), respectively. Most organic semiconductors have band gaps in the range of 2–3 eV, resulting in a high injection barrier for at least one of the charge carriers, when a single electrode material is employed. The combination of two organic semiconductors in a bilayer or blend, as in the bulk heterojunction configuration, can provide a great opportunity to overcome this limitation. If the materials are warily chosen, the work function of the electrodes could line up properly with the HOMO of one semiconductor for hole injection and with the LUMO of the other for electron injection.

This chapter concentrates on organic bipolar transistors, including details about the basic operation principles, device configurations, and processing methods, and describing the various strategies that have been applied to achieve ambipolar transport. Touching upon small molecule-based FETs and the hybrid approach, the main focus will be on polymer-based bipolar transistors since they can provide one of the ultimate solutions for simple, low-cost fabrication of flexible bipolar FETs.

16.2
Basics of Organic Field-Effect Transistors

16.2.1
Operation Principles of FETs

A field-effect transistor is basically a three-terminal electrical switch that is able to modulate the flow of current. The three terminals are (Figure 16.1a): the gate electrode, which is separated from the semiconducting layer by a dielectric layer, and the so-called source and drain electrodes, which are in direct contact with the semiconductor layer. The position of these two electrodes defines the geometrical characteristics of the device, that are, the channel width (W) and length (L). By applying voltage to the gate it is possible to control the current flowing between the source and the drain in the semiconducting active layer. The gate electrode is most commonly a heavily doped silicon wafer, but also a metal or a conducting polymer can serve as the gate electrode. Gate dielectrics can be oxides, such as SiO_2 (especially when doped silicon is used as gate) and Al_2O_3, or polymeric insulators, such as PMMA (poly(methyl methacrylate)) and many others recently synthetized [16, 17]. The organic semiconducting layer can be applied by vapor deposition (sublimation) or solution processing. The vacuum sublimed small molecule layers generally yield better transistor performance due to higher crystallinity and controllable morphology, although solution processing of small molecules and polymer has the potential for low cost and high volume production. The source and drain electrodes for charge injection are usually fabricated using high work function metals (Au, Ag, Pd, Pt) or conducting polymers (PEDOT:PSS).

16.2.1.1 Unipolar FETs

Figure 16.1 illustrates the typical operation regimes of a unipolar FET where only one type of charge carriers (holes or electrons) can accumulate in the channel depending on the nature of the semiconductor. In operation conditions, a voltage is applied to the drain (V_d) and gate (V_g) electrode while the source electrode is grounded ($V_s = 0$). The difference between the potential at source and drain is the source–drain voltage (V_{ds}), while the potential difference between the source and the gate is called the gate voltage (V_g). The gate itself is isolated from the source and drain by an insulator (dielectric); therefore in ideal conditions no current flows through the dielectric. The charge-injecting electrode is the source, which injects electrons or holes into the channel depending on the polarity with respect to the gate electrode. When a positive voltage is applied to the gate electrode, electrons accumulate at the semiconductor–dielectric interface. The opposite situation occurs for a negatively applied gate voltage; positive charges accumulate at the semiconductor–dielectric interface. When a voltage higher than the threshold voltage (V_{th}) is reached, there will be the formation of a conductive channel from the source to the drain, allowing current to flow. The threshold voltage is mostly due to deep traps in the active layer, which reduce the contribution of induced charges to the current flow. These deep traps first have to be filled, making the effective gate voltage $V_g - V_{th}$. At higher gate voltage more charge carriers will accumulate, causing an extension of the conductive channel. In this manner the gate voltage controls the conductivity between the source and the drain. The transistor can operate in two regimes, the linear and the saturation regimes. In the so-called linear

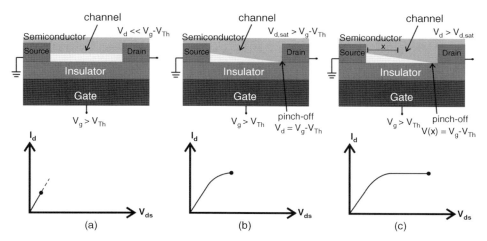

Figure 16.1 Operating regimes of a unipolar FET: schematic of a typical FET and graph representation of the linear regime (a), the pinch off point (b), and the saturation regime (c). (Reprinted with permission from Ref. [18]).

regime, when a small source–drain bias is applied $\ll$ ($V_{sd} \ll V_g$) (Figure 16.1a) and $V_g > V_{th}$, current starts flowing in the channel and it is directly proportional to the V_{ds}. When V_{ds} is further increased and reaches the pinch off point ($V_{ds} = V_g - V_{th}$), a depletion region forms next to the drain electrode, since the difference between the gate voltage and the local potential (V_x) is below the V_{th} (Figure 16.1b). This narrow depletion region allows the flow of a space-charge limited saturation current ($I_{ds,sat}$) from the pinch off point to the drain. Further increasing V_{ds} (Figure 16.1c) expands the depletion region, shortens the channel, and the current saturates at $I_{ds,sat}$ since the potential drop between the pinch off point and the source electrode stays approximately constant.

16.2.1.2 Bipolar FETs

The current in FETs with bipolar nature is composed of both electrons and holes, with proportion depending on the bias conditions. The typical operating regimes for a bipolar FET are illustrated in Figure 16.2 [19]. In regime 1 when $|V_g-V_{th}| > |V_{ds}|$ and they are both positive, the current is dominated by electrons, similarly to the unipolar electron transport. While in regime 4 when $|V_g-V_{th}| > |V_{ds}|$, but both voltages are negative, the current is carried by holes and the bipolar FET operates as unipolar p-type FET. In regime 6, when V_g-V_{th} is positive and V_{ds} is negative, the

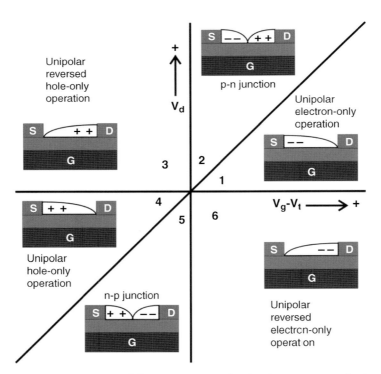

Figure 16.2 Illustration of all operating regimes for a bipolar FET. (Reprinted with permission from Ref. [20]).

effective gate potential becomes positive throughout the whole channel and the bipolar FET behaves as it was a unipolar n-type FET where the source and drain electrodes are inverted. Likewise, when $V_g - V_{th}$ is negative and V_{ds} is positive, the effective gate potential is negative in the channel and the bipolar FET operates as a unipolar p-type FET where the source and drain electrodes are inverted.

Bipolarity only occurs in regimes 2 and 5 where $V_{ds} > V_g - V_{th} > 0$ and $V_{ds} < V_g - V_{th} < 0$, respectively. Under these bias conditions a unipolar transistor operates in the saturation regime and charges cannot be accumulated in the pinched off part of the channel. However, under these conditions, the effective gate potential changes sign at a certain position in the channel in a bipolar FET. As a result, holes and electrons accumulate at the opposite sides of the FET and are separated by a narrow transition region acting as a p–n junction.

16.2.2
Current–Voltage Characteristics

16.2.2.1 Unipolar FET

The current–voltage characteristics of a unipolar p-type FET in linear and saturation regime are illustrated in Figure 16.3 [16]. Figure 16.3a shows the typical output characteristics, where the drain current is plotted against the source–drain voltage at different constant gate voltages. The linear and saturation regimes are obtained at low and high V_{ds}, respectively. Figure 16.3b illustrates the drain current versus the gate voltage (transfer characteristic) in the linear regime ($V_{ds} \ll V_g$) both on a semilogarithmic and a linear scale. The semilogarithmic plot provides the so-called onset voltage (V_{on}) where the drain current starts to increase sharply. From the linear plot the linear field-effect mobility can be extracted since the gradient of the current increase is directly

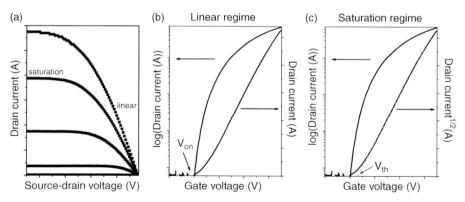

Figure 16.3 Typical current–voltage characteristics of a unipolar n-type FET: (a) output characteristics showing the linear and saturation regimes; (b) transfer characteristics in the linear regime indicating the onset voltage (V_{on}); (c) transfer characteristics in the saturation regime indicating the threshold voltage (V_{th}).

proportional to the mobility according to the following equation:

$$\mu_{\text{lin}} = \frac{\partial I_{ds}}{\partial V_g} \times \frac{L}{WCV_{ds}}. \tag{16.1}$$

Figure c represents the transfer characteristics in the saturation regime. In this plot the extrapolation of the linear fit gives the threshold voltage (V_{th}) while the square root of the drain current can be used to calculate the saturation mobility depending on the gate voltage according to Eq. (16.2):

$$\mu_{\text{sat}} = \left(\frac{\partial \sqrt{I_{ds,\text{sat}}}}{\partial V_g}\right)^2 \times \frac{2L}{WC}. \tag{16.2}$$

Another important feature of the transfer curves is the $I_{\text{on}}/I_{\text{off}}$ ratio that is defined as the ratio of the drain current in the on-state at a given gate voltage and the drain current in the off state. To obtain excellent switching properties, this ratio has to be maximized.

16.2.2.2 Bipolar FETs

The transfer and output curves of a bipolar FET are more complex [16]. The output curve, illustrated in Figure 16.4a, shows the superposition of the standard saturation behavior for one of the charge carriers at higher V_g and a superlinear current increase at lower V_g and high V_{ds} due to the injection of the opposite charge carriers. The transfer curve (Figure 16.4b) has a unique V-shape where the left arm represents the hole transport and the right arm corresponds to the electron transport.

16.2.3
Device Configurations

The different transistor geometries can strongly influence the device performance. The most commonly used configurations are top contact/bottom gate (TC/BG),

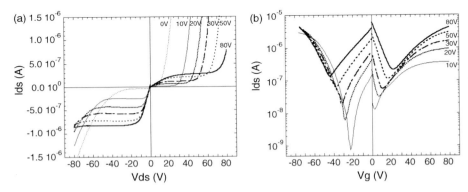

Figure 16.4 (a) Transfer and (b) output characteristics of a bipolar FET. (Reprinted with permission from Ref. [21]).

Figure 16.5 Typical FET configurations: (a) top contact/bottom gate (TC/BG), (b) bottom contact/bottom gate (BC/BG) and (c) bottom contact/top gate (BC/TG).

bottom contact/bottom gate (BC/BG), and bottom contact/top gate (BC/TG) (Figure 16.5). The main difference among these configurations is the position of the charge injecting electrodes with respect to the gate electrode. In the BC/BG geometry, the charges can be directly injected into the channel at the interface of the semiconductor and the dielectric while in the two other geometries, TC/BG and BC/TG, the electrodes are separated from the channel by the semiconductor. In the last two structures the current is larger since charges can be injected not only from the edges of the electrodes but also from the overlapping area with the gate electrode [22]. The properties of the interface between the electrodes and the semiconductor and between the dielectric and the semiconductor are very critical and can greatly affect the device performance.

16.2.4
Role of the Injecting Electrodes

In order to obtain reasonable FET performance, the injection of electrons and holes from the source electrode into the semiconductor has to be ensured. Good ohmic contact between the metal electrode and the semiconductor can be achieved when the work function of the metal is closely aligned with the LUMO or the HOMO of the material for the injection of electrons or holes, respectively [23]. Without proper alignment, a potential barrier will form, depending on the difference between the work function of the electrode and the energy levels of the semiconductor. This will lead to poor injection and increased resistance of the transistor. Non ohmic contacts result in irregularities in the linear regime of the output characteristics of the transistor (i.e., primary suppression and superlinear current increase). High-contact resistance has similar effect due to a source–drain voltage drop at the contacts. The potential barrier can be reduced by choosing the appropriate injecting electrode or introducing dipoles at the metal surface with self-assembled monolayers that are able to modify (increase or decrease) the metal work function [24].

The position of the injecting electrode also affects the contact resistance. Contrary to BC/BG, in BC/TG and TC/BG configurations, where the source and drain electrodes overlap with the gate electrode, the contact resistance is lower since the charges are injected from a larger area. Moreover, evaporated top metals can introduce surface states in the semiconductor facilitating or hindering the charge injection [25].

16.2.5
Applications: Inverters and Light-Emitting Transistors

The development of low cost, good performing bipolar FETs is highly desirable for CMOS inverter technology and light-emitting field-effect transistors (LE-FETs). Several groups have already demonstrated discrete inverters [13, 26, 27] and simple circuits [28] that are crucial building blocks in chip design for logic functions such as NAND, NOR, and NOT. The most basic component for these circuits is the voltage inverter, which inverts the incoming signal into an outgoing signal. CMOS inverters in silicon-based technology are composed of a p-type and an n-type transistor. As we will discuss in the following paragraphs, also bipolar transistors can be used for the purpose. Figure 16.6a shows the schematic of a complementary inverter [16]. In operation, the p- and n-type transistors are connected at the drain and the gate electrodes serving as output (V_{out}) and input (V_{in}) nodes, respectively. The source of the load transistor is connected to the power supply (V_{supply}), while the source of the driver transistor is grounded. The state of the n- and p-type transistors can be varied by changing the input voltage. An example of the quasi-static transfer curves of an inverter is shown in Figure 16.6b where a sharp inversion of the input signal can be observed with the maximum voltage gain [16]. An important characteristic of complementary inverters is that when the output is in a steady logic state ($V_{out} = 0$ or V_{supply}), only one of the transistor is "on" and the other is in "off" state. This reduces the current flow to a minimum equivalent to the leakage current through the transistor. Only during switching both transistors are "on" for a very short time. This considerably reduces the power consumption of these circuits with respect to other circuit architectures. Therefore, the advantages of

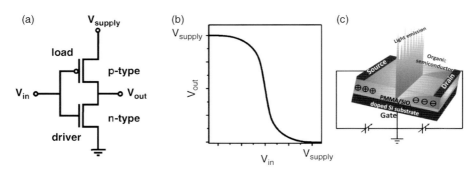

Figure 16.6 (a) Illustration of a complementary inverter structure. (b) Transfer characteristics of a voltage inverter. (c) Schematic of the LE-FET. (Reprinted with permission from Ref. [16, 20])

CMOS inverters are that they can operate at relatively high speed with very little power loss and exhibit good noise margins in both low and high states.

Light-emitting field-effect transistors combine the switching properties of transistors with the electroluminescence of light-emitting diodes. They have been demonstrated recently for several classes of materials including inorganic semiconductors [29], carbon nanotubes [30], and organic semiconductors. LE-FETs provide planar nanometer-sized light sources that can be easily integrated in silicon, glass, or plastic substrates. Organic semiconductors often show strong photoluminescence [31] and electroluminescence [32], making them ideal active layers for this new class of devices. LE-FETs can be realized by using a p–n junction within the transistor channel forming a radiative recombination interface for electrons and holes. The requirements for good performing LE-FETs are: (i) balanced electron and hole mobilities, (ii) control over the position of the recombination zone, (iii) high current densities. Figure 16.6c shows a schematic of the LE-FET and its working mechanism [33].

16.3
Bipolar Field-Effect Transistors

16.3.1
Single-Component Bipolar FETs

From the electronic point of view organic semiconductors should in principle conduct equally both positive and negative carriers. However for many years the vast majority of known organic semiconductors have displayed either positive (p-channel) or, in smaller number, negative (n-channel) transport properties. Recently, the crucial role played by traps for the electron current has been understood. These traps are present as impurities of the semiconductor and on the surface of several dielectrics commonly used in the fabrication of FETs (hydroxyl, silanol, and carbonyl groups). Therefore, only with an appropriate treatment of the dielectric surface and by using pure materials processed in an inert atmosphere, it was possible to observe electron transport in several polymers, previously considered only p-type [34]. The use of bipolar materials allows for the fabrication of complementary-like circuits through the use of a single component that functions both as p-channel and/or as n-channel. This significantly reduces fabrication complexity avoiding the necessity of micropatterning of the p-channel and n-channel.

A challenge in fabricating ambipolar transistors from a single component consists in getting efficient injection of both charge carriers from the same electrode. Optimal charge injection takes place when the work function of the metal electrode lines up with the HOMO level of the semiconductor for hole injection and with the LUMO level for electron injection. Since most of the standard organic semiconductors have band gaps of 2–3 eV, the injection of at least one carrier will be contact limited for any given electrode material. However, one may also consider realizing asymmetric electrodes by choosing two different materials for each contact.

Separately engineering distinct contacts for hole–electron injection in bipolar OFETs will facilitate more efficient injection of the two types of carriers, but at the same time it will make the device realization more complex.

The issue of the injection for the same metal can be solved by using polymer with a narrow band gap (i.e., lower than 1.8 eV). This lowers the injection barrier for both charge carriers and efficient ambipolar transport can take place. This concept was first demonstrated by Klapwijk and coworkers [13], who investigated a few wide-band gap polymers, (poly[2-methoxy-5-(3′,7′-dimethyloctyloxy)]-p-phenylene vinylene (OC1C10-PPV) and related PPVs, poly(2,5-thienylene vinylene) (PTV), and P3HT) and attributed the lack of ambipolar transistor action to the presence of large injection barriers. To confirm this point they reduced the barrier using a small bandgap polymer, poly(3,9-di-t-butylindeno[1,2-b] fluorene) (PIF), which has a band gap energy of 1.55 eV. Typical output characteristics of a field-effect transistor based on PIF in combination with gold electrodes demonstrated operation both in the hole-enhancement mode and in the electron-enhancement mode. They also pointed out that the purity of the material is an important requirement to achieve ambipolar operation.

The same concept was further investigated by Sirringhaus and coworkers [35] who synthetized a series of regioregular polyselenophene-based polymers, which, with respect to polythiophene, present a reduced band-gap. The lower lying LUMO in these polymers results in an improved electron transport due to an enhanced electron injection from the metal electrodes and to reduced susceptibility of electrons to traps states and oxidation. Top-gate, bottom contact (TG/BC) transistor configurations with gold source–drain electrodes were used for all polymers. The best polymer, poly(3,3″-di-n-decylterselenophene) (PSSS-C10) showed clean ambipolar transport characteristics with similar hole and electron saturation and linear mobilities of >0.01 cm^2 (V^{-1}s^{-1}) (PMMA as gate dielectric) (Figure 16.7a). While the saturation mobility values for holes and electrons were similar, some reversible hysteresis was systematically observed in the transfer characteristics in the electron transport regime but not in the hole transport regime, thus indicating the presence of a larger number of shallow traps for electrons than for holes. This is a common issue in bipolar semiconductors, which makes the observation of n-type regime more difficult and in general limits the value of the electron mobility. Using this polymer, complementary-like inverters based on two identical TG/BC ambipolar transistors were fabricated, with a common gate as input and a common drain as output, eliminating the need for semiconductor patterning. Despite the general fact that none of these TFTs can be fully switched off, the authors obtained very high switching gain in this inverter, with an absolute value as high as 86 (Figure 16.7b). This is one of the higher gain values reported so far in inverters composed of bipolar organic FETs.

For the realization of narrow band gap polymers, a useful synthetic strategy consists in copolymerizing an electron-rich and an electron-poor unit, to obtain a donor–acceptor structure. An example of this alternating donor–acceptor architecture has been reported by Jenekhe [36] with the synthesis of the copolymer PNIBT consisting of an electron-donating dialkoxybithiophene and

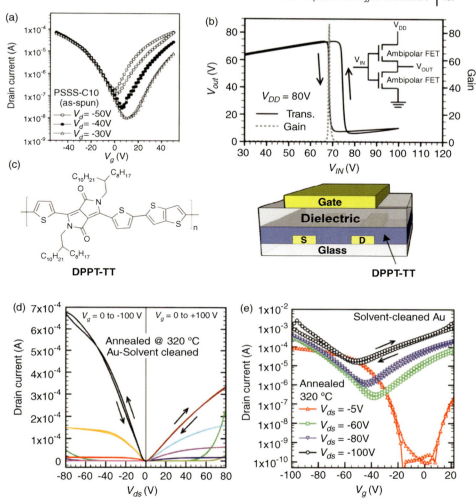

Figure 16.7 (a) Transfer characteristics of an as-spun PSSS-C10 ambipolar FET with channel length (L) of 40 μm and channel width (W) of 2 cm. The hole and electron mobilities for this device are both ∼0.026 cm^2 (V^{-1} s^{-1}). (b) The transfer characteristic and the corresponding gain (in absolute value) of a complementary-like inverter comprises two identical as-spun PSSS-C10 TG/BC ambipolar FETs. The inset shows the inverter circuit configuration. (c) Chemical structure of the DPPT-TT polymer and schematic diagram of the top gate, bottom contact OFET structure. The output and transfer (d and e) characteristics of a typical OFET based on DPPT-TT thin films annealed at 320 °C. (Adapted with permission from Ref. [35]).

electron-accepting naphthalene bisimide. High-mobility ambipolar transistors and high-gain complementary-like inverters were fabricated and exhibit electron and hole mobilities as high as 0.04 and 0.003 cm^2 (V^{-1} s^{-1}), respectively, and output voltage gains as high as 30.

In the previous example, and in most of the ambipolar polymers reported, there is an unbalanced charge transport, the hole mobility being always higher than the electron mobility. In order to solve this problem it is possible to either optimize the device or introduce a stronger electron withdrawing unit in the donor–acceptor structure of the polymer. Following this strategy to improve the electron transport characteristics, Cho et al. introduced a more deficient building block, benzothiadiazole (BZT), in the donor–acceptor alternating copolymer consisting of DPP and two unsubstituted thiophene rings [37]. The resulting polymer, PDTDPP-*alt*–BTZ, exhibits balanced hole and electron mobilities of 0.061 and 0.054 cm^2 (V^{-1} s^{-1}), respectively. These values were obtained using a bottom gate, top contact configuration with Ag source–drain electrodes. The authors observed that when different work function metals were used as the source and drain electrodes, it was possible to improve the n-type and p-type performances. The mobilities obtained from the optimized FET with two different electrodes (Au for source electrode and Al for drain electrode) were 0.097 cm^2 (V^{-1} s^{-1}) for the hole and 0.089 cm^2 (V^{-1} s^{-1}) for the electron. The well-balanced bipolar nature of PDTDPP-*alt*–BTZ FETs makes the materials suitable for the realization of CMOS-like inverters. The steepness of the inverter curve indicated the maximum gain of ∼35, which is a relatively good value for single-component-based inverters fabricated from organic semiconductors.

This demonstrates that an accurate design of the polymers combined with a careful optimization of the device architecture, charge injection, and thin film quality enable to reach high, balanced electron and hole mobilities. A further improvement in the device performance has been reported by Sirringhaus and coworkers with the synthesis of a diketopyrrolopyrrole polymer, DPPT-TT (Figure 16.7c), with balanced hole and electron field-effect mobilities both exceeding 1 cm^2 (V^{-1} s^{-1}) [11]. The narrow band gap and the position of the HOMO and LUMO energy levels were designed to enable efficient bipolar charge injection and transport. For the realization of the devices, the authors used a bottom contact top gate configuration with PMMA as dielectric. The drawback of this system is that in order to observe high electron mobility an annealing treatment of the devices at 320 °C was necessary. This treatment allows eliminating electron-trapping impurities in the DPPT-TT films and improves the contact of the polymers with the electrodes. After the optimization of the annealing temperature and the contact electrode functionalization, the authors realized a complementary-like voltage inverter by combining two identical ambipolar DPPT-TT OFETs with a common gate as the input voltage (V_{in}) and a common drain as the output voltage (V_{out}). The inverter static-voltage-transfer characteristic exhibited a good symmetry with a switching voltage nearly half of that of the power supply (V_{DD}), little hysteresis, and a relatively high gain (absolute value of gain > 20) comparable with previously reported complementary-like inverters based on bipolar OFETs. These results show that DPPT-TT based OFETs are promising candidates for applications in ambipolar devices and integrated circuits, as well as model systems for fundamental studies of bipolar charge transport in conjugated polymers. However, the high annealing temperature required to reach high electron mobility is a limiting factor for the application and implementation in circuit. Research is therefore focused on synthesizing polymers that can provide balanced bipolar behavior, with high hole and electron mobility, at room temperature.

In a similar system, Winnewisser and coworkers investigated the possibility to combine both ambipolar behavior and light-emitting properties [38]. They reported on a narrow band gap diketopyrrolopyrrole-based polymer, BBTDPP1, with ambipolar charge transport properties and near-infrared light emission from top-gate as well as bottom-gate transistor structure. By recording the electroluminescence with an infrared sensitive camera system, they showed that the recombination zone is moving through the transistor channel. This is an indication that the transistor is working in a truly bipolar mode: both a hole and an electron accumulation layer coexist and recombination occurs at the point in the channel where the two meet, depending on the relative gate and source drain voltage applied.

16.3.2
Bilayer Bipolar FETs

The possibility to obtain bipolar behavior in organic field-effect transistors by combining layers of n-type and p-type materials was first demonstrated in systems where both layers were deposited by evaporation. Dodabalapur and coworkers combined layers of two small molecules, the hole-conducting R-hexathienylene (R-6T) and the electron-conducting fullerene C_{60}, and observed both hole and electron transport in these devices, although with lower mobilities than those of the single component [39]. Both the relative position of HOMO and LUMO levels of the two materials as well as the deposition order were found to be important for achieving bipolar characteristics. Using this approach different combinations of hole conducting and electron conducting small molecules have been explored [40–42]. Air-stable, ambipolar transistors based on copper-hexadecafluoro-phthalocyanine (FCuPc), as an efficient electron conductor, and 2,5-bis(4-biphenylyl)-bithiophene (BP2T), as a hole conductor, were reported by Wang *et al*. Figure 16.8 shows the output (a) and transfer (b) characteristics of such a device. Using this material combination, bipolar mobilities of up to $0.04 \, cm^2 \, (V^{-1} s^{-1})$ for holes and $0.036 \, cm^2 \, (V^{-1} s^{-1})$ for electrons are possible under ambient conditions. Inverters fabricated with these transistors showed gains as high as 13 and good noise margins [43].

Dinelli *et al*. showed that by using layered structures of small molecules it is also possible to obtain light emission together with balanced ambipolar transport. For this application, α,ω-dihexyl-quaterthiophene (DH4T) and N,N'-ditridecylperylene-3,4,9,10-tetracarboxylic diimide (P13) were used for the p- and n-type layers, respectively (Figure 16.8c). The mobility values reported are as large as $3 \times 10^{-2} \, cm^2 \, (V^{-1} s^{-1})$, among the highest values reported for bipolar bilayer OLETs. Morphological analysis by confocal PL microscopy indicates that "growth" compatibility is required in order to form a continuous interface between the two organic films. The authors pointed out that this compatibility is crucial in controlling the quality of the interface and the resulting optoelectronic properties of the OLETs. Therefore, they concluded that the optimum performance is not necessarily achieved by

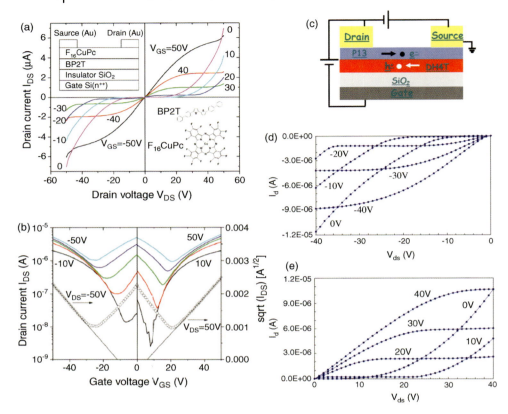

Figure 16.8 (a) Typical output characteristics of bipolar OFETs based on FCuPc and BP2T bilayers for positive and negative gate biases. A schematic cross-section of the device structure and the molecular structures of FCuPc and BP2T is given in the insets. (b) Typical transfer characteristics for positive and negative gate biases. The solid lines show the logarithmic drain current versus gate voltage for various V_{ds}, and the open symbols show the square root of the drain current. (c) Schematic of an OLET device based on a DH4T–P13 bilayer. (d) Output curves (I_d vs. V_{ds}) obtained from a DH4T–P13 device operating in a p-type configuration or in an n-type configuration. V_{gs} values are reported beside each curve. I_d = drain current; V_{ds} = drain–source voltage; V_{gs} = gate–source voltage. (Reprinted with permission from Ref. [43]).

employing materials with the highest mobility values in the single-layer devices, but by combining materials that are compatible and able to form a defined interface when evaporated one on top of the other [44].

In the bilayer architecture, depending on the device configuration and on the materials used, charge accumulation and transport of holes and electrons can occur in different layers. However, at least one of the two accumulation zones will form at the interface between the two organic layers and therefore charge transport will depend on the quality of this interface. In bilayer devices realized by evaporation, as shown in the previous examples, an accurate control of the growth conditions is necessary. In order to reach balanced bipolar behavior, intermixing between the two layers and rough interfaces need to be avoided, since they limit charge transport. This is the reason why almost all the examples of bilayer ambipolar transistors are based on sublimed small molecules thin films. There are very few reports on bipolar bilayer transistors based on solution-processed semiconductors [45, 46]. This is due to the well-known difficulty in fabricating well-defined smooth bilayer structures by depositing one component on top of the other, without damaging the layer underneath and without intermixing. A possible solution to this issue has been proposed by Heeger *et al.* who realized FETs using regioregular poly(3-hexylthiophene) (rr-P3HT) and [6,6]phenyl C_{61} butyric acid methyl ester (PCBM) separated by an intermediate layer of titanium oxide (TiO_x). All layers were processed from solution. In this multilayer FET structure, the TiO_x layer enables the bipolar properties by electronically separating the p- from the n-channel. The comparison of the data with and without the TiO_x confirms its role as an electron transport and hole blocking material. The optimized device shows good hole and electron mobilities, 8.9×10^{-3} cm^2 (V^{-1} s^{-1}) in the *n*-channel mode and 5.7×10^{-3} cm^2 (V^{-1} s^{-1}) in the p-channel mode [47].

An alternative approach for the realization of bipolar organic thin-film transistors with a bilayer structure has been demonstrated by Hashimoto and coworkers by using a contact-film-transfer method. They successfully fabricated bipolar FET and inverters based on a bilayer structure of P3HT and PCBM using a simple solution-based, contact-film-transfer method (Figure 16.9a). The transistors exhibited balanced electron and hole mobilities of 2.1×10^{-2} and 1.1×10^{-2} cm^2 (V^{-1} s^{-1}), respectively. Complementary inverters based on two identical bipolar transistors showed good performance with a gain of 14 (Figure 16.9d). These results indicate that the contact-film-transfer method provides a facile way to construct multi-layered structures with well-defined smooth interfaces and to fabricate complex organic electronic devices. The authors pointed out that further improvement could be achieved by using other organic materials with higher mobilities [48].

Jenekhe and coworkers demonstrated the possibility to obtain well-defined n–p polymer–polymer heterojunctions for bipolar FETs using sequential spin coating from solutions. By selecting polymers soluble in orthogonal solvents, it was possible to deposit one directly on top of the other, without using film transfer and lamination [12]. The polymer used for the n-type layer poly(benzobisimidazo-benzophenanthroline), BBL, is soluble in methane sulfonic acid but not in chlorinated aromatic solvents whereas the p-type polymers P3HT, poly(benzobisthiazole-*alt*-3-octylquarterthiophene) PBTOT, and poly(thiazolothiazole) PSOTT are soluble in the latter solvents (Figure 16.10a). Bipolar charge transport in the n–p heterojunctions had electron and hole mobilities of $\sim$0.001–0.01 cm^2 (V^{-1} s^{-1}), which are similar to those measured in the corresponding single-layer OFETs. This demonstrates that the deposition of the second layer does not affect the performance of the first one. The authors also tracked the field-effect mobilities of electrons and holes in the n–p heterojunction FETs in air for more than 6 months recording good stability (Figure 16.10d).

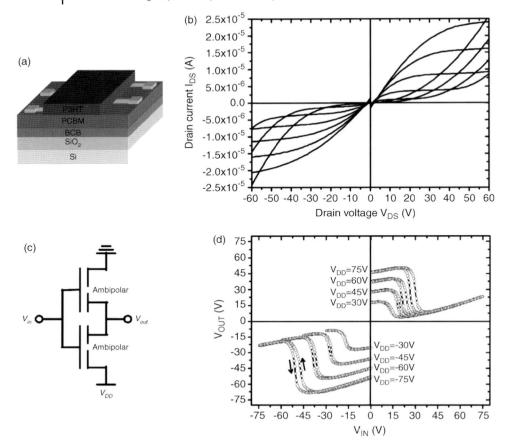

Figure 16.9 (a) Schematic structure of the bilayer realized by contact-film-transfer method. (b) Output characteristics of the P3HT/PCBM bilayer transistor. (c) Schematic representation of the electrical connections for the inverter based on two identical ambipolar transistors and (d) transfer curves. (Reprinted with permission from Ref. [48]).

Moreover, the authors showed the application of these devices in logic circuits. Digital logic circuits that perform a logic calculation of binary information (represented by 0 and 1) have played an essential role in the development of current information technology. These circuits are fabricated by integration of multiple field-effect transistors. Sharp signal switching of the circuits can be obtained from complementary circuits that consist of p- and n-type transistors or bipolar transistors. Each transistor is selectively turned on and off based on the voltages at the terminal electrodes, inducing current flow through a certain pathway of the circuit and resulting in the targeted output voltage, which is close to either the supplied voltage (V_{dd}; representing signal 1) or ground (representing signal 0) of the logic operation. For example, the input signal is inverted after NOT gate operation (from 0 to 1, or from 1 to 0). The output signal of NAND gate is 0 only when two input signals are 1;

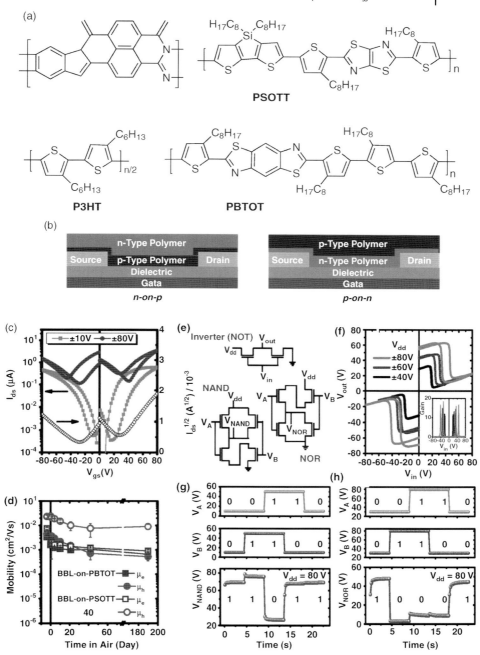

Figure 16.10 (a) Molecular structures of the n-channel (BBL) and p-channel polymer semiconductors (P3HT, PBTOT, and PSOTT). (b) Schematics of bipolar field-effect transistors based on n/p polymer heterojunctions. (c) Transfer curves of a BBL-on-PBTOT bilayer

otherwise the output is 1. The NOR operation results in 1 only when the input signals are both 0. Complementary NOT gate requires one n-channel transistor and one p-channel transistor, and complementary NAND and NOR gates require two n-channel transistors and two p-channel transistors each. The advantage of using bipolar transistors in logic gate is that it is possible to use one type of transistor, although the total number of transistors is the same as in circuits with unipolar transistors (Figure 16.10e). Circuit diagrams of an inverter (NOT-gate) consisting of two transistors, and two-input NAND and NOR circuits with four transistors each, are shown in Figure 16.10e. The circuits consist of identical bipolar FETs from BBL-on-PBTOT n–p heterojunctions with the same geometric factors.

Complementary logic gates, including NOT, NAND, and NOR circuits, based on the polymeric bilayer bipolar OFETs exhibited sharp switching, with a gain as high as 16–18 (Figure 16.10f). Some degree of hysteresis is likely due to the threshold voltage difference of n- and p-channel modes in the individual OFETs. In the logic circuits of NAND and NOR gates, good switching characteristics were also observed, as shown in the output voltages plotted with the corresponding input voltages V_A and V_B (Figure 16.10g and h). High and low voltages at the terminals represent signals 1 and 0, respectively. These results demonstrate that bipolar OFETs can be used for designing and constructing various complementary circuits. Because bipolar transistors and circuits are fabricated by simple solution process, the devices presented here could ultimately be printable allowing for the effective use of this approach on a large area.

16.3.3
Bulk Heterojunction Bipolar FETs

Another way to obtain bipolar performance is to blend n- and p-type organic semiconductors, combining the advantageous properties of the two components in a single structure. Blending organic materials may lead to heterostructured bipolar FETs when an interpenetrating (percolation) network of the two materials is obtained. This can be achieved by coevaporation or solution processing, if the two materials are soluble in a common solvent.

In comparison to single-component films, bipolar semiconducting blends are much more challenging. The necessity to have percolation networks for both electrons and holes, each one conducted by one component of the blend, imposes strict requirements for the supra-molecular organization of the two materials. On one side an extended intermixing of the two semiconductors at the molecular level will

transistor. (d) Hole and electron mobilities in the BBL on-PBTOT and BBL-on-PSOTT transistors as a function of time in air. Data points before day 0 represent the mobilities in inert conditions. (e) Circuit diagrams of a complementary inverter, and NAND and NOR logic gates. (f) Voltage transfer characteristics of an inverter. (g, h) Output voltages of complementary logic gates and the truth tables with corresponding input voltages V_A and V_B: (g) NAND and (h) NOR gates. The transistors in the circuits are based on BBL-on-PBTOT heterojunctions. (Reprinted with permission from Ref. [12]).

be negative, since it will allow electron–hole recombination; and on the other side macroscopic phase segregation can determine low percolation of the carriers between source and drain electrodes and it can cause problems for the charge injection. In conclusion, the main challenge is controlling the phase separation degree between the two components of the blend during the deposition process. This can be achieved by warily tuning the processing temperature, the solution viscosity, and the evaporation rate of the solvent. The objective is to adjust the miscibility and degree of crystallinity of the individual compounds in order to obtain in the blend continuous percolation pathways for both electrons and holes.

16.3.3.1 Coevaporated Blends

Although the bilayer approach yields some impressive device characteristics and interesting insight into the electronic properties of organic semiconductor interfaces, it is still an issue to deposit two layers one top of the other. Lamination and transfer methods are not compatible with printing technology and sequential depositions directly from solution are challenging due to the need of orthogonal solvents for the deposition of the two layers. An alternative method is to use blends of n- and p-channel materials to realize bipolar transport in a single layer. For the blend approach, both coevaporated and solution-processed films are feasible.

Rost *et al.* showed that coevaporating $N,N8$-ditridecylperylene-3,4,9,10-tetracarboxylic diimide PTCDI-$C_{13}H_{27}$ (P13) and quinquethiophene (5T) (Figure 16.11a),

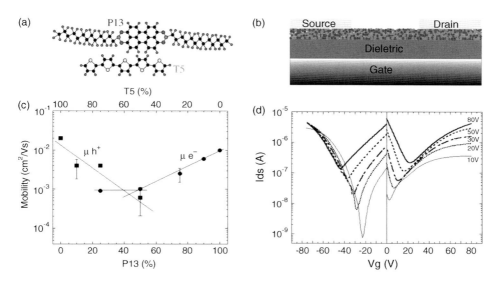

Figure 16.11 (a) Molecular structure of 5T PTCDI-$C_{13}H_{27}$ and (b) device structure of bipolar field-effect transistor consisting of a coevaporated thin film of 5T and PTCDI-$C_{13}H_{27}$. (c) Electron (filled circles) and hole (filled squares) field-effect mobilities for different bulk-heterojunction compositions. (d) Transfer characteristics of the coevaporated 5T/PTCDI-$C_{13}H_{27}$ thin-film transistor for negative and positive gate biases. (Reprinted with permission from Ref. [21]).

with equal fractions results in good bipolar characteristics with hole and electron mobilities of 10^{-4} cm^2 (V^{-1}s^{-1}) and 10^{-3} cm^2 (V^{-1}s^{-1}), respectively, which were smaller than those for the pure materials [49]. This is understandable, as explained above, since in the blend there is an interpenetrating network of n-channel and p-channel materials. In the presence of a not optimized morphology of this network, electron–hole recombination may occur with consequent losses in carrier population. Despite these limitations, it is important to underline that this was the first report of a bipolar light-emitting transistor (LE-FET): the observation of pronounced ambipolar conduction over a wide range of bias conditions was accompanied by light emission.

Loi *et al.* later investigated the impact of the relative fractions of each material on hole and electron mobilities in the same system and found the expected increase of hole mobility with an increasing fraction of 5T and equally an enhanced electron mobility with an increasing PTCDI-C$_{13}$H$_{27}$ fraction (Figure 16.11c) [21]. Balanced hole and electron mobilities for this system are obtained with a ratio of PTCDIC$_{13}$H$_{27}$/5T of 2 : 3. The authors were also able to detect electroluminescence from this system that correlates with the voltage applied at the drain–source and gate electrode. Moreover, by using laser scanning confocal microscopy, they found the relation between the working characteristics of the transistors (p-channel, n-channel, ambipolar) and the supra-molecular organization of the thin film. Other coevaporated blends resulting in bipolar transport include pentacene/PTCDI-C$_{13}$H$_{27}$ [50] and pentacene/fluorinated pentacene [51].

16.3.3.2 Polymer–Small Molecule Blends

Due to the easy processing, thin films of solution processable polymer–small molecule composites are more appealing for the realization of integrated circuits on large area by printing techniques. As for the coevaporated systems, the composition and microstructure of the film will affect the performances of the devices. For the polymer–small molecule blends, these characteristics can be tuned by the choice of solvents and spin coating parameters [52], until the optimal conditions are reached for balanced bipolar transport.

The first bipolar FETs based on a blend were fabricated by Tada *et al.* [53, 54] by mixing the electron conducting dye N,N-bis(2,5-di-*tert*-butylphenyl)-3,4,9,10- perylene dicarboximide, whose chemical structure is similar to PTCDI-C$_{13}$H$_{27}$ (shown in Figure 16.11a), with p-type poly(3-dodecylthiophene) in chloroform and spin casting this mixture on a Si/SiO$_2$ substrate with prepatterned Ti/Au electrodes. Although bipolar behavior was observed, the effective mobilities were extremely low (10^{-7} cm^2 (V^{-1}s^{-1}) for holes and 10^{-9} cm^2 (V^{-1}s^{-1}) for electrons).

The first demonstration of a polymer blend transistor with appreciable ambipolar mobilities was accomplished by Meijer *et al.* by blending poly(2-methoxy-5-(3,7-dimethyloctoxy)- *p*-phenylene vinylene) (OC$_1$C$_{10}$-PPV) and PCBM (Figure 16.12a) [13]. Here, the hole and electron mobilities reached 7×10^{-4} cm^2 (V^{-1}s^{-1}) and 3×10^{-5} cm^2 (V^{-1}s^{-1}), respectively. A representation of the interpenetrating network and cross-section of the field-effect transistor is presented in Figure 16.12a. For both materials the charge injection comes from

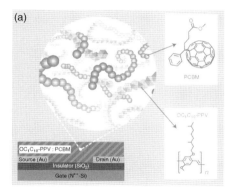 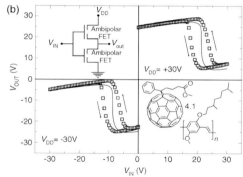

Figure 16.12 (a) Schematic of the cross-section of the FET geometry; the molecular structures of PCBM and OC$_1$C$_{10}$-PPV and a representation of the interpenetrating networks of the two semiconductors. (b) Transfer characteristics of CMOS-like inverters based on two identical OC$_1$C$_{10}$-PPV: PCBM field-effect transistors. Depending on the polarity of the supply voltage, V_{DD}, the inverter works in the first or the third quadrant. A schematic representation of the electrical connections in the inverter is given in the inset. (Reprinted with permission from Ref. [13]).

the gold contact: the HOMO level of OC$_1$C$_{10}$-PPV, at -5.0 eV, is aligned with the work function of gold (about -5.1 eV), resulting in an ohmic contact for hole injection from gold into the OC$_1$C$_{10}$-PPV network. Instead the mismatch in energy levels between the gold work function and the LUMO level of the PCBM results in an injection barrier of about 1.4 eV for electron injection into the PCBM network. However, the authors claimed that this injection barrier can be significantly reduced to 0.76 eV, due to the formation of a strong interface dipole layer at the Au–PCBM interface.

Typical output characteristics of a field-effect transistor based on the OC$_1$C$_{10}$-PPV: PCBM blend, in combination with Au electrodes, demonstrated operation in both the hole-enhancement and electron-enhancement modes. For high negative V_g, the transistor is in the hole-enhancement mode and its performance is identical to a unipolar transistor based on OC$_1$C$_{10}$-PPV, with a field-effect mobility of 7×10^{-4} cm^2 (V^{-1} s^{-1}). At positive V_g, the transistor operates in the electron-enhancement mode, with a field-effect mobility of 3×10^{-5} cm^2 (V^{-1} s^{-1}). This mobility is two orders of magnitude lower than the electron mobility in a PCBM transistor [55] due to the interpenetrating network of the two components. For this kind of system, a better matching of the energy levels of the n-type semiconductors with the electrode metal work function could lead to a more balanced bipolar transport in the blend. Inverters based on two identical bipolar transistors were realized, demonstrating CMOS-like operation (Figure 16.12b). A gain of 10 for the OC$_1$C$_{10}$-PPV: PCBM blend inverter was easily achieved, in combination with a good noise margin.

Other examples of bipolar behavior have been reported for blends of poly(2-methoxy-5-(2c-ethylhexyloxy)-1,4- phenylenevinylene) (MEH-PPV) and C$_{60}$ [56],

MEH-PPV and PCBM [57], CuPc and poly(benzobisimidazo-benzophenanthroline) [58], and P3HT and PCBM [59].

The best results obtained until now with this class is reported by Shkunov et al. from bipolar blends of thieno[2,3-b]thiophene terthiophene polymer and phenyl C_{61} butyric acid methyl ester (Figure 16.13) [27]. The authors studied the effect of different surface treatments and obtained the highest field-effect mobility for bipolar blends on OTS-treated substrates in saturation regime, with electron mobility reaching 9×10^{-3} cm^2 (V^{-1}s^{-1}) and hole mobility reaching 4×10^{-3} cm^2 (V^{-1}s^{-1}). CMOS-like inverters have been built on a single substrate using two identical transistors. As shown in the circuit schematic, a common gate has been used for both the transistors (inset in Figure 16.13c). The

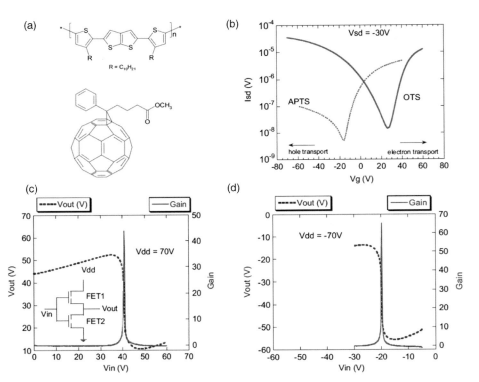

Figure 16.13 (a) Chemical structures of thieno [2,3-b]thiophene terthiophene polymer and PCBM. (b) Transfer characteristics of FETs with APTS-modified (dotted line) and OTS-modified (solid line) SiO$_2$ surfaces. The polymer/PCBM blend composition is identical in both cases. The devices operated in electron-enhancement mode with offset voltages −16 and 27 V for APTS and OTS transistors, respectively. (c, d) Transfer characteristics of CMOS-like inverters are shown as dashed lines. The corresponding gain curves are thin solid lines. Inverters are operational in the first quadrant (d) or in the third quadrant (c). The inset of (c) shows the schematic configuration for these inverters based on two identical transistors. V_{in}: input voltage; V_{out}: output voltage; V_{dd}: supply voltage. (Reprinted with permission from Ref. [27]).

devices are operational in two quadrants: with positive supply (V_{DD}) and input (V_{in}) voltages, the inverters work in the first quadrant with a maximum gain of 45, whereas under negative bias, the inverters operate in the third quadrant and exhibit a gain of ∼65, which is the highest value reported for inverters made with bipolar blends.

16.3.3.3 Hybrid Blends

Reports on semiconducting polymer-inorganic material-based bipolar FETs are rare to date, most probably due to the difficulty in controlling the morphology of a complex active layer. Aleshin *et al.* have shown that embedding ZnO nanoparticles into poly[9,9-bis-(2-ethylhexyl)-9*H*-fluorene-2,7-diyl] (PFO) matrix can serve as active layer for light-emitting unipolar and bipolar transistors depending on the concentration ratio of the PFO:ZnO [60]. PFO is one of the most promising and widely used conjugated polymers for LEDs, while ZnO is a nontoxic, environmentally stable, and solution processable inorganic n-type semiconductor. Recently ZnO based FETs have been reported with very high mobilities up to 7.2 cm^2 (V^{-1} s^{-1}) in combination with high on/off ratio [61].

The major challenge in fabricating bipolar FETs arises from the mismatch of the electrode work function and the energy levels of the semiconductors. The authors have proposed to use two different metal electrodes to overcome this limitation. Gold electrodes have been used to inject holes into the HOMO levels of the PFO and aluminum electrodes to facilitate electron injection into the LUMO of the ZnO. Due to a large injection barrier, gold electrodes are unable to sufficiently inject electrons to the higher lying LUMO level of the ZnO.

The structure used for this device is composed of an n+ silicon substrate and 200 nm thermally grown SiO$_2$ as gate dielectric with thermally evaporated Au and Al electrodes. The device is finished by spin-casting or drop casting the blend of PFO and ZnO from chloroform with different concentration ratios. After deposition, the films were dried at 80 °C in N$_2$ for 15 min. The output and transfer characteristics of such a device are shown in Figure 16.14. The best results are produced with concentration ratio of 1: 0.2 between PFO and ZnO nanoparticles. The hole and electron enhancement modes are evident from Figure 16.14a and b. However, the characteristic S-shape of the output curves strongly suggests contact resistance problems. The transfer characteristic for the same device in Figure 16.14c shows both electron and hole accumulation regimes with a small reversible hysteresis. Electron and hole field-effect mobilities were found to be ∼0.021 and ∼0.029 cm^2 (V^{-1} s^{-1}), respectively. The on/off ratio for this device was calculated to be in the range of ∼10^3 for V_g ∼ 20 V. As the concentration of the ZnO nanoparticles increases the transfer plot of the PFO:ZnO FET becomes asymmetric indicating the transition to the n-type unipolar regime. The optical output characteristics (electroluminescence (EL) intensity versus V_{ds}) of the ambipolar PFO: ZnO FET, at different gate voltages, are shown in Figure 16.14d. The EL intensity is increasing by enhancing the source–drain voltage and both for negative and positive gate voltages. The onset of the light EL emission at V_{ds} ∼ −5 V is nearly constant and independent of the ZnO nanoparticles concentration and the gate voltage polarity. The

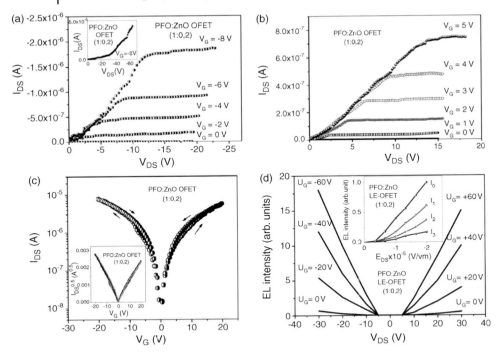

Figure 16.14 (a) Output characteristics of the PFO:ZnO FET in hole enhancement mode; (inset) output I–V characteristics for the same sample at high $V_{ds} > 20$ V and $V_G = -20$ V; (b) output characteristics in electron enhancement mode. (c) Transfer characteristic for $V_{ds} = -10$ V; (inset) square root of I_D versus V_g for the same sample; (d) EL intensity versus V_{ds} at different V_g (inset): EL intensity versus electric field for different spectral regions: I_0 — integral; I_1 — 600–830 nm; I_2 — 450–620 nm; I_3 — 300–400 nm. (Reprinted with permission from Ref. [60]).

inset in Figure 16.14d shows the EL intensity versus the electric field at different spectral regions. Clearly, the EL emission takes place mainly in the green spectral region and reaches ~40% of the integral EL intensity, whereas in the blue region reaches only ~20%. These results demonstrate that polymer composite thin films such as PFO and semiconducting nanoparticles can also serve as multifunctional devices fabricated by compatible techniques for printed technologies.

16.3.3.4 Polymer–Polymer Blends

Blends of semiconducting polymers can also be used as active layers to achieve bipolar charge transport. Since controlling the morphology of polymer blends is challenging, there are only few reports based on polymer–polymer composite bipolar devices.

Babel et al. investigated the dependence of bipolar carrier transport on thin-film morphology [14]. They studied two series of binary polymer–polymer blends of the n-type poly(benzobisimidazobenzophenanthroline) (BBL) with the p-type semiconductors poly[(thiophene-2,5-diyl)-alt-(2,3-diheptylquinoxaline-5,8-diyl)] (PTHQ$_x$) and poly(10-hexyl-phenoxazine-3,7-diyl-alt-3-hexyl-2,5-thiophene) (POT).

Atomic force microscopy (AFM) was used to investigate the thin-film blend morphologies of BBL/PTHQ$_x$ blend films as a function of the composition; the images are shown in Figure 16.15a. There are two distinct phases (lighter and darker) present in the images. The light features in Figure 16.15a correspond to PTHQ$_x$ since their size increases with increasing PTHQ$_x$ concentration. In the case of 10 wt% PTHQ$_x$ the phase-separated domain size is ∼50 nm, which increases to ∼300 nm in the 80 wt% PTHQ$_x$ blend. By increasing more the PTHQ$_x$ concentration, an interpenetrating two-phase bicontinuous network structure can be observed. These data were supported by transmission electron microscopy (TEM) measurements confirming that PTHQ$_x$ tends to form spherical aggregates randomly dispersed in the BBL matrix. The BBL/POT blends show similar phase separated morphology. The size of the light features increases by increasing the concentration of POT indicating the origin of this phase.

The compositional dependence of the charge transport was investigated in simple bottom contact/bottom gate FET configuration. The binary blends were spin coated from methanesulfonic acid (MSA) followed by drying and annealing in vacuum at 60 °C. Using 10–90 wt% PTHQ$_x$ in the BBL/PTHQ$_x$ blend, only

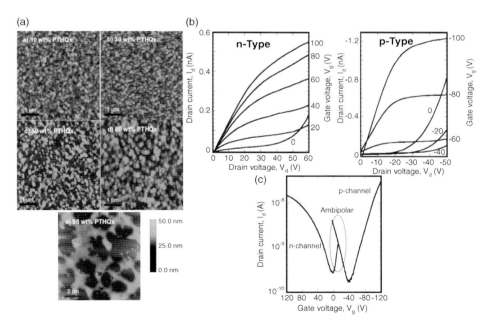

Figure 16.15 (a) AFM topography images of 10, 30, 50, 80, and 98 wt% of BBL/PTHQ$_x$ blends. (b) Bipolar operation of the 98 wt% PTHQx blend in electron enhancement and hole enhancement mode. Transfer characteristics of a 98 wt% PTHQx blend. (Reprinted with permission from Ref. [14]).

unipolar electron transport could be observed. The calculated saturation field-effect electron mobility was relatively constant at 1.0×10^{-3} cm^2 (V^{-1}s^{-1}) almost over the whole blend-composition range. It is interesting to note that the electron mobility did not decrease by adding the second component, if compared with the mobility in the BBL FET. In most of the cases the mobility sharply decreases by one or two orders of magnitude in the presence of another component in the active layer. However, when the PTHQ$_x$ concentration was increased to 90 wt%, a drop in the electron mobility was observed, indicating that the electron mobility in the BBL/PTHQ$_x$ blends is independent of the composition until a certain threshold in concentration is reached.

This constant electron mobility can be explained by considering the blend thin-film morphology. The AFM images show that even at high PTHQ$_x$ concentrations (50–80 wt%) PTHQ$_x$ exists as separate domains in the BBL matrix while BBL forms an interconnected matrix facilitating the electron transport. Since there is no sign of hole transport in these blends, it can be concluded that the single crystalline domains of PHTQ$_x$ are not connected. Ambipolar transport was observed in these blends only above 90 wt% PTHQ$_x$. The output and transfer curves of such a device are shown in Figure 16.15, where the 98 wt% PTHQx blend transistor operating in electron and hole enhancement mode is shown. At low gate voltages and high drain voltages the typical nonlinear increase in the drain current appears in both modes due to the formation of the p–n junctions in the channel. From the transfer curves, the saturation field-effect electron and hole mobilities were calculated to be 1.4×10^{-5} cm^2 (V^{-1}s^{-1}) and 1×10^{-4} cm^2 (V^{-1}s^{-1}), respectively. The hole field-effect mobility is still comparable to the single component PTHQ$_x$ device mobility, while the electron mobility in the blend at this concentration is about two orders of magnitude lower compared to the BBL-only device mobility. According to the authors, the decrease in the electron mobility of these bipolar FETs may be due to the smaller amount of BBL in the blend.

The BBL/POT blends showed similar trend: only electron transport was observed until a threshold concentration (50 wt%) of POT. Above this concentration bipolar transport appeared with electron and hole mobilities of 6×10^{-4} cm^2 (V^{-1}s^{-1}) and 1.2×10^{-6} cm^2 (V^{-1}s^{-1}), respectively for 80 wt% of POT. Both the electron and hole mobilities decreased in this blend with respect to the single-component devices, which can be explained by the variation of the morphology. The same authors investigated binary blends of poly(9,9-dioctylfluorene) and regioregular poly(3-hexylthiophene). The polymers were found to be phase separated and to exhibit only hole transport [62].

The lack of the bipolar transport in binary blends of polymers for most of the compositions shows that controlling the thin-film morphology is challenging. Therefore, the key issue is to realize an interpenetrating and bicontinuous networks of binary polymer blends in order to establish ambipolar charge transport.

Another example of all-polymer bulk heterojunction bipolar FETs was demonstrated recently by Szendrei et al. [15]. The limited number of polymer blend bipolar FETs is due to the scarcity of high-performing n-type polymers. The recent

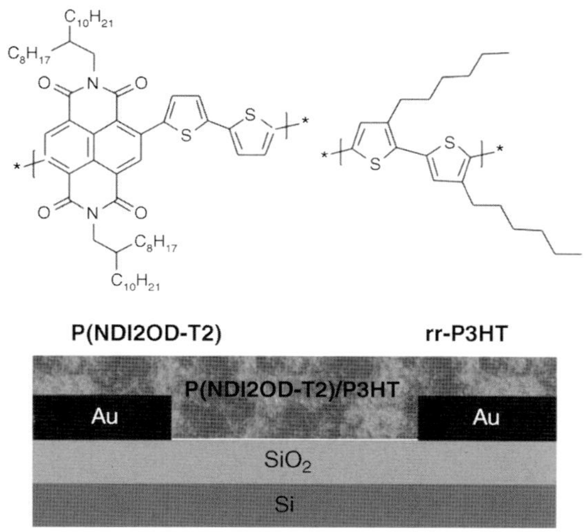

Figure 16.16 Chemical structure of the semiconducting polymers P(NDI2ODT2) and rr-P3HT and illustration of the bottom-gate/bottom-contact bipolar FET architecture. (Reprinted with permission from Ref. [15]).

discovery of the n-type polymer poly{[N,N'-bis(2-octyldodecyl)-naphthalene-1,4,5,8-bis(dicarboximide)-2,6-diyl]-alt-5,5'-(2,2'-bithiophene)} (P(NDI2OD-T2)) [8] exhibiting large electron mobility in ambient conditions opened up the way toward the development of solution processed efficient polymer-based bulk heterojunctions. Using P(NDI2OD-T2) as the n-type component and regioregular poly(3-hexylthiophene) (rr-P3HT), the authors demonstrated a bipolar FET with high and balanced hole and electron mobilities. The chemical structures of P(NDI2OD-T2) and P3HT are shown in Figure 16.16, together with the device configuration used in this study to test the electrical properties of the polymer blend. The devices are fabricated in a bottom gate/bottom contact (Au) configuration where the polymer blend (typically in the proportion of 1 : 1 by weight) is spin coated from 1,2-orthodichlorobenzene (ODCB) and annealed overnight at 110 °C in a vacuum oven.

Figure 16.17 shows the output characteristics of the all-polymer bulk heterojunction FETs. The ambipolar nature in both electron enhancement (Figure 16.17a) and hole enhancement (Figure 16.17b) modes is evident. At high positive gate voltages (V_g) these transistors function as only electrons are accumulated at the semiconductor–insulator interface, similarly to the unipolar P(NDI2OD-T2) FETs. For lower V_g the devices show the typical nonlinear increase in current at high V_{ds} due to the injection of both charge carriers in the channel. However, at low V_{ds} the output curves show clear indications of contact resistance for electron injection due to the injection barrier between the gold electrodes and the LUMO level of P(NDI2OD-T2).

A similar behavior is observed for hole transport when applying negative V_g and V_{ds} biases (Figure 16.17b). However, the current is far less limited for hole

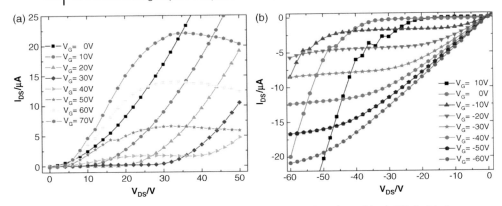

Figure 16.17 Typical output characteristics of the bipolar polymer blend FETs in (a) electron enhanced mode and (b) hole enhancement mode. (Reprinted with permission from Ref. [15]).

injection due to the almost ohmic nature of the contact between Au and the HOMO level of rr-P3HT.

Figure 16.18a and b shows the transfer characteristics of these polymer bipolar FETs for positive and negative V_{ds}, respectively. In both cases the transfer shows a symmetric shape indicating the presence of balanced electron and hole populations in the channel. The saturation field-effect electron and hole mobilities were calculated to be 4×10^{-3} cm^2 (V^{-1} s^{-1}) at $V_{ds} = +30$ V and a p-type mobility of 2×10^{-3} cm^2 (V^{-1} s^{-1}) at $V_{ds} = -30$ V. These mobilities are the highest balanced mobilities reported so far for solution processed all polymer bulk heterojunction bipolar FETs. The balanced FET mobilities indicate the presence of sufficient percolation pathways for both charge carriers in these polymer blends.

To provide evidence for the presence of good interpenetrating bicontinuous network, the surface morphology of the blend was investigated by AFM. Figure 16.19

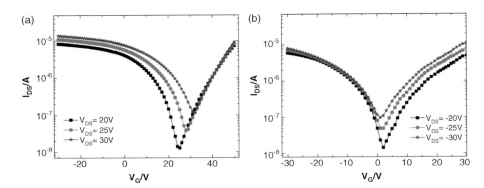

Figure 16.18 Typical transfer characteristics of the bipolar polymer blend FETs in (a) electron enhancement mode and (b) hole enhancement mode. (Reprinted with permission from Ref. [15]).

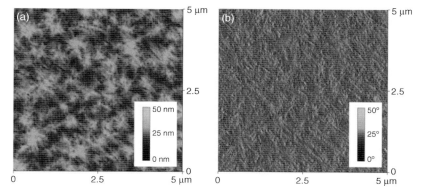

Figure 16.19 AFM (a) topography and (b) phase images of the P(NDI2OD-T2)/rr-P3HT blends. (Reprinted with permission from Ref. [15]).

shows the AFM topography (a) and phase (b) of the P(NDI2OD-T2)/rr-P3HT active layer. The P(NDI2OD-T2)/rr-P3HT blend thin-film surface is quite smooth, characterized by small protrusions and an RMS roughness of 3.9 nm. The phase image (Figure 16.19b) shows a very fine network of the two materials with two different phases, suggesting phase segregation between P(NDI2OD-T2) and rr-P3HT. This fine network proves the presence of efficient percolation pathways for both charge carriers, supporting the good bipolar nature of this polymer–polymer blend.

Despite the challenge of controlling the morphology of polymer composite the excellent bipolar performance of these blends may hold promise for easy, cheap, and solution-processed bipolar optoelectronic applications.

16.4
Perspectives

In this chapter we have reviewed the state of art in the fabrication of bi-polar field effect transistors by using different approaches. We have observed that, in the past few years, the synthesis of polymers with bi-polar characteristics has progressed considerably, also thanks to the understanding of the detrimental role of impurities and trapping (especially electron trapping) at the semiconductor–oxide interface.

A better understanding of the role of the supra-molecular order for the realization of continuous percolation paths for both charge carriers has brought noticeable improvement in the performances of polymer–polymer composites. Further improvements with this last approach will be achieved when methods for the control of the composite morphology will be developed. Moreover, the synthetic efforts toward better n- and p-type polymers will also be fundamental in the achievement of high bipolar mobilities and on–off ratios in polymer-based composites. In the meantime, it is possible that interesting results will come from emerging fields, as the organic–inorganic hybrid composites.

References

1 Lilienfeld, E.J. (1928) US patent 1900018.
2 Horowitz, G. (1998) *Adv. Mater.*, **10**, 365–377.
3 Voss, D. (2000) *Nature*, **407**, 442–444.
4 Rotzoll, R., Mohapatra, S., Olariu, V., Wenz, R., Grigas, M., Dimmler, K., Shchekin, O., and Dodabalapur, A. (2006) *Appl. Phys. Lett.*, **88**, 123502.
5 Sirringhaus, H., Kawase, T., and Friend, R.h. (2001) *MRS Bull.*, **26**, 539–543.
6 Gelinck, G.H., Huitema, H.E.A., van Veenendaal, E., Cantatore, E., Schrijnemakers, L., van der Putten, J.B.P.H., Geuns, T.C.T., Beenhakkers, M., Giesbers, J.B., Huisman, B.-H., Meijer, E.J., Benito, E.M., Touwslager, F.J., Marsman, A.W., van Rens, B.J.E., and de Leeuw, D.M. (2004) *Nat. Mater.*, **3**, 106–110.
7 Bronstein, H., Chen, Z., Ashraf, R.S., Zhang, W., Du, J., Durrant, J.R., Shakya Tuladhar, P., Song, K., Watkins, S.E., Geerts, Y., Wienk, M.M., Janssen, R.A.J., Anthopoulos, T., Sirringhaus, H., Heeney, M., and McCulloch, I. (2011) *J. Am. Chem. Soc.*, **133**, 3272–3275.
8 Yan, H., Chen, Z., Zheng, Y., Newman, C., Quinn, J.R., Dotz, F., Kastler, M., and Facchetti, A. (2009) *Nature*, **457**, 679–686.
9 Marks, T.J. (2010) *MRS Bull.*, **35**, 1018–1027.
10 Sze, S.M. and Ng, K.K. (2007) *Physics of Semiconductor Devices*, John Wiley & Sons, Inc.
11 Chen, Z., Lee, M.J., Shahid Ashraf, R., Gu, Y., Albert-Seifried, S., Meedom Nielsen, M., Schroeder, B., Anthopoulos, T.D., Heeney, M., McCulloch, I., and Sirringhaus, H. (2012) *Adv. Mater.*, **24**, 647.
12 Kim, F.S., Ahmed, E., Subramaniyan, S., and Jenekhe, S.A. (2010) *ACS Appl. Mater. Interfaces*, **2**, 2974–2977.
13 Meijer, E.J., de Leeuw, D.M., Setayesh, S., van Veenendaal, E., Huisman, B.-H., Blom, P.W.M., Hummelen, J.C., Scherf, U., and Klapwijk, T.M. (2003) *Nat. Mater.*, **2**, 678–682.
14 Babel, A., Zhu, Y., Cheng, K.-F., Chen, W.-C., and Jenekhe, S.A. (2007) *Adv. Funct. Mater.*, **17**, 2542–2549.

15 Szendrei, K., Jarzab, D., Chen, Z., Facchetti, A., and Loi, M.A. (2009) *J. Mater. Chem.*, **20**, 1317–1321.
16 Zaumseil, J. and Sirringhaus, H. (2007) *Chem. Rev.*, **107**, 1296–1323.
17 Ortiz, R.P., Facchetti, A., and Marks, T.J. (2009) *Chem. Rev.*, **110**, 205–239.
18 Bisri, S.Z. (2011) *Light-emitting transistors towards current-induced amplified spontaneous emission in organic single crystals*. Doctoral dissertation. Tohoku University, Sendai, Japan
19 Smits, E.C.P., Anthopoulos, T.D., Setayesh, S., van Veenendaal, E., Coehoorn, R., Blom, P.W.M., de Boer, B., and de Leeuw, D.M. (2006) *Phys. Rev. B*, **73**, 205316.
20 Bisri, S.Z. (2008) *Optoelectronic studies of organic single crystal transistors*. Master thesis. Tohoku University, Sendai, Japan
21 Loi, M.A., Rost-Bietsch, C., Murgia, M., Karg, S., Riess, W., and Muccini, M. (2006) *Adv. Funct. Mater.*, **16**, 41–47.
22 Street, R.A. and Salleo, A. (2002) *Appl. Phys. Lett.*, **81**, 2887.
23 Ishii, H., Sugiyama, K., Ito, E., and Seki, K. (1999) *Adv. Mater.*, **11**, 605–625.
24 de Boer, B., Hadipour, A., Mandoc, M.M., van Woudenbergh, T., and Blom, P.W.M. (2005) *Adv. Mater.*, **17**, 621–625.
25 Lei, C.H., Das, A., Elliott, M., Macdonald, J.E., and Turner, M.L. (2004) *Synth. Met.*, **145**, 217–220.
26 Anthopoulos, T.D., de Leeuw, D.M., Cantatore, E., van 't Hof, P., Alma, J., and Hummelen, J.C. (2005) *J. Appl. Phys.*, **98**, 054503.
27 Shkunov, M., Simms, R., Heeney, M., Tierney, S., and McCulloch, I. (2005) *Adv. Mater.*, **17**, 2608–2612.
28 Anthopoulos, T.D., Setayesh, S., Smits, E., Cölle, M., Cantatore, E., de Boer, B., Blom, P.W.M., and de Leeuw, D.M. (2006) *Adv. Mater.*, **18**, 1900–1904.
29 Walters, R.J., Bourianoff, G.I., and Atwater, H.A. (2005) *Nat. Mater.*, **4**, 143–146.
30 Misewich, J.A., Martel, R., Avouris, P., Tsang, J.C., Heinze, S., and Tersoff, J. (2003) *Science*, **300**, 783–786.

31 Burroughes, J.H., Bradley, D.D.C., Brown, A.R., Marks, R.N., Mackay, K., Friend, R.H., Burns, P.L., and Holmes, A.B. (1990) *Nature*, **347**, 539–541.
32 Friend, R.H., Gymer, R.W., Holmes, A.B., Burroughes, J.H., Marks, R.N., Taliani, C., Bradley, D.D.C., Santos, D.A.D., Bredas, J.L., Logdlund, M., and Salaneck, W.R. (1999) *Nature*, **397**, 121–128.
33 Zaumseil, J., Donley, C.L., Kim, J.-S., Friend, R.H., and Sirringhaus, H. (2006) *Adv. Mater.*, **18**, 2708–2712.
34 Chua, L.-L., Zaumseil, J., Chang, J.-F., Ou, E.C.-W., Ho, P.K.-H., Sirringhaus, H., and Friend, R.H. (2005) *Nature*, **434**, 194–199.
35 Chen, Z., Lemke, H., Albert-Seifried, S., Caironi, M., Nielsen, M.M., Heeney, M., Zhang, W., McCulloch, I., and Sirringhaus, H. (2010) *Adv. Mater.*, **22**, 2371–2375.
36 Kim, F.S., Guo, X., Watson, M.D., and Jenekhe, S.A. (2010) *Adv. Mater.*, **22**, 478–482.
37 Cho, S., Lee, J., Tong, M., Seo, J.H., and Yang, C. (2011) *Adv. Funct. Mater.*, **21**, 1910–1916.
38 Bürgi, L., Turbiez, M., Pfeiffer, R., Bienewald, F., Kirner, H., and Winnewisser, C. (2008) *Adv. Mater.*, **20**, 2217–2224.
39 Dodabalapur, A., Katz, H.E., Torsi, L., and Haddon, R.C. (1996) *Appl. Phys. Lett.*, **68**, 1108.
40 Sakamoto, Y., Suzuki, T., Kobayashi, M., Gao, Y., Fukai, Y., Inoue, Y., Sato, F., and Tokito, S. (2004) *J. Am. Chem. Soc.*, **126**, 8138–8140.
41 Rost, C., Karg, S., Riess, W., Loi, M.A., Murgia, M., and Muccini, M. (2004) *Appl. Phys. Lett.*, **85**, 1613.
42 Shi, J.W., Wang, H.B., Song, D., Tian, H.K., Geng, Y.H., and Yan, D.H. (2007) *Adv. Funct. Mater.*, **17**, 397–400.
43 Wang, H., Wang, J., Yan, X., Shi, J., Tian, H., Geng, Y., and Yan, D. (2006) *Appl. Phys. Lett.*, **88**, 133508.
44 Dinelli, F., Capelli, R., Loi, M.A., Murgia, M., Muccini, M., Facchetti, A., and Marks, T.J. (2006) *Adv. Mater.*, **18**, 1416–1420.
45 Liu, C. and Sirringhaus, H. (2010) *J. Appl. Phys.*, **107**, 014516.
46 Liu, C. and Sirringhaus, H. (2010) *Org. Electron.*, **11**, 558–563.
47 Cho, S., Yuen, J., Kim, J.Y., Lee, K., Heeger, A.J., and Lee, S. (2008) *Appl. Phys. Lett.*, **92**, 063505.
48 Wei, Q., Tajima, K., and Hashimoto, K. (2009) *ACS Appl. Mater. Interfaces*, **1**, 1865–1868.
49 Rost, C., Karg, S., Riess, W., Loi, M.A., Murgia, M., and Muccini, M. (2004) *Appl. Phys. Lett.*, **85**, 1613.
50 Unni, K.N.N., Pandey, A.K., Alem, S., and Nunzi, J.-M. (2006) *Chem. Phys. Lett.*, **421**, 554–557.
51 Inoue, Y., Sakamoto, Y., Suzuki, T., Kobayashi, M., Gao, Y., and Tokito, S. (2005) *Jpn. J. Appl. Phys.*, **44**, 3663–3668.
52 Arias, A.C., Corcoran, N., Banach, M., Friend, R.H., MacKenzie, J.D., and Huck, W.T.S. (2002) *Appl. Phys. Lett.*, **80**, 1695.
53 Tada, K., Harada, H., and Yoshino, K. (1997) *Jpn. J. Appl. Phys.*, **36**, L718–L720.
54 Tada, K., Harada, H., and Yoshino, K. (1996) *Jpn. J. Appl. Phys.*, **35**, L944–L946.
55 Anthopoulos, T.D., Tanase, C., Setayesh, S., Meijer, E.J., Hummelen, J.C., Blom, P.W.M., and de Leeuw, D.M. (2004) *Adv. Mater.*, **16**, 2174–2179.
56 Hayashi, Y., Kanamori, H., Yamada, I., Takasu, A., Takagi, S., and Kaneko, K. (2005) *Appl. Phys. Lett.*, **86**, 052104.
57 Naber, R.C.G., Tanase, C., Blom, P.W.M., Gelinck, G.H., Marsman, A.W., Touwslager, F.J., Setayesh, S., and de Leeuw, D.M. (2005) *Nat. Mater.*, **4**, 243–248.
58 Babel, A., Wind, J.D., and Jenekhe, S.A. (2004) *Adv. Funct. Mater.*, **14**, 891–898.
59 Cho, S., Yuen, J., Kim, J.Y., Lee, K., and Heeger, A.J. (2006) *Appl. Phys. Lett.*, **89**, 153505.
60 Aleshin, A.N., Shcherbakov, I.P., Petrov, V.N., and Titkov, A.N. (2011) *Org. Electron.*, **12**, 1285–1292.
61 Pal, B.N., Trottman, P., Sun, J., and Katz, H.E. (2008) *Adv. Funct. Mater.*, **18**, 1832–1839.
62 Babel, A. and Jenekhe, S.A. (2003) *Macromolecules*, **36**, 7759–7764.

17
Nanostructured Conducting Polymers for Sensor Development

Yen Wei, Meixiang Wan, Ten-Chin Wen, Tang-Kuei Chang, Gaoquan Shi, Hongxu Qi, Lei Tao, Ester Segal, and Moshe Narkis

17.1
Introduction

Since the discovery that a conjugated polyacetylene (PA) could achieve high electronic conductivity upon doping with iodine in 1977 [1], "conducting polymers" or "synthetic metals" have been established as a new class of functional materials. For this landmark discovery, Professors A. G. MacDiarmid, A, J. Heeger, and H. Shirakawa were awarded the Nobel Prize in Chemistry in 2000 [2]. In the past several decades, fundamental research on the synthesis of new materials, structural characterization, solubility and processability, structure–property relationship and conduction mechanism of the conducting polymers as well as their technological applications have been widely done and significant progress has been made [3]. As compared with other materials, conducting polymer not only shows unique electrical and redox properties via special doping processes, which are quite different from inorganic semiconductors, but also possesses light-weight and flexible characteristics of common polymers. Moreover, conducting polymers with tailored nanostructures not only have the unique properties of conducting polymers as mentioned above, but also have associated characteristics of nanomaterials, such as large surface, size, and quantum effects. These excellent properties give the nanostructured conducting polymers an important place in the field of nanosciences and nanotechnology.

The aim of this chapter is to provide a brief overview of various synthetical approaches to nanostructured conducting polymers and the development of these polymers for chemical and biological sensor applications. The references cited in this chapter are meant only to highlight specific aspects of conducting polymer nanostructures and their sensor applications. Some excellent books and reviews are available on more comprehensive descriptions of the fundamental aspects of conducting polymers and their nanostructures [3–5].

Semiconducting Polymer Composites: Principles, Morphologies, Properties and Applications, First Edition.
Edited by Xiaoniu Yang.
© 2012 Wiley-VCH Verlag GmbH & Co. KGaA. Published 2012 by Wiley-VCH Verlag GmbH & Co. KGaA.

17.2
Conducting Polymers and Their Nanostructures

As compared with other materials, conducting polymers have some unique characteristics. First of all, their structures consist of highly π-conjugated polymeric chain, leading to special electrical properties [6]. However, the π-conjugated polymer chain often results in insolubility in any organic solvent or aqueous solution, limiting their applications in technology. For instance, polyaniline (PANI) is one of the most promising conducting polymers from the point of view of application, but only limited commercial applications have been found due to poor solubility in common solvents. Thus insufficient processability of conducting polymers is still a challenge for practice application in technology. Another is of a transition of π-conjugated polymer from insulator to metal produced by a "doping" process. However, the "doping" process used in conducting polymers significantly differs from that of traditional inorganic semiconductors [4]. The main differences in the doping process between the conducting polymers and the inorganic semiconductors are summarized as follows.

i) The doping in conducting polymers is an oxidation (p-type doping) or reduction (n-type doping) process, rather than atom replacement in inorganic semiconductors. The doped π-conjugated polymer (i.e., conducting state) can be consequently reversed to insulator (i.e., insulating state) by a de-doping process. Such de-doping process never takes place in an organic semiconductor [3]. Interestingly, the doping/de-doping process in conducting polymers is reversible, which results in the conductivity of conducting polymers covering whole insulator–semiconductor–metal region.

ii) As in inorganic semiconductors, p-doping (withdrawing electron from polymeric chain) or n-doping (adding electron onto polymeric chain) in conducting polymers can take place. However, the doping process in conducting polymers is consequently accompanied by the incorporation of counterions (e.g., action for p-doping or anion for n-doping) into polymer chain to satisfy electrical nature. This is completely different from that of inorganic semiconductors, which are without counterions in the doped inorganic semiconductor. Thereby, the polymeric chain of conducting polymers actually consists of π-conjugated chain and incorporated counterions that result in the electrical properties of conducting polymers being affected by both the chain structure (i.e., π-conjugated length) and the nature of dopants [5]. Except for redox doping, proton doping discovered in PANI is another useful doping concept in conducting polymers [7]. The proton doping does not involve a change in the number of electrons associated with the polymer chain [7]. This is quite different from the redox doping (e.g., oxidation or reduction doping) where the partial addition (reduction) or removal (oxidation) of electrons to or from the system of the polymer backbone takes place [8]. Besides, photo-doping and charge-injection doping have also been employed [9].

iii) The degree of doping in conducting polymers is much greater than that in inorganic semiconductors. For instance, the doping degree in the doped PANI can be achieved up to as high as 50% [9]. The electrical and transport properties of conducting polymers caused by the special doping process are therefore expected to be different from either metal or inorganic semiconductors. For a semiconductor, for instance, electrons or holes are assigned as the charge carries and the transport properties are not only dominated by the minion charge carriers but also controlled by the width of the energy gap [5]. For a conducting polymer, on the other hand, solitons [10], polarons [11], and bipolarons [12] have been proposed to interpret conduction mechanism of conducting polymers via a doping process. The metal-like nature of conducting polymers has been demonstrated by a metal-like conductivity (10^2–10^3 S cm^{-1}) at room temperature, the optical properties, the thermo-electrical power, and the magnetic susceptibility [3, 5]. Temperature dependence of the conductivity of conducting polymers, as measured by a four-probe method, only shows a semiconductor behavior and obeys variable range hopping (VRH) model proposed by Mott [13] due to a large intercontact resistance in the interfebrile, intergranular, or intercrystalline regions of conducting polymers. A metallic temperature dependence of the conductivity for the doped PA [14], polypyrrole (PPy) [15], and polythiophene (PTH) [16] has been observed by using voltage shorted compaction (VSC) method proposed by Coleman [17]. However, the measurement of an intrinsic metallic temperature dependence of conductivity for conducting polymers is still desired because VSC method is only a qualitative method. Therefore to eliminate intercontact resistance in the interfebrile, intergranular, or intercrystalline regions are a key to show intrinsic metal-like nature of conducting polymers.

Applications of conducting polymers in technology are related to their properties as shown in Figure 17.1. In the semiconducting region, for instance, conducting polymer-based electronic devices such as Schottky rectifier, field-effect resistor, light-emitting diode (LED), and solar cell can be fabricated in the same way as inorganic semiconductors. In the metal-like region, conducting polymers are excellent candidates as electromagnetic interference (EMI) shielding, microwave absorbing materials, and conductive textiles. Combination of the reversible redox with a metal-like conductivity leads to the application of conducting polymers as rechargeable batteries and supper-capacitors. The color change induced by a doping/de-doping process can be used in the manufacturing of multi-chromic displays or electrochromic windows. Especially, the electrical properties sensitive to doping/de-doping process resulting in reversible change in the conductivity can be used to fabricate drug-releasing agents, gas separation membrane, and chemical or biochemical sensors. Detailed reviews on promising applications of conducting polymers are available in several handbooks on conducting polymers [3, 5].

Although conducting polymers have received great attention in electronic or electrochemical devices for displays, energy storage devices, actuators, and sensors [3, 5], as mentioned above, the interchange rate is usually slow (i.e., a few

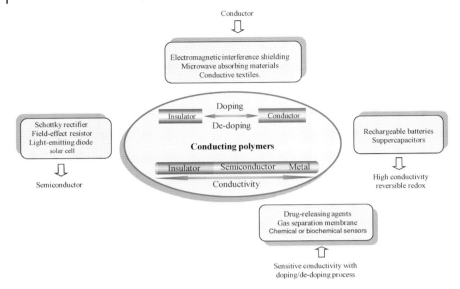

Figure 17.1 Some unique properties of conducting polymers and their relevance to technological applications.

hundred milliseconds) due to the rate-determining process of counterion transport into the polymer layer for charge balance so that fast charge/de-charge capability for application of conducting polymer-based sensors, electrochromic, superchromic, and supercapacitors are desired. One approach to increase the interchange rate is of tailoring the polymer structure to introduce molecular-scale porosity [18]. Another approach is to create nanostructures (e.g., nanotubes, nanowires, or random nanoporous structures) to provide intrinsically high surface area, leading to high charge/discharge capacities and short diffusion distances for ion transport [19]. Therefore, conducting polymer nanostructures plays an important role in the field of nanosciences and nanotechnology. Like bulk conducting polymers, the application of conducting polymer nanostructures in technology is also related to their electrical and transport properties. Thereby to measure intrinsic electrical and transport properties of conducting polymer nanostructures is a challenge. Presently, however, the pellet conductivity, as measured by the standard two- or four-probe method, has been commonly presented as the electrical properties for the conducting polymer nanostructures. Obviously, thus resultant electrical properties are often not intrinsic because of large interchain and interfiber, or intertube contact resistance involved [5]. That is why temperature dependence of the pellet conductivity, as measured by four-probe method, for conducting polymer nanostructures only shows semiconductor behavior [20] and obeys the variable range hopping model [13]. Martin [21], for the first time, measured the conductivity of a single nanofiber synthesized by a hard-template method that shows room-temperature conductivity of single nanofiber or nanotube is enhanced by one or two orders compared with that of the pellet nanotubes or nanofibers. Moreover, a size effect on the conductivity, which means the conductivity increases with decrease of the diameter, was observed. However, the conductivity was calculated by assuming that the number and diameter of the tubes were known, and that the resistance of the membrane as the template was neglected. The conductivity measured by this method may not be precise because of not knowing the number and the diameter of tubes or fibers in membrane. Thereby the measurement of the electrical properties for a single nanostructure (e.g., nanofibers or nanotube) has received great attention. For instance, Long *et al.* [22] recently measured the electrical properties of a single PANI nanotube by the four-probe method. The results showed the size effect on conductivity. However, the temperature dependence of the conductivity still exhibits a semiconductor behavior and obeys a 1D-VRH model [13]. Additionally, Zhang *et al.* [23] reported that the transport properties of single poly(3,4-ethylenedioxythiophene) (PEDOT) nanotube (50–100 nm in diameter) synthesized by a reverse emulsion polymerization also showed classical semiconductor behavior following the 3D-VRH model [13]. Similar size effect on conductivity for conducting polymer nanostructures has also been reported [24]. Elimination of such intercontact resistance between individual fibers and tubes to show pure metal-like behavior for conducting polymer nanostructures is still a big challenge.

The interfacial electronic states of the electron transport properties in conducting polymer nanostructures are a key factor in determining the performances of the devices. The current-sensing atomic force microscopy (CS-AFM) with a conducting tip is a technology at present to study electrical properties of conducting polymer films. It allows for not only the topographical and current images to be obtained simultaneously but also the current–voltage (I–V) traces to be recorded on selected spots of the image. This technique provides important information for the studies of doping distributions by obtaining two-dimensional current images and nanoscale electrical properties by measuring the I–V characteristics [25]. Park and coworkers used this technique in studying nanoscale electrical properties of the conducting polymers by measuring I – V characteristics [26]. Han *et al.* [27] also reported recently that the morphological and electrical properties of PEDOT films, as measured by CS-AFM, are very sensitive to the experimental conditions (e.g., preparation methods, solvents, and electrolytes). Besides, electronic transport properties of PEDOT, PPy, and PANI have also been reported [28].

17.3
Synthetical Methods for Conducting Polymer Nanostructures

Synthesis of conducting polymer nanostructures is a basic and important issue for the fundamental understanding and applications. Some important and common approaches are reviewed in the following sections focusing on hard- and soft-template methods as well as electrospinning technology, for the preparation of conducting polymer nanostructures. Figure 17.2 gives a schematic summary of those common methods for the synthesis of conducting polymers with nanostructures.

17.3.1
Hard-Template Method

Hard-template and soft-template methods are major approaches to prepare conducting polymer nanostructures [5]. As illustrated in Figure 17.2, in the hard-template method, a porous membrane is required as the template, which guides the growth of the nanostructures within the pore of the membrane [21], leading to a well-controlled morphology and size of the nanostructures. This is a major advantage of the hard-template method as compared with other methods. On the other hand, the template often needs to be removed after polymerization in order to obtain pure nanostructures. The post-process of template removal not only results in a complex preparation process, but also often destroys the nanostructures. As a result, some simplifications of hard-template method have been reported [29]. For instance, a new approach of oxygen plasma etching to effectively expose the ends of a gold nanowire template, synthesized by using a polycarbonate membrane as a hard template, has been reported by Martin and coworkers [29a]. Wan group also discovered that octahedral or spherical cuprous oxide (Cu_2O) could be used as a template to synthesize corresponding hollow octahedrons or microspheres of PANI [29b]. Interestingly, the morphology and conductivity of the resultant hollow microstructures are controlled by the molar

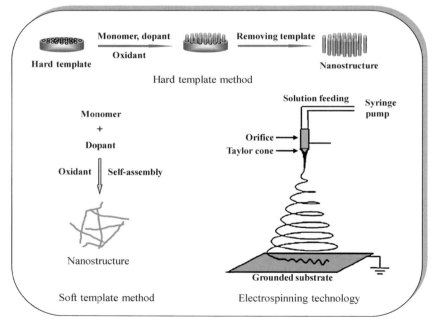

Figure 17.2 Schematic illustration of common methods, such as hard-template, and soft-template methods as well as electrospinning technology, for the preparation of conducting polymer nanostructures.

ratio of ammonium peroxydisulfate (APS) to aniline when the octahedral cuprous oxide as the hard-template [29c]. This method eliminates the need for post-synthesis template removal because the soluble Cu^{2+} ions produced by the reaction of Cu_2O with APS could be readily washed away. This method is expected to be applicable to other conducting polymer nanostructures because many conducting polymers are polymerized by using APS as oxidant.

It is well known that a porous membrane as the hard-template plays a role of controlling morphology of the resulting nanostructures. Alumina membranes (Al_2O_3) fabricated by an electrochemical approach [30] and PC membranes by a "track-etch" method [31] are widely used as commercial membranes. In addition, porous silicates such as MCM-41, carbon nanotubes, electrospun polymer fibers, highly oriented pyrolytic graphite (HOPG), and DNA have also been employed as the hard-templates to synthesize conducting polymer nanostructures [32]. Besides, nanofiber seedings such as PANI, hexapeptide, and V_2O_5 nanofibers have been employed to prepare 1D nanofibers of conducting polymers as the hard-template [33]. On the other hand, most of the pores in membranes or seeding fibers are of tubular or spherical shape, whereas membranes with a more complex morphology are lacking. Wan group [34] recently developed a novel approach of aniline/citric acid (CA) salt as the hard-template through a gas/solid reaction to synthesize brain-like nanostructures of PANI with many convolutions (140–170 nm in average diameter). This method not only differs from the traditional polymerization in acidic solution, but also provides a novel brain-like morphology.

The hard-template method usually includes chemical and electrochemical methods. Chemical hard-template method is accomplished by simply immersing the membrane into a solution of the desired monomer and its oxidizing agent, and then allows monomer polymerization within the pores reserving as "nanoreactor" [21]. By controlling the polymerization time, tubules with thin walls for short polymerization times or thick walls, even fibers, for long polymerization times can be produced [21]. However, the diffusion of monomer and oxidant into those pores in the membrane is generally limited by capillary phenomenon produced by nanoscaled diameter of the pores. A "molecular anchor," which interacts with the material being pre-deposited on the pore wall, is generally used to overcome capillary phenomenon [21]. For an electrochemical hard-template synthesis, a metal film coated on one surface of the membrane is required in order to carry out electrochemical polymerization of the desired polymer within the pores of the membrane [35]. As compared with chemical hard-template method, although synthesis of the electrochemical hard-template is more complex and expensive, the morphology and diameter of the resulting nanostructures are easily controlled by changing current density, applied voltage, and polymerization time. However, large scaled synthesis for both chemical and electrochemical hard-template method is almost impossible due to the limitation of template size.

Martin [21] proposed a mechanism for the growth of the nanostructures prepared by the hard-template method. However, the mechanism is insufficient in explaining the growth of partially filled nanotubes by an electrochemical template synthesis [36]. Recently, Lee and coworkers [37] investigated the electrochemical synthetic mechanisms of conductive polymer nanotubes in a porous alumina template by using PEDOT as a model compound. They found the growth mechanism is affected by the applied potentials, showing that the mechanism of the PEDOT nanotubes formed at the potentials below 1.4 V is dominated by the electrochemically active site works, while the mechanism is based on diffusion and reaction kinetics works at the potentials above 1.4 V.

17.3.2
Soft-Template Method

Soft-template method is another powerful approach to synthesize conducting polymer nanostructures. As compared with hard-template method, this method does not need a porous membrane and the growth of the nanostructures is a self-assembly process as shown in Figure 17.2 [5]. Moreover, Shi and coworkers [38] recently synthesize a variety of PPy microstructures (e.g., bowls, cups, and bottles) by using "soup bubbles" as the soft-template. In the soft-template method, hydrogen bonding, π–π stacking, van der Waals forces, and electrostatic interactions are usually served as the driving forces for the self-assembly of nanostructures [39]. As a result, the soft-template method is simpler than the hard template method because of elimination of the membrane as a hard template and the post-synthesis template removal. However, the absence of the hard-template in the soft-template method often results in poor controllability in morphology and size of the resulting nanostructures as compared with hard-template method. Nevertheless, the soft-template method utilizes molecular interactions as the driving forces might open a new door for preparing complex three dimensional (3D) micro/nanostructures via self-assembling rather than mostly one-dimensional (1D) nanostructures as by the hard template method. Hagner et al. [40] recently developed a polyelectrolyte-based preparative route to prepare PEDOT nanowires by using $FeCl_3$ as an oxidant in the presence of a poly(acrylic acid) (PAA). They found the morphology is strongly affected by the molar ratio of PAA to EDOT. When the molar ratio of PAA to EDOT is increased up to 4 : 1, for instance, a large quantity of flower-like nanostructures is obtained, while cage-like in shape was observed when the molar ratio is decreased down to 1 : 10. A cooperation effect of PAA as the template, the strong $\pi - \pi$ interactions between the PEDOT coated on the individual PAA chains, and electrostatic attraction between positive charged PEDOT chains and negative charged PAA polymer chain have been proposed to explain variation in morphology of the PEDOT-PAA nanostructures with the molar ratios [40].

17.3.3
Electrospinning Technology

The electrospinning technique patented in the 1930s [41] has recently received intense interest in the preparation of polymer fibers with length of 100 m or longer and diameters in the range of 30–2000 nm. As shown in Figure 17.2, a high electrical field in this technology is generally applied between a polymer fluid contained in a glass syringe with a capillary tip and a metallic collection screen. When the applied field reaches a critical value, the charge overcomes the surface tension of the deformed drop of the suspended polymer solution from on the tip of the syringe and a jet is produced. The dry fibers are accumulated on the surface of the collection screen, resulting in a nonwoven mesh of nano- to microdiameter fibers. The morphology and diameter of the fibers are influenced by preparation parameters such as the applied voltage, solution concentration, polymer molecular weight, solution surface tension, dielectric constant of the solvent, and solution conductivity [42]. As compared with other methods, long fibers and nonwoven meshes of polymers can be fabricated by the electrospinning technique. However, a soluble polymer with a properly high viscosity is required. Pure conducting polymer nanofibers are difficult to be fabricated by the electrospinning method due to a low viscosity of the conducting polymers dissolved in organic solvent. Thereby, only composite nanofibers of conducting polymers with other nonconductive polymers are fabricated by using electrospinning method [43]. In order to overcome this problem, electrospun polymer fibers are used as templates to prepare tubular materials with controlled dimensions [44]. In this route, the electrospun polymer nanofibers are used as the template and conducting polymers are then coated on the surface of the core fibers by *in situ* deposition polymerization from solution containing monomer and oxidant. By using this method, coaxial conducting polymer composite nanofibers and tubular conducting polymers after dissolving the core polymer can be fabricated. The selection of the core polymer to be used as the fiber template is therefore critical to the process of the tubular materials. Another key requirement is that the decomposition temperature of the core fibers should not exceed the threshold to cause structural damage to the outer shell material [45].

17.4
Typical Conducting Polymer Nanostructures

Despite significant efforts, the types of conducting polymers are limited to PHT, PPy, PA, PANI, PEDOT, polyphenylenes (PP), poly(*p*-phenylene vinylene) (PPV), and a few others. Here, some significant results on PANI, PPy, and PEDOT conducting polymers and their nanostructures are briefly reviewed as follows.

17.4.1
Polyaniline (PANI)

Alan G. MacDiarmid and his colleagues reported that aniline monomer in an acid aqueous solution (e.g., 1.0 M HCl) could be chemically oxidized by APS to form a green powder with a conductivity as high as ~ 3 S cm^{-1}, which was latter called the emeraldine salt (ES) form of PANI [46]. PANI is easy to synthesize and at a low cost and has great potential for modification of the molecular structure, and undergoes a special proton doping mechanism compared with other conducting polymers. In particular, its chemical and physical properties are controlled by both the oxidation state and protonation levels [5]. These unique properties have led to PANI being one of the most excellent conducting polymers and most promising for potential applications. The base form of PANI consists of an alternating reduced and oxidized repeat unit chain that is divided into the completely reduced form (the "leukoemeraldine" base form, LEB, and $y=1$), the half-oxidized form (the emeraldine base form, EB, and $y=0.5$), and the completed oxidation state (the "pernigraniline" form, PEN, and $y=0$), where the oxidation states are varied with the of value y between 1 and 0, respectively [5c]. This indicates that there is more room to modify the molecular structure that is different from other conducting polymers. In addition, PANI was the first sample of a conducting polymer doped by protonation, which results in a change from the insulating EB to the conducting ES form [46]. As mentioned before, the proton doping does not involve a change in the number of electrons associated with the polymer backbone and results in the formation of a delocalized poly-semiquinone radical cation [46, 47]. Besides, PANI has some advantages of low cost, easy synthesis, high yield, and changeable physical properties (e.g., optical, electrical, and magnetic properties) and their corresponding effects (e.g., Schottky and thermochromic effect and photoemission effect) [5c]. However, insufficient thermal stability (i.e., PANI cannot be melt-processed like most thermoplastics), poor mechanical properties, limited solubility, and synthesis in an acidic condition are obvious disadvantages of PANI. Many attempts including modification of the polymer with various ring or N-substitutes, post-treatment of the polymer with fuming sulfuric acid, and self-doped methods have been used to improve the processability of PANI have been reviewed [5]. However, to improve the solubility in organic solvent and the processability of PANI remains a challenge in commercial applications.

The novel reversible doping/de-doping process is a basic requirement in some applications (e.g., rechargeable batteries and sensors). However, de-proton doping process often results in the loss of electrical properties that limit the commercial applications of PANI in technology. Moreover, the conducting state of PANI (i.e., the emeraldine salt form, ES) can be only synthesized in an acidic solution [9]. Since the activity of biomolecules is retained only in a neutral condition, the acidic condition is unfavorable for biomaterials, limiting application of PANI as the biosensors. Thus, the use of enzymes as chemical catalysts in the synthesis of PANI has attracted great interest [48]. The enzymatic approach is environmentally benign, can offer a higher degree of control over the kinetics of the reaction, and has the potential of product high yield. A major drawback of enzymatic polymerization, however, has been that as soon as polymer begins to form in aqueous solutions, it precipitates out and only very low molecular weight polymers (i.e., oligomers) are formed [49]. To address this problem and to improve processability, a variety of modified enzymatic polymerizations have been employed [50]. Furthermore, the toxicity of the aniline monomer is a concern with regard to an environmentally friendly synthesis (i.e., green chemistry), and needs to be addressed for commercial applications.

Till date, liquid-crystalline phases [51], colloidal particles [52], and structure-directing molecules [53] as the soft-template have been employed to synthesize PANI nanostructures. Based on the traditional synthesis method of PANI, in particular, some simple approaches such as interfacial polymerization [54], mixed reactions [55], dilute polymerization [56] and ultrasonic irradiation [57] have also been employed to synthesize PANI. The interfacial polymerization method only allows the oxidative polymerization of aniline to take place at the interface of the organic/water phases and the product directly enters into the water phase, which could facilitate environmentally friendly processing.

Wan group accidentally discovered that PANI microtubes could be prepared by a common *in situ* doping polymerization in the presence of β-naphthalene sulfuric acid (β-NSA) as the dopant without using any membrane template [58]. This approach was called a "template-free" method because it did not use a porous membrane as the hard template. After discovering this new method, they extensively studied its universality, controllability, and self-assembly mechanism by changing the polymeric chain length, the polymerization method, the dopant structure, and reaction conditions. A detailed discussion is given in Ref. [5]. In particular, the diameter of the synthesized PANI nanostructures by the template-free method could be adjusted by the dopant structure, the molar ratio of dopant to aniline, and reaction conditions that have been reviewed in details [5c]. Furthermore, a micelle composed of dopant, dopant/aniline salt, or even aniline itself has been proposed as a soft-template in the formation of the PANI micro/nanostructures by a self-assembly process [59], as shown in Figure 17.3. The micelle model has been demonstrated by means of dynamic light scattering (DLS) [59] and freeze-fracture transmission electronic microscopy (FFTEM) [60]. The growth of the template-free synthesized nanostructures is a self-assembly process by an accretion and elongation process [61]. In particular, the template-free method has been further simplified, called simplified template-free method (STFM), by using APS as both oxidant

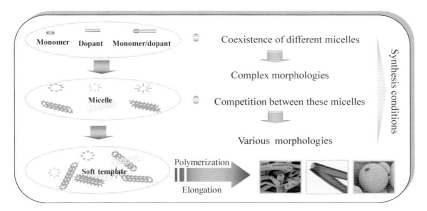

Figure 17.3 Micellar soft-templates for self-assembling conducting polymer nanostructures via a "template-free" process.

and dopant because H_2SO_4 as a dopant could be produced by a reaction of aniline with APS [62]. The STFM has been further proven by the formation of PANI nanofibers with a well-controlled diameter using different redox potential of oxidants in the absence of acidic dopant [63]. It indicates that change in the redox potential of the oxidants is a powerful way of controlling the diameter of the self-assembled PANI nanofibers. To the best of our knowledge, STFM is the most facile approach to synthesize PANI nanostructures at the current time because it not only avoids the use of a hard template and post-synthesis template removal, but also simplifies the reagents required. It is reasonable to believe that STFM is able to expand to synthesis of other conducting polymer nanostructures because APS is a common oxidant for almost all conducting polymers discovered so far.

Since the formation of the micelle-soft-template is strongly affected by the nature of polymeric chain and dopant as well as polymerization conditions, the structure of micelle-soft templates formed in a reaction solution can vary [5c]. Moreover, the micelle-soft template and the molecular interactions as the driving forces coexist in the reaction solution, resulting in cooperation between them that might be employed to complex micro/nanostructures of PANI via the self-assembly process. This prediction has been confirmed by the formation of hollow rambutan-like spheres [64], hollow dandelion-like microstructure [65] and hollow cube box-like 3D microstructures of PANI [66] as shown in Figure 17.4. These complex 3D micro/nanostructures are self-assembled from 1D nanofibers and show electrical and supper-hydrophobic properties. The trick is to use perfluorooctane sulfuric acid (PFOSA) or perfluorosebacic acid (PFSEA) as the dopant, which has doping,

Figure 17.4 Three-dimensional multifunctional microstructures of PANI self-assembled from 1D nanofibers by a cooperative effect of micelles as soft-templates and molecular interactions as driving forces [64–67].

soft-template, and hydrophobic functions at the same time. Another sample is rose-like 3D microstructures of PANI, which are self-assembled from 2D plates constructed with 1D nanofibers, synthesized by using APS as both oxidant and dopant at a high relative humidity (80%) [67], where micelle soft-template cooperated with a hydrogen bonding produced by a related high humidity as the driving forces [68]. From above results, one can see that cooperation of micelle soft-template and molecular interaction as the driving force is a powerful approach to synthesize complex 3D micro/nanostructures self-assembled from 1D or 2D nanostructures.

Kaner and coworkers [69], for the first time, found that when a porous film made of PANI nanofibers is illuminated (e.g., a camera flash), the generated heat will lead to rapid "melting" to form a smooth and continuous film. Based on this finding, they prepared the asymmetric PANI nanofiber films by flash welding that has been developed as monolithic actuators [70]. Kaner and coworkers [71] also developed an interfacial rout to prepare PANI composite nanometers or films by simply mixing nanofibers with other water-soluble functional materials such as nanoparticles, or polymers in water. Based on the above-given idea, Au, Pt, and Pd salts can be reduced by the emeraldine base form of PANI to produce metal nanoparticles inside or on the surface of the nanofibers, while PANI is oxidized to a higher oxidation state (i.e., pernigraniline) and can then be reduced back to the emeraldine oxidation state [72]. By the repetition of this process or adjustment of concentration, temperature and reaction time, the size and density of the deposited nanoparticles can be controlled. Furthermore, a combination of different metals can be introduced into the same fibers by sequentially exposing the fibers to different metal ions. Gallon *et al.* [72] also found that palladium nanoparticles supported on PANI nanofibers prepared by exposing PANI nanofibers to a palladium salt can be used as an active catalyst for Suzuki coupling between aryl chlorides and phenylboronic acid in water at relatively low temperatures. The PANI nanofibers play a multifunctional role of reducing agent to produce Pd nanoparticles, a stabilizer to prevent the agglomeration of Pd nanoparticles and a water-dispersible catalyst support at the same time. Recently, aligned PANI nanofibers synthesized by a step-wise electrochemical process [74] and PANI arrays synthesized by dilute chemical oxidative polymerization [75] have been reported.

17.4.2
Poly(3,4-ethylenedioxythiophene) (PEDOT)

PEDOT is a unique conducting polymer with a small band gap, high optical transparency, and conductivity in the doped state [76], whose molecular structure is depicted in Figure 17.3. Poly(3,4-ethylenedioxythiophene)-poly(styrenesulfonate), known as PEDOT-PSS, is today probably the most successful conductive polymer in commercial applications [77]. It is obvious that transparency is required for optoelectronic devices including light-emitting diodes and photovoltaic cells. Although metal oxides (e.g., indium tin oxide (ITO)) are widely used as conventional transparent electrode, they are limited by the abundance of such metals in earth and the rigidity of the metal oxides [78]. Transparent electrode made of cheap and transparent thin films with high conductivity and high mechanical flexibility is therefore highly desired. PEDOT-PSS emerged as a promising conducting polymer to replace ITO in the optoelectronic application because of its high transparency in the visible light range, solution processability, high mechanical flexibility, and good thermal stability [76]. Although PEDOT-PSS has been extensively used in the optoelectronic devices as the electrode material [79], its conductivity is much lower than that of ITO [80], adversely affecting the application of PEDOT-PSS in many aspects. Much effort has been made to improve the conductivity of PEDOT- PSS by adding high-boiling-point polar organic compounds into the PEDOT-PSS aqueous solution or treating the PEDOT-PSS film with polar solvent (e.g., ethylene glycol or dimethyl sulfoxide) [81]. Moreover, the conductivity can be significantly enhanced by adding anionic surfactant into the PEDOT: PSS aqueous solution and the conductivity enhancement was attributed to the effect of the anionic surfactant on the conformation of the conductive PEDOT chains, which means that the PEDOT chain follows the structure of the PSS chain in water, while the anionic surfactant replaces PSS as the counteranions to PEDOT in water [82].

Ionic liquids are organic/inorganic salts with a good chemical stability, low flammability, negligible vapor pressure, and high ionic conductivity that has been used in the synthesis of self-assembled nanostructures [83], electrochemical synthesis [84], and as electrolytes in conducting polymer-based electrochemical devices [85]. For instance, the conductivity of the PEDOT-PSS could be enhanced by adding ionic liquids as the additives and the enhancement of the conductivity is affected by the nature of the used ionic liquids [86]. Till now, interfacial polymerization, Al_2O_3, and electrospun nanofibers as the template, reverse cylindrical micelle-mediated interfacial polymerization, and V_2O_5 seeding approach have been widely employed to prepare PEDOT nanostructures [87].

17.4.3
Polypyrrole (PPy)

As shown in Figure 17.3, PPy is one of the most important conducting polymers because of its high stability, electronic conductivity, ion exchange capacity, and biocompatibility [88]. 1D nanostructured PPy has also been synthesized by using hard template or soft-template [89] and magnetic ionic liquid [90]. Recently, it was reported that a super-hydrophilic PPy nanofiber network could electrochemically be synthesized in an aqueous solution by using phosphate buffer solution (PBS) in the absence of templates, surfactants, and structure-directing molecules. This is a simple, environment-friendly and one-step approach to fabricate a PPy film with unique nanostructure. It was also proposed that the presence of the hydrogen bonding between phosphate and PPy oligomers is essential to produce the 1D nanofibers, and the electrostatic interactions between ionic dopant and PPy oligomers lead to an irregular nanostructure.

Alignment of nanostructures, providing the ordering at a large scale, is one of the key factors to obtain the high performance of functional nanomaterials. However, most resultant conducting polymer nanostructures synthesized at the current time are disordered. Although hard-template method is a common and efficient approach to synthesize alignment nanostructures of conducting polymers, costs of template are high and handling processes are relative complex. Especially removing the template often destroys the alignment and the property of nanostructures. Wei and coworkers [92] recently exposed a facile method, which is electrochemical polymerization combined with a chemical interfacial polymerization, to synthesize PPy nanofiber arrays. The biphasic method ensures a very low concentration of pyrrole monomer in the polymerization process, while electropolymerization provides good controllability. In addition, a new one-step electrochemical strategy for direct formation of oriented PPy nanowire array has been recently reported [93]. They proposed that the formation of the oriented PPy nanofibers is related to the water oxidation that leads locally to the formation of O_2 nanobubbles, which protect the PPy film against hydroxyl radicals (i.e., against overoxidation) allowing the electropolymerization of pyrrole to continue growing nanowires.

17.5
Multifunctionality of Conducting Polymer Nanostructures

Multifunctional nanocomposites are a special class of materials, which are originated from suitable combinations of two or more nanomaterials, have received great attention due to their unique physical properties, and have wide application potential in diverse areas. The design and synthesis of multi-functionalized micro/nanostructures are therefore necessary for realizing their practical applications in nanodevices. Functional inorganic nanoparticle (e.g., magnetic ferric oxides or optical titanium dioxide, TiO_2) as the hard-template is a common method to prepare 1D-multifunctionalized nanostructures of conducting polymers [94]. Moreover, soft template-guided method is another common approach to synthesize 1D-nanostructured conducting polymers [95]. However, both methods are difficult to prepare 2D- or 3D-nanostructured conducting polymers due to the limitation of the porous morphology of the hard-template. For a soft-template, as mentioned above, the formation of micelle soft-template is strongly affected by the nature of polymeric chain and dopant as well as polymerization conditions [5c], resulting in a variety of the possible structures of micelle soft-templates. Coordination of such characters of micelle soft-template with molecular interaction as the driving forces could be used to synthesize complex and multi-functionalized 3D-micro/nanostructures assembled from 1D nanostructures via a self-assembly process. Electrosuperhydrophobic hollow rambutan-like spheres, hollow dandelion-like, and hollow cube box-like 3D microstructures self-assembled from 1D nanofibers are typical samples of using cooperative effect as the powerful approach [64–66]. Moreover, the micelle soft-template associated with other approaches, such as hard-template, functional dopant induced method, and electromagnetic interaction as the driving force can be used to a variety of multifunctional micro/nanostructures of conducting polymers [5c]. For example, the electromagnetic functional nanotubes or nanofibers of PANI have been synthesized by a micelle soft-template associated with ferric oxides (Fe_3O_4 or γ-Fe_2O_3) as the hard-template [96]. Electromagnetic functional core-shell structured micro/nanostructures of PANI or PPy, where spherical carbonyl ion magnet is served as the core, whereas the self-assembled PANI or PPy nanofibers formed on the surface of the core are used as the shell, have been also synthesized by a coordination of micell soft-template with carbonyl ion magnet as hard-template [97]. Interestingly, cage-like electromagnetic functionalized microstructures of PANI, where the octahedron $CoFe_2O_4$ magnet as the hard-template while the self-assembled PANI nanofibers around the magnet through an electromagnetic interaction have been prepared by a micelle soft-template associated with the electromagnetic interaction [98]. In general, the magnetic properties of the electromagnetic functionalized composite nanostructures increase with an increase in the content of the magnetic particles, whereas the conductivity decreases with an increase in the content of magnetic nanoparticles as a result of insulating the magnetic particles [96]. However, a maximum conductivity has been observed in the above core–shelled PANI-carbonyl ion micro/nanostructures [97] or cage-like PANI-$CoFe_2O_4$ micro/nanostructures [98]. The reason may be due to decreasing the intercontact resistance resulted from the conductive nanofibers coated on the surface of the magnetic carbonyl ion [97] or electromagnetic interaction between the spin-polar and magnetic $CoFe_2O_4$ nanoparticles [98]. However, the magnetic micro- or nanoparticles in all the above samples are pre-prepared and serve as the hard templates or additives.

As is well known, $FeCl_3$ is widely used as the oxidant for the preparation of both conducting polymers and magnetic ferric oxides (e.g., Fe_3O_4 or γ-Fe_2O_3). On the other hand, the preparation condition for conducting polymers and ferric oxides is significantly different. For ferric oxides, a basic condition is required. In contrast, polymerization of conducting polymers (e.g., PANI) is usually carried out in an acidic condition. Based on the facts described above, one is unable to synthesize conducting polymers accompanied by the formation of ferric oxides to form electromagnetic functionalized conducting polymers at the same time. However, the electromagnetic functional PANI-γ-Fe_2O_3 fibers have been prepared by a coordination reaction of $FeCl_3$ as oxidant for both conducting polymers and ferric oxides [99]. The trick to this method was to adjust the concentration of Fe^{3+} and Fe^{2+} and the molar ratio of aniline to APS. Interestingly, the electrical properties are controlled by the concentration of Fe^{3+} ions, whereas the ferromagnetic properties are adjusted by the concentration of the Fe^{2+} ions [99]. This method not only is the simplest approach to prepare electromagnetic PANI nanostructures at the current time, but also might be expanded to other electromagnetic conducting polymers because $FeCl_3$ is a common oxidant for conducting polymers.

Electro-optic composite nanostructures or chiral nanostructures of conducting polymers have received great attention in the application of electro-optic nanodevices [100]. Coordination of micelle soft-template with functional dopant induced method is a simple and an efficient approach to prepare electro-optical nanostructures of conducting polymers. However, the dopant selected in this method must be of doping, soft-template and bring about optic function at the same time. Electrophotoisomerization nanotubes of PANI, by using this method, have been self-assembled in the presence of azobenzene sulfuric acid (ABSA) as the dopant [101], where the SO_3H group of the ABSA plays a role of dopant and soft-template at the same time, while the photoisomerization function of the PANI-ABSA nanotubes is induced by the ABSA. Similarity, chiral PANI nanotubes induced by chiral camphorsulfonic acid (CSA) or PCA have been prepared [102] as the chiral dopant.

High electrical, thermal conductivities and unique optical properties of gold nanostructures have been widely applied as substrates for surface-enhanced Raman spectroscopy [103], efficient catalysts for chemical reactions [104], and as probes for detecting nuclei acids, proteins, and biologically relevant small molecules [105]. Conducting polymer nanostructures containing Au nanoparticles have recently received great attention [106]. Many novel three-dimensional conducting polymer nanostructures containing Au nanostructures synthesized through a one-step process have been reported. For instance, Shi and coworkers [107] reported three-dimensional dendritic Au nanostructures (3D-DGNs), which were synthesized at room temperature by an interfacial reaction between $HAuCl_4$ in aqueous solution and EDOT in organic solution, and can be used as substrate for Raman scattering. Moreover, PEDOT-silica colloidal composite has been synthesized by a template approach using mesoporous silica spheres (MSS) [108]. Wolf and coworkers [109] also reported a series of conjugated polymer-*meso*-porous silica composites synthesized by a template method.

Although multi-functionality of conducting polymers could be realized by a variety of methods as described above, facile, efficient, and controllable synthesis for multi-functional conducting polymer nanostructures is still in need. Moreover, the previous papers paid most attention to multi-functionality, whereas the study of its size effect on the physical properties is lacking.

17.6
Conducting Polymer-Based Sensors

Among various applications of conducting polymer nanostructures as electronic devices, conducting polymer nanostructures have shown promise as sensor materials because of their reversible electrical properties induced by doping/de-doping process [3], mechanical flexibility, and ease of preparation [110]. It has been demonstrated that the controllable geometry and high surface-to-volume ratio associated with the nanowires play an important role in improving response and increased sensitivity [111]. These properties can be utilized and developed into optical and chemical sensors as well as biosensors [112] as shown in Figure 17.3. In general, high selectivity and short response time are required in the conducting polymer-based sensors. The ultra-thin-film composite membrane is a promising and versatile approach for sensor design. Two types of sensors based on ultra-thin film can be conceptualized. The first type is of molecule-recognition chemistry directly quilted into the ultra-thin film, whereas the second type is based on a function of "pre-filter" that would transport the analyses molecules into an internal sensing solution [113]. Organic thin-film transistors (OTFTs) are field-effect devices with organic or polymer thin-film semiconductors as channel material that can act as multi-parametric sensor with remarkable response repeatability, and as semi-conducting polymer-based sensing circuits [114]. Here, various conducting polymer-based sensors, such as gas, pH, and biosensors are reviewed in the following sections.

17.6.1
Gas Sensors

One-dimensional (1D) nanomaterials, such as nanowires, nanorods, nanotubes, and nanofibers are of interest in the sensor applications because of their ultra-small size and high surface area features, which offer great promise for gas sensors and biosensors [115]. Recently, PANI nanofibers synthesized by an interfacial polymerization were found to respond much faster than the conventional film for both acid doping and de-doping. [54]. Interestingly, the nanofiber films show essentially no thickness dependence (0.2–2.5 μm) in their performance. Virji et al. [116] also reported the electrical resistance of the nanofibers as a function of time when the materials are exposed to hydrochloric acid (HCl), ammonia (NH_3), hydrazine (N_2H_4), chloroform ($CHCl_3$), and methanol (CH_3OH). They found that in all cases the response time and extent of response are significantly better for PANI nanofiber films than that observed for PANI films produced by conventional method although each gas produces a different response mechanism consisting of protonation, de-protonation, reduction, swelling, and conformational alignment, respectively. Moreover, Gao et al. [117] reported that the PANI nanotubes synthesized by using electrospun poly(vinyl alcohol) (PVA) fiber as the template can be also used as gas (e.g., NH_3, N_2H_4, and $(C_2H_5)_3N$) sensors. PANI nanowires prepared by chemical synthesis [118] or by electrospinning technique [119] have been incorporated into interdigitized electrodes to prepare gas sensors with an excellent sensitivity. PANI nano-framework–electrode junctions (PNEJs) have been also prepared by electrochemical polymerization at low and constant current [120]. In addition, the gas sensors of PEDOT nanorods (conductivity ~ 72 S cm^{-1}) synthesized by interfacial polymerization using a reverse micelle as a soft-template response to NH_3 and HCl vapor concentration as low as 10 ppm and 5 ppm have been also reported, in which sensors display good reproducibility and reversibility in response [121]. These results indicate that the high surface area, small nanofiber diameter, and porous nature of the nanofibers give significantly better performance in both sensitivity and response time.

The detection of NH_3 in air is of interest for environmental monitoring and process control. PANI can be used as NH_3 sensors due to following facts. First, it has a high affinity for NH_3, resulting from the similarity of the coordinative roles of nitrogen atoms in PANI and NH_3 [122]. Another is that PANI exhibits p-type semiconductor characteristics. Consequently, electron-supplying NH_3 gases reduce the charge-carrier concentration and the conductivity of PANI [122]. Based on the above concept, a nonlithographic deposition process to form single PANI-PEO nanowire (100 nm in diameter) chemical sensors showed a rapid and reversible resistance change upon exposure to NH_3 gas at concentrations as low as 0.5 ppm [122]. Besides, PANI has been used as a sensing material for a variety of toxic gases such as CO and NO_2 [123].

Although conducting polymer nanostructures have advantages of facile preparation, design flexibility, high and tunable conductivity, and environmental stability, they also have some shortcomings such as low mechanical strength and low sensitivity for gas sensing. Much effort has been made to the synthesis and fabrication of organic–inorganic hybrid materials, expecting to find new properties or applications over their single components. Hybrid materials of conducting polymers and semiconductor metal oxide have been widely studied as promising sensing materials due to their unique sensing abilities [124]. Wu *et al.* [125] recently reported that core-shell hybrid materials with SnO_2 hollow spheres as the core and PPy as the shell prepared by an *in situ* chemical oxidative polymerization of pyrrole in the presence of pre-prepared SnO_2 hollow spheres exhibited fast response and high sensitivity to NH_3 at room temperature with a detection limit at the ppm level.

Control of surface wettability has received considerable attention because of its promising applications in intelligent devices based on reversible switching between super-hydrophobic and super-hydrophilic [126]. The wettability of conducting polymers has recently received attention because it is a crucial parameter for controllable separation and for conducting polymer-based sensors [127]. For instance, PANI coated on the surface of fabrics by *in situ* doping polymerization in the presence of perfluorosebacic acid (PFSEA) as the dopant represents fast and reversible switching from super-hydrophobic to super-hydrophilic triggered by ammonia gas [128]. It was found that change in the normalized contact angle is a function of the concentration of ammonia gas, which indicates that the ammonia-responsive reversible wettability allows PANI-PFSEA-coated fabric to be used as a sensor to monitor ammonia gas [128]. The above results reveal a new concept of preparing conducting polymer nanostructure-based sensors with a fast response, stability, flexibility, and good reproducibility by changing wettability from super-hydrophobic to super-hydrophilic by a doping/de-doping process. However, a high sensitivity and reversibility, and *in situ* measurement of the conducting polymer sensors guided by reversible wettability are still desired.

Although palladium metal or its alloys are used as hydrogen sensors because of their fast and reversible response, they require elevated temperatures to work well and are inhibited by oxygen [129]. Janata and coworkers [130] recently showed that a field effect transistor with two layers, that is, palladium and PANI, can be used as a good sensor for hydrogen. Kaner and coworkers [129c] also reported camphorsulfonic acid (CSA) doped PANI nanofibers as hydrogen sensors. These results suggest that PANI nanofibers have the potential of being a good room-temperature hydrogen sensor in a dry atmosphere.

Hydrogen peroxide (H_2O_2) is widely used in the industry for bleaching and cleaning, as well as disinfecting, releasing contaminants in the surrounding environment in large quantities [131]. H_2O_2 is also a major reactive oxygen species in living organisms and plays an important role as a second messenger in cellular signal transduction. Oxidative damages resulting from the cellular imbalance of H_2O_2 and other reactive oxygen species are connected to aging and severe human diseases such as cancers and cardiovascular disorders [132]. Furthermore, H_2O_2 is one of the products of reactions mediated by almost all oxides [133]. Therefore, molecular probes for H_2O_2 are of broad interest to the research community in environmental sciences and biochemistry as well as in clinical assays and screening [134]. Conducting polymers such as PPy and PANI give rise to electronic transitions in the UV–visible region, particularly in doped states [135]. These characters are similar to those in nanoparticle systems, indicating that the optical properties of the conjugated polymers may find potential applications in various chemical analyses [135b]. More recently, an optical sensing probe has been developed based on a fact that the absorbance of the oxidation product of pyrrole is proportional to the concentration of hydrogen peroxide, when H_2O_2 is used as an oxidant for pyrrole in the presence of a surfactant, sodium dodecyl sulfate, and Fe (II) in a slightly acidic aqueous solution [136].

Besides, since H_2S reacts rapidly with many metal salts (e.g., $CuCl_2$) to form a metal sulfide precipitate and generate a strong acid (i.e., HCl), based on these experimental facts, Virji et al. [73] reported that when the PANI composite film mixed with $CuCl_2$ is exposed to H_2S vapor, it becomes doped by the HCl generated as a byproduct of CuS formation within a few seconds that leads to the sensitivity of PANI nanofibers to H_2S vapor.

Since hydrazine is a strong reducing agent, both doped and de-doped, the emeraldine oxidation state of PANI changes to its fully reduced leukoemeraldine base oxidation state by hydrazine, which is accompanied with a decrease in the conductivity [137]. The decrease in conductivity associated with change in oxidation state can be therefore used to develop PANI hydrazine sensors. Interestingly, incorporation of an additive, such as glucose, urea, oxygen, chloride, and fluorinated alcohol (e.g., hexafluoroisopropanol, HFIP), into PANI could increase the response of PANI to hydrazine [138].

With the recent developments in nanoscience and nanotechnology, there is a pressing need for flexible, mechanically robust, and environmentally stable chemical vapor sensors with a high efficiency and low power consumption. This type of sensors can be used for real-time sensing of chemical warfare stimulants in a battlefield by monitoring the resistance changes in soldiers' clothing [139]. Wei et al. [140] recently reported a novel concept for developing a new class of multifunctional chemical vapor sensors with a low power consumption, high sensitivity, good selectivity, and excellent environmental stability by partially coating perpendicularly aligned carbon nanotube arrays with appropriate flexible polymer films. This might have important implications for potential use of aligned carbon nanotubes and polymer composites as a new class of very promising multifunctional materials and devices.

17.6.2
pH Sensors

The measurement of the solution pH is an important task that is required in clinical diagnostics, environmental and industrial control, and various branches of modern science and technology. Since the response of the electrode is provided by the difference in the electrochemical potentials of hydrogen ions inside and those outside the H^+-permeable glass membrane, its theoretical sensitivity is limited by the prediction of the Nernst equation, which is 59 mV per pH unit for the theoretical limit of sensitivity of such electrodes [141]. Among the conducting polymers, PANI represents one of the most promising materials for pH sensing, due to special proton doping mechanism or electrochemical oxidation of aniline [5c]. For regular PANI, the slopes of electrode potential versus pH were reported to be near Ernestinal [142]. Interestingly, the potentiometer response as high as 70 mV/pH unit has been reported for self-doped PANI [143]. Orlov et al. [141] further reported that self-doped PANI-modified both glassy carbon and screen-printed carbon electrodes by dip-coating exhibited a fully reversible potentiometer response of approximately 90 mV/pH unit over the range from pH 3 to 9. The higher sensitivity of self-doped PANI toward pH is beneficial for biosensors.

As a sensing element, moreover, PANI was used in different pH transducers including electrochemical, optical, and gravimetric ones [144]. However, PANI loses its electrochemical activity in solutions of pH greater than 4 [5c] so that adaptation of PANI to neutral pH is an important issue. In order to extend its conductivity to neutral pH, many attempts have been made to modify the properties of PANI. The first attempt is to synthesize self-doped PANI by sulfuric acid groups introduced on the PANI backbone through sulfonation of the emeraldine and leukoemeraldine forms of PANI [5a]. Self-doped PANI also can be synthesized by copolymerization of aniline and metanilic acid or by homopolymerization of metanilic acid [145]. Another strategy extended to neutral pH is to synthesize PANI in the presence of anionic polyelectrolyte with a sulfonate group [146]. Therefore, several amperometric sensor devices, which are responsive to glucose, urea, and NADH, have been developed using the modified PANI that can be operated in pH 7 buffer solution [147]. Since the conductivity of the PPy film is also sensitive to pH of the solution, PPy as a pH-sensitive transducer for enzyme and the PPy derivatives as good pH response has also been reported [148].

Acetic acid is a component of fermentation products (e.g., wines, vinegar, and soy sauces) and the tats and flavor of the fermented products are affected by the concentration of acetic acid. In addition, acetic acid molecules are also relevant to disease diagnostics and are used as a biomarker in the breath analysis of patients due to their noninvasiveness and rapid detection [149]. The accurate and rapid determination of acetic acid concentrations is therefore necessary in industrial food laboratories and disease diagnostics [150]. Conducting polymers can be used as acetic acid sensing materials because of their conductivity and physical properties of being influenced by factors such as polaron number and charge transfer to adjacent molecules [151]. Jang and coworkers [152] reported PANI-modified conducting PPy nanotubes as chemoreceptive vapor sensors for acetic acid detection where sensitivity and response of the sensors were found to be a function of the number of amine spacers and the analyses concentration.

17.6.3
Biosensors

Conducting polymers such as PANI, PPy, and PEDOT have been utilized as the immobilization matrix for enzymes in glucose sensors due to inherent charge transport properties and biocompatibility of the conducting polymers [153]. The majority of glucose biosensors are based on amperometric detection with glucose oxidase (GOx) enzyme electrodes [154]. Since conducting PANI is only synthesized from acidic solutions, most of the PANI–based biosensors are fabricated by the physical adsorption of the enzyme/host species after the polymerization that often results in poor sensitivities due to low amount of loading [155]. Recently, acidic polyelectrolyte, for example, poly(styrene sulfonate) (PSS), has proven to be a suitable microenvironment for the formation of conducting PANI because the pH at the acidic polyelectrolyte surfaces is much lower than that of bulk aqueous medium [156]. Based on the above idea, Kanungo et al. [157] reported microtubule sensors for glucose, urea, and triglyceride fabricated based on the doped PANI-PSS within the pores of track-etched PC membranes as the hard-template. The sensor response for urea and triglyceride is enhanced by a factor of 10^2 times compared to PANI -based sensors, where the enzymes were immobilized by physical adsorption after the polymerization. Moreover, the sensors based on urea and triglyceride were found to have a higher linear range of response, better sensitivity, improved multiple use capability, and faster response time compared to the sensors based on PANI.

Traditionally, the interaction of p-doped conducting polymers with DNA has been attributed to interact with positively charged molecules. However, Ocampo et al. [158] recently reported that polythiophene derivatives are able to prevent DNA digestion, which suggested that the formation of such polymer-DNA complexes is based not only on electrostatic interactions but also on other kinds of interactions (i.e., hydrogen bonds, stacking, Van Der Waals, charge transfer). Zanuy and Alemán [159] evaluated the ability of pyrrole and thiophene to interact with the methylated analogues of DNA bases (9-methyladenine (mA), 9-methylguanine (mG), 1-methylcytosine (mC), and 1-methylthymine (mT)) through specific hydrogen-bonding interactions.

Gene analysis plays an ever-increasing role in a number of areas related to human health such as diagnosis of infectious diseases, genetic mutations, drug discovery, forensics, and food technology [160]. Currently widely used gene array technologies have shortcomings arising from limited tagging efficiency, hazardous waste disposal, and complex multi-step analysis [160]. A new generation of gene sensors that are fast, reliable, and cost-effective needs to be developed. Electrochemical gene sensors are regarded as particularly suitable for direct and fast bio-sensing since they can convert the bio-recognition event (e.g., DNA hybridization) into a direct electrical signal [161]. Conducting polymers possess a delocalized electronic structure that is sensitive to changes in the polymeric chain environment and other perturbations of chain conformation that may be used as active substrates in DNA hybridization [160].

Field-effect transistors (FETs) have attracted increasing interest as primary candidates for fabricating state-of-the art sensor platforms because they are capable of achieving high current amplification and sustaining an enhanced signal-to-noise ratio [162]. One-dimensional (1D) conducting polymer nanomaterials have attracted much attention as promising building blocks for FET sensor applications because of their anisotropic electronic properties, high surface area, and small dimensions [163]. However, there are some limitations: one is that conducting polymer nanomaterials need a delicate and time-consuming process for adhering to a patterned electrode because the detection of biological species is commonly carried out in solution. Another is that the interaction between polymer transducers and bio-receptors has a major impact on the quality and performance of biosensors [164]. Although microelectronics technology can be used to fabricate interdigitized microelectrode elements and miniaturize the devices, chemical cross-talk between neighboring devices can occur if the immobilization of the enzyme that imparts specificity to the sensor is not localized to the designated microelectrode pair. Sangodkar *et al.* [165] reported the fabrication of PANI-based microsensors and microsensor arrays for the estimation of glucose, urea, and triglycerides. In this method, microelectronics technology was used to produce gold interdigitized microelectrodes on oxidized silicon wafers, and PANI deposition and enzyme immobilization were achieved electrochemically. The trick of this method is to direct enzyme immobilization to the chosen microelectrodes by controlling electrochemical potential, preventing enzyme from contacting with other microelectrodes. This has enabled the immobilization of three different enzymes on three closely spaced microelectrodes, resulting in a sensor array that can analyze a sample containing a mixture of glucose, urea, and so on in a single measurement using a few microliters of the sample. This strategy is quite general and can be extended to other enzyme-substrate systems to eventually produce an "electronic tongue." Meyer *et al.* [166] have recently described an array of 400 sensors, each independently addressable to obtain a two-dimensional concentration profile of glucose by immobilizing glucose oxidase on all the sensor elements.

17.6.4
Artificial Sensors

It is commonly known that human tongue is able to distinguish four basic types of tastes (e.g., sweet, salty, sour, and bitter) [167]. Although the human tongue and nose can sufficiently respond to chemical substances, they cannot directly contact with many toxic chemical substances. Artificial sensors, such as electronic tongue or electronic nose, are the possible alternatives for detecting toxic and unpleasant substances. The use of artificial sensors for evaluating tartans has attracted attention, because it is an important tool to improve quality control in the food and beverage industry. Conducting polymers and mixtures with satiric acid have been reported as artificial sensors. As mentioned earlier, a microtubule sensor array prepared in the presence of PSS can be used for the detection of glucose, urea, and triglyceride [165]. Ferreira *et al.* [168] also reported a sensor array made up of nanostructured LB films as an electronic tongue. However, one key requirement for electronic tongues is the need to have a good performance for the different basic tastes. Contractor and coworkers [169] described the fabrication of microtubular biosensors and sensor arrays of PANI as an "electronic tongue" for glucose, urea, and triglycerides. The response of the microtubular sensor for glucose is higher by a factor of more than 10^3. Moreover, an artificial tongue composed of four sensors (e.g., salty, sour, sweet, and bitter) made from ultra-thin films deposited onto gold interdigitized electrodes has also been reported [170]. Additionally, the use of artificial materials with comparable recognition properties has been proposed for designing bio-mimetic sensors [171]. This approach has overcome some limitations of biosensors such as those due to the availability or the cost of biocomponents for a particular analysis. Malitesta *et al.* [171] reported biosensors based on poly(o-phenylenediamine) (PPD) electropolymerized by using a glucose template, showing that the presence of the template in the synthesis medium plays a fundamental role in determining the molecular recognition capabilities of the imprinted polymer.

17.7
Summary and Outlook

In comparison with other nanomaterials, conducting polymer nanostructures not only possess highly π-conjugated polymer chain, metal-like conductivity and reversible chemical, electrochemical and physical properties controlled by doping/de-doping process, but also have characteristics of surface, size and quantum effect of nanomaterials as well as light-weight and flexibility of common polymers. These unique properties led to a wide range of applications as nanodevices and sensors and enabled conductive polymers to hold an important position in the field of nanoscience and nanotechnology.

The reversible electrical and redox properties are basic requirements for some applications (e.g., rechargeable battery, hydrogen storage, tissue engineering scaffolds, super capacitors and sensors). Conducting polymers work well in this regard. However, the de-doping process results in decrease in conductivity, which is one of unfavorable features of conducting polymer nanostructures. Design and synthesis of new types of conducting polymers are required in order to overcome this and other disadvantages. Although a variety of methods have been explored, the synthesis of conducting polymer nanostructures with multi-functionality, with uniform, mono-disperse and tailored morphology and size, and with evenly oriented nanostructure arrays, is still needed. In particular, searching and developing a simple, low-cost, universal, large scaled, and reproducible synthetic method for the preparation of conducting polymer nanostructures remain to be a challenge. Moreover, the transport properties of conducting polymer nanostructures even in single nanotube or nanofiber form still only exhibit semiconductor behavior due to large inter-contact resistance between fibers or nanotubes. Therefore, the elimination of such contact-resistance is a key issue in improving intrinsic transport properties and understanding the conduction mechanism of conducting polymers and their nanostructures.

For the construction of electronic devices on a molecular or nanoscale level, additionally, there are a number of problems that need to be addressed. One of the main problems is the nature of electrical contacts between the individual molecules or the molecular assemblies and the metal leads. Establishing the electrical contact is a challenge in the development of molecular electronic devices and characterization of their electronic transport properties across the contacts. Although research on conducting polymer-based sensors is very active at present, improvement in sensitivity, responsiveness, and reproducibility of the conducting polymer nanostructures as sensing materials and development of real-time measurement method for conducting polymer-based sensors are still required in order for practical applications. The above issues with respect to conducting polymers and their nanostructures need more fundamental, technological, and translational investigations in the future.

Acknowledgments

MXW and YW thank the Chinese Academy of Sciences for the support via an Outstanding Oversea Scholar Award. YW acknowledges the support by the Ministry of Science and Technology of China via the National 973 Program (Grant No. 2011CB935700) and the National Natural Science Foundation of China via Key Program (No. 21134004). TCW and YW are most grateful to National Cheng Kung University (NCKU) for the support through the NCKU Project of Promoting Academic Excellence & Developing World Class Research Centers (No. HUA98-12-3-140).

References

1 Shirakawa, H., Louis, E.J., MacDiarmid, A.G., Chiang, C.K., and Heeger, A.J. (1977) *J. Chem. Soc. Chem. Commun.*, 578–580.
2 http://www.nobel.se/chemistry/laureaters/2000/idex.html.
3 Skotheim, T.A. (ed.) (1986 and 1998) *Handbook of Conducting Polymers*, Marcel Dekker, New York.
4 Nalwa, H.S. (ed.) (2004) *Encyclopedia of Nanoscience and Nanotechnology*, American Scientific Publishers, Stevenson Ranch, CA.
5 (a) Wan, M.X. (2008) *Conducting Polymers with Micro or Nanometer Structure*, Tsinghua University Press, Beijing and Springer, Berlin, Heidelberg; (b) Wan, M.X. and Nalwa, H.S. (eds) (2004)

Conducting polymer nanofibers, in *Encyclopedia of Nanoscience and Nanotechnology*, American Scientific Publishers, Stevenson Ranch, CA, pp. 153–169; (c) Wan, M.X. (2009) *Macromol. Rapid Commun.*, **30**, 963–975; (d) Li, C., Bai, H., and Shi, G.Q. (2009) *Chem. Soc. Rev.*, **38**, 2397–2409; (e) Wei, Y., Li, B.S., Fu, C.K., and Qi, H.X. (2010) *Acta Polym. Sin.*, **12**, 1399–1405; (f) Lu, X.F., Wang, C., and Wei, Y. (2009) *SMALL*, **5** (21), 2349–2370; (g) Lu, X.F., Zhang, W.J., Wang, C., Wen, T.-C., and Wei, Y. (2011) *Prog. Polym. Sci.*, **36** (5), 671–712.

6 Stenger-Smith, J.D. (1998) *Progr. Polym. Sci.*, **23**, 57.

7 (a) Chiang, J.C. and MacDiarmid, A.G. (1986) *Synth. Met.*, **13**, 193; (b) MacDiarmid, A.G., and Epstein, A.J. (1989) *Faraday Discuss. Chem. Soc.*, **88**, 317.

8 Seymour, R.B. (1981) *Conductive Polymers*, Plenum Press, New York, pp. 23–47.

9 MacDiarmid, A.G. (2001) *Angew. Chem. Int. Ed.*, **40**, 2581.

10 Su, W.P., Schrieffer, J.R., and Heegel, A.J. (1979) *Phys. Rev. Lett.*, **42**, 1698.

11 (a) Kaufman, H., Colaneri, N., Scott, J.C., and Street, G.B. (1984) *Phys. Rev. Lett.*, **53**, 1005; (b) Stafstrom, S., Bredas, J.L., Epstein, A.J., Woo, H.S., Tanner, D.B., Huang, W.S., and MacDiarmid, A.G. (1987) *Phys. Rev. Lett.*, **59**, 1464.

12 (a) Brazovski, S.A. and Kirova, N.N. (1981) *Sov. Phys. JETP Lett.*, **33**, 4; (b) Briad, J.L., Chance, R.R., and Silbey, R. (1981) *Mol. Cryst. Liq. Cryst.*, **77**, 319.

13 Mott, N.F. and Davis, E.A. (1979) *Electronic Processes in Noncrystalline Solid*, 2nd edn, Clarendon Press, Oxford, p. 32.

14 Wan, M.X., Wang, P., Cao, Y., Qian, R.Y., Wang, F.S., Zhao, X.J., and Gong, Z. (1983) *Solid State Commun.*, **47**, 759.

15 (a) Wang, P., Wan, M.X., Bi, X.T., Yao, Y.X., and Qian, R.Y. (1984) *Acta Phys. Sin.*, **33**, 1771; (b) Bi, X.T., Yao, Y.X., Wan, M.X., Wang, P., Xiao, K., Yang, Q.Y., and Qian, R.Y. (1985) *Makromol. Chem.*, **186**, 1101.

16 Cao, Y., Wang, P., and Qian, R.Y. (1985) *Makromol. Chem.*, **186**, 1093.

17 Coleman, L.B. (1978) *Rev. Sci. Instrum.*, **49**, 58.

18 (a) Groenendaal, L., Zotti, G., Aubert, P.-H., Waybright, S.M., and Reynolds, J.R. (2003) *Adv. Mater.*, **15**, 855–879; (b) Cirpan, A., Argun, A.A., Grenier, C.R.G., Reeves, B.D., and Reynolds, J.R. (2003) *Mater. J. Chem.*, **13**, 2422–2428.

19 (a) Arico, A.S., Bruce, P., Scrosati, B., Tarascon, J.-M., and van Schalkwijk, W. (2005) *Nat. Mater.*, **4**, 366–377; (b) Long, J.W., Dunn, B., Rolison, D.R., and White, H.S. (2004) *Chem. Rev.*, **104**, 4463–4492; (c) Xia, Y., Yang, P., Sun, Y., Wu, Y., Mayers, B., Gates, B., Yin, Y., Kim, F., and Yan, H. (2003) *Adv. Mater.*, **15**, 353–389; (d) Zhang, D. and Wang, Y. (2006) *Mater. Sci. Eng. B.*, **134**, 9–19; (e) Arico, A.S., Bruce, P., Scrosati, B., Tarascon, J.-M., and van Schalkwijk, W. (2005) *Nat. Mater.*, **4**, 366–377; (f) Long, J.W., Dunn, B., Rolison, D.R., and White, H.S. (2004) *Chem. Rev.*, **104**, 4463–4492.

20 (a) Huang, J. and Wan, M.X. (1999) *J. Polym. Sci., Part A: Polym. Chem.*, **37**, 1277; (b) Wei, Z.X. and Wan, M.X. (2003) *Adv. Mater.*, **15**, 136; (c) Qiu, H.J., Wan, M.X., Matthews, B., and Dai, L.M. (2001) *Macromolecules*, **34**, 675; (d) Wan, M.X. and Li, J.L. (2003) *Polym. Adv. Technol.*, **14**, 320.

21 Martin, C.R. (1994) *Science*, **266**, 1961.

22 (a) Long, Y.Z., Chena, Z.J., Wang, N.L., Zhang, Z.M., and Wan, M.X. (2003) *Physica. B.*, **325**, 208; (b) Long, Y.Z., Chen, Z.J., Zheng, P., Wang, N.L., Zhang, Z.M., and Wan, M.X. (2003) *J. Appl. Phys.*, **93**, 2962.

23 Zhang, X.Y., Lee, J.-S., Lee, G.S., Cha, D.-K., Kim, M.J., and Yang, D.J., and Manohar, S.K. (2006) *Macromolecules*, **39**, 470–472.

24 (a) Lee, H.J., Jin, Z.X., Aleshin, A.N., Lee, J.Y., Goh, M.J., Akagi, K.Y., Kim, S., Retho, D.W., and Park, Y.W. (2004) *J. Am. Chem. Soc.*, **126**, 16722; (b) Duvail, J.L., Fernandez, P., Louarn, V.G., Molinie, P., and Chauvet, O. (2004) *J. Phys. Chem. B.*, **108**, 18552–18556; (c) Martin, C.R., Van Dyke, L.S., Cai, Z., and Liang, W.J. (1990) *Am. Chem. Soc.*, **112**, 8976; (d) Liang, W. and Martin, C.R. (1990) *J. Am. Chem. Soc.*, **112**, 9666.

25 (a) Saha, S.K., Su, Y.K., Lin, C.L., and Jaw, D.W. (2004) *Nanotechnology*, **15**, 66;

(b) I-Z, C., Mechler, A., Carter, S.A., and Lal, R. (2004) *Adv. Mater.*, **16**, 385; (c) Wu, C.G. and Chang, S.S. (2005) *Phys. J. Chem. B.*, **109**, 825.

26 (a) Lee, H.J. and Park, S.-M. (2005) *J. Phys. Chem. B.*, **108**, 1590; (b) Han, D.-H. and Park, S.M. (2004) *J. Phys. Chem. B.*, **108** (139), 21; (c) Lee, H.J. and Park, S.-M. (2004) *J. Phys. Chem. B.*, **108**, 16365; (d) Lee, H.J. and Park, S.M. (2005) *J. Phys. Chem. B.*, **109**, 13247; (e) Park, S.M. and Lee, H.J. (2005) *Bull. Korean Chem. Soc.*, **26**697; (f) Hong, S.Y., Jung, Y.M., Kim, S.B., and Park, S.M., *J. Phys. Chem. B.*, **109**, 3844; (g) Park, J.G., Lee, S.H., Kim, B., Saha, S.K., Lin, C.L., and Jaw, D.W. (2004) *Nanotechnology*, **15**, 66.

27 Han, D.H., Kim, J.W., and Park, S.M. (2006) *J. Phys. Chem. B.*, **110**, 14874–14880.

28 (a) Kiriy, N., Jahne, E., Adler, H.J., Schneider, M., Kiriy, A., Gorodyska, G., Minko, S., Jehnichen, D., Simon, P., Fokin, A.A., and Stamm, M. (2003) *Nano Lett.*, **3**, 707; (b) Merlo, J.A. and Frisbie, C.D. (2004) *J. Phys. Chem. B.*, **108**, 19169; (c) Kima, B.H., Park, D.H., Joo, J., Yu, S.G., and Lee, S.H. (2005) *Synthetic Met.*, **150**, 279.

29 (a) Yu, S.F., Li, N.C., Wharton, J., and Martin, C.R. (2003) *Nano Lett.*, **3**, 815; (b) Zhang, Z., Sui, J., Zhang, L., Wan, M., Wei, Y., and Yu, L. (2005) *Adv. Mater.*, **17**, 2854; (c) Zhang, Z., Deng, J., Sui, YuJ., Wan, L.M., and Wei, Y. (2006) *Macromol. Chem. Phys.*, **207**, 763.

30 Foss, C.A.J., Hornyak, G.L., Stockert, J.A., and Martin, C.R. (1993) *Adv. Mater.*, **5**, 135.

31 (a) Foss, C.A.J., Hornyak, G.L., Stockert, J.A., and Martin, C.R. (1994) *J. Phys. Chem.*, **98**, 2963; (b) Tonucci, R.J., Justus, B.L., Campillo, A.J., and Ford, C.E. (1992) *Science*, **258**, 783.

32 (a) Cao, L., Chen, H.Z., Zhou, H.B., Zhu, L., Sun, J.Z., Zhang, X.B., Xu, J.M., and Wang, M. (2003) *Adv. Mater.*, **15**, 909; (b) Dong, H., Prasad, S., Nyame, V., and Jones, W.E. (2004) *Chem. Mater.*, **16**, 371; (c) Ma, Y.F., Zhang, J.M., Zhang, G.J., and He, H.X. (2004) *J. Am. Chem. Soc.*, **126**, 7097; (d) Noll, J.D., Nicholson, M.A., Vanpatten, P.G., Chung, C.W., and Myrick, M.L. (1998) *J. Electrochem. Soc.*, **145**, 3320.

33 (a) Zhang, X.Y., Goux, W.J., and Manohar, S.K. (2004) *J. Am. Chem. Soc.*, **126** (14), 4502–4503; (b) Huang, J., Virji, S., Weiller, B.H., and Kaner, R.B. (2003) *J. Am. Chem. Soc.*, **125**, 314; (c) Von, B.M., Friedhoff, P., Biernat, J., Heberle, J., Mandelkow, E.M., and Mandelkow, E. (2000) *Proc. Natl. Acad. Sci. U.S.A.*, **97**, 5129; (d) Bailey, J.K., Pozarnsky, G.A., and Mecartney, M.L. (1992) *J. Mater. Res.*, **7**, 2530.

34 Zhu, Y., Li, J., Wan, M., Jiang, L., and Wei, Y. (2007) *Macromol. Rapid Commun.*, **28**, 1339.

35 Van Dyke, L.S. and Martin, C.R. (1990) *Langmuir*, **6**, 1123–1132.

36 (a) Cho, S.I., Kwon, W.J., Choi, S.J., Kim, P., Park, S.A., Kim, J., Son, S.J., Xiao, R., Kim, S.H., and Lee, S.B. (2005) *Adv. Mater.*, **17**, 171–175; (b) Cho, S.I., Choi, D.H., Kim, S.H., and Lee, S.B. (2005) *Chem. Mater.*, **17**, 4564–4566.

37 Xiao, R., Cho, S., Liu, R., and Lee, S.B. (2007) *J. Am. Chem. Soc.*, **129**, 4483–4489.

38 (a) Qu, L.T., Shi, G.Q., Chen, F.E., and Zhang, J.X. (2003) *Macromolecules*, **36**, 1063; (b) Qu, L.T. and Shi, G.Q. (2003) *Chem. Commun.*, **2**, 206.

39 Boal, A.K., Ihan, F., DeRouchey, J.E., Albrecht, T.T., Russell, T.P., and Rotello, V.M. (2000) *Nature*, **404**, 746.

40 Sun, X.P. and Hagner, M. (2007) *Macromolecules*, **40** (24), 8537–8539.

41 Formhals, A. (1934) *U.S. Patent*1,975,504.

42 Fong, H., Chun, I., and Reneker, D.H. (1999) *Polymer*, **40**, 4585.

43 MacDiarmid, A.G., Jones, W.E., Jr., Norris, I.D., Gao, J., Johnson, A.T., Pinto, N.J., Hone, J., Han, B., Ko, F.K., Okuzaki, H., and Llaguno, M. (2001) *Synth. Met.*, **119**, 27.

44 (a) Bognitzki, M., Hou, H., Ishaque, Frese., Hellwig, M.T., Schwarte, M., Wendorff, C.A.J.H., and Greiner, A. (2000) *Adv. Mater.*, **12**, 637; (b) Hou, H., Jun, Z., Reuning, A., Schaper, A., Wendorff, J.H., and Greiner, A. (2002) *Macromolecules*, **35**, 2429; (c) Caruso, R.A., Schattka, J.H., and Greiner, A. (2001) *Adv. Mater.*, **13**, 1577.

45 Hong, D., Sudhindra, P., Verrad, N.E., and Wayne, J.J. (2004) *Chem. Mater.*, **16**, 371.

46 Huang, W.S., Humphrey, B.D., and MacDiarmid, A.G. (1986) *J. Chem. Soc.*, **82**, 2385.

47 (a) MacDiarmid, A.G., Chiang, J.C., Richter, A.F., and Epstein, A.J. (1987) *Synth. Met.*, **18**, 285; (b) Wan, M.X., and Yang, J.P. (1995) *J. Appl. Polym. Sci.*, **55**, 399.

48 Akkara, J.A., Kaplan, D.L., John, V.J., and Tripathy, S.K. (1996) *Polymeric Materials Encyclopedia*, vol. **3** (ed. J.C. Salamone), CRC Press, Boca Raton, FL, pp. 2116–2125.

49 Saunders, B.C., Holmes-Siedle, A.G., and Stark, B.P. (1964) *Peroxidase*, Butterworths, London.

50 (a) Liu, W., Kumar, J., Tripathy, S., Senecal, K.J., and Samuelson, L. (1999) *J. Am. Chem. Soc.*, **121**, 71–78; (b) Samuelson, L.A., Anagnostopoulos, A., Alva, K.S., Kumar, J., and Tripathy, S.K. (1998) *Macromolecules*, **31**, 4376; (c) Premachandran, R., Banerjee, S., John, V.T., McPherson, G.L., Akkara, J.A., and Kaplan, D.L. (1997) *Chem. Mater.*, **9**, 1342; (d) Bruno, F., Akkara, J.A., Samuelson, L.A., Kaplan, D.L., Marx, K.A., Kumar, J., and Tripathy, S.K. (1995) *Langmuir*, **11**889.

51 (a) Hulvat, J.F. and Stupp, S.I. (2003) *Angew. Chem.*, **42**, 778; (b) Huang, L.M., Wang, Z.B., Wang, H.T., Cheng, X.L., Mitra, A., and Yan, Y.S. (2002) *J. Mater. Chem.*, **12**, 388.

52 (a) Wang, D. and Caruso, F. (2001) *Adv. Mater.*, **13**, 350; (b) Bartlett, P.N., Birkin, P.R., Ghanem, M.A., and Toh, C.S. (2001) *J. Mater. Chem.*, **11**, 849.

53 Wang, X., Liu, N., Yan, X., Zhang, W.J., and Wei, Y. (2005) *Chem. Lett.*, **34**, 42.

54 (a) Huang, J., Virji, S., Weiller, B.H., and Kaner, R.B. (2003) *J. Am. Chem. Soc.*, **125**, 314–315; (b) Huang, J. and Kaner, R.B. (2004) *J. Am. Chem.Soc.*, **126**, 851–855; (c) Huang, J.X. and Kaner, R.B. (2004) *Nat. Mater.*, **3**, 783–786; (b) Li, D. and Xia, Y.N. (2004) *Nat. Mater.*, **3**, 753–754; (d) Huang, J.X. and Kaner, R.B. (2004) *Nat. Mater.*, **3**, 783–786.

55 Huang, J. and Kaner, R.B. (2004) *Angew. Chem.*, **43**, 5817.

56 (a) Zhang, Z.M., Wei, Z.X., Zhang, L.J., and MaWan, M.X. (2005) *Acta Mater*, **53**, 1373; (b) Zhang, L.X., Zhang, L.J., Wan, M.X., and Wei, Y. (2006) *Synth. Met.*, **156**, 454; (c) Chiou, N.R. and Epstein, A.J. (2005) *Adv. Mater.*, **17**, 1679.

57 (a) Liu, H., Hu, X.B.,Wang, J.Y., and Boughton, R.I. (2002) *Macromolecules*, **35**, 9414; (b) Zhang, L.J. and Wan, M.X. (2005) *Thin Solid Films*, **477**, 24.

58 (a) Wan, M.X., Shen, Y.Q., Huang, J. (1998) invs.: CH 98109916.5; (b) Zhang, L.J., Wan, M.X., and Wei, Y. (2006) *Macromol. Rapid. Commun.*, **27**, 366.

59 Wei, Z.X., Zhang, Z., and Wan, M.X. (2002) *Langmuir*, **18**, 917.

60 Huang, K., Wan, M.X., Long, Y.Z., Chen, Z.J., and Wei, Y. (2005) *Synth. Met.*, **155**, 495.

61 (a) Kim, B.J., Oh, S.G., Han, M.G., and Im, S.S. (2000) *Langmuir*, **16**, 5841; (b) Harada, M. and Adachi, M. (2000) *Adv. Mater.*, **12**, 839.

62 (a) Konyushenko, E.N., Stejskal, J., Sedecaron, I., Trchova, M., Sapurina, I., Cieslar, M., and Prokeš, J. (2006) *Polym. Int.*, **55**, 31; (b) Trchova, M., Syědĕnková, I., Konyushenko, E.N., Stejskal, J., Holler, P., and CÄirić-Marjanović, G. (2006) *J. Phys. Chem. B*, **110**, 9461.

63 Ding, H.J., Wan, M.X., and Wei, Y. (2007) *Adv. Mater.*, **19**, 465.

64 Zhu, Y., Hu, D., Wan, M.X., Lei, J., and Wei, Y. (2007) *Adv. Mater.*, **19**, 2092.

65 Zhu, Y., Li, J.M., Wan, M., and Jiang, X.L. (2008) *Macromol. Rapid. Commun.*, **29**, 239.

66 Zhu, Y., Li, J., Wan, M., and Jiang, L. (2008) *Polymer*, **49**, 3419.

67 Zhu, Y.H., Wan, M., and Jiang, L. (2008) *Macromol. Rapid Commun.*, **29**, 1689.

68 (a) Taylor, R.S., Dang, L.X., and Garrett, B.C. (1996) *J. Phys. Chem.*, **100**, 11720; (b) Chen, B., Siepmann, J.I., and Klein, M.L. (2002) *J. Am. Chem. Soc.*, **124**, 12232.

69 (a) Huang, J.X. and Kaner, R.B. (2004) *Nat. Mater.*, **3**, 783–786; (b) Li, D. and Xia, Y.N. (2004) *Nat. Mater.*, **3**, 753–754; (c) Huang, J.X. and Kaner, R.B. (2004) *Nat. Mater.*, **3**, 783–786.

70 Baker, C.O., Shedd, B., Innis, P.C., Whitten, P.G., Spinks, G.M., Wallace,

G.G., and Kaner, R.B. (2008) *Adv. Mater.*, **20**, 155–158.

71 (a) Virji, S., Fowler, J.D., Baker, C.O., Huang, J.X., Kaner, R.B., and Weiller, B.H. (2005) *Small*, **1**, 624–662. (b) Masdarolomoor, F., Innis, P.C., Ashraf, S., Kaner, R.B., and Wallace, G.G. (2006) *Macromol. Rapid Commun.*, **27**, 1995–2000.

72 (a) Tseng, R.J., Huang, J.X., Ouyang, J., Kaner, R.B., and Yang, Y. (2005) *Nano Lett.*, **5**, 1077–1080; (b) Tseng, R.J., Baker, C.O., Shedd, B., Huang, J.X., Kaner, R.B., Ouyang, J.Y., and Yang, Y. (2007) *Appl. Phys. Lett.*, **90** (5), 3101; (c) Gallon, B.J., Kojima, R.W., Kaner, R.B., and Diaconescu, P.L. (2007) *Angew. Chem., Int. Ed.*, **46**, 7251–7254.

73 Virji, S., Fowler, J.D., Baker, C.O., Huang, J.X., Kaner, R.B., and Weiller, B.H. (2005) *Small*, **1**, 624–627.

74 (a) Liang, L., Liu, J., Windisch, C.F., Exarhos, G.J., and Lin, Y.H. (2002) *Angew. Chem., Int. Ed.*, **41**, 3665–3668; (b) Liu, J., Lin, Y.H., Liang, L., Voigt, J.A., Huber, D.L., Tian, Z.R., Coker, E., McKenzie, B., and McDermott, M.J. (2003) *Chem. Eur. J.*, **9**, 605–611.

75 Chiou, N.R., Liu, C.M., Guan, J.J., Lee, L.J., and Epstein, A.J. (2007) *Nat. Nanotechnol.*, **2**, 354–357.

76 (a) Groenendaal, L.B., Jonas, F., Freitag, D., Pielartzik, H., and Reynolds, J.R. (2000) *Adv. Mater.*, **12**, 481; (b) Groenendaal, L.B., Zotti, G., Aubert, P.H., Waybright, S.M., and Reynolds, J.R. (2003) *Adv. Mater.*, **15**, 855; (c) Ha, Y.H., Nikolov, N., Pollack, S.K., Mastrangelo, J., Martin, B.D., and Shashidhar, R. (2004) *Adv. Funct. Mater.*, **14**, 615.

77 (a) Kirchmeyer, S. and Reuter, K. (2005) *J. Mater. Chem.*, **15** (21), 2077; (b) Groenedaal, L.B., Jonas, F., Freitag, D., Pielartzik, H., and Reynolds, J.R. (2000) *Adv. Mater.*, **12** (7), 481.

78 (a) Chipman, A. (2007) *Nature (London)*, **449**, 131; (b) Zhang, F., Johansson, M., Andersson, M.R., Hummelen, J.C., and Inganäs, O. (2002) *Adv. Mater.*, **14**, 662.

79 Cao, Y., Yu, G., Zhang, C., Menon, R., and Heeger, A.J. (1997) *Synth. Met.*, **87**, 171.

80 Groenendaal, L., Jonas, F., Freitag, D., Peilartzik, H., and Reynolds, J.R. (2000) *Adv. Mater.*, **12**, 481.

81 (a) Kim, J.Y., Jung, J.H., Lee, D.E., and Joo, J. (2002) *Synth. Met.*, **126**, 311; (b) Pettersson, L.A.A., Ghosh, S., and Inganäs, O. (2002) *Org. Electron.*, **3**, 143; (c) Nardes, A.M., Janssen, R.A.J., and Kemerink, M. (2008) *Adv. Funct. Mater.*, **18**, 865; (d) Ouyang, J., Xu, Q., Chu, C.W., Yang, Y., Li, G., and Shinar, J. (2004) *Polymer*, **45**, 8443; (e) Ouyang, J., Chu, C.W., Chen, F.C., Xu, Q., and Yang, Y. (2005) *Adv. Funct. Mater.*, **15**, 203.

82 (a) Fan, B., Mei, X.G., and Ouyang, J.Y. (2008) *Macromolecules*, **41**, 5971–5973; (b) Fan, B., Mei, X., and Ouyang, J.Y. (2008) *Macromolecules*, **41**, 5971–5973; (c) Nardes, A.M., Janssen, R.A.J., and Kemerink, M. (2008) *Adv. Funct. Mater.*, **18**, 865.

83 (a) Antonietti, M., Kuang, D., Smarsly, B., and Zhou, Y. (2004) *Angew. Chem. Int. Ed.*, **43**, 4988; (b) Wei, D., Kvarnstrom, C., Lindfors, T., and Ivaska, A. (2006) *Electrochem. Commun.*, **8** (10), 1563–1566.

84 (a) Randriamahazaka, H., Plesse, C., Teyssie, D., and Chevrot, C. (2005) *Electrochim. Acta*, **50**, 1515; (b) Pringle, J.M., Forsyth, M., MacFarlane, D.R., Wagner, K., Hall, S.B., and Officer, D.L. (2005) *Polymer*, **46**, 2047–2058.

85 (a) Lu, W., Fadeev, A.G., Qi, B.H., Smela, E., Mattes, B.R., Ding, J., Spinks, G.M., Mazurkiewicz, J., Zhou, D.Z., Wallace, G.C., MacFarlane, D.R., Forsyth, S.A., and Forsyth, M. (2002) *Science*, **297**, 983–987; (b) Ding, J., Zhou, D., Spinks, G., Wallace, G., Forsyth, S., Forsyth, M., and MacFarlane, D. (2003) *Chem. Mater.*, **15** (12), 2392; (c) Marcilla, R., Alcaide, F., Sardon, H., Pomposo, J.A., Pozo-Gonzalo, C., and Mecerreyes, D. (2006) *Electrochem. Commun.*, **8**, 482–488.

86 Döbbelin, M., Marcilla, R., Salsamendi, M., Pozo-Gonzalo, C., Carrasco, PedroM., Pomposo, JoseA., and Mecerreyes, D. (2007) *Chem. Mater.*, **19**, 2147–2149.

87 (a) Mumtaz, M., de Cuendias, A., Putaux, J.L., Cloutet, E., and Cramail, H. (2006) *Maromol. Rapid Commun.*, **27**, 1446; (b) Han, M.G., and Foulger, S.H. (2005) *Chem. Commun.*, 1; (c) Abidian, M.R., Kim, D.H., and Martin, D.C., *Adv. Mater.* (2006) **18**, 405; (d) Jang, J., Chang, M., and Yoon, H. (2005) *Adv. Mater.*, **17**, 1616; (e) Zhang, X., MacDiarmid, A.G., and Manohar, S.K. (2005) *Chem. Commun.*, 5328.

88 Zang, J., Li, C.M., Bao, S.J., Cui, X.Q., Bao, Q.L., and Sun, C.Q. (2008) *Macromolecules*, **41** (19), 7053–7057.

89 (a) Xia, Y., Yang, P., Sun, Y., Wu, Y., Mayers, B., Gates, B., Yin, Y., Kim, F., and Yan, H. (2003) *Adv. Mater.*, **15**, 353; (b) Zhang, X. and Manochar, S.K. (2004) *J. Am. Chem. Soc.*, **126**, 12714; (c) Zhou, Q., Li, C.M., Li, J., Cui, X.Q., and Gervasio, D. (2007) *J. Phys. Chem. C*, **111** (30), 11216–11222; (d) Zhong, W.B., Liu, S.M., Chen, X.H., Wang, Y.X., and Yang, W.T. (2006) *Macromolecules*, **39** (9), 3224–3230; (e) Goren, M. and Lennox, R.B. (2001) *Nano Lett.*, **1**, 735; (f) Jang, J. and Yoon, H. (2005) *Langmuir*, **21**, 11484; (g) Zhang, X., Zhang, J., Liu, Z., and Robinson, C. (2004) *Chem. Commun.*, 1852; (h) Shi, W., Ge, D., Wang, J.Z., Jiang, L., and Zhang, R.Q. (2007) *Macromol. Rapid Commun.*, **27**, 926.

90 Kim, J.Y., Kim, J.T., Song, E.A., Min, Y.K., and Hamaguchi, H.O. (2008) *Macromolecules*, **41**, 2886.

91 Zang, J., Li, C.M., Bao, S.J., Cui, X.Q., Bao, QL., and Sun, C.Q. (2008) *Macromolecules*, **41** (19), 7053–7057.

92 Li, M., Wei, Z., and Jiang, L. (2008) *J. Mater. Chem.*, **18**, 2276–2280.

93 Catherine Debiemme-Chouvy (2009) *Electrochem. Comm.*, **11**, 298–301.

94 (a) Gurunathan, K. and Trivedi, D.C. (2000) *Mater. Lett.*, **45**, 262; (b) Xia, H. and Wang, Q. (2002) *Chem. Mater.*, **14**, 2158; (c) Zhang, L.J. and Wan, M.X. (2003) *J. Phys. Chem. B*, **107**, 6748; (d) Zhang, L. and Wan, M. (2003) *J. Phys. Chem. B*, **107**, 6748.

95 (a) Gudiksen, M.S., Lauhon, L.J., Wang, J., Smith, D.C., and Lieber, C.M. (2002) *Nature*, **415**, 617; (b) Park, S., Lim, J.H., Chung, S.W., and Mirkin, C.A. (2004) *Science*, **303**, 348; (c) Nicewarner-Pena, S. R., Freeman, R.G., Reiss, B.D., He, L., Pena, D.J., Walton, I.D., Cromer, R., Keating, C.D., and Natan, M.J. (2001) *Science*, **294**, 137; (d) Park, S., Lim, J.H., Chung, S.W., and Mirkin, C.A. (2004) *Science*, **303**, 348; (e) Park, S., Lim, J.H., Chung, S.W., and Mirkin, C.A. (2004) *Science*, **303**, 348.

96 (a) Zhang, Z.M. and Wan, M.X. (2003) *Synth. Met.*, **132**, 205; (b) Zhang, Z.M., Wan, M.X., and Wei, Y. (2006) *Nanotechnology*, **16**, 2827.

97 (a) Li, X., Shen, J.Y., Wan, M.X., Chen, Z.J., and Wei, Y. (2007) *Synth. Met.*, **157**, 575. (b) Li, X. and Wan, M.X. (2006) *J. Phys. Chem. B*, **110**, 14623.

98 Ding, H., Liu, X.M., Wan, M., and Fu, S.Y. (2008) *J. Phys. Chem. B*, **112**, 9289.

99 Zhang, Z.M., Deng, J.Y., Shen, J.Y., Wan, M.X., and Chen, Z.J. (2007) *Macromol. Rapid Commun.*, **28**, 585.

100 (a) Foss, C.A.J., Hornyak, G.L., Stockert, J.A., and Martin, C.R. (1993) *Adv. Mater.*, **5**, 135; (b) Foss, C.A.J., Hornyak, G.L., Stockert, J.A., and Martin, C.R. (1994) *J. Phys. Chem.*, **98**, 2963.

101 Huang, K. Wan, M.X. (2002) *Chem. Mater.*, **14**, 3486.

102 (a) Zhang, L. and Wan, M.X. (2005) *Thin Solid Films*, **477**, 24; (b) Yang, Y.S., and Wan, M.X. (2002) *J. Mater. Chem.*, **12**, 897–901.

103 (a) Ren, B., Picardi, G., Pettinger, B., Schuster, R., and Ertl, G. (2005) *Angew. Chem. Int. Ed.*, **44**, 139; (b) Wang, H., Levin, C.S., and Halas, N.J. (2005) *J. Am. Chem. Soc.*, **127**, 14992; (c) Lu, L., Randjelovic, I., Capek, R., Gaponik, N., Yang, J., Zhang, H., and Eychmueller, A. (2005) *Chem. Mater.*, **17**, 5731.

104 (a) Guzman, J. and Gates, B.C. (2004) *J. Am. Chem. Soc.*, **126**, 2672; (b) Hutchings, G.J. (2005) *Catal. Today*, **100**, 55; (c) Ma, S., Yu, S., and Gu, Z. (2006) *Angew. Chem., Int. Ed.*, **45**, 200.

105 (a) Taton, T.A., Mirkin, C.A., and Letsinger, R.L. (2000) *Science*, **289**, 1757; (b) Hirsch, L.R., Jackson, J.B., Lee, A., Halas, N.J., and West, J.L. (2003) *Anal. Chem.*, **75**, 2377; (c) Storhoff, J.J., Lucas, A.D., Garimella, V., Bao, Y.P., and Müller, U.R. (2004) *Nat. Biotechnol.*, **22**, 883.

106 (a) Li, X., Li, Y., Tan, Y., Yang, C., and Li, Y. (2004) *J. Phys. Chem. B*, **108**, 5192; (b) Pillalamarri, S.K., Blum, F.D., Tokuhiro, A.T., and Bertino, M.F. (2005) *Chem. Mater.*, **17**, 5941; (c) Huang, J. and Kaner, R.B. (2004) *J. Am. Chem. Soc.*, **126**, 851; (d) Jang, J., Bae, J., and Park, E. (2006) *Adv. Mater.*, **18**, 354; (e) Hormozi Nezhad, M.R., Aizawa, M., Porter, L.A., Jr., Ribbe, A.E., and Buriak, J.M. (2005) *Small*, **11**, 1076.

107 Lu, G., Li, C., and Shi, G.Q. (2007) *Chem. Mater.*, **19** (14), 3433–3440.

108 Kelly, T.L., Yamada, Y., Che, S.P.Y., Yano, K., and Wolf, M.O. (2008) *Adv. Mater.*, **20**, 2616–2621.

109 Kelly, TimothyL., Che, SaraP.Y., Yamada, Yuri, Yano, Kazuhisa, and Wolf, Michael O. (2008) *Langmuir*, **24** (17), 9809–9815.

110 (a) Nuraje, N., Su, K., Yang, N., and Matsui, H. (2008) *ACS Nano*, **2**, 502; (b) Li, B. and Lambeth, D.N. (2008) *Nano Lett.*, **8**, 3563.

111 (a) Li, D., Jiang, Y.D., Wu, Z.M., Chen, X.D., and Li, Y.R. (2000) *Sens. Actuators, B*, **66**, 125–127; (b) Matsuguchi, M., Io, J., Sugiyama, G., and Sakai, Y. (2002) *Synth. Met.*, **128**, 15–19; (c) Huang, J.X., Virji, S., Weiller, B.H., and Kaner, R.B. (2003) *J. Am. Chem. Soc.*, **125**, 314–315.

112 (a) Pringsheim, E., Zimin, D., and Wolfbeis, O.S. (2001) *Adv. Mater.*, **13**, 819–822; (b) Janata, J. and Josowicz, M. (2003) *Nature Mater.*, **2**, 19–24; (c) Gao, M., Dai, L., and Wallace, G.G. (2003) *Eletroanalysis*, **15**, 1089–1094.

113 Ballarin, B., Brumlik, C.J., Lawson, D.R., Liang, W., Van Dyke, L.S., and Martin, C.R. (1992) *Anal. Chem.*, **64** (21), 2647–2651.

114 Torsi, L. (2000) *Sens. Actuators, B*, **67**, 312–316.

115 (a) Ramanathan, K., Bangar, M.A., Yun, M., Chen, W., Myung, N.V., and Mulchandani, A. (2005) *J. Am. Chem. Soc.*, **127**, 496–497; (b) Mieszawska, A.J., Slawinski, G.W., and Zamborini, F.P. (2006) *J. Am. Chem. Soc.*, **128**, 5622–5623; (c) Lu, C., Qi, L., Yang, J., Tang, L., Zhang, D., and Ma, J. (2006) *Chem. Commun.*, **33**, 3551–3553; (d) Zhang, X., Goux, W.J., and Manohar, S.K., *J. Am. Chem. Soc.* (2004) **126**, 4502–4503;

(e) Bai, H. and Shi, G. (2007) *Sensors*, **7**, 267–307.

116 Virji, S., Huang, J., Kaner, R.B., and Weiller, B.H. (2004) *Nano Lett.*, **4** (3), 491–496.

117 Gao, Y., Li, X., Gong, J., Fan, B., Su, Z., and Qu, L. (2008) *J. Phys. Chem. C*, **112**, 8215–8222.

118 Sharma, S., Nirke, C., Pethkar, S., and Athawale, A.A. (2002) *Sens. Actuators, B*, **85**, 131.

119 Anderson, M.R., Mattes, B.R., Reiss, H., and Kaner, R.B. (1991) *Science*, **252**, 1412.

120 Wang, J., Chan, S., Carlson, R.R., Luo, Y., Ge, G., Ries, R.S., Heath, J.R., and Tseng, H.R. (2004) *Nano Lett.*, **4**, 1693.

121 Jang, J., Chang, M., and Yoon, H. (2005) *Adv. Mater.*, **17**, 1616.

122 Liu, H., Kameoka, J., Czaplewski, D.A., and Craighead, H.G. (2004) *Nano Lett.*, **4** (4), 671–675.

123 (a) Nicho, M.E., Trejo, M., Garcia-Valenzuela, A., Saniger, J.M., Palacios, J., and Hu, H. (2001) *Sens. Actuators, B*, **76**, (b) Chabukswar, V.V., Pethkar, S., and Athawale, A.A. (2001) *Sens. Actuators, B*, **77**, 657–663; (c) Xie, D., Jiang, Y., Pan, W., Li, D., Wu, Z., and Li, Y. (2002) *Sens. Actuators, B*, **81**, 158–164.

124 (a) Hosono, K., Matsubara, I., Murayama, N., Woosuck, S., and Izu, N. (2005) *Chem. Mater.*, **17**, 349; (b) Geng, L., Zhao, Y., Huang, X., Wang, S., Zhang, S., and Wu, S. (2007) *Sens. Actuators, B*, **120**, 568.

125 Zhang, J., Wang, S., Xu, M., Wang, Y., Xia, H., Zhang, S., Guo, X., and Wu, S. (2009) *J. Phys. Chem. C*, **113** (5), 1662–1665.

126 (a) Fu, Q., Rao, G.V., Rama, S.B., Basame, D.J., Keller, K., Artyushkova, J.E., Fulghum, G., and López, P. (2004) *J. Am. Chem. Soc.*, **126**, 8904; (b) Yu, X., Wang, Z., Jiang, Y., Shi, F., and Zhang, X. (2005) *Adv. Mater.*, **17**, 1289; (c) Minko, S., Müller, M., Motornov, M., Nitschke, M., Grundke, K., and Stamm, M. (2003) *J. Am. Chem. Soc.*, **125**, 3896; (d) Lahann, J., Mitragotri, S., Tran, T.N., Kaido, H., Sundaram, J., Choi, I.S., Hoffer, S., Somorjai, G.A., and Langer, R. (2003) *Science*, **299**, 371.

127 (a) Isaksson, J., Tengstedt, C., Fahlman, M., Robinson, N., and Berggren, M. (2004) *Adv. Mater.*, **16**, 316; (b) Xu, L., Chen, W., Mulchandani, A., and Yan, Y. (2005) *Angew. Chem. Int. Ed.*, **44**, 6009.

128 Zhu, Y., Li, J.M., Wan, M.X., and Jiang, L. (2007) *Macromol. Rapid. Commun.*, **28**, 2230.

129 (a) Zuttel, A., Nutzenadel, Ch., Schmid, G., Chartouni, D., and Schlapbach, L. (1999) *J. Alloys Compd.*, 472–475; (b) Watari, N., Ohnish, S., and Ishi, T. (2000) *J. Phys.: Condens. Matter*, **12**, 6799–6823; (c) Virji, S., Kaner, R.B., Weiller, B.H. (2006) *J. Phys. Chem. B*, **110**, 22266–22270.

130 Domansky, K., Baldwin, D.L., Grate, J.W., Hall, T.B., Josowicz, M., and Janata, J. (1998) *Anal. Chem.*, **70**, 473–481.

131 (a) Price, D., Worsfold, P.J., Mantoura, R.F.C., and Fauzi, C. (1992) *Trends Anal. Chem.*, **11**, 379; (b) Anglada, J.M., Aplincourt, Ph., Bofill, J.M., and Cremer, D. (2002) *Chem. Phys. Chem.*, **3**, 215.

132 (a) Chang, M.C.Y., Pralle, A., Isacoff, E.Y., and Chang, C.J. (2004) *J. Am. Chem. Soc.*, **126**, 15392; (b) Miller, E.W., Albers, A.E., Pralle, A., Isacoff, E.Y., and Chang, C.J. (2005) *J. Am. Chem. Soc.*, **127**, 16652.

133 Barman, T.E. (ed.) (1974) *The Enzyme Handbook*, Springer, Berlin.

134 Wolfbeis, O.S., Durkop, A., Wu, M., and Lin, Z.H. (2002) *Angew. Chem. Int. Ed.*, **41**, 4495.

135 (a) Nalwa, H.S. (ed.) (1997) *Handbook of Organic Conductive Molecules and Polymers*, vol. **3**, John Wiley & Sons Ltd., West Sussex, England, (b) Hong, S.-Y., Jung, Y.M., Kim, S.B., and Park, S.-M. (2005) *J. Phys. Chem. B*, **109**, 3844; (c) Hung, C.C., Wen, T.-C., and Wei, Y. (2010) *Mater. Chem. Phys.*, **122** (2–3), 392–396.

136 Wang, H. and Su-Moon, P. (2007) *Anal. Chem.*, **79** (1), 240–245.

137 (a) Hsu, C.H., Peacock, P.M., Flippen, R.B., Manohar, S.K., and MacDiarmid, A.G. (1993) *Synth. Met.*, **60**, 233. (b) Yano, J., Terayama, K., Yamasaki, S., and Aoki, K. (1998) *Electrochim. Acta*, **44**, 337.

138 (a) Volotovsky, V., Soldatkin, A.P., Shul'ga, A.A., Rossokhaty, V.K., Strikha, V.I., and El'skaya, A.V. (1996) *Anal. Chim. Acta*, **322**, 77; (b) Soldatkin, A.P., Volotovsky, V., El'skaya, A.V., Jaffrezic-Renault, N., and Martelet, C. (2000) *Anal. Chim. Acta*, **403**, 25; (c) Xu, W., McDonough, R.C., Langsdorf, B., Demas, J.N., and DeGraff, B.A. (1994) *Anal. Chem.*, **66**, 4133; (d) Jae, H.S., Hyo, L.L., Sung, H.C., Ha, J., Nam, H., and Geun, S.C. (2004) *Anal. Chem.*, **76**, 4217; (e) Virji, S., Kaner, R.B., and Weiller, B.H. (2005) *Chem. Mater.*, **17**1256.

139 Dai, L. (2004) *Intelligent Macromolecules for Smart Devices: From Materials Synthesis to Device Applications*, Springer-Verlag, New York.

140 Wei, C., Dai, L., Roy, A., and Tolle, T.B. (2006) *J. Am. Chem. Soc.*, **128**, 1412–1413.

141 Orlov, Andrey V., Karpachova, Galina P., and Wang, J. (1999) *Anal. Chem.*, **71**, 2534–2540.

142 (a) Wan, Q.J., Zhang, X.J., Zhang, C.G., and Zhou, X.Y. (1997) *Chemical J. Chin. Univ. -Chinese*, **18**, 226–228; (b) Lindino, C.A., and Bulhoes, L.O.S. (1996) *Anal. Chim. Acta*, **334**, 317–322.

143 Karyakin, A.A., Bobrova, O.A., Lukachova, L.V., and Karyakina, E.E. (1996) *Sens. Actuators, B*, **33**, 34–38.

144 (a) Grummt, U.W., Pron, A., Zagorska, M., and Lefrant, S. (1997) *Anal. Chim. Acta*, **357**, 253–259; (b) Pringsheim, E., Terpetschnig, E., and Wolfbeis, O.S. (1997) *Anal. Chim. Acta*, **357**, 247–252; (c) Zhou, X.Y., Cha, H.Y., Yang, C., and Zhang, W.M. (1996) *Anal. Chim. Acta*, **329**, 105–109.

145 (a) Karyakin, A.A., Strakhova, A.K., and Yatimirsky, A.K. (1994) *J. Electroanal. Chem.*, **371**, 259; (b) Krishnamoorthy, K., Contractor, A.Q., and Kumar, A. (2002) *Chem. Commun.*, 240.

146 Austria, G., Jang, G.W., MacDiarmid, A.G., Doblhofer, K., and Zhang, C. (1991) *Ber. Bunsen-Ges. Phys. Chem.*, **95**, 1381.

147 (a) Castillo-Ortega, M.M., Rodriguez, D.E., Encinas, J.C., Plascencia, M., Mendez-Velarde, F.A., and Olayo, R. (2002) *Sens. Actuators, B*, **B85** (1–2), 19; (b) Tatsuma, T., Ogawa, T., Sato, R., and Oyama, N. (2001) *J. Electroanal. Chem.*, **501** (1–2), 180; (c) Raitman, O.A., Katz, E., Bueckmann, A.F., and Willner, I. (2002) *J. Am. Chem. Soc.*, **124** (22), 6487.

148 (a) Nishizawa, M., Matsue, T., and Uchida, I., *AMI. Che*, **1002** (64), 2042–2044; (b) Aquino-Binag, C.N., Kumar, N., Lamb, R.N., and Pigram, P.J. (1996) *Chem. Mater.*, **8** (11), 2579–2585.

149 Cui, Y., Wei, Q., Park, H., and Lieber, C.M. (2001) *Science*, **293**, 1289.

150 (a) Joo, J., Lee, J.K., Lee, S.Y., Jang, K.S., Oh, E.J., and Epstein, A.J. (2000) *Macromolecules*, **33**, 5131; (b) Sadki, S., Schottland, P., Brodie, N., and Sabouraud, G. (2000) *Chem. Soc. Rev.*, **29**, 283; (c) Zhong, Z., Qian, F., Wang, D., and Lieber, C.M. (2003) *Nano Lett.*, **3**, 343.

151 Hernández, R.M., Lee, R., Steve, S., Stranick, S., and Mallouk, E.T. (2004) *Chem. Mater.*, **16**, 3431.

152 Ko, S. and Jang, J. (2007) *Biomacromolecules*, **8**, 182.

153 (a) Zhu, L., Yang, R., Zhai, J., and Tian, C. (2007) *Biosens. Bioelectron.*, **23**, 528; (b) Kros, A., Nolte, R.J.M., and Sommerdijk, N.A.J.M. (2002) *Adv. Mater.*, **14**, 1779; (c) Geetha, S., Rao, C.R.K., Vijayan, M., and Trivedi, D.C. (2006) *Anal. Chim. Acta*, **568**, 119.

154 (a) Zayats, M., Katz, E., Baron, R., and Willner, I. (2005) *J. Am. Chem. Soc.*, **127**, 12400; (b) Ekanayake, E., Preethichandra, D.M.G., and Kaneto, K. (2007) *Biosens. Bioelectron.*, **23**, 107; (c) Li, J. and Lin, X. (2007) *Biosens. Bioelectron.*, **22**, 2898.

155 Kanungo, M., Kumar, A., and Contractor, A.Q. (2003) *Anal. Chem.*, **75**, 5673–5679.

156 (a) Manning, G.S. (1979) *Acc. Chem. Res.*, **12**, 443; (b) Manning, G.S. (1988) *J. Chem. Phys.*, **89**, 3722.

157 Kanungo, M., Kumar, A., and Contractor, A.Q. (2003) *Anal. Chem.*, **75**, 5673–5679.

158 Ocampo, C., Armelin, E., Estrany, F., del Valle, L.J., Oliver, R., Sepulcre, F., and Alemán, C. (2007) *Macromol. Mater. Eng.*, **292**, 85.

159 Zanuy, D. and Alemán, C. (2008) *J. Phys. Chem. B*, **112**, 3222–3230.

160 Peng, H., Soeller, C., and Travas-Sejdic, J. (2007) *Macromolecules*, **40**, 909–914.

161 (a) Katz, E. and Willner, I. (2003) *Electroanalysis*, **15**, 913–947; (b) Peng, H., Soeller, C., Vigar, N., Kilmartin, P.A., Cannell, M.B., Bowmaker, G.A., Cooney, R.P., Travas-Sejdic, J. (2005) *J. Biosens. Bioelectron.*, **20**, 1821–1828; (c) Wang, J. (1999) *J. Chem. Eur.*, **5**, 1681–1685.

162 (a) Janata, J. and Josowicz, M. (2003) *Nat. Mater.*, **2**, 19; (b) Freeman, R., Gill, R., and Willner, I. (2007) *Chem. Commun.*, 3450; (c) Star, A., Joshi, V., Han, T.-R., Altoé, M.V.P., and Grüner, G. (2004) *Org. Lett.*, **6**, 2089.

163 (a) Yoon, H., Kim, J.H., Lee, N., Kim, B.G., and Jang, J. (2008) *Chem. BioChem*, **9**, 634; (b) Wanekaya, A.K., Bangar, M.A., Yun, M., Chen, W., Myung, N.V., and Mulchandani, A. (2007) *J. Phys. Chem. C*, **111**, 5218; (c) Jang, J., Chang, M., and Yoon, H. (2005) *Adv. Mater.*, **17**, 1616; (d) Yoon, H., Chang, M., and Jang, J. (2007) *Adv. Funct. Mater.*, **17**, 431.

164 Yoon, H., Ko, S., and Jang, J. (2008) *J. Phys. Chem. B*, **112**, 9992–9997.

165 Sangodkar, H., Sukeerthi, S., Srinivasa, R.S., Lal, R., and Contractor, A.Q. (1996) *Anal. Chem.*, **68**, 779–783.

166 Meyer, H., Drewer, H., Grundig, B., Camman, K., Kakerow, R., Manoli, Y., Mokowa, W., and Rospert, M. (1995) *Anal. Chem.*, **67**, 1164.

167 Durán, L. and Costell, E. (1999) *Food Sci. Technol. Int.*, 299.

168 Ferreira, M., Riul, A., Jr., Wohnrath, K., Fonseca, F.J., Oliveira, O.N., Jr., and Mattoso, L.H. C. (2003) *Anal. Chem.*, **75**, 953.

169 Sukeerthi, S. and Contractor, A.Q. (1999) *Anal. Chem.*, **71**, 2231.

170 Riul, A., Jr., Santos, D.S., Wohnrath, K., Tommazo, R. Di, Carvalho, A.C.P.L.F., Fonseca, F.J., Oliveira, O.N., Jr., Taylor, D.M., and Mattoso, L.H.C. (2002) *Langmuir*, **18**, 239.

171 Malitesta, C., Losito, I., and Zambonin, P.G. (1999) *Anal. Chem.*, **71**, 1366–1370.

Index

a

absorption coefficients 372, 378
absorption intensity ratio, time
  dependence 272
acceptor materials, properties 336
acceptor polymers 399, 418
acene/polymer system 228
acetone 16, 316, 318
acetonitrile 16, 280
acetophenone 8, 17, 232
adiabatic energy 85, 89
air–film interface 199
all-polymer blends
– charge separation efficiency 409
– efficiency 399
all-polymer solar cells 399, 400, 403, 407,
  409, 419, 421
– charge transport 418, 419
– development 400
– device structure 400, 401
– donor/acceptor combinations 404–406
– interfacial charge separation 407–410
– low fill factor 407
– morphology 411–418
ammonium peroxydisulfate (APS) 495
amplified spontaneous emission (ASE) 443
aniline monomer, toxicity 498
anodic aluminum oxidation (AAO) 183, 387
anodic protection mechanism 270–274
anodization process 389
anticorrosion agents 269
anticorrosion polyaniline coating, test
  coating 281, 282
Arrhenius equation 271
artificial sensors 511, 512
atomic force microscopy (AFM) 9, 39, 56, 135,
  174, 307, 411, 481, 485
atom transfer radical polymerization
  (ATRP) 302

Auger deexcitation (AD) 78
azobenzene sulfuric acid (ABSA) 505

b

bandgaps 66, 74, 335, 339, 340, 342, 364, 366,
  373, 391, 402, 428
– strategies for reducing 340
band transport mobility 79, 80
"benchmark" system 240
benzo[1,2-b;3,4-b]dithiophene (BDT) 340,
  342–344
– polymers 342
γ-benzyl-L-glutamate N-carboxyanhydride
– ring-opening polymerization 321
BHJ. see bulk heterojunction (BHJ)
BHSCs. see bulk heterojunction solar cells
  (BHSCs)
binary polymer blend, phase diagram 195
binding energy 69, 75, 76, 81, 93, 96, 98, 100,
  101, 109, 129, 138, 372, 403, 429
binodal curves 195
biosensors 510, 511
bipolar charge transport 471
bipolar FET. see bipolar field-effect transistors
bipolar field-effect transistors
– bilayer 469–474
– bulk heterojunction 474, 475
– coevaporated blends 475, 476
– hybrid blends 479, 480
– n–p polymer–polymer heterojunctions
  471
– operating regimes 460
– polymer–polymer blends 480–485
– polymer–small molecule blends 476–479
– schematics 473
– single-component 465–469
– transfer and output curves 462
bipolar polymer blend FETs, typical output
  characteristics 484

Semiconducting Polymer Composites: Principles, Morphologies, Properties and Applications, First Edition.
Edited by Xiaoniu Yang.
© 2012 Wiley-VCH Verlag GmbH & Co. KGaA. Published 2012 by Wiley-VCH Verlag GmbH & Co. KGaA.

bipolar transistors. *see also* bipolar field-effect transistors
– complementary metal oxide–semiconductor (CMOS) technology 457
– current–voltage characteristics 461
–– bipolar FETs 462
–– unipolar FET 461, 462
– device configurations 462, 463
– field-effect transistor (FET) 457, 458, 459
–– bipolar 460, 461
–– unipolar 459, 460
– injecting electrodes, role 463, 464
– inverters/light-emitting transistors 464, 465
3,6-bis(5-(benzofuran-2-yl)thiophen-2-yl)-2,5-bis(2-ethylhexyl)pyrrolo[3,4-c]pyrrole-1,4-dione (DPP(TBFu)$_2$) 9
4,4-bis(2-ethylhexyl)dithieno[3,2-b:2′,3′-d]silole and N-octylthieno[3,4-c]pyrrole-4,6-dione (PDTSTPD) 20
bis(1,2,5-thiadiazolo)-p-quinobis(1,3-dithiole) (BTQBT) 80
6,13-bis(triisopropylsilylethynyl) pentacene (TIPS-pentacene) 227, 231
– amorphous polystyrene blend 232
– blends, SIMS data 229
blend films
– absorption spectra 177
– with embedded P3HT nanowires 207–213
–– from conjugated block copolymers 212
–– electrospun nanowires from conjugated polymer blends 212, 213
–– P3AT nanowires 208, 209
– photoluminescence spectra 178
– with vertical stratified structure 194–207
–– charge carrier mobility 201–204
–– crystallization-induced vertical phase segregation 206, 207
–– environmental stability 201
–– OTFTs, semiconducting and insulating layers, one-step formation 198–201
–– polymer blends, patterned domains 201–204
–– polymer blends, phase behavior 194–198
Boltzmann constant 266
bond occupation probability 245
σ-bond polymers 90
– quasi-one-dimensional band dispersion 90
–– using hexatriacontane (n-CH$_3$(CH$_2$)$_{32}$CH$_3$) vacuum deposited on metal surfaces 90
–– using LB films of Cd arachidate {Cd^{2+}[CH$_3$(CH$_2$)$_{18}$COO$^-$]$_2$}, 91
boron dipyrromethene (bodipy) derivatives 315

bottom-gate (BCBG) OFET structure 235
Bragg scattered 448
buckminsterfullerene (C$_{60}$) 332
bulk heterojunction (BHJ) 2
– blend materials, self-assembly 349
– characteristics 177
– composite nanostructures, optimization 173–182
– devices 373
–– performances 181
– donor–acceptor interface 172
– 3D representation 41
– ESSENCIAL method 174, 176, 180, 181
– film 411
– layer 305
– PV cells 173
– solar cells 2, 332, 334, 336 (*see also* bulk heterojunction solar cells (BHSCs))
–– active layer 346
–– developments in 339
–– hybrid, photocurrent generation mechanism 372
–– loss mechanisms 338
–– optimization 351
–– performance 334
–– photoexcitation in 192
–– structural complexity 241
– structures 373
–– characteristics 177
bulk heterojunction solar cells (BHSCs) 2, 7, 192, 371, 384
BZT-coated surface 202

c

cadmium selenide (CdSe) 436
– nanoparticles 378
– NP-based devices 374
– P3HT blended devices 376
– quantum dots (QDs) 364, 378, 379
–– absorption spectra 363
–– NCs 374, 387, 391
–– photoluminescence (PL) graphs 363
– tetrapods (TPs) 375, 376
Cahn's theory 196
camphorsulfonic acid (CSA)-doped PANI 274, 507
capillary forces 387
carbon–carbon cross-coupling reactions 314
carbon nanotubes (CNTs) 24, 26, 207, 391
carrier drift length 337
cathodic electrodeposition 389
cathodic protection mechanism 274–277
CdS/P3HT NW-based hybrid solar cell device 384

CdTe/polymer systems 379
charge carriers 465
– injection 432, 433
– mobility 193, 219, 224
– transport 233, 260
– – and recombination 433
charge hopping 84–86
– parameters determine mobility 85
charge relaxation processes 333
charge separation
– efficiency 407, 409, 410, 412, 419
– interfacial 407–410
– Onsager–Braun theory 129–131
– process, defect states influencing 374
– role of charge transfer states 131–133
– theory of field and temperature
    dependence 129
charge transfer 337, 403
– behavior to TiO$_2$ 390
– binding energies 130
– complex 260, 261
– – C-PCPDTBT and Si-PCPDTBT 21
– at D–A interfaces 333
– Dexter mechanism 429
– interaction 195
– parameters influencing, separation of charge
    transfer states 133
– – charge carrier mobilities 138
– – electric field 137, 138
– – electronic coupling/reorganization
    energies 138
– – energetic disorder 138
– – excess energy 133
– – morphology 133–137
– – temperature 138
– photoinduced 333, 334, 340
– in polymer/fullerene composites 125–129
– – diffusion-limited charge transfer 128, 129
– – Marcus theory 126–128
– – quantum-mechanical considerations 128
– – theory 125
– in related heterojunction structure 97
– relaxed state 129
– states, role 131–133
– strong electric field dependence 409
charge transport mechanisms 209, 214, 232
– variable range hopping 235
chemical hard-template method 495
chemical interaction 29, 30, 278
chiral camphorsulfonic acid (CSA) 505
chlorobenzene 22, 42, 44, 51, 135, 211, 235,
    307, 349, 375, 386
chloroform 7, 9, 10, 12, 16, 96, 210, 284, 413,
    479, 506

1-chloronaphthalene 15
ClAlPc dipole layer 102
cohesive energy density (CED) 3
coil–rod–coil block copolymers 316
colloidal synthetic methods 362
complementary inverter structure 464
complementary metal oxide–semiconductor
    (CMOS) technology 457
composite materials
– carbon black–polymer composites 162, 163
– carbon nanotube (CNT)–polymer
    composites 158
– semiconductor–dielectric composites 161
– semiconductor–insulator composite 156
– Si-SiO$_2$ composites 157
– structure 156–159
– σ(x) dependence, observations and
    interpretations 159
– – critical behavior of σ(x) 161–165
– – percolation threshold 159–161
– W-Al$_2$O$_3$ composite 163
conduction band (CB) 68–70, 74, 76, 336,
    371, 428
conduction pathway 225, 228, 229, 231
conductive additives 282, 283
conductive AFM (C-AFM) techniques 307
conductive atomic force microscopy
    (C-AFM) 210
conductivity 258, 428
conductor-like screening model for real
    solvents (COSMO-RS) 4
confocal PL microscopy 469
conjugated block copolymers 212, 418
conjugated–insulating block copolymers 299–
    322
– conjugated–insulating rod–rod block
    copolymers 320–322
– oligo/polythiophene rod–coil block
    copolymers 300–308
– polyfluorenes (PFs) 313–318
– poly(p-phenylene vinylene) block
    copolymers 308–312
– semiconducting rod–coil systems 319, 320
conjugated–insulating rod–coil polymers 319
conjugated–insulating rod–rod block
    copolymers 319, 320
conjugated polymer-based photovoltaic cells,
    device architectures 172
conjugated polymer blends
– all-polymer solar cells 399–421
– – charge transport 418–420
– – interfacial charge separation 407–410
– – key issues affect 407–420
– – morphology 411–418

- device achievements to date 403–407
- material considerations 401–403
- polymer photophysics, and device operation 400, 401
conjugated polymer:fullerene bulk heterojunction solar cells 7
conjugated polymers 1, 71, 113, 192, 209, 252, 261, 339, 366, 367, 399
- anostructured arrays formation 387
- blending 26 (see also conjugated polymer blends)
- chemical doping 1
- with deep LUMOs 403
- electronic structure and excited states 108
- electron mobility 419
- electrospinning 213
- energy levels 375
-- of polaron 73
- excited state dynamics 113
-- role of disorder in energy transfer 113, 114
-- singlet exciton energy transfer 114, 115
-- triplet exciton dynamics 115–123
- exciton delocalized along 110
- matrix 381
- nanowires and nanorods, with luminescent 438
- optical properties 508
- in situ synthesis of ligand-free semiconductors 381, 382
- utilized in hybrid solar cells 374
-- with respective HOMO–LUMO levels 368–370
π-conjugated polymer semiconductors 208
conjugated polymers synthesis 192
conjugated polymer systems 299
contact-film transfer method, bilayer realized by
- output characteristics of P3HT/PCBM bilayer transistor 472
- transfer curves 472
controlled radical polymerization (CRP) methods 302
copper indium disulfide (CIS) nanoparticle 381
copper phthalocyanine 1
corrosion potential 278
corrosion protection mechanisms 277
cost-effective photovoltaics 331
Coulomb interactions 107, 267, 332, 364, 366
counter diode 337
counterion-induced processability 284
crystalline–crystalline system 226
crystalline insulating polymer 262

current-sensing atomic force microscopy (CS-AFM) 493
current–voltage characteristics 407
cyclic voltammetry (CV) 366
cyclohexanone (CHN) 211
cyclooctene 302
cyclopentadithiophene-based copolymers 341
cyclopentadithiophene benzothiadiazole copolymer (CDT-BTZ) 192

d

density of gap states (DOGS) 101
- in pentacene on VL-increased HOPG system 102
density of states (DOS) 82, 102, 111, 113, 114, 128, 252, 266, 267, 428
Dexter transfer 429
2D grazing incidence X-ray scattering (2D GISAXS/GIWAXS) 347
DH4T:P3HT system 225, 226
diblock copolymers
- self-assembly behavior 183
- synthesis 304
3,6-dibromofluorene 313
1,2-dichlorobenzene 7, 8, 350
o-dichlorobenzene (ODCB) 209, 254
dielectric bisbenzocyclobutene (BCB) derivative 198
differential scanning calorimetry (DSC) 228
diffusion coefficient 123, 124, 244, 245, 262
2,8-difluoro-5,11-bis(triethylsilylethynyl) anthradithiophene (diF-TESADT) 228
dihalide fluorene monomers 314
α,ω-dihexyl-quaterthiophene (DH4T) 469
- critical concentration 224
diiodoalkanes 20
1,8-diiodooctane (DIO) 20
diketopyrrolopyrrole-based acceptors 403
dilute Harrison's solution (DHS) 277
dimethylformamide 16
dimethylsiloxane (DMS) 205
dioctyl phosphate (DOPH)-doped PANI 278
diode efficiency 432
dip-pen nanolithography (DPN) 202, 203
direct percolation pathway 373
distributed Bragg reflector (DBR) 445
distributed feedback (DFB) lasers 448
dithienosilole (DTS) 341, 342
4,7-dithien-5-yl-2,1,3-benzodiathiazole (DTBT) 340
divinyltriphenylamine (DVTPA) monomer 387
dodecylbenzenosulfonic acid (DBSA)-doped polyaniline 284, 285

doping 72, 191, 252, 253, 258, 260
– in conducting polymers 490, 491
– with functional dopants 279
– with iodine 489
– oxidation/reduction 490
– photo-doping and charge-injection 490
– of polythiophene/IP composite 265
– proton doping 498
– p-type 490
– reversible 498
double-screw extruder 284
DPP-based polymers 345
DPPT-TT based OFETs 467, 468
dye-sensitized solar cells (DSSCs) 391
dynamic light scattering (DLS) 377, 499

e
EB/ER blend-coated steel–copper couple
– scheme 275
– visual observations 276
effective mass approximation 364
Einstein relation, for diffusive motion 85
electrical conductivity 66, 145, 146, 164, 191, 283, 284, 287
electroluminescence 131, 132, 191, 436, 437, 439, 445, 465, 469, 476, 479
electromagnetic interference (EMI) shielding 491
electron-accepting polymers 366, 399, 400, 403, 417, 418
electron blocking layer (EBL) 371, 391, 432
electron-conducting fullerene C_{60} 469
electron-donating dialkoxybithiophene 466
electron field-effect mobility 244, 245
electron–hole recombination 475
electronic structure
– and control of π-electron density distribution 93
– – annealing-induced increment, of molecular tilt angle 101
– – conjugated poly(3-hexylthiophene) (P3HT) thin films 93–98
– – NEXAFS spectra 97, 98
– – P3HT thin-film conformation 95
– – PIES spectra 95, 96, 98, 100
– – UPS spectra 96, 99
– energy levels, described by molecular orbitals and 70
– essential properties of delocalized π-electron system 71, 72
– evolution, from single molecule 67
– excited state dynamics in conjugated polymers 113–123

– and excited states of conjugated polymers 108
– – nature of excited states 108–113
– from single atom to polymer chain 70
electronic tongue 511, 512
electron mobility 178, 179, 238, 240, 243, 418, 469, 482, 483
electron–phonon coupling 66, 84
– in conjugated polymers 107
electron transfer rates 85
electrospinning technique 496
electrospun nanowires 212, 213
– from conjugated polymer blends 212
– field-effect hole mobilities 213
electrostatic damage 282
electrostatic interaction polyaniline/silica hybrid 289
EL intensity 479
elongated fullerenes 391
emeraldine salt (ES) 497
emulsion polymerization 284, 286, 293
energy autarkic miniaturized systems 392
energy band diagram, from bulk values of CdSe QDs 365
energy band dispersion 67, 76, 79, 92
energy barriers 58, 374
energy conversion 1, 53, 331
energy level alignment (ELA) 65, 66
– impact of charge injection 102
– at interface 73, 74
– role of interface dipole layer 101–103
energy transfer 108
– relevance to device performance 123, 124
– role of disorder 113
– – Gaussian disorder model (GDM) 113
– – Miller–Abrahams jump rates 114
– – relaxation 114
– singlet exciton 114, 115
engineering polymers 224, 225
ESSENCIAL process for fabricating polymer solar cells 175–182
ethanedithiol (EDT) 380
ethanol 16, 288
2-ethylhexyl moiety 342
2-ethyl-2-oxazoline 302
excitons 1, 54, 107–111, 115, 123, 124, 133, 172, 176, 332–334, 371, 401, 429, 436
– diffusion 54, 124, 171, 173, 182, 333, 372, 412, 414
– dissociation efficiency 174, 180, 346, 372, 380, 401
external quantum efficiency (EQE) 8, 335

f

Fabry–Perot dye laser cavity 447
Fermi–Dirac distribution function 74, 252, 267
Fermi level 373, 374
field-effect charge carrier mobility 220
field-effect mobility 193, 200, 204, 205, 238, 239, 242, 321
field-effect transistor (FET) 212, 213, 246, 251, 457
– bipolar 462 (*see also* bipolar FET)
– configurations 463
– geometry, schematic of 477
– mobility 256
– n-p heterojunction 471
– performance 253, 262
– unipolar 459, 460 (*see also* unipolar FET)
fill factor (FF) 7, 10, 15, 41, 180, 335, 337, 339, 350, 403, 407, 410
flat panel displays (FPDs) 1, 152, 192
flexible coil-like polymers 300
Flory–Huggins theory 194
fluorene–dithienylbenzothiadiazole copolymer 403
fluorene monomer 313
fluorescence resonance energy transfer efficiency (FRET) 115, 117, 316, 317
focused ion beam (FIB) lithography 183
Förster energy transfer 429, 436, 437
free charge carriers 109, 129, 133, 135, 332, 373, 380
free energy 4, 6, 176, 222, 223, 269
freeze-fracture transmission electronic microscopy (FFTEM) 499
fullerene
– acceptor 240
– aggregation 18, 22, 244
– bilayer diffusion 241
– charge transfer in composites 125
– clustering process 241, 243
– intercalation 348
– molecular weight, effects 240
– new derivative 345
– polymer–fullerene miscibility 28
– polymers, blend morphology 233
– solar cells, optimization 338
– solubility 7

g

Gaussian disorder model (GDM) 113, 118
germanium 191
Gibbs free energy 5, 195
Gilch route 309
glass transition temperatures 23, 31, 44, 184, 230, 265, 415

graphene–polymer 161
grazing incident X-ray diffraction (GIXD) 230

h

Hansen solubility parameters (HSPs) 6–9, 22, 33
Heck coupling reactions 309, 310
hexadecylamine (HDA) 362, 376, 386
hexafluoroisopropanol (HFIP) 508
high-density polyethylene (HDPE) 225
highest occupied molecular orbital (HOMO) state 66, 234, 337, 342, 366, 401, 428
– hole–vibration coupling 84, 89
– HOMO–LUMO gap 66, 101, 367
highly oriented pyrolytic graphite (HOPG) 495
hole blocking layer (HBL) 432
hole transporting layer (HTL) 432
hoping distance 266
hopping mobility 67, 72, 84, 85, 89, 90
Horner–Wadsworth–Emmons condensation reactions 309
hot injection method 362
Huang–Rhys factors 85
hybrid solar cells 53, 361, 367, 374, 381, 383, 385, 386
– conjugated polymers 366
– density–voltage characteristic 379
– development novel approaches 381–390
– – less toxic semiconductor NCs, utilization 381
– – nanostructured donor–acceptor phases 384–390
– – one-dimensional structured donor–acceptor nanostructures, utilization 382–384
– – *in situ* synthesis of ligand-free semiconductors 381, 382
– photoactive composite films 366
– photoactive layer 367
– state of art 374–380
– typical device structures 371
– *vs.* pure OPVs 390, 391
hydrogen bonding 3, 6, 195, 288, 496, 501, 502, 510
– interaction 3, 195, 510

i

ideality factor 337
incident photon to electron conversion efficiency (IPCE) 13, 16, 335
indene-C_{60} bisadduct (ICBA) 345, 346
indium tin oxide (ITO) 58, 352, 371, 401, 501
– composite–MgAg diodes 434

– PF/ZnO nanorods 439
inkjet-printed semiconductor 231
inkjet printing 15, 211, 228, 231, 367
in situ chemical polymerization method 280
insulating matrix 262, 267
insulating polymers (IPs) 251, 267
– mechanical characteristics 194, 214
– passive effect 194
integrated circuits 1, 233, 468, 476
– fabrication 457
interface dipole layer, role 101
– E_F shift 101
– pentacene/ClAlPc(ML)/HOPG system 101
–– ClAlPc dipole layer 101, 102
–– relaxed positive polaron 102
–– ultralow-density DOGS reaching to E_F 103
–– UPS measurements 102
– VL shift 101
interfacial electron–hole pairs 409
internal electric field 372
intrinsically conducting polymer (ICPs)/composites 269–289
– anticorrosion application 269–282
–– conducting composite coating, matrix resin 278, 279
–– oxidation of metals 278
–– oxygen reduction 278
–– processing methods 279, 280
–– protection mechanism 270–278
–– reduction potential 277, 278
– antistatic coating 282–288
–– ICPs processing 284–288
–– processable ICPs synthesis 283, 284
– application 269
– corrosion protection mechanisms 277
– dopant ions 275
– metal anticorrosion coating 270
– negative effect 279
– paints, products 279
– in polymer matrix 279, 280
– redox states 274
– reduction potentials 273
intrinsic bandgap 252
intrinsic conductivity 252
inverse photoemission spectroscopy (IPES) 66, 69, 101
inverted cells 401
ionization potentials 132, 252
isotropic factor 197

k
kinetic energy 76

l
lab-on-a-chip systems 449
leukoemeraldine 498, 508, 509
light-emitting field-effect transistors (LE-FETs) 464, 465, 476
linear operating regime 221
liquid–liquid phase separation 222
low-bandgap polymers 341
lowest unoccupied molecular orbital (LUMO) state 66, 234, 345, 346, 366, 401, 428, 458

m
Marangoni flow 232
Marcus theory 126–128
MDMO-PPV/PCBM
– BHJ subcells 354
– as model for amorphous donor system 42–48
–– optimum blending ratio 347
MEH-PPV polymer 374
melamine–urea resin 287
melt mixing, disadvantage 285
16-mercaptohexadecanoic acid (MHA) 202
mesitylene 17
mesoporous silica spheres (MSS) 505
metal corrosion 269
metallic conduction, in polyacetylene 1
metal-oxidesemiconductor field-effect transistor (MOSFET) 193
metastable atom electron spectroscopy (MAES) 78, 94
metastable deexcitation spectroscopy (MDS) 78
metastable impact electron spectroscopy (MIES) 78
metastable quenching spectroscopy (MQS) 78
methanesulfonic acid (MSA) 481
[2-methoxy-5-(3′,7′-dimethyloctyloxy-*p*-phenylene vinylene)] (MDMO-PPV) 236
methylene chloride (CH_2Cl_2) 210
micellar soft-templates 499
micellar triblock polymers 316
micelle formation 306
micelle-soft-templates 500
microchip laser 449
Miller–Abraham equation 266
MIM (metal/insulator/metal) devices 336
miscibility 22, 23, 34, 40, 227, 339
miscibility–thermodynamic relationships 5, 6
mixed mesitylene (MS) 7, 8
mobile carriers 419
mobility gap 238
molecular beam deposition 362

Monte Carlo simulations 4, 122, 136, 137, 265, 267
morphology control, importance 40–42
morphology imaging 48–50
Mott–Wannier excitons 332
multiple trap and release (MTR) models 209
mutual solubility regimes 8

n

NAND/NOR gates 474
nanocomposites (NCs) 382, 383, 430, 432, 503
nanofibrillar network in PS matrix, formation 210
nanoimprinted polymer solar cells, device performance 417
nanoimprint lithography (NIL) molds 173, 183, 184, 186, 389, 415, 417
nanopolyaniline-based coating 279
nanoscale electrical properties, measurements 56–60
nanostructured conducting polymers 489
– doping, degree 491
– doping process 490
– electromagnetic interference (EMI) 491
– electron transport properties 493
– electro-optic composite nanostructures/chiral nanostructures 504
– electrospinning technology 494
– inorganic semiconductors 490
– light-emitting diode (LED) 491
– multifunctionality 503–505
– polyaniline (PANI) 497–501
– poly(3,4-ethylenedioxythiophene) (PEDOT) 501, 502
– polypyrrole (PPy) 502, 503
– properties 492
– Schottky rectifier 491
– sensors 505
– – artificial sensors 511, 512
– – biosensors 510, 511
– – gas sensors 506–508
– – pH sensors 509, 510
– – synthetical methods 493
– – electrospinning technique 496, 497
– – hard-template method 494–496
– – soft-template method 496
nanostructured donor/acceptor blends, formation using sacrificial PDA 416
nanowires 438
β-naphthalene sulfuric acid (β-NSA) 499
naphtho[1,2-c:5,6-c]bis[1,2,5]-thiadiazole (NT) 343
NBDAE acceptor 318
n-channel (BBL), molecular structures 473
NC–polymer hybrid solar cells 391
NC-type semiconductors, energy values 365
near-edge X-ray absorption fine structure spectroscopy (NEXAFS) 94
neutral polyaniline (EB) 277
neutron reflectivity (NR) measurements 228, 230
nitrobenzene (NtB) 16
nitroxide mediated polymerization (NMP) 272, 302, 303
NP percolation pathways 373
nuclear magnetic resonance (NMR) 376
nucleation, and growth mechanism 195, 222

o

octanedithiol (ODT) 18–21, 349
octylamine 362, 385
octyltrichlorosilane (OTS) 201
o-dichlorobenzene 22, 42
oleic acid (OA) 17, 376
oligo(phenylene vinylene) (OPV) 310
oligothiophenes 110, 224, 226, 228, 245, 300, 301
o-oxylene 7
open-circuit potential (OCP) 271
open-circuit voltage (V_{OC}) 7, 41, 61, 335, 336, 340, 374, 383, 402, 403
optical density (OD) 7, 135
OPV-b-PEO polymers
– aggregation behavior 312
– UV/fluorescence spectra 312
organic field-effect transistors (OFETs) 209, 219, 221, 342
– bipolar 466, 468, 470, 474
– mobilities 220, 227, 231
– polymer/fullerene ambipolar 232–245
– rr-P3HT block copolymers 307
– unipolar films 224
organic light-emitting diodes (OLEDs) 65, 107, 111, 123, 124, 220, 267, 427, 430, 431, 432, 435, 438
– applications 308
– cadmium selenide (CdSe) 436
– color emission 433–441
– design 312
– efficiency 432
– – injection of charge carriers 432, 433
– – transport and recombination of charge carriers 433
organic photovoltaic cells (OPVs) 2, 299, 399, 427
– solar cells 233
organic semiconductor lasers 442
– applications and developments 448–451

– chips 450
– materials 443–446
– resonators 446–448
– working principle 441–443
organic semiconductors 1, 7, 9, 26, 65, 103, 119, 133, 224, 232, 332, 431, 465
organic solar cells (OSCs) 2, 9, 65, 123, 124, 174, 371
– device optimization 332, 338
– properties 332
organic tandem cell 353
organic thin-film solar cells
– quantum efficiency 333
– working mechanism, broken down 333
organic thin-film transistor (OTFT) 192, 506
– application 198
– array 203, 204
– based on semiconducting/insulating blends, preparation strategies 194
– consisting of crystalline–crystalline blend 206
– developments, based on semiconducting blends 192
– electric characteristics 204
– environmental stability 201
– fundamental principle, and operating mode 193, 194
– one-step formation of semiconducting and insulating layers 198–201
– water-gated 205
1,2-orthodichlorobenzene (ODCB) 483
oxide film growth rates 271
oxygen scavenging protection model 277

p

PANI nano-framework–electrode junctions (PNEJs) 506
parallel plate capacitor 336
passivation potential 273
PBDTT–DPP polymer 353, 354
p-channel polymer semiconductors, molecular structures 473
p-channel transistor 212
PEG-b-PPV insulating–conjugated block polymers 312
PEGylated polyphenylene ethynylene (PEG–OPE) 320
pendant group polymers 92, 93
– band dispersion of upper valence bands 93
– UPS and PIES spectra for thin films 93
penning ionization (PI) 78
– bandgap solid by metastable He (He$^+$) impact 78

penning ionization electron spectroscopy (PIES) 65
– ARUPS measurements of rubrene 83
– detection of molecular orientation using 79
– determination of intermolecular band dispersion 80
– HOMO band dispersion in pentacene 82, 83
– metastable He, use 78
– photoelectron takeoff angle 81
– spectra of monolayerequivalent OTiPc film on HOPG surface 80
pentacene
– DOGS in 102
– He I UPS spectra 82
– HOMO band dispersion in 82
– hopping mobility 89
– precursors, production 227
– relaxed positive polaron 102
– TIPS-pentacene 228, 231, 232
– – blends 230
– – SIMS data 229
percolation
– behavior, interparticle conduction by tunneling 154–156
– continuum 152–154
– lattice 146–152
– – Gaussian distribution of local conductances 149
– – global specific conductivity 148
– – lattice-occupied-by-spheres model 151, 152
– – length scale of "percolating" system 147
– – links–nodes–blobs (LNB) model 148
– – resistance of whole backbone network 148
– – two-dimensional 147
– theory 145, 225
– threshold 207
perfluorooctane sulfuric acid (PFOSA) 500
perfluorosebacic acid (PFSEA) 500, 507
pernigraniline 501
PFB:F8BT blends 409
– chemical composition maps 413
– resonant soft X-ray scattering profiles 413
PFO:ZnO FET, output characteristics of 480
phase diagrams 30–32
– binary 227
– Flory–Huggins phase diagram 29
– liquid–liquid phase region 254
– P3HT:PC$_{61}$BM 31
– P3HT/PE (1/9) blend and xylene 206
– of polymer–fullerene–solvent system 13
– ternary 32, 33, 195, 196
– tricomponent systems 255

phase-directing agent (PDA) 414
phase separation 194, 197, 223
– domains 402
– stages 196
[6,6]-phenyl-C_{61}-butyric acid methyl ester ($PC_{61}BM$) 7, 9, 42, 43, 173, 233–236, 238, 240–245, 367, 373, 375, 471
– aggregate 179
– bilayer structure 471
– clusters, crystallization 350
– composite films, image 45
– crystallites 233
– electroluminescence spectra 132
– electron diffraction pattern 43
– energy levels 375
– FET geometry 477
– fluorine copolymer:PCBM blend 137
– MDMO-PPV/PCBM blend films, image 47
– mobility of holes and electrons 236
– PCBM ([6,6]-phenyl-C_{61}-butyric acid methyl ester 375
– PCBM–P3HT-based organic solar cells 384
– phase segregation 348
– P3HT:[60]PCBM
– – bilayer film fabricated on SiO_2:Si substrate 242
– – blends, image 234
– – 2D concentration profile (C/C_0) 244
– – DSC heating thermogram 238
– – field-effect mobility of holes and electrons 237, 239, 240, 243
– P3HT/PCBM bulk heterojunction 53
– P3HT/PCBM film, TEM images 351
– weight ratio 177
– – of PCBM to P3HT 177
[6,6]-phenyl-C_{71}-butyric acid methyl ester 7, 9, 240, 241
phosphate buffer solution (PBS) 502
– QDs polymer 380
photoactive CdS NR-P3HT hybrid films 382
photoactive composite film 378
photoactive layer
– parameters, determining morphology creation 42
– volume characterization 50–56
photocurrent generation mechanism 372
photocurrent–voltage characteristics in 335
photoexcitations 107
photoinduced charge transfer 334
photoluminescence 412
– quenching 24–26
photon absorption processes 333
photooxidation 443, 451
photovoltaic cells 337, 390

– applications, semiconducting polymer composite thin films 171
planar microcavity 447
plasticization effect 287
plastic lab-on-a-chip system 450
plastic solar cells 338
P(NDI2OD-T2)/rr-P3HT blends 485
Poisson distribution 86
polaron 66, 72, 73, 86, 111, 118, 120, 491, 509
polar solvents 283
poly (o-ethoxyaniline) 270
polyacetylene (PA) 191, 489
– conducting polymers/synthetic metals 489
poly(alkyl methacrylate) 284
poly(3-alkylthiophene)s (P3ATs) 30, 301
– nanowires 208, 209
polyaniline (PANI) 213, 269, 271, 497–501
– ABSA nanotubes 505
– chiral camphorsulfonic acid (CSA) 505
– DOPH-containing epoxy resin 279
– electroactivity 288
– electronic tongue 512
– electrophotoisomerization nanotubes 505
– film 270
– nanofibers 501, 506
– nano-framework–electrode junctions (PNEJs) 506
– nanotube 493, 506
– three-dimensional multifunctional microstructures 500, 501
poly(o-anisidine) 270
poly(benzobisimidazobenzophenanthroline) (BBL)/$PTHQ_x$ blend films
– AFM topography images 481
– ambipolar transport 482
– thin-film blend morphologies, atomic force microscopy (AFM) 481
poly(benzobisthiazole-alt-3-octylquarterthiophene) (PBTOT) 471
poly(γ-benzyl-L-glutamate)-b-poly[2-(dimethylamino)ethyl methacrylate] block copolymers 319, 320
poly[2,6-(4,4-bis-(2-ethylhexyl)-4H-cyclopenta [2,1-b;3,4-b']-dithiophene)-alt-4,7-(2,1,3-benzothiadiazole)] (PCPDTBT) 7, 8, 341
– thin film 378
poly(2,5-bis(3-tetradecyllthiophen-2-yl)thieno [3,2-b]thiophene) (pBTTT) 347
poly(3-butylthiophene) (P3BT) 30
– a-PS composite 254, 256, 263
– – conductivity 258
– – ODCB tricomponent systems phase diagram 254
– chemical structure 252

– insulating polymer composites 258
– nanowires 253, 259
– PMMA composites 258
– PS composites 203
– – bulk heterojunction in 263
– – conductivity 258
– – film 209
poly(ε-caprolactone) (PCL) 213
poly(9,9-dialkyfluorene-*alt*-triarylamine) (TFB) 198
poly(3,3‴-didodecylquaterthiophene) (PQT-12) 192, 201, 252
polydimethylsiloxane (PDMS) layer 205
poly(2,7-(9,9-di-*n*-octylfluorene-*alt*-benzothiadiazole)) (F8BT) 26, 111, 112, 407, 408
poly(2,7-(9,9-di-*n*-octylfluorene)-*alt*-(1,4-phenylene-((4-*sec*-butylphenyl)imino)-1,4-phenylene)) (TFB) 26, 27, 198, 199
poly(9,9-dioctylfluorene) (PFO) 212, 444
poly[2,7-(9,9-dioctylfluorene)-*alt*-5,5-(40,70-di-2-thienyl-20,10,30-benzothiadiazole)] (PFDTBT) 31
poly(9,9-dioctylfluorene-cobithiophene) (F8T2) 192
poly(dioctylfluorene-*co*-dimethyltriarylamine) 231
poly(3,9-di-t-butylindeno[1,2-b] fluorene) (PIF) 466
poly(3-dodecylthiophene) (P3DDT) 30
polyesters 282, 300
polyethylene (PE) 71, 206, 251, 300
– derivatives 300
poly(3,4-ethylenedioxythiophene) (PEDOT) 285, 493, 501, 502
– PAA nanostructures 496
poly(3,4-ethylenedioxythiophene):poly(styrene sulfonate) (PEDOT:PSS) layer 174, 180, 371, 401
poly(ethylene oxide) (PEO) 213
poly(2-ethyl-2-oxazoline) (PEOXA) 302
polyfluorene-*b*-poly(*N*-isopropylacrylamide) rod–coil block copolymers 317
polyfluorene-*b*-poly(stearyl acrylate) block copolymers 315
polyfluorene polymers blends 407–409, 412–414, 416
polyfluorenes (PFs) 313, 439
– chemical structures 313
– derivatives 444
– preparation 313
– rod–coil block copolymers 315
– synthetic methods 314

poly(10-hexyl-phenoxazine-3,7-diyl-*alt*-3-hexyl-2,5-thiophene) (POT) 481
poly(3-hexylselenothiophene) (P3HS) 30
poly(3-hexylthiophene) (P3HT) 173, 192, 224, 251, 252, 335
– diblock copolymers 212
– nanowires
– – charge transport 209
– – electronic properties 210
– – preparation 209
poly(3-hexylthiophene)-*b*-poly(γ-benzyl-L-glutamate) films 321
poly(3-hexylthiophene)-*b*-poly(3-cyclohexylthiophene) (P3HT-*b*-P3cHT) 212
poly(3-hexylthiophene)-*b*-poly(2-(dimethylamino)ethyl methacrylate) rod–coil block copolymers 304
poly(3-hexylthiophene)-*b*-poly(2-ethyl-2-oxazoline) (P3HT–PEOXA) diblock copolymers 305
poly(3-hexylthiophene)-*b*-poly(4-vinylpyridine) 320
poly(3-hexylthiophene-2,5-diyl) (P3HT) 7
– based hybrid solar cells 391
– bilayer structure 471
– crystals 176
– diblock copolymers, synthesis 302
– nanofibers 383
– P3HT:F8TBT blend solar cells
– – photocurrent transient measurements on 419
– – short-circuit photocurrent 420
– P3HT:F8TBT cell, EQE spectrum 407
– P3HT/F8TBT nanoimprinted devices 417
– P3HT–PBLG diblock copolymer 322
– P3HT/PCBM
– – based devices, performance 350
– – blends 410
– – film 351
– – solar cells 339
– P3HT:[60]PCBM blends 235, 236
– – field-effect mobility 237
– – molecular structures 234
– – optical microscope images 234
– P3HT:PC$_{61}$BM films 12
– P3HT:[60]PCBM OFET mobility measurements 241
– P3HT:[70]PCBM system 241
– P3HT/PE blend, phase diagram 206
– P3HT–PE diblock copolymer synthesis 305
– P3HT–PE polymer blend/xylene system 206
– P3HT/PMMA films 200

- P3HT/polyethylene composite, phase evolution 256
- P3HT– PS–PS-*graft*-fullerene rod–coil polymer 307
- polymers 178, 179
- SEM micrographs 182
- triblock copolymers synthesis 304
poly(*N*-isopropylacrylamide)s (PNIPAMs) 315
polylactide (PLA)–PPV rod–coil block copolymers 311
polymer-based composites 430
polymer 3,6-bis(5-(benzofuran-2-yl)thiophen-2-yl)-2,5-bis(2-ethylhexyl)pyrrolo[3,4-c]pyrrole-1,4-dione (DPP(TBFu)$_2$) 9
polymer blend transistor 476
polymer-bridged bithiophene poly[2,6-(4,4-bis-(2-ethylhexyl)-4H-cyclopenta[2,1-b;3,4-b0]-dithiophene)-alt-4,7-(2,1,3-benzothiadiazole)] (PCPDTBT) 8
- PCBM device, efficiency 341
polymer brushes 387
polymer composites, in photonic applications 430
- organic light-emitting diodes 430–441
- organic semiconductor lasers 441–451
polymer/fullerene ambipolar OFETs 232–245
- polymer:fullerene bilayer diffusion 241–245
- polymer:fullerene blend morphology 233–241
polymer/fullerene BHJ photovoltaic device 334
polymer:fullerene bilayer diffusion 241–245
- modeling fullerene diffusion 243–245
polymer:fullerene blend morphology 233–241
- blend composition 236–238
- fullerene molecular weight effect 240, 241
- solvent and polymer molecular weight 235, 236
- temperature/time-dependent annealing 238–240
polymer:fullerene blends 241, 245
polymer–fullerene miscibility 28–30
polymer/fullerene solar cells 354
polymer–fullerene–solvent system 13
polymeric light-emitting diodes (PLEDs) 192
polymeric semiconductors 191, 205, 219
polymer matrix 210, 231, 264, 286
- electrical conductivity 283
- ICPs in
-- doping with functional dopants 279, 280
-- side groups 279
-- in situ polymerization 280

polymer–nanocrystal hybrid solar cells 383
polymer/NPs system 380
- composite materials 361
polymer/PCBM systems 390
polymer photovoltaic (PV) cells 171
polymer/polymer blends 411, 480
polymer–polymer interaction 196
polymer/polymer interface 197
polymer–polymer miscibility 26, 27
polymer solar cell devices, optimization 346–354
- bilayer structure 333
- bulk heterojunction structure 333
- device architectures 351–354
-- inverted device structures 352
-- tandem structure 353, 354
- morphology control 346–351
-- blending ratio 347
-- solvent effects 347–350
-- thermal annealing 350
- power conversion efficiency (PCE) 331
polymer solar cells (PSCs) 39, 53, 174, 175, 181, 267, 331, 335, 336, 338, 344, 417
- devices, optimization 346–354 (*see also* polymer solar cell devices, optimization)
- performance 353
polymer–solvent interaction 348
polymer tandem solar cells 354
polymethacrylates 300
poly[2-methoxy-5-(3′,7′-dimethyloctyloxy)-1,4-phenylene vinylene] (MDMO-PPV) 42
poly-[2-methoxy-5-(2′-ethylhexyloxy)-1,4-(1-cyanovinylene)phenylene (MEH-PPV) 236, 237
poly(2-methoxy-5-(2′-ethyl-hexyloxy)-1,4-phenylene vinylene (MEH-PPV) 57, 212, 236, 251, 311, 366, 380, 432, 433, 449, 477
poly(methyl acrylate) (PMA) 212
poly(methyl methacrylate) (PMMA) 199, 224, 271, 443, 458
- nanodomains 184–186
poly(methyl methacrylate)-*co*-4-(2-acryloyloxyethylamino)- 7-nitro-2,1,3-benzooxadiazole-*block*-polyfluorene-*block*-poly(methyl methacrylate)- *co*-4-(2-acryloyloxyethylamino)-7-nitro-2,1,3-benzooxadiazole (P(MMA-*co*-NBDAE)- *b*-PF-*b*-P(MMA-*co*-NBDAE)) 316, 318
polymethylthiophene (PMT) 272
poly(1-naphthylamine) (PNA)/poly(vinyl alcohol) (PVA) 279
poly[*N*-9″-heptadecanyl-2,7-carbazole-*alt*-5,5-(4′,7′-di-2-thienyl-2′,1′,3′-benzothiadiazole)] (PCDTBT) 13

poly(3-octylthiophene) (P3OT) 388
polyolefins 282
poly(oxa-1,4-phenylene-1,2-(1-cyanovinylene)-2,5-dialkoxy-1,4-phenylene-1,2-(2-cyanovinylene)-1,4-phenylene) (PCNEPV) 412
poly (2,5-bis(3-alkylthiophen-2-yl)thieno[3,2-b]thiophene) (PBTTT) 192
poly(o-phenylenediamine) 270
polyphenylenes (PP) 497
poly(p-phenylene vinylene)s (PPVs) 308, 339, 403, 497
– conventional polymer light-emitting diodes 446
– PPV$_7$-b-PLA$_{72}$ thin film 311
– PPV rod–coil block copolymers 311
– synthetic routes 309
polypyrrole (PPy) 270, 502, 503
polypyrrole aluminum flake composite (PAFC) 277
polystyrene (PS) 202, 224, 251
polystyrene-b-poly(3-hexylthiophene) (PS-b-P3HT) 307
polystyrene (PS):polythiophene blends 13
polystyrene segment 307
poly(styrene sulfonate) (PSS) 510
poly(thienylene vinylene) (PTV) 343
poly(2,5-thienylene vinylene) (PTV) 466
polythiophene (PTH) 260, 491
poly[(thiophene-2,5-diyl)-alt-(2,3-diheptylquinoxaline-5,8-diyl)] (PTHQ$_x$) blend films 481
– AFM topography images 481
– ambipolar transport 482
– thin-film blend morphologies, atomic force microscopy (AFM) 481
poly(3- (2,5-dioctylphenyl)thiophene) film 442
polythiophene/insulating polymer composites
– doped by molecular dopant, conductivity 260–261
– enhanced conductivity/mobility mechanisms 261–267
– – crystallinity and molecular ordering 261, 262
– – matrix at interface, bulk 3-D interface and reduced polarization 263, 264
– – perspective 267
– – reduced polaron–dopant interaction 265–267
– – "self-encapsulation" effect 262
– – "zone refinement" effect 264, 265
– at low doping level, enhanced conductivity 258–260

– morphological requirement, enhanced electrical conductivity 251–267
– phase evolution and morphology 253–255
– uses 214
polythiophene/IP composite 265
polythiophene:P(NDI2OD-T2) solar cells 409
polytriarylamine (PTAA) 228
poly(vinyl alcohol) (PVA) fiber 506
polyvinylcarbazole (PVK) 438
poly(9-vinylcarbazole) (PvCz) 93
poly(2-vinylnaphthalene) (PvNp) 93
porous AAO membrane 387
porous TiO$_2$ nanotubes 390
postproduction annealing 233
power conversion efficiency (PCE) 7, 171, 331, 375, 418
– key factors determining 41
PQT-12/PS composite 205
printing technologies 191, 352
processing constraints 403
PS-b-PMMA block copolymer 183, 184, 186
PTCDI-C$_{13}$H$_{27}$ 475
push–pull driving forces 340
pyridine ligand 376, 379

q

quantitative correlation 381
quantum dots (QDs) 362, 364, 378, 379, 433
– absorption spectra 363
– blue, atomic force microscope phase image 436
– energy bandgap 435
– LED, normalized electroluminescence spectra 437
– NCs 374, 387, 391
– photograph 436
– photoluminescence (PL) graphs 363
– RGB CdSe 437
quasi-one-dimensional band dispersion, along polymer chains
– π-conjugated polymer chain 91, 92
– σ-bond polymers 90, 91
quenching
– of delayed fluorescence 123
– donor–acceptor interface, excitons 107
– exciton quenching 123, 135, 136
– fluorescence 412
– light emission by 433
– into phase separation and 198
– photoluminescence 24, 26, 132, 138, 382
– roomtemperature solvent evaporation 226
– into two-phase region 222

r

radio-frequency identification (RFID) 220, 457
– tags 192
recombination processes 307, 407
reduction potentials 273
regioregular poly(3-alkylthiophene)s, synthesis procedures 301
regioregular poly(3-hexylthiophene) (rr-P3HT) 182, 471, 483
relaxation energies 85
reorganization energy 86–90
resonant soft X-ray scattering (R-SoXS) 412
resorcinol formaldehyde (RF)-cured composite coatings 279
reversible addition–fragmentation chain transfer polymerization (RAFT) 302
RGB components 435
Rieke method, for synthesis polymers 301
rigid rod polymers 300
ring-opening cationic polymerization 302
ring-opening metathesis polymerization (ROMP) 302
rod–coil block copolymers 302, 304, 305, 310
– classes 300
– self-assembling properties 312
– syntheses 302
rod-rod copolymers 299
– diblock copolymer 322
– diblock copolymer synthesis 321
– triblock copolymer synthesis 321
roll-to-roll processing 367
rr-P3HT-b-PMMA copolymers 303, 308
rr-P3HT-b-PS copolymers 303
rr-P3HT diblock copolymers synthesis 303

s

scanning electron microscope (SEM) 271, 347
scanning reference electrode technology (SRET) 274
scanning tunneling microscope (STM)
– tunneling injection 445
secondary ion mass spectrometry (SIMS) 228
– measurements 229, 231
self-assembled monolayers (SAMs) 204
"self-encapsulation" effect 262
semiconducting diblock copolymer 207, 300
semiconducting nanocrystal/conjugated polymer composites
– applications in hybrid polymer solar cells 361–392
– composite materials 361–367
– – colloidal semiconductor nanocrystals 361–366
– – conjugated polymers 366, 367
– device structure 367–374
– – device principle 371–373
– – donor acceptor pairs, band alignment and choice 373, 374
– – photoactive layer 367–371
– hybrid PV and OPV vs. PV technologies 391, 392
– hybrid solar cells, state of art 374–380
– hybrid solar cells vs. pure OPVs 390, 391
semiconducting polymers (SPs) 53, 173, 194, 251, 252, 262, 264, 265, 267, 300
– IP composites 260, 267
– – charge transport 260, 264
– – conductivity 260
– – FET mobility 261
– – morphological requirements 262
– – phase evolution routes 257
– – phase separation 264
– nanostructure, fabrication
– – atomic force microscopy (AFM) 174
– – exciton diffusion efficiency 171
– – nanoimprint mold fabrication 183–186
– – photoluminescence (PL) 174
– – SEM micrographs 182
– – sub-20 nm scale 182
– – X-ray photoelectron spectroscopy (XPS) 174
semiconductor, band diagram 428
semiconductor–dielectric interfaces 198, 235, 242, 262, 459
semiconductor–insulator interfaces 263
semiconductor nanocrystals (NCs) 361, 362
semiconductor nanowires 212
semi-interpenetrating networks 287
sensors 505
– artificial 511, 512
– biosensors 510, 511
– development 489
– gas 506–508
– pH 509, 510
short-circuit current density (J_{SC}) 7
Siegrist polycondensation 310
silicon-based p–n junction solar cells 331
silicone film 174, 176, 181
silicon mold 389
silicon phthalocyanine derivative (SiPc) 23
silicon processing capabilities 2
silicon technology 457
simplified template-free method (STFM) 499
single-junction OPVs 338
single-walled carbon nanotubes (SWCNTs) 207
SiO_2 nanopillar structure 185

size-tunable inorganic core 377
small-molecule organic semiconductors 219
small polaron binding energy 86–90
sodium dodecylsulfate (SDS) 284
soft X-ray microscopy 412
solar cells 307
– efficiency 391
– PCE values 392
solar spectrum 240, 339, 353, 366, 367, 402
sol–gel process 287, 288
solubility 3, 4
solution-based processes 226
solution/dispersion method 285
solution-processed bulk heterojunction films 411
solution-processed materials 222
solvents 9–21
– blends 15, 16
– impact on solid-state morphology 10–14
– – AFM images 10, 11
– non-halogenic solvents 14, 15
– parameters of different key solvents for OPV 22
– poor solvents 16, 17
– processing additives 18–21
– – alkane dithiols, chain length efficiency 18
– – functional end groups of 1,8-di(R)octane 18
– – j–V characteristics, PCPDTBT/PC$_{71}$BM composite films 19
– – processing, self-assembly of bulk heterojunction blend materials 19
– solution concentration 21
source–drain voltage drop 463
space-charge limited current (SCLC) model 178, 179
spin coating 2, 197
spinodal curves 195
spinodal decomposition 222
stainless steel (SS) samples 270
state-of-the-art polymer solar cells
– fullerene/polymer solar cells, optimization 338–354
– – absorption enhancement 340–344
– – benzo[1,2-b;3,4-b]dithiophene 342, 343
– – conjugated side chains 343, 344
– – design rules 339
– – fluorene-like donor units 340–342
– – HOMO and LUMO energy levels, fine-tuning 344, 345
– – mobility and solubility improvement 345
– – new fullerene derivative 345, 346
– – new materials design 339–346
– – polymer solar cell devices, optimization 346–354
– working mechanism 332–338
– – bulk heterojunction structures 332–334
– – device parameters 334–338
– – fill factor 337, 338
– – open-circuit voltage 335, 336
– – organic solar cells, unique properties 332
– – short-circuit current density 334, 335
– – theoretical efficiency 334–338
steric interactions 410
Stille coupling reactions 309, 310
stimulated emission (SE) 441
stress–elongation curves 208
structure–property relationships 300
surface charge density 220
surface energy 23, 24
surface-sensitive techniques 228
Suzuki coupling reactions 309, 310

t

tandem solar cells, chemical structures 353
temperature-dependent solubility 7
template-free method 499
tetra-fluorotetracyanoquinodimethane (F4TCNQ) 260, 261
tetrahydrofuran (THF) 13, 302
– ring-opening cationic polymerization 302
– toluene solvent systems 315
thermal annealing (TA) 173, 350
thiazolo[5,4-d]thiazole (TTZ) 343
thieno[3,4-b]thiophene (TT) 344
thieno [2,3-b]thiophene terthiophene polymer
– chemical structures 478
thienopyrroledione (TPD) 342
– synthesis 343
thin-film photovoltaic devices, polymer solar cells 336
thin-film transistor (TFT) 204, 388
– based on polythiophene/insulating polymer composites with enhanced charge transport 191–214
– blend films with embedded P3HT nanowires 207–213
– – from conjugated block copolymers 212
– – electrospun nanowires from conjugated polymer blends 212, 213
– – P3AT nanowires 208, 209
– blend films with vertical stratified structure 194–207
– – charge carrier mobility 201–204
– – crystallization-induced vertical phase segregation 206, 207
– – environmental stability 201

– – polymer blends, patterned domains 201–204
– – polymer blends, phase behavior 194–198
– field-effect characteristics in 191
– OFETs
– – device operation 220–222
– – polymer/fullerene ambipolar 232–245
– – unipolar films 224–232
– OTFTs
– – fundamental principle and operating mode 193, 194
– – semiconducting/insulating blends, preparation strategies 194
– – semiconducting/insulating layers, one-step formation 198–201
– semiconducting organic molecule/polymer composites 219–246
– small-molecule/polymer film morphology 222, 223
thiophene-based oligomers 300
thiophene–phenylene–thiophene (TPT) unit 345
thiophenes, oligomers 301
third-generation PV technologies 392
three-dimensional dendritic Au nanostructures (3D-DGNs) 505
threshold voltage 221
TiO_2 nanopillars 388
TiO_2 nanotubes 389, 390
TiO_2/P3HT hybrid solar cell 388
titanyl phthalocyanine 79
toluene 7, 16
top contact/bottom gate (TC/BG) configurations 463
top-gate, bottom contact (TG/BC) transistor configurations 466
transistor channel 221
transition metal-mediated condensation reactions 314
transition metal-mediated cross-coupling reactions 314
transition metal-mediated polycondensation reactions 314
transmission electron microscopy (TEM) 233, 347, 411
transparent conducting oxide (TCO) film 431
trap-assisted recombination 419
1,2,4-trichlorobenzene (TCB) 375
triethylsilylethynyl anthradithiophene (TESADT)
– PMAA blend 232
– PMMA blend 200
– poly(α-methylstyrene) (PαMS) composite 264, 265

trioctylphosphine (TOP) 362
trioctylphosphine oxide (TOPO) 362
tris(8-hydroxy-quinoline) aluminum (Alq3) thin film 430
two-component water-based epoxy resin coating system 279
two-dimensional (2D) lamellar crystals 208

u

ultraviolet photoelectron spectroscopy (UPS) 65, 67, 366
– angle-resolved UPS (ARUPS), measuring 77
– binding energy 77
– electronic structure probed by 75
– feature of HOMO 76
– kinetic energy and wave vector, for observed photoelectron 76, 77
– three-step model, general assumptions 76
– use thin films 76
unbalanced charge transport 347
unipolar FET
– n-type, typical current–voltage characteristics 461
– operating regimes 459
– p-type, current–voltage characteristics 461
unipolar systems 245
unmixing process 196
upper critical solution temperature (UCST) system 195
UV–Vis region 346
UV–Vis spectra 272, 363

v

vacuum level energy (E_{vac}) 76
valance band (VB) 371
vanadium oxides 352
van der Waals forces 496
vapor liquid solid (VLS) process 388
variable range hopping (VRH) models 209
– charge transport mechanism 235
vertical cavity surface emitting lasers (VCSELs) 447
vertical stratified blend films, preparation 194
vinylene bonds 309
voltage-controlled device 193. *see also* organic thin-film transistor (OTFT)
voltage shorted compaction (VSC) method 491

w

water-based polyaniline 286
water-gated OTFTs 205
weak donor–strong acceptor strategy 344

weakly bound ligands 377
Wessling methods 308, 309
wide-angle X-ray diffraction 227

x

X-ray absorption spectra 412
X-ray microscopy 414
X-ray photoelectron spectroscopy (XPS) technique 75, 228, 270, 366
xylene 16
– MEH–PPV dissolved 443
p-xylene 7, 14

z

zero-loss energy-filtered transmission electron microscopy 411
ZnO nanoparticles 479
ZnO nanorods, defects 439
ZnO nanowires 438
– cotton-like product consisting 438
– LED device structure, schematic diagram 439
ZnO–P3HT hybrid solar cell device, TEM micrograph 382
"zone refinement" effect 264, 265